Lecture Notes in Artificial Intelligence 9329

Subseries of Lecture Notes in Computer Science

More information about this series at http://www.springer.com/series/1244

Manuel Núñez · Ngoc Thanh Nguyen
David Camacho · Bogdan Trawiński (Eds.)

Computational Collective Intelligence

7th International Conference, ICCCI 2015
Madrid, Spain, September 21–23, 2015
Proceedings, Part I

 Springer

Editors
Manuel Núñez
Universidad Complutense de Madrid
Madrid
Spain

Ngoc Thanh Nguyen
Wrocław University of Technology
Wrocław
Poland

David Camacho
Computer Science Department
Universidad Autónoma De Madrid
Madrid
Spain

Bogdan Trawiński
Wrocław University of Technology
Wrocław
Poland

ISSN 0302-9743 ISSN 1611-3349 (electronic)
Lecture Notes in Artificial Intelligence
ISBN 978-3-319-24068-8 ISBN 978-3-319-24069-5 (eBook)
DOI 10.1007/978-3-319-24069-5

Library of Congress Control Number: 2015948851

LNCS Sublibrary: SL7 – Artificial Intelligence

Springer Cham Heidelberg New York Dordrecht London
© Springer International Publishing Switzerland 2015

Printed on acid-free paper

Springer International Publishing AG Switzerland is part of Springer Science+Business Media
(www.springer.com)

Preface

This volume contains the proceedings of the 7th International Conference on Computational Collective Intelligence (ICCCI 2015), held in Madrid, Spain, September 21–23, 2015. The conference was co-organized by the Universidad Complutense de Madrid, Spain, the Universidad Autónoma de Madrid, Spain, and Wrocław University of Technology, Poland. The conference was run under the patronage of the IEEE SMC Technical Committee on Computational Collective Intelligence.

Following the successes of the 1st ICCCI (2009) held in Wrocław, Poland, the 2nd ICCCI (2010) in Kaohsiung, Taiwan, the 3rd ICCCI (2011) in Gdynia, Poland, the 4th ICCCI (2012) in Ho Chi Minh City, Vietnam, the 5th ICCCI (2013) in Craiova, Romania, and the 6th ICCCI (2014) in Seoul, South Korea, this conference continued to provide an internationally respected forum for scientific research in the computer-based methods of collective intelligence and their applications.

Computational Collective Intelligence (CCI) is most often understood as a sub-field of Artificial Intelligence (AI) dealing with soft computing methods that enable making group decisions or processing knowledge among autonomous units acting in distributed environments. Methodological, theoretical, and practical aspects of computational collective intelligence are considered as the form of intelligence that emerges from the collaboration and competition of many individuals (artificial and/or natural). The application of multiple computational intelligence technologies such as fuzzy systems, evolutionary computation, neural systems, consensus theory, etc., can support human and other collective intelligence, and create new forms of CCI in natural and/or artificial systems. Three subfields of the application of computational intelligence technologies to support various forms of collective intelligence are of special interest but are not exclusive: Semantic Web (as an advanced tool for increasing collective intelligence), social network analysis (as the field targeted to the emergence of new forms of CCI), and multiagent systems (as a computational and modeling paradigm especially tailored to capture the nature of CCI emergence in populations of autonomous individuals).

The ICCCI 2015 conference featured a number of keynote talks and oral presentations, closely aligned to the theme of the conference. The conference attracted a substantial number of researchers and practitioners from all over the world, who submitted their papers for the main track and 10 special sessions.

The main track, covering the methodology and applications of computational collective intelligence, included: knowledge integration; data mining for collective processing; fuzzy, modal, and collective systems; nature-inspired systems; language processing systems; social networks and Semantic Web; agent and multi-agent systems; classification and clustering methods; multi-dimensional data processing; web systems; intelligent decision making; methods for scheduling; and image and video processing.

The special sessions, covering some specific topics of particular interest, included: Collective Intelligence in Web Systems - Web Systems Analysis, Computational Swarm Intelligence, Cooperative Strategies for Decision Making and Optimization, Advanced Networking and Security Technologies, IT in Biomedicine, Collective Computational Intelligence in an Educational Context, Science Intelligence and Data Analytics, Computational Intelligence in Financial Markets, Ensemble Learning, and Big Data Mining and Searching.

We received in total 186 submissions. Each paper was reviewed by 2–4 members of the International Program Committee of either the main track or one of the special sessions. We selected the 110 best papers for oral presentation and publication in two volumes of the Lecture Notes in Artificial Intelligence series.

We would like to express our thanks to the keynote speakers, Francisco Herrera, B. John Oommen, Guy Theraulaz, and Jorge Ufano, for their world-class plenary speeches. Many people contributed towards the success of the conference. First, we would like to recognize the work of the Program Committee Co-chairs and special sessions organizers for taking good care of the organization of the reviewing process, an essential stage in ensuring the high quality of the accepted papers. The Workshops and Special Sessions Chairs deserve a special mention for the evaluation of the proposals and the organization and coordination of the work of the 10 special sessions. In addition, we would like to thank the PC members, of the main track and of the special sessions, for performing their reviewing work with diligence. We thank the Organizing Committee Chairs, Liaison Chairs, Publicity Chair, Special Issues Chair, Financial Chair, Web Chair, and Technical Support Chair for their fantastic work before and during the conference. Finally, we cordially thank all the authors, presenters, and delegates for their valuable contribution to this successful event. The conference would not have been possible without their support.

It is our pleasure to announce that the conferences of ICCCI series continue a close cooperation with the Springer journal Transactions on Computational Collective Intelligence, and the IEEE SMC Technical Committee on Transactions on Computational Collective Intelligence.

Finally, we hope and intend that ICCCI 2015 significantly contributes to the academic excellence of the field and leads to the even greater success of ICCCI events in the future.

September 2015

Manuel Núñez
Ngoc Thanh Nguyen
David Camacho
Bogdan Trawiński

ICCCI 2015 Conference Organization

General Chairs

Ngoc Thanh Nguyen	Wrocław University of Technology, Poland
Manuel Núñez	Universidad Complutense de Madrid, Spain

Steering Committee

Ngoc Thanh Nguyen	Wrocław University of Technology, Poland (Chair)
Piotr Jędrzejowicz	Gdynia Maritime University, Poland (Co-Chair)
Shyi-Ming Chen	National Taiwan University of Science and Technology, Taiwan
Adam Grzech	Wrocław University of Technology, Poland
Kiem Hoang	University of Information Technology, VNU-HCM, Vietnam
Lakhmi C. Jain	University of South Australia, Australia
Geun-Sik Jo	Inha University, South Korea
Janusz Kacprzyk	Polish Academy of Sciences, Poland
Ryszard Kowalczyk	Swinburne University of Technology, Australia
Toyoaki Nishida	Kyoto University, Japan
Ryszard Tadeusiewicz	AGH University of Science and Technology, Poland

Program Committee Chairs

David Camacho	Universidad Autónoma de Madrid, Spain
Costin Bădică	University of Craiova, Romania
Piotr Jędrzejowicz	Gdynia Maritime University, Poland
Toyoaki Nishida	Kyoto University, Japan

Organizing Committee Chairs

Jesús Correas	Universidad Complutense de Madrid, Spain
Sonia Estévez	Universidad Complutense de Madrid, Spain
Héctor Menéndez	Universidad Autónoma de Madrid, Spain

Liaison Chairs

Geun-Sik Jo	Inha University, South Korea
Attila Kiss	Eötvös Loránd University, Hungary
Ali Selamat	Universiti Teknologi Malaysia, Malaysia

Special Sessions and Workshops Chairs

Alberto Núñez   Universidad Complutense de Madrid, Spain
Bogdan Trawiński   Wrocław University of Technology, Poland

Special Issues Chair

Jason J. Jung   Chung-Ang University, South Korea

Publicity Chair

Antonio González   Bilbao Center for Applied Mathematics, Spain

Financial Chair

Mercedes G. Merayo   Universidad Complutense de Madrid, Spain

Web Chair

Luis Llana   Universidad Complutense de Madrid, Spain

Technical Support Chair

Rafael Martínez   Universidad Complutense de Madrid, Spain

Special Sessions

WebSys: Special Session on Collective Intelligence in Web Systems – Web Systems
Analysis
Organizers: Kazimierz Choroś and Maria Trocan
CSI: Special Session on Computational Swarm Intelligence
Organizers: Urszula Boryczka, Mariusz Boryczka, and Jan Kozak
CSDMO: Special Session on Cooperative Strategies for Decision Making and
Optimization
Organizers: Piotr Jędrzejowicz and Dariusz Barbucha
ANST: Special Session on Advanced Networking and Security Technologies
Organizers: Vladimir Sobeslav, Ondrej Krejcar, Peter Brida, and Peter Mikulecky
ITiB: Special Session on IT in Biomedicine
Organizers: Ondrej Krejcar, Kamil Kuca, Teodorico C. Ramalho, and Tanos C.C.
Franca
ColEdu: Special Session on Collective Computational Intelligence in an Educational
Context
Organizers: Danuta Zakrzewska and Marta Zorrilla
SIDATA: Special Session on Science Intelligence and Data Analytics.
Organizers: Attila Kiss and Binh Nguyen

CIFM: Special Session on Computational Intelligence in Financial Markets
Organizers: Fulufhelo Nelwamondo and Sharat Akhoury
EL: Special Session on Ensemble Learning
Organizers: Piotr Porwik and Michał Woźniak
BigDMS: Special Session on Big Data Mining and Searching
Organizers: Rim Faiz and Aymen Elkhlifi

Program Committee (Main Track)

Abulaish, Muhammad	Jamia Millia Islamia, Kingdom of Saudi Arabia
Bădică, Amelia	University of Craiova, Romania
Barbucha, Dariusz	Gdynia Maritime University, Poland
Bassiliades, Nick	Aristotle University of Thessaloniki, Greece
Bielikova, Maria	Slovak University of Technology in Bratislava, Slovakia
Byrski, Aleksander	AGH University Science and Technology, Poland
Calvo-Rolle, José Luis	University of A Coruña, Spain
Camacho, David	Universidad Autónoma de Madrid, Spain
Capkovic, Frantisek	Slovak Academy of Sciences, Slovakia
Ceglarek, Dariusz	Poznan High School of Banking, Poland
Chiu, Tzu-Fu	Aletheia University, Taiwan
Chohra, Amine	Paris-East University (UPEC), France
Choros, Kazimierz	Wrocław University of Technology, Poland
Colhon, Mihaela	University of Craiova, Romania
Czarnowski, Ireneusz	Gdynia Maritime University, Poland
Davidsson, Paul	Malmö University, Sweden
Do, Tien V.	Budapest University of Technology and Economics, Hungary
Elci, Atilla	Aksaray University, Turkey
Ermolayev, Vadim	Zaporozhye National University, Ukraine
Faiz, Rim	IHEC - University of Carthage, Tunisia
Florea, Adina Magda	Polytechnic University of Bucharest, Romania
Gaspari, Mauro	University of Bologna, Italy
Gónzalez, Antonio	Universidad Autónoma de Madrid, Spain
Ha, Quang-Thuy	Vietnam National University, Vietnam
Hoang, Huu Hanh	Hue University, Vietnam
Hong, Tzung-Pei	National University of Kaohsiung, Taiwan
Huang, Jingshan	University of South Alabama, USA
Hwang, Dosam	Yeungnam University, South Korea
Istrate, Dan	UTC, Laboratoire BMBI, France
Ivanovic, Mirjana	University of Novi Sad, Serbia
Jędrzejowicz, Joanna	University of Gdansk, Poland
Jozefowska, Joanna	Poznan University of Technology, Poland
Jung, Jason	Yeungnam University, South Korea
Kefalas, Petros	The University of Sheffield International Faculty - CITY College, Greece

Kim, Sang-Wook	Hanyang University, South Korea
Kisiel-Dorohinicki, Marek	AGH University of Science and Technology, Poland
Koychev, Ivan	University of Sofia "St. Kliment Ohridski", Bulgaria
Kozierkiewicz-Hetmanska, Adrianna	Wrocław University of Technology, Poland
Krejcar, Ondrej	University of Hradec Kralove, Czech Republic
Kulczycki, Piotr	Polish Academy of Science, Poland
Kuwabara, Kazuhiro	Ritsumeikan University, Japan
Leon, Florin	Technical University "Gheorghe Asachi" of Iasi, Romania
Li, Xiafeng	Texas A&M University, USA
Lu, Joan	University of Huddersfield, UK
Lughofer, Edwin	Johannes Kepler University Linz, Austria
Meissner, Adam	Poznan University of Technology, Poland
Menéndez, Hector	Universidad Autónoma de Madrid, Spain
Mercik, Jacek	Wrocław School of Banking, Poland
Michel, Toulouse	Université de Montréal, Canada
Moldoveanu, Alin	Polytechnic University of Bucharest, Romania
Mostaghim, Sanaz	Otto von Guericke University of Magdeburg, Germany
Nalepa, Grzegorz J.	AGH University of Science and Technology, Poland
Neri, Filippo	University of Napoli Federico II, Italy
Nguyen, Ngoc Thanh	Wrocław University of Technology, Poland
Nguyen, Linh Anh	University of Warsaw, Poland
Nguyen Trung, Duc	Yeungnam University, South Korea
Nguyen Tuong, Tri	Yeungnam University, South Korea
Núñez, Manuel	Universidad Complutense de Madrid, Spain
Núñez, Alberto	Universidad Complutense de Madrid, Spain
Ponnusamy, Ramalingam	Madha Engineering College, India
Precup, Radu-Emil	Polytechnic University of Timisoara, Romania
Quaresma, Paulo	Universidade de Evora, Portugal
Ratajczak-Ropel, Ewa	Gdynia Maritime University, Poland
Selamat, Ali	Universiti Teknologi Malaysia, Malaysia
Stoyanov, Stanimir	University of Plovdiv "Paisii Hilendarski", Bulgaria
Takama, Yasufumi	Tokyo Metropolitan University, Japan
Trawiński, Bogdan	Wrocław University of Technology, Poland
Treur, Jan	Vrije Universiteit Amsterdam, The Netherlands
Unold, Olgierd	Wrocław University of Technology, Poland
Wierzbowska, Izabela	GMU, Poland
Žagar, Drago	University of Osijek, Croatia
Zakrzewska, Danuta	Technical University of Lodz, Poland
Zamfirescu, Constantin-Bala	"Lucian Blaga" University of Sibiu, Romania
Zdravkova, Katerina	FINKI, FYR of Macedonia, Macedonia

Additional Reviewers

Abro, Altaf Hussain	Hwang, Dosam	Mollee, Julia
Acs, Zoltan	Kaczor, Krzysztof	Molnár, Bálint
Camacho, Azahara	Kluza, Krzysztof	Ramírez-Atencia, Cristian
Cañizares, Pablo C.	Kontopoulos, Efstratios	Rácz, Gábor
Eriksson, Jeanette	Lócsi, Levente	Salmeron, José L.
Hajas, Csilla	Malumedzha, Tendani	Thilakarathne, Dilhan

Invited Talks

Text Classification Using Novel "*Anti*-Bayesian" Techniques

B. John Oommen

School of Computer Science, Carleton University, Canada.
Also an *Adjunct Professor* with the University of Agder in Grimstad, Norway.
oommen@scs.carleton.ca

The problem of Text Classification has been studied for decades, and this problem is particularly interesting because the features are derived from syntactic or semantic indicators, while the classification, in and of itself, is based on statistical Pattern Recognition (PR) strategies. Thus, all the recorded TC schemes work using the fundamental paradigm that once the statistical features are inferred from the syntactic/semantic indicators, the classifiers themselves are the well-established ones such as the Bayesian, the Naïve Bayesian, the SVM etc. and those that are neural or fuzzy. In this paper, we shall demonstrate that by virtue of the skewed distributions of the features, one could advantageously work with information latent in certain "non-central" quantiles (i.e., those distant from the mean) of the distributions. We, indeed, demonstrate that such classifiers exist and are attainable, and show that the design and implementation of such schemes work with the recently-introduced paradigm of Quantile Statistics (QS)-based classifiers. These classifiers, referred to as Classification by Moments of Quantile Statistics (CMQS), are essentially "Anti"-Bayesian in their modus operandi. Being a Plenary/Keynote talk, we will concentrate and survey the new "Anti"-Bayesian paradigm. We shall show that by using it, we can obtain optimal or near-optimal results by working with a very few (sometimes as small as two) points distant from the mean.

Big Data: Technologies and Algorithms to Deal with Challenges

Francisco Herrera

Dpto. de Ciencias de la Computación e Inteligencia Artificial,
Universidad de Granada, Spain
herrera@decsai.ugr.es

In this age, big data applications are increasingly becoming the main focus of attention because of the enormous increment of data generation and storage that has taken place in the last years, in science, business, . . . This situation becomes a challenge when huge amounts of data are processed to extract knowledge because the data mining techniques are not adapted to the new space and time requirements. We must consider the new paradigms to develop scalable algorithms. At this conference we will introduce briefly the technologies that have emerged strongly in recent years (Hadoop ecosystem, Spark, . . .) and the libraries such as Mahout, ML lib, . . . We will discuss some applications, and we will focus the attention on the steps to create a learning algorithms that was winner for the ECBDL'14 big data competition, processing an extremely imbalanced big data bioinformatics problem.

Collective Information Processing in Fish Schools: From Data to Computational Models

Guy Theraulaz

Centre de Recherches sur la Cognition Animale, CNRS, UMR 5169,
Université Paul Sabatier, France
guy.theraulaz@univ-tlse3.fr

Swarms of insects, schools of fish and flocks of birds display an impressive variety of collective behaviors that emerge from local interactions among group members. These puzzling phenomena raise a variety of questions about the interactions rules that govern the coordination of individuals' motions and the emergence of large-scale patterns. While numerous models have been proposed, there is still a strong need for detailed experimental studies to foster the biological understanding of such collective motion. I will present the methods that we used to characterize interactions among individuals and build models for animal group motion from data gathered at the individual scale. Using video tracks of fish shoal in a tank, we determined the stimulus/response function governing an individual's moving decisions from an incremental analysis at the local scale. We found that both attraction and alignment interactions are present and act upon the fish turning speed, yielding a novel schooling model whose parameters are all estimated from data. We also found that the magnitude of these interactions changes as a function of the swimming speed of fish and the group size. The consequence being that groups of fish adopt different shapes and motions: group polarization increases with swimming speed while it decreases as group size increases. The phase diagram of model also revealed that the relative weights of attraction and alignment interactions play a key role in the emergent collective states at the school level. Of particular interest is the existence of a transition region in which the school exhibits multistability and intermittence between schooling and milling for the same combination of individual parameters. In this region the school becomes highly sensitive to any kind of perturbations that can affect the behavior of just a single fish.

Trading and Poker: Using Computers to Take Intelligent Decisions

Jorge Ufano

INVINCO, Spain
jufano@clasesdebolsa.com

In order to be a successful poker player and/or a trader in stock markets it is necessary to acquire a specific set of skills. Despite the fact that both poker and trading involve a high degree of chance, it is crucial to use statistical models to analyse, interpret and quantify patterns that repeat themselves. For example in trading we analyse patterns related to price, volume, seasonal and sentiment trends while in poker we study patterns related to betting, sequences, ranges in different boards and psychological tendencies such as 'tells'. Therefore, the collection and generation of intelligence through complex software tools is a must. For example, we can use tools such as Tradestation, ProRealTime and Ninja Trader for trading, and Poker Tracker, Holdem Resources and Flopzilla in the world of poker.

The key for success in poker and trading is to gain a small edge over the rest of players and be able to exploit it. Once we have developed our own method and strategy with a positive mathematical expectation, we need to put it in practice many times. These small advantages will in turn generate highly probable earnings, an important factor considering that the variance of results in these games is significant. In addition, a proper management and assess of risk is also fundamental to succeed.

One of the characteristics that make both poker and trading such fantastic games is that they reward the best players in the long term, while still giving almost any player a chance to succeed in the short term. In this talk I will review the main tools and methodologies used by players and traders and will illustrate, with examples, the differences and similarities.

Contents – Part I

Ontologies and Information Extraction

Formal Models and Simulation

Neural Networks, SMT and MIS

Contents – Part II

ITiB: Special Session on IT in Biomedicine

ColEdu: Special Session on Collective Computational Intelligence in Educational Context

SIDATA: Special Session on Science Intelligence and Data Analytics

CIFM: Special Session on Computational Intelligence in Financial Markets

EL: Special Session on Ensemble Learning

BigDMS: Special Session on Big Data Mining and Searching

Text Classification Using Novel "*Anti*-Bayesian" Techniques

B. John Oommen[1]($\boxtimes$), Richard Khoury[2], and Aron Schmidt[2]

[1] School of Computer Science, Carleton University, Ottawa K1S 5B6, Canada
oommen@scs.carleton.ca
[2] Department of Software Engineering, Lakehead University,
Thunder Bay P7B 5E1, Canada
{rkhoury,aschmid1}@lakeheadu.ca

Abstract. This paper presents a non-traditional "*Anti*-Bayesian" solution for the traditional Text Classification (TC) problem. Historically, all the recorded TC schemes work using the fundamental paradigm that once the statistical features are inferred from the syntactic/semantic indicators, the classifiers themselves are the well-established statistical ones. In this paper, we shall demonstrate that by virtue of the skewed distributions of the features, one could advantageously work with information latent in certain "non-central" quantiles (i.e., those distant from the mean) of the distributions. We, indeed, demonstrate that such classifiers exist and are attainable, and show that the design and implementation of such schemes work with the recently-introduced paradigm of Quantile Statistics (QS)-based classifiers. These classifiers, referred to as Classification by Moments of Quantile Statistics (CMQS), are essentially "Anti"-Bayesian in their *modus operandi*. To achieve our goal, in this paper we demonstrate the power and potential of CMQS to describe the *very* high-dimensional TC-related vector spaces in terms of a limited number of "outlier-based" statistics. Thereafter, the PR task in classification invokes the CMQS classifier for the underlying multi-class problem by using a linear number of pair-wise CMQS-based classifiers. By a rigorous testing on the standard 20-Newsgroups corpus we show that CMQS-based TC attains accuracy that is comparable to the best-reported classifiers. We also propose the potential of fusing the results of a CMQS-based method with those obtained from a traditional scheme.

Keywords: Text classification · Quantile statistics (QS) · Classification by the moments of QS (CMQS)

1 Introduction

This paper presents a non-traditional and totally novel solution to the problem of Text Classification (TC). TC is the challenge of associating a given unknown

The authors are grateful for the partial support provided by NSERC, the Natural Sciences and Engineering Research Council of Canada.

B. John Oommen—*Chancellor's Professor; Fellow: IEEE* and *Fellow: IAPR*. This author is also an *Adjunct Professor* with the University of Agder in Grimstad, Norway.

© Springer International Publishing Switzerland 2015
M. Núñez et al. (Eds.): ICCCI 2015, Part I, LNAI 9329, pp. 1–15, 2015.
DOI: 10.1007/978-3-319-24069-5_1

text document with a category selected from a predefined set of categories (or classes) based on its content. As opposed to this, statistical Pattern recognition (PR) is the process by which unknown *statistical* feature vectors are categorized into groups or classes based on their *statistical* components [3]. The field of statistical PR has been so well developed that it is not necessary for us to survey the field here. Suffice it to mention that all the recorded TC schemes work using the fundamental paradigm that once the statistical features are inferred from the *syntactic or semantic* indicators, the classifiers themselves are the well-established *statistical*, neural or fuzzy ones such as the Bayesian, Naïve Bayesian, Linear Discriminant, the SVM, the Back-propagation etc.

The TC problem has been studied since the 1960's [13], but it has taken a special importance in recent years as the sheer amount of text available has increased super-exponentially – thanks to the internet, text-based communications such as e-mail, tweets and text messages, and the numerous book-digitization projects that have been undertaken by the various publishing houses. Over the decades, many approaches[1] have been proposed to accomplish this goal. When it concerns classification and PR, the TC problem is particularly interesting both from an academic and a research perspective. This is because, whereas the features in TC are derived from *syntactic or semantic* indicators, the classification, in and of itself, is based on *statistical*, neural or fuzzy strategies.

The goal of this paper is to show that we can achieve TC using "Anti"-Bayesian quantile statistics-based classifiers which only use information contained in, let us say, non-central quantiles (which are sometimes outliers) of the distributions, and that it can do this by operating with a philosophy that is totally contrary to the acclaimed Bayesian paradigm. Indeed, the fact that such a classification can be achieved is, strictly speaking, not easy to fathom.

To motivate this paper and to place its contribution the right context, we present the following simple example. Consider the problem of distinguishing a document that belongs to one of two classes, namely, *Sports* or *Business*. It is obvious that one can trivially distinguish them if we merely considered those words which occurred frequently in one class and not the other, for example, "football" and "basketball" *versus* "dollars" and "euros". Our hypothesis is that it is not *merely* these truly "distinguishing" words that possess "discriminating" capabilities. We intend to demonstrate that there are "outliers" quantiles of the words which occur in both categories, and which also can be used to achieve the classification. Hopefully, this would be both a pioneering and remarkable result.

It should, first of all, be highlighted that we do not intend to obtain a classification that *surpasses* the behavior of the scheme that involves a Bayesian strategy invoking the truly "distinguishing" words. Attempting to do this would be tantamount to accomplishing the impossible, because the Bayesian approach

---

[1] Due to space limitations, it is impossible to survey the field of TC here. The unabridged version of this paper [8] contains a more detailed survey of the field and includes the preliminaries of the Vector Space model, the Bag-of-Words (BoW), the Term Frequency (TF), the Term Frequency-Inverse Document Frequency (TFIDF) weighting schemes, and the Cosine Similarity metric etc. [11,12].

maximizes the *a posteriori* probability and it thus yields the optimal hallmark classifier. What we endeavor to do is to show that if we use the above-mentioned non-central quantiles and work within an "Anti"-Bayesian paradigm using only *these* quantile statistics, we can obtain accuracies comparable to this optimal hallmark! Indeed, we demonstrate that a near-optimal solution can be obtained by invoking counter-intuitive features *when they are coupled with* a counter-intuitive PR paradigm.

As a backdrop, we note that the basic concept of traditional *parametric* classification is to model the classes based on the assumptions related to the underlying class *distributions*, and this has been historically accomplished by performing a learning phase in which the moments, i.e, the mean, variance etc. of the respective classes are evaluated. However, there have been some families of indicators (or distinguishing quantifiers) that were until recently, noticeably, *uninvestigated* in the PR literature. Specifically, we refer to the use of phenomena that have utilized the properties of the *Quantile Statistics* (QS) of the distributions. This has led to the "Anti"-Bayesian methodology alluded to.

1.1 Contributions of this Paper

The novel contributions of this paper are:

- To demonstrate that text and document classification can be achieved using an "Anti"-Bayesian methodology;
- To show that this "Anti"-Bayesian PR can be achieved using syntactic information that that has not been used in the literature before, namely the information contained in the symmetric quantiles of the distributions, and which are traditionally considered to be "outlier"-based;
- To show that the results of our "Anti"-Bayesian PR is not highly correlated with the results of any of the traditional TC schemes, implying that one can use it in conjunction with a traditional TC scheme for an ensemble-based classifier;
- To suggest that a strategy that incorporates the fusion of the features and methodology proposed here and the distinct ones from the state-of-the-art has great potential. This is an avenue that we will explore in future research.

As in the case of the quantile-based PR results, to the best of our knowledge, the pioneering nature and novelty of these TC results hold true.

2 Background: Traditional Text Classifiers

Apart from the methods presented above, many authors have also looked at ways of enhancing the document and class representation by including not only words but also bigrams, trigrams, and n-grams in order to capture common multi-word expressions used in the text [4]. Likewise, character n-grams can be used to capture more subtle class distinctions, such as the distinctive styles of different authors for authorship classification. While these approaches have, so far, considered ways to enrich the representation of the text in the word vector, other authors have attempted to augment the text itself by adding extra information into it, such as synonyms of the words taken from a thesaurus, be it a specialized custom-made one for a project such as the affective-word thesaurus built in [7], or, more commonly, the more general-purpose linguistic ontology, *WordNet* [5].

Adding another generalization step, it is increasingly common to enrich the text not only with synonymous words but also with synonymous *concepts*, taken from domain-specific ontologies [17] or from Wikipedia [1]. Meanwhile, in an opposing research direction, some authors prefer to simplify the text and its representation by reducing the number of words in the vectors, typically by grouping synonymous words together using a Latent Semantic Analysis (LSA) system or by eliminating words that contribute little to differentiating classes as indicated by a Principal Component Analysis (PCA) [6]. Other authors have looked at improving classification by mathematically transforming the sparse and noisy category word space into a more dense and meaningful space. A popular approach in this family involves Singular Value Decomposition (SVD), a projection method in which the vectors of co-occurring words would project in similar orientations, while words that occur in different categories would be projected in different orientations.This is often done before applying LSA or PCA modules to improve their accuracy. Likewise, authors can transform the word-count space to a probabilistic space that represents the likelihood of observing a word in a document of a given category. This is then used to build a probabilistic classifier, such as the popular Naïve-Bayes' classifier [10], to classify the text into the most probable category given the words it contains.

An underlying assumption shared by all the approaches presented above is that one can classify documents by comparing them to a representation of what an average or typical document of the category should look like. This is immediately evident with the BOW approach, where the category vector is built from average word counts obtained from a set of representative documents, and then compared to the set of representative documents of other categories to compute the corresponding similarity metric. Likewise, the probabilities in the Naïve-Bayes' classifier and other probability-based classifiers are built from a corpus of typical documents and represent a general rule for the category, with the underlying assumption that the more a specific document differs from this general rule, the less probable it is that it belongs to the category. The addition of information from a linguistic resource such as a thesaurus or an ontology is also based on this assumption, in two ways. First, the act itself is meant to add words and concepts that are missing from the specific document and thus make it more like a typical document of the category. Secondly, the development of these resources is meant to capture general-case rules of language and knowledge, such as "these words are typically used synonymously" or "these concepts are usually seen as being related to each other."

The method we propose in this paper is meant to break away from this assumption, and to explore the question of whether there is information usable for classification outside of the norm, at "the edges (or fringes) of the word distributions", which has been ignored, so far, in the literature.

3 CMQS-Based Text Classifiers

3.1 How Uni-dimensional"Anti"-Bayesian Classification Works

We shall first describe how uni-dimensional "Anti"-Bayesian classification works, and then proceed to explain how it can be applied to TC, which, by definition, involves PR in a highly multi-dimensional feature space. Classification by the Moments of Quantile Statistics[2], (CMQS) is the PR paradigm which utilizes QS in a pioneering manner to achieve optimal (or near-optimal) accuracies for various classification problems. Rather than work with "traditional" statistics (or even sufficient statistics), the authors of [14] showed that the set of *distant* quantile statistics of a distribution do, indeed, have discriminatory capabilities. Thus, as a *prima facie* case, they demonstrated how a generic classifier could be developed for any uni-dimensional distribution. Then, to be more specific, they designed the classification methodology for the Uniform distribution, using which the analogous classifiers for other symmetric distributions were subsequently created. The results obtained were for symmetric distributions, and the classification accuracy of the CMQS classifier exactly attained the optimal Bayes' bound. In cases where the symmetrtic QS values crossed each other, one invokes a *dual* classifier to attain the same accuracy.

Unlike the traditional methods used in PR, one must emphasize the fascinating aspect that CMQS is essentially "Anti"-Bayesian in its nature. Indeed, in CMQS, the classification is performed in a counter-intuitive manner i.e., by comparing the testing sample to a few samples *distant* from the mean, as opposed to the Bayesian approach in which comparisons are made, using the Euclidean or a Mahalonibis-like metric, to *central* points of the distributions. Thus, opposed to a Bayesian philosophy, in CMQS, the points against which the comparisons are made are located at the positions where the Cumulative Distribution Function (CDF) attains the percentile/quantile values of $\frac{2}{3}$ and $\frac{1}{3}$, or more generally, where the CDF attains the percentile/quantile values of $\frac{n-k+1}{n+1}$ and $\frac{k}{n+1}$.

In [9], the authors built on the results from [14] and considered various symmetric and *asymmetric* uni-dimensional distributions within the exponential family such as the Rayleigh, Gamma, and Beta distributions. They again proved that CMQS had an accuracy that attained the Bayes' bound for symmetric distributions, and that it was very close to the optimal for asymmetric distributions.

---

[2] The authors of [14], [9] and [15] (cited in their chronological order) had initially proposed their theoretical and experimental results as being based on the *Order*-Statistics of the distributions. This was later corrected in [16], where they showed that their results were, rather, based on their *Quantile* Statistics.

3.2   TC: A Multi-dimensional "Anti"-Bayesian Problem

Any problem that deals with TC must operate in a space that is very high dimensional primarily the because cardinality of the BOW can be very large. This, in and of itself, complicates the QS-based paradigm. Indeed, since we are speaking about the quantile statistics of a distribution, it implicitly and explicitly assumes that the points can be *ordered*. Consequently, the multi-dimensional generalization of CMQS, theoretically and with regard to implementation, is particularly non-trivial because there is no well-established method for achieving the ordering of multi-dimensional data specified in terms of its uni-dimensional components.

To clarify this, consider two patterns, $\mathbf{x}_1 = [x_{11}, x_{12}]^T = [2, 3]^T$ and $\mathbf{x}_2 = [x_{21}, x_{22}]^T = [1, 4]^T$. If we only considered the first dimension, x_{21} would be the first QS since $x_{11} > x_{21}$. However, if we observe the second component of the patterns, we can see that x_{12} would be the first QS. It is thus, clearly, not possible to obtain the ordering of the *vectorial* representation of the patterns based on their individual components, which is the fundamental issue to be resolved before the problem can be tackled in any satisfactory manner for multi-dimensional features. One can only imagine how much more complex this issue is in the TC domain – when the number of elements in the BOW is of the order of hundreds or even thousands.

To resolve this, multi-dimensional CQMS operates with a paradigm that is analogous to a Naïve-Bayes' approach, although it, really, is of an *Anti*-Naïve-Bayes' paradigm. Using such a *Anti*-Naïve-Bayes' approach, one can design and implement a CMQS-based classifier. The details of this design and implementation for two and multi-dimensions (and the associated conclusive experimental results) have been given in [15]. Indeed, on a deeper examination of these results, one will appreciate the fact that the higher-dimensional results for the various distributions do not necessarily follow as a consequence of the lower uni-dimensional results. They hold by virtue of the factorizability of the multi-dimensional density functions that follow the *Anti*-Naïve-Bayes' paradigm, and the fact that the d-dimensional QS-based statistics are concurrently used for the classification in every dimension.

3.3   Design and Implementation: "Anti"-Bayesian TC Solution

"Anti"-Bayesian TC Solution: The Features. Each class is represented by two BOW vectors, one for each CMQS point used. For each class, we compute the frequency distribution of each word in each document in that class, and generate a frequency histogram for that word. While the traditional BOW approach would then pick the average value of this histogram, our method computes the area of the histogram and determines the two symmetric QS points. Thus, for example, if we are considering the $\frac{2}{7}$ and $\frac{5}{7}$ QS points of the two distributions, we would pick the word frequencies that encompass the $\frac{2}{7}$ and $\frac{5}{7}$ of the histogram area respectively. The reader must observe the salient characteristic of this strategy: By working with such a methodology, for each word in the BOW, we represent

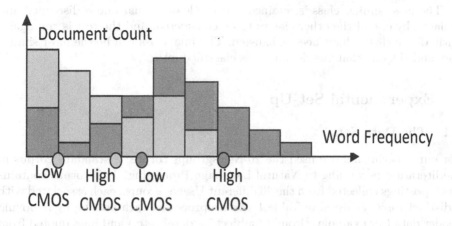

Fig. 1. Example of the QS-based features extracted from the histogram of a lower class (light grey) and of a higher class (dark grey), and the corresponding lower and higher CMQS points of each class.

the class by two of its non-central cases, rather than its average/median sample. This renders the strategy to be "Anti"-Bayesian!

For further clarity, we refer the reader to Figure 1. For any word, the histograms of the two classes are depicted in light grey for the lower class, and in dark grey for the higher class. The QS-based features for the classes are then extracted from the histograms as clarified in the figure.

"Anti"-Bayesian TC Solution: The Multi-Class TC Classifier. Let us assume that the PR problem involves C classes. Since the "Anti"-Bayesian technique has been extensively studied for two-class problems, our newly-proposed multi-class TC classifier operates by invoking a sequence of $C-1$ pairwise classifiers. More explicitly, whenever a document for testing is presented, the system invokes a classifier that involves a pair of classes from which it determines a winning class. This winning class is then compared to another class until all the classes have been considered. The final winning class is the overall best and is the one to which the testing document is assigned.

"Anti"-Bayesian TC Solution: Testing. To classify an unknown document, we compute the cosine similarity between it and the features representing pairs of classes. This is done as follows: For each word, we mark one of the two groups as the high-group and the other as the low-group based on the word's frequency in the documents of each class, and we take the high CMQS point of the low-group and the low CMQS point of the high-group, as illustrated in Figure 1. We build the two class vectors from these CMQS points, and we compute the cosine similarity [8] between the document to classify each class vector.

The most similar class is retained and the least similar one is discarded and replaced by one of the other classes to be considered, and the test is run again, until all the classes have been exhausted. The final class will be the most similar one, and the one that the document is classified into.

4   Experimental Set-Up

4.1   The Data Sets

For our experiments, we used the 20-Newsgroups corpus, a standard corpus in the literature pertaining to Natural Language Processing. This corpus contains 1,000 postings collected from the 20 different Usenet groups, each associated with a distinct topic, as listed in Table 1. We preprocessed each posting by removing header data (for example, "from", "subject", "date", etc.) and lines quoted from previous messages being responded to (which start with a '>' character), performing stop-word removal and word stemming, and deleting the postings that became empty of text after these preprocessing phases.

Table 1. The topics from the "20-Newsgroups" used in the experiments.

comp.graphics	alt.atheism	sci.crypt	misc.forsale
comp.sys.mac.hardware	talk.religion.misc	sci.electronics	rec.autos
comp.windows.x	talk.politics.guns	sci.med	rec.motorcycles
comp.os.ms-windows.misc	talk.politics.mideast	sci.space	rec.sport.hockey
comp.sys.ibm.pc.hardware	talk.politics.misc	soc.religion.christian	rec.sport.baseball

In every independent run, we randomly selected 70% of the postings of each newsgroup to be used for training, and retained the remaining 30% for testing.

4.2   The Histograms/Features and Benchmarks Used

We first describe the process involved in the construction of the histograms and the extraction of the Quantile-based features.

Each document in the 20-Newsgroups dataset was preprocessed by word stemming using the Porter Stemmer algorithm and by a stopword removal phase. It was then converted to a BOW representation. The documents were then randomly assigned into training or testing sets.

The word-based histograms (please see Figure 2) were then computed for each word in each category by tallying the observed frequencies for that word in each training document in that category, where the area of each histogram was the total sum of all the columns. The CMQS points were determined as those points where the cumulative sum of each column was equal to the CMQS moments when normalized with the total area. For further clarification, we present an example of

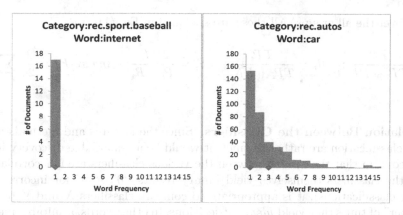

Fig. 2. The histograms and the $\frac{1}{3}$ and $\frac{2}{3}$ QS points for the two words "internet" and "car" from the categories "rec.sport.baseball" and "rec.autos".

two histograms[3] in Figure 2 below. The $\frac{1}{3}$ and $\frac{2}{3}$ QS points of each histogram are marked along their horizontal axes. In this case, the markings represent the word frequencies that encompass the $\frac{1}{3}$ and $\frac{2}{3}$ areas of the histograms respectively. The histogram on the left depicts a less significant word for its category while the histogram on the right depicts a more significant word for its category. Note that in both histograms the first CMQS point is located at unity. To help clarify the figure, we mention that for the word "internet" in "rec.sport.baseball", both the CMQS points lie at unity - i.e., they are on top of each other.

To compare the various methods used, we have developed three benchmarks for our system: A BOW classifier which involved the TFs and invoked the cosine similarity measure, a BOW classifier with the TFIDF features, and a Naïve-Bayes' classifier. Since they are well-established classifiers, their details are omitted – they are found in [8].

The Metrics Used. In every testing case, we used the respective data to train and test our classifier and each of the three benchmark schemes. For each news-group i, we counted the number of *True Positives* (TP_i) of postings correctly identified by a classifier as belonging to that group, the number of *False Negatives*, (FN_i) of postings that should have belonged in that group but were misidentified as belonging to another group, and the number of *False Positives* (FP_i) of postings that belonged to other groups but were misidentified as belonging to this one. The Precision P_i is the proportion of postings assigned in group i that are correctly identified, and the Recall R_i is the proportion of postings belonging in the group that were correctly recognized. The F score is an average of these two metrics for each group, and the *macro-F1* is the average of the F

---

[3] The documents used in this test were very short, which explains why the histograms are heavily skewed in favour of lower word frequencies.

scores over the all groups. All these are specified in Eq. (1).

$$P_i = \frac{TP_i}{TP_i + FP_i}; \quad R_i = \frac{TP_i}{TP_i + FN_i}; \quad F_i = \frac{2P_iR_i}{P_i + R_i}; \quad macro\text{-}F1 = \frac{1}{20}\sum_{i=1}^{20} F_i. \tag{1}$$

Correlation Between the Classifiers. Since the features and methods used in the classification are rather distinct, it would be a remarkable discovery if we could confirm that the results between the various classifiers are not correlated. Since the classifiers themselves yield binary results ('0' or '1' for incorrect or correct classifications), it is appropriate to compare classifiers X and Y by the "number" of times they yield *identical* decisions. In other words, a suitable metric for evaluating how any two classifiers X and Y yield identical results is given by Eq. (2) below:

$$\text{ClassifierSim}_{X,Y} = \frac{Pos_X Pos_Y + Neg_X Neg_Y}{Pos_X Pos_Y + Pos_X Neg_Y + Neg_X Pos_Y + Neg_X Neg_Y}, \tag{2}$$

where $Pos_X Pos_Y$ and $Neg_X Neg_Y$ are the count of cases where the classifiers X and Y both return identical decisions '1' or '0' respectively, and where '0' and '1' represent the events of a classifier classifying a document incorrectly or correctly respectively. Analogously, $Pos_X Neg_Y$ and $Neg_X Pos_Y$ are the counts of cases where X returns '1' and Y returns '0' and vice-versa respectively. Although this is a statistical measure of the relative similarities between the classifiers, we shall refer to this as their mutual "correlation".

5   Experimental Results

5.1   The Results Obtained: "Anti"-Bayesian TF Scheme

The experimental results that we have obtained for the "Anti"-Bayesian scheme that used only the TF criteria are briefly described below. We performed 100 tests, each one using a different random 70%/30% split of training and testing documents. We then evaluated the results of each classifier by computing the Precision, Recall, and F-score of each newsgroup, whence we computed the *macro-F1* value for each classifier over the 20-Newsgroups. The average results we obtained, over all 100 tests, are summarized in Table 2.

The results show that for *half* of the CMQS pairs, the "Anti"-Bayesian classifier performed as well as and sometimes even better than the BOW classifier.

Figure 3 displays the plots of the correlation between the different classifiers for the 100 classifications achieved, where in the case of the "Anti"-Bayesian scheme, the method used the TF features. The reader should observe the uncorrelated nature of the classifiers when the CMQS points are non-central [8].

Table 2. The *macro-F1* score results for the 100 classifications attempted and for the different methods. In the case of the "Anti"-Bayesian scheme, the method used the TF features.

Classifier	CMQS Points	*macro-F1* Score
"Anti"-Bayesian	1/2, 1/2	0.709
	1/3, 2/3	0.662
	1/4, 3/4	0.561
	1/5, 4/5	0.465
	2/5, 3/5	0.700
	1/6, 5/6	0.389
	1/7, 6/7	0.339
	2/7, 5/7	0.611
	3/7, 4/7	0.710
	1/8, 7/8	0.288
	3/8, 5/8	0.686
	1/9, 8/9	0.264
	2/9, 7/9	0.515
	4/9, 5/9	0.713
	1/10, 9/10	0.243
	3/10, 7/10	0.631
BOW		0.604
BOW-TFIDF		0.769
Naïve-Bayes		0.780

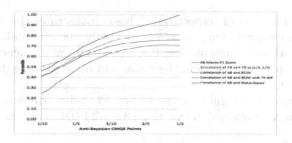

Fig. 3. Plots of the correlation between the different classifiers for the 100 classifications achieved. In the case of the "Anti"-Bayesian scheme, the method used the TF features.

5.2   The Results Obtained: "Anti"-Bayesian TFIDF Scheme

The results of the "Anti"-Bayesian scheme when it involves TFIDF features are shown in Table 3.

1. The results show that for *all* CMQS pairs, the "Anti"-Bayesian classifier performed much better than the traditional BOW classifier. For example, while the BOW had a *macro-F1* score of 0.604, the corresponding index for the CQMS pairs $\langle \frac{1}{3}, \frac{2}{3} \rangle$, was significantly higher, i.e., 0.747. Further, the *macro-F1* score indices for $\langle \frac{1}{4}, \frac{3}{4} \rangle$, $\langle \frac{3}{7}, \frac{4}{7} \rangle$ and $\langle \frac{4}{9}, \frac{5}{9} \rangle$ were consistently higher

Table 3. The *macro-F1* score results for the 100 classifications attempted and for the different methods. In the case of the "Anti"-Bayesian scheme, the method used the TFIDF features.

Classifier	CMQS Points	*macro-F1* Score
"Anti"-Bayesian	1/2, 1/2	0.742
	1/3, 2/3	0.747
	1/4, 3/4	0.746
	1/5, 4/5	0.742
	2/5, 3/5	0.745
	1/6, 5/6	0.736
	1/7, 6/7	0.729
	2/7, 5/7	0.747
	3/7, 4/7	0.744
	1/8, 7/8	0.720
	3/8, 5/8	0.746
	1/9, 8/9	0.712
	2/9, 7/9	0.745
	4/9, 5/9	0.744
	1/10, 9/10	0.705
	3/10, 7/10	0.748
BOW		0.604
BOW-TFIDF		0.769
Naïve-Bayes		0.780

　　－ 0.746, 0.744 and 0.744 respectively. This demonstrates the validity of our counter-intuitive paradigm – that we can truly get a remarkable accuracy even though we are characterizing the documents by the syntactic features of the points quite distant from the mean and more towards the extremities of the distributions.

2. In all the cases, the values of the Macro-*F1* index was only slightly less than the indices obtained using the BOW-TFIDF and the Naïve-Bayes approaches.

　　Figure 4 displays the plots of the correlation between the different classifiers for the 100 classifications achieved, where in the case of the "Anti"-Bayesian scheme, the method used the TFIDF features. The reader should again observe the uncorrelated nature of the classifiers for non-central CMQS points. This correlation increases as the feature points become closer to the mean/median.

　　To continue the analysis, it would be good to examine if the two "Anti"-Bayesian classifiers are relatively uncorrelated in and of themselves. Thus, if a particular pair of CMQS points yielded distinct classification decisions using the two schemes, and if they, all the same, yielded comparable accuracies, the potential of the paradigm is shown to be significantly more. This is precisely what we embark on achieving now – i.e., examining the correlation (or lack thereof) of the "Anti"-Bayesian TF and TFIDF schemes. This correlation is depicted graphically in Figure 5 whence the trends in the correlation with the increasing values of the CMQS points is clear.

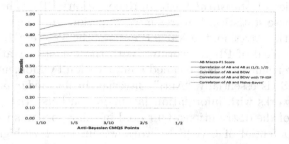

Fig. 4. Plots of the correlation between the different classifiers for the 100 classifications achieved. In the case of the "Anti"-Bayesian scheme, the method used the TFIDF features.

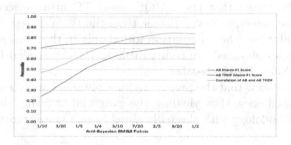

Fig. 5. The correlation between the two "Anti"-Bayesian classifiers for the 100 classifications when they utilized the TF and the TFIDF features respectively.

When the CMQS points are close to the mean or median, the correlation is quite high (for example, 0.842). This is not surprising at all, since in such cases, the "Anti"-Bayesian classifier reduces to become a Bayesian classifier. Also, when the CMQS points are far from the mean or median, the correlation is quite high (for example, 0.659 for the CMQS points $\langle \frac{2}{9}, \frac{7}{9} \rangle$). This is quite surprising because although both schemes are "Anti"-Bayesian in their philosophy, the lengths of the documents play a part in determining the decisions that they individually make because the IDF values account for document lengths.

The unabridged version of the paper [8] also describes how the various classifiers can be fused. This discussion is omitted here in the interest of space.

6 Conclusions

In this paper we have considered the problem of Text Classification (TC), which is a problem that has been studied for decades. From the perspective of classification, problems in TC are particularly fascinating because while the feature extraction process involves *syntactic or semantic* indicators, the classification uses the principles of *statistical* Pattern Recognition (PR). The state-of-the-art in TC uses these statistical features in conjunction with the well-established methods such as the Bayesian, the Naïve Bayesian, the SVM etc. Recent research has advanced the field of PR by working with the Quantile Statistics (QS) of the features. The resultant scheme called Classification by Moments of Quantile Statistics (CMQS) is essentially "Anti"-Bayesian in its *modus operandus*, and advantageously works with information latent in "outliers" (i.e., those distant from the mean) of the distributions. Our goal in this paper was to demonstrate the power and potential of CMQS to work within the *very* high-dimensional TC-related vector spaces and their "non-central" quantiles. To investigate this, we considered the cases when the "Anti"-Bayesian methodology used both the TD and the TFIDF criteria.

Our PR solution for C categories involved $C-1$ pairwise CMQS classifiers. By a rigorous testing on the well-acclaimed data set involving the 20-Newsgroups corpus, we demonstrated that the CMQS-based TC attains accuracy that is comparable to and sometimes even better than the BOW-based classifier, even though it essentially uses the information found only in the "non-central" quantiles. The accuracies obtained are comparable to those provided by the BOW-TFIDF and the Naïve Bayes classifier too!

Our results also show that the results we have obtained are often uncorrelated with the established ones, thus yielding the potential of fusing the results of a CMQS-based methodology with those obtained from a more traditional scheme.

References

1. Alahmadi, A., Joorabchi, A., Mahdi, A.E.: A new text representation scheme combining bag-of-words and bag-of-concepts approaches for automatic text classification. In: Proceedings of the 7th IEEE GCC Conference and Exhibition, Doha, Qatar, pp. 108–113, November 2014
2. Debole, F., Sebastiani, F.: Supervised term weighting for automated text categorization. In: Proceedings of the 18th ACM Symposium on Applied Computing, Melbourne, USA, pp. 784–788, March 2003
3. Duda, R.O., Hart, P.E., Stork, D.G.: Pattern Classification. A Wiley Interscience Publication (2006)
4. Dumoulin, J.: Smoothing of n-gram language models of human chats. In: Proceedings of the Joint 6th International Conference on Soft Computing and Intelligent Systems (SCIS) and 13th International Symposium on Advanced Intelligent Systems (ISIS), Kobe, Japan, pp. 1–4, November 2012
5. Lu, L., Liu, Y.-S.: Research of english text classification methods based on semantic meaning. In: Proceedings of the ITI 3rd International Conference on Information and Communications Technology, Cairo, Egypt, pp. 689–700, December 2005
6. Madsen, R.E., Sigurdsson, S., Hansen, L.K., Larsen, J.: Pruning the vocabulary for better context recognition. In: Proceedings of the 17th International Conference on Pattern Recognition, Cambridge, UK, vol. 2, pp. 483–488, August 2004
7. Ning, Y., Zhu, T., Wang, Y.: Affective-word based chinese text sentiment classification. In: Proceedings of the 5th International Conference on Pervasive Computing and Applications (ICPCA), Maribor, Slovenia, pp. 111–115, December 2010

8. Oommen, B.J., Khoury, R., Schmidt, A.: Text Classification Using "Anti"-Bayesian Quantile Statistics-based Classifiers. Unabridged version of this paper. Submitted for publication
9. Oommen, B.J., Thomas, A.: Optimal Order Statistics-based "Anti-Bayesian" Parametric Pattern Classification for the Exponential Family. Pattern Recognition **47**, 40–55 (2014)
10. Qiang, G.: An effective algorithm for improving the performance of naïve bayes for text classification. In: Proceedings of the Second International Conference on Computer Research and Development, Malaysia, pp. 699–701, May 2010
11. Salton, G., McGill, M.: Introduction to Modern Information Retrieval. Mc-Graw Hill Book Company, New York (1983)
12. Salton, G., Yang, C.S., Yu, C.: Term weighting approaches in automatic text retrieval. Technical Report, Ithaca, NY, USA (1987)
13. Sebastiani, F.: Machine Learning in Automated Text Categorization. ACM Computing Surveys **34**, 1–47 (2002)
14. Thomas, A., Oommen, B.J.: The Fundamental Theory of Optimal "Anti-Bayesian" Parametric Pattern Classification Using Order Statistics Criteria. Pattern Recognition, 376–388 2013
15. Thomas, A., Oommen, B.J.: Order Statistics-based Parametric Classification for Multi-dimensional Distributions. Pattern Recognition, 3472–3482 (2013)
16. Thomas, A., Oommen, B.J.: Corrigendum to Three Papers that deal with "Anti"-Bayesian Pattern Recognition. Pattern Recognition, 2301–2302 (2014)
17. Wu, G., Liu, K.: Research on text classification algorithm by combining statistical and ontology methods. In: Proceedings of the International Conference on Computational Intelligence and Software Engineering, Wuhan, China, pp. 1–4, December 2009

Multi-agent Systems

ADELFE 3.0 Design, Building Adaptive Multi Agent Systems Based on Simulation a Case Study

Wafa Mefteh[1,2](✉), Frederic Migeon[2], Marie-Pierre Gleizes[2],
and Faiez Gargouri[1]

[1] MIRACL Laboratory, University of Sfax, Sfax, Tunisia
wafa.mefteh07@gmail.com, migeon@irit.fr, faiez.gargouri@fsegs.rnu.tn
[2] IRIT Laboratory, University of Paul Sabatier, Toulouse, France
gleizes@irit.fr

Abstract. ADELFE is a methodology dedicated to applications characterized by openness and the need of the system adaptation to an environment. It was proposed to guide the designer during the building of an Adaptive Multi-Agent System (AMAS). Given that designing such systems is not an easy task, this leads us to provide means to bring the AMAS design to a higher stage of automation and confidence thanks to Simulation. ADELFE 3.0 is a new version of ADELFE based on a Simulation Based Design approach in order to assist the designer of AMAS and make his task less difficult. This paper focuses on a practical example of building an AMAS using ADELFE 3.0.

Keywords: Adaptive multi-agent systems · ADELFE methodology · Simulation · ADELFE3.0

1 Introduction

Some complex systems are qualified by the heterogeneity and diversity of actors involved, the large mass of data, by the distribution of the manipulated information and also by the dynamic of the environments in which they are immersed. Modeling such complex systems requires the use of efficient techniques. In recent years, several research works are interested in the development of new techniques best suited to this kind of problems. The AMAS (Adaptive Multi-Agent System) theory [1] [2] was proposed to model complex systems. This theory has shown that for a system immersed in a dynamic environment, a system in cooperative interactions with its environment is functionally adequate. The ADELFE methodology [4] [5] was proposed to guide the AMAS designers through an approach based on the RUP (Rational Unified Process). However, the AMAS theory stipulates that the designer must find all Non Cooperative Situations that an agent may encounter or create. For each of these situations, it must give the actions to be performed to ensure the agent to come back and stay in a cooperative state with others and itself. The problem lies in the definition of the

© Springer International Publishing Switzerland 2015
M. Núñez et al. (Eds.): ICCCI 2015, Part I, LNAI 9329, pp. 19–28, 2015.
DOI: 10.1007/978-3-319-24069-5_2

cooperative behaviour that is complete and accurate enough to give the agent any means to avoid what are called Non Cooperative Situations (NCS). Building such self-organizing systems is not an easy task. Our objective was to provide agents behaviours to self-design. The goal is to help the designer, to discharge him from the inherent difficulty in search of cooperative behaviour of agents and accelerate design time. For this, this we studied the contribution of simulation to design these systems. Simulation enables to improve and test the behaviour of an agent but also the behaviour of the collective. The ADELFE methodology was enriched by engineering processes and tools to achieve the "living design" that integrates modelling, programming and simulation techniques. For this, we dened three challenges in order to enrich ADELFE by engineering processes and tools to achieve the Simulation-Based Design; *Firstly*, we help the designer and discharge him from the difficulty of searching for the cooperative behaviour of agents (search for Non-Cooperative Situations, anticipating problems ...). Thus, we dene iterative activities that enable (i) the design of a preliminary version of the cooperative behaviour that (ii) is tested according to predefined situations and give information on what goes right and wrong in order (iii) for the designer to be able to analyze what is going wrong and to start a new design cycle of the cooperative behaviour. *Secondly*, we automate the design and provide tools that can determine what kinds of Non Cooperative Situations have been encountered by the agent and what kinds of cooperative behaviour cure the malfunctioning. *Finally*, we enable the designer to provide only a basic behaviour for the agent that could adapt to the Non Cooperative Situations encountered during runtime according to analysis and adaptation of its cooperative behaviour automatically. This "Living Design" of the agent, as we call it, defines the most efficient mechanisms of adaptation obtained so far. In [7], we presented in details the different contributions integrated in ADELFE3.0 and compared ADELFE3.0 to ADELFE2.0. The experimentation results [7] indicates that ADELFE 3.0 is better than ADELFE 2.0 for the all verified aspects. This paper gives an example of application of ADELFE3.0 in order to present how it helps the designer and make his task less difficult. The objective of this paper is also to serve as a guide for a designer who wants to use ADELFE3.0 with another case study. For this, we begin by presenting briefly ADELFE3.0 and more details can be found in [7]. Then, we present the case study the Red Donkey Game resolved using ADELFE3.0.

1.1 ADELFE 3.0

Initially ADELFE (figure 1) includes five phases that were initially inspired from the RUP and gathers twenty one activities, producing twenty six work products [5]. Each activity contains a set of steps which are carried out by different roles. Each step uses one or more input documents and produces one or more outputs. The different documents (inputs or outputs) have different types. ADELFE was described using the FIPA template [3]. ADELFE was proposed to guide AMAS designers. This design process takes into account the specific aspects of AMAS, including those relating to cooperation. With ADELFE3.0 [7] (figure 2),

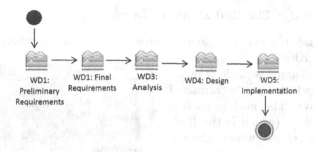

Fig. 1. ADELFE 2.0

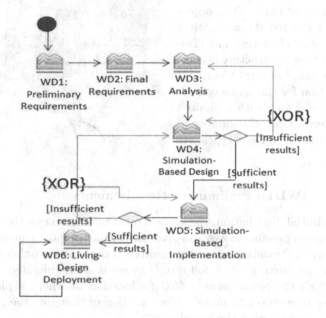

Fig. 2. ADELFE 3.0

we integrated a Simulation Based Design (SBD) approach. Our goal is twofold. *Firstly*, as usual, we want to shorten development time and lower costs by detecting as soon as possible misconceived parts of the system. *Secondly*, we would like to benet as well as to enhance the adaptation capabilities of the Multi-Agent System by means of simulation-based techniques. So the ADELFE methodology was enriched by an engineering process and tools to achieve the Simulation-Based Design. Design, implementation and deployment phases are concerned. This issue includes techniques for modelling, programming and simulating. This technique of SBD permits to visualize information about the system that is running from models used in the specification of the system. With ADELFE, simulation helps in the design of the agent behaviour as well as in the validation activities and the construction of the self-design of the agents.

1.2   Case Study: The Red Donkey Game

The Red Donkey is a classical game originating from Poland where it is known as Klotski (or Klocki).

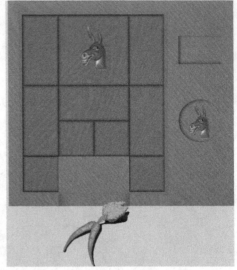

The original game is made of many wood blocks which can be arranged in various ways. The goal is to free the "master bloc" (which is the Red-Donkey) by successive shifts of the elements. The player is not allowed to remove blocks; he may only slide blocks horizontally and vertically. The objective is to exit the red donkey with a minimum number of moves. The classic Red-Donkey game structure is composed of 10 pieces. But our objective is to find a solution for this game considering a very large area with unlimited number of pieces. Only the number of empty cells and the pieces types are known.

1.3   Phase 1 (WD1): Preliminary Requirements

From a given initial distribution, the system must be able to exit the red donkey with the minimum possible moves and give the number of moves made to achieve the given solution. Considering that the number of the pieces is unknown, according to the user requirements, the following keywords are highlighted: *Board* (the ground in which the pieces move), *Cell* (a location in where a piece can be shifted), *Piece* (the object to move), *Move* (action of changing the pieces place without losing contact with the board).

1.4   Phase 2 (WD2): Final Requirements

We extract the following limits:

1. For a piece, a move is only possible if it does not lose contact with the board.
2. Minimize the number of moves.

We define three use cases as the following: *Set Initial Configuration* (the user sets the initial distribution of the pieces), *Calculate solution* (the user ask the system to calculate a solution), *Manage* (The user can stop, play, reset and restart the game). The user is the only actor interacting with the system. Based on the previous steps, the system can be designed as a multi-agent system. Indeed, the result given by the system is realized by the participation of all the entities of the system. Each entity is autonomous and has a well dened role.

1.5   Phase 3 (WD3): Analysis

The pieces are active entities because: *they are autonomous* (they decide themselves to move), *they have a local goal* (which is moving to a better place when
it is possible), *they have a local vision about their environment* (each piece can
perceive its neighbors (other pieces or free cells which are just behind it and also
the door for the red donkey) and they can interact between themselves). The
Cells (occupied and free) are passive entities because they cant change their state
by themselves. Figure 3 presents the entities structure of the system. Each piece
is able to move and check if this movement can cause a Non Cooperative Situation (NCS). The pieces, which are able to move, negotiate between themselves in
order to let the most critical piece to move. After each move, a piece checks, with
local knowledge, if the last move creates a problem or not. If yes, the problem
must be corrected and this situation is saved in order to anticipate it next time.
The interactions between entities are represented in figure 4. The next activity
is to verify the AMAS adequacy. It permits the us to decide if the system can
be built as an AMAS or not. ADELFE provides a decision tool which helps us
to take this decision. Using the decision function dened in the AMAS adequacy
tool to verify the Global AMAS adequacy, we obtain the result that our system
can be developed as AMAS. Indeed, the global task of the system cannot be
fully specified because though the result we want to achieve is known, the steps
(moves) must be realized to get this result are unknown because of the dynamic
of the environment. The solution requires a correlated activity of several components and several trials are required before giving an optimal solution. The
environment evolves because it hasnt the same structure during the execution
of the system. With each movement, pieces are facing a new environment with
new empty places. A conceptual distribution is useful to solve the global task
of the system. The system includes a considerable number of entities and it is
non-linear because the entities are connected and the behaviour of an entity is
influenced by the behaviour of the others. Active entities (pieces) are considered
as agents because each piece takes its own decision (to move or not). A piece is
able to change its state itself and moves when it is possible to a better place.
Pieces interact between themselves and negotiate the best move. We define two

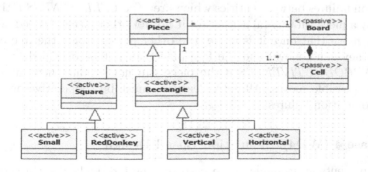

Fig. 3. Entities Structure Diagram

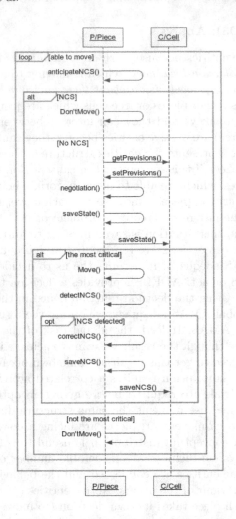

Fig. 4. Entities Interaction Diagram

cooperation failures between entities which are: *The USELESSNESS* if the move realized by a piece isn't useful neither for it nor for other pieces. Indeed, a useless move is a move that once it is made, the other pieces beside the free cells are blocked (they can't move) and the free place can not be occupied by another piece. *UNPRODUCTIVENESS* if the piece can not produce new information about the possible moves. All active entities must be cooperative agents and detect cooperation failures.

1.6   Phase 4 (WD4): Simulation Based Design

Figure 5 presents an example of piece agent (the red donkey) architecture and an inner state related to its behaviour. With the simulation tool SeSAm, we use

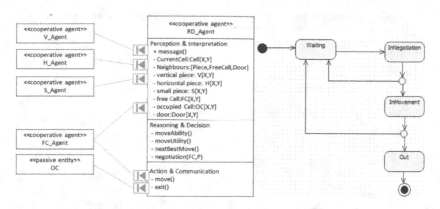

Fig. 5. (left) An example of a piece agent (the red donkey) architecture. (right) The inner state related to the behaviour of the P_Agent

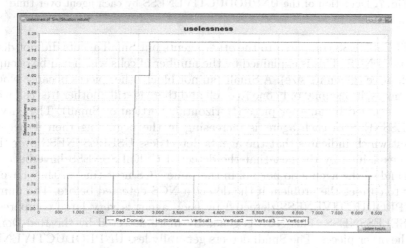

Fig. 6. Detection of the USELESSNESS by each agent over time

the S-DLCAM model to develop four types of S-DLCAM [6] agents: Red Donkey, Vertical, Horizontal and Small. We give a preliminary behaviour (a non complete behaviour) to the agents. With Sesam, an agent behaviour is developed as an activity diagram which is composed of a set of activities and transitions between these activities. The S-DLCAM agent behaviour is composed of three types of activities: nomonal activities, NCS-detection activities and NCS-correction activities. To develop the agent of our system, we need to redefine some activities then, we execute the simulated system. The analysis of the experimentation results are given by Sesam in figure 6 and figure 7. These results indicate the number of NCS (USELESSNESS and UNPRODUCTIVENESS) detected by each agent during the same game. The horizontal axis presents the number of the step. The vertical axis presents the number of detected USELESSNESS (and UNPRODUCTIVENESS). The Red-Donkey agent detects a large number

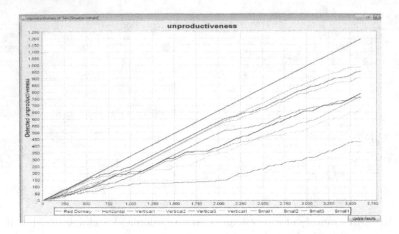

Fig. 7. Detection of the UNPRODUCTIVENESS by each agent over time

of USELESSNESS compared to the other agents but Small agents did not detect this type of NCS. This is explained by the number of cells which can be occupied by each piece (given its size). A Small can not block other pieces because when a Small moves, it occupies only one free cell and there is still another free cell which can be occupied by another piece (Horizontal, Vertical or Small). The curve of USELESSNESS of each agent is increasing in the beginning then it becomes constant which indicates that the agents detect less USELESSNESS over time. This is explained by the fact that the detected USELESSNESS have been corrected and so the agents do not fall in the same NCS next time or they have the ability to correct the problem if the detect a NCS detected before. The number of UNPRODUCTIVENESS detected by each agent is very large compared to the USELESSNESS. This NCS is generally more detected by the Red-Donkey than the other pieces. The Small detects generally less UNPRODUCTIVENESS compared to the others. This is explained by the fact that the more the piece is large the less it finds a possibility to move. The curves of UNPRODUCTIVE-NESS of all agents are increasing. Indeed, given that there are only two free cells so at maximum only two moves are possible at each step. We observe that all detected NCS are corrected automatically. According to these results, we the design can be validated because the agents are not blocked and they are able to overcome the problem when a NCS is detected.

1.7 Phase 5 (WD5): Simulation Based Implementation

The objective of this phase is to develop the system and also to verify that it is implemented exactly as described in the design phase. The system has been developed using the Jade platform. With JADE, the agent behaviour can be implemented as a set of activities and transitions (which is very suitable for AMAS agents because it allows us to focus principally on the behaviour of each agent). The execution of the system shows that it behaves exactly as expected in the design simulations. The agents are able to detect the NCS and correct them. The system gives always a solution but the number of steps differs from a game to another. Figure 8 presents the progress of the system in terms of number of steps made by each game over time. The progress curve has a decreasing shape. This indicates that our system is improving based on its experience. It learns from the previous games (by detecting Non Cooperative Situations and correct them) in order to self-adapt and better play the next games (by avoiding the problems previously detected). The implementation is so validate and the system can be deployed.

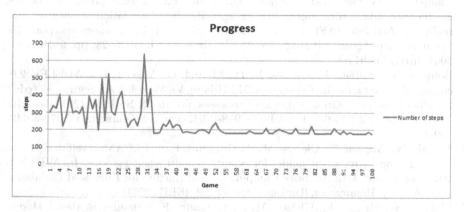

Fig. 8. Progress of the system in terms of number of steps

1.8   Phase 6 (WD6): Living-Design Deployment

At this phase, we have only to install and activate the system. The system is able to adapt to its environment. The agents self-design during their lives by detecting, avoiding and correcting themselves the encountered problems (Non Cooperative situations) and the number of steps made by the system to exit the Red-Donkey from the board is fixed to the minimum.

2   Conclusion

Adaptive Multi Agent Systems (AMAS) are complex and self-organizing systems with emergent functionality. ADELFE was proposed to guide the designer of such systems. Building such systems is not an easy task because the designer has to exhaustively find all Non Cooperative Situations in order to obtain a coherent global behaviour. ADELFE3.0 integrates a Simulation Based Design approach in order to assist the designer in the detection and the correction of the Non Cooperative Situations that an agent may encounter during its life. This approach gives to the agent the ability to self-design. This paper gave an example of application of ADELFE3.0 [7].

References

1. Gleizes, M.-P., Georg, J.-P., Glize, P.: Conception de systmes adaptatifs fonction-nalit mergente: la thorie des AMAS. Revue dIntelligence Articielle **17**(4/2003), 591–626 (2003)
2. Glize, P.: LAdaptation des Systmes Fonctionnalit mergente par Auto-Organisation Cooprative. Habilitation diriger des recherches, University of Paul Sabatier, Toulouse, France, June 2001
3. Cossentino, M.: Design Process Documentation Template. Experimental, 2011 IEEE Foundation for Intelligent Physical Agents, June 2011. http://www.pa.org/
4. Camps, V., Gleizes, M.-P., Bernon, C., Glize, P.: La conception de systmes multiagents adaptatifs: contraintes et spcicits. In: Association Francaise dIntelligence Articielle (AFIA). Atelier Methodologie et Environnements pour les Systmes Multi-Agents - Plate-forme AFIA, Grenoble, June 25–28, pp. 9–18, June 2001. http://aa.lri.fr/
5. Bonjean, N., Mefteh, W., Gleizes, M.-P., Maurel, C., Migeon, F. : ADELFE 2.0. chapitre/chapter dans. In: Cossentino, M., Hilaire, V., Molesini, A., Seidita, V. (eds.) Handbook on Agent-Oriented Design Processes, pp. 19–63. Springer (2014). doi:10.1007/978-3-642-39975-6_3, Print ISBN 978-3-642-39974-9, Online ISBN 978-3-642-39975-6
6. Mefteh, W., Migeon, F., Gleizes, M.-P., Gargouri, F.: S-DLCAM: a self-design and learning cooperative agent model for adaptive multi agent systems. In: WETICE 2013 Conference, 11th Adaptive Computing (and Agents) for Enhanced Collaboration (ACEC), Hammamet, Tunisia, June 17–20. IEEE (2013)
7. Mefteh, W., Migeon, F., Gleizes, M.-P., Gargouri, F. : Simulation Based Design for Adaptive Multi-Agent Systems with the ADELFE Methodology. International Journal of Agent Technologies and Systems **7**(1): 1–16 (2015)

Multi Agent Model Based on Chemical Reaction Optimization for Flexible Job Shop Problem

Bilel Marzouki[2(✉)] and Olfa Belkahla Driss[1,2]

[1] Stratégies d'Optimisation et Informatique IntelligentE, High Institute of Management, University of Tunis, 41, Street of Liberty Bouchoucha-City CP-2000-Bardo, Tunis, Tunisia
olfa.belkahla@isg.rnu.tn
[2] Higher Business School of Tunis, University of Manouba, Manouba, Tunisia
marzouki.bilel@gmail.com

Abstract. The Flexible Job Shop Problem (FJSP) is an extension of classical job shop problem such that each operation can be processed on different machine and its processing time depends on the machine used. This paper proposes a new multi-agent model based on the meta-heuristic Chemical Reaction Optimization (CRO) to solve the FJSP in order to minimize the maximum completion time (makespan). Experiments are performed on benchmark instances proposed in the literature to evaluate the performance of our model.

Keywords: Scheduling · Flexible job shop · Multi agent system · Chemical reaction optimization

1 Introduction

The Job Shop Problem (JSP) is considered one of the most difficult scheduling problems. This problem consists on assigning a set of operations on a set of machines such as each operation must be processed on one machine. The Flexible Job Shop Problem (FJSP) is an extension of the classical job shop problem (JSP) where each operation can be processed on different machines and its processing time depends on the machine used, thus FJSP is harder than JSP. FJSP is classified, as most of scheduling problems, NP-Hard in complexity theory [9]. Therefore the approximate methods are the most used to solve this problem. The flexible job shop problem has been studied for the first time by [4]. They have developed a polynomial algorithm to solve FJSP.

Meta-heuristics are the most used methods to solve the FJSP. Several researches are made based on tabu search as [3], which resolved the resource allocation problem using the rules of priority. The durations are varied resources functions, the assignment problem is solved then we get the classical job shop problem which is solved by a tabu search method. The neighborhood function used allows to permute two critical operations, then a reallocation of these critical operations are performed at predefined time intervals, this process is repeated until reaching the stop criterion which is the maximum number of iterations. Tabu search has been also used by [5], [6], [14], [17] to solve the FJSP.

M. Núñez et al. (Eds.): ICCCI 2015, Part I, LNAI 9329, pp. 29–38, 2015.
DOI: 10.1007/978-3-319-24069-5_3

[12], [19] and [20] used genetic algorithms in their researches to solve the FJSP. [15] have proposed a genetic hybridization algorithm and Hill Climbing approach to solve the FJSP as its objective is to minimize the makespan, the maximum load and the total load resources. [8] presented a mathematical model and heuristic approaches (integrated and hierarchical) for FJSP to solve real size problems. For the integrated approach, they used an algorithm that uses tabu search called ITS (Integrated approach with Tabu Search heuristic) and another algorithm that uses simulated annealing named ISA (Integrated approach with Simulated Annealing heuristic) for the allocation and sequencing problems consecutively. For the hierarchical approach, they used algorithms based on tabu search and simulated annealing named HSA / SA, HSA / TS, HTS / TS and HTS / SA. An approach named AIA was proposed by [2] for the FJSP and based on natural immune system.

Distributed methods are always used; we find the approach of [7] to solve the FJSP and it is also based on a tabu search method, this approach presented three classes of agents: Job agents, the resource agents responsible for satisfying the resource constraints and an Interface agent containing the core of the tabu search. [1] have developed a multi-agent approach based on the combination of tabu search and genetic algorithm method. This work proposes a new approach using two classes of agents: resource agents and Interface agent. Another multi-agent approach has been introduced in [16] based on the combination of genetic algorithm and tabu search, where a Scheduler agent applies a genetic algorithm for a global exploration of the search space whereas a local search technique is used by a set of cluster agents to guide the research in promising regions of the search space. So, distributed methods used for the flexible job shop problem show their effectiveness, that's why, we propose in this paper a multi-agent model based on a new meta-heuristic "Chemical Reaction Optimization" (CRO) for solving the FJSP to minimize the maximum completion time (makespan). In fact, CRO is proposed in 2010 by [13] to optimize combinatorial problems. Due to its ability to escape from the local optimal, it has been applied for solving many scheduling problem such as grid scheduling [10], multi-objective optimization of Flexible Job shop scheduling with maintenance activity constraints [11] where three minimization objectives are considered simultaneously: the maximum completion time, the total workload of machines and the workload of the critical machine, etc. Experimental comparisons demonstrated that CRO is one of the most powerful optimization algorithms.

The remainder of this paper is organized as follows. The details of the FJSP are presented in section 2. The CRO meta-heuristic is defined in section 3. We describe our proposed model namely Multi-Agent model based on Chemical Reaction Optimization for Flexible Job Shop Problem (MACRO—FJSP) in section 4. We give some experimental results in section 5. A conclusion and perspectives are given in section 6.

2   Problem Formulation

The Flexible Job Shop Problem can be formulated as follows [18]:
- A set of n jobs to be performed on m machines M_k, $k = 1, 2, ..., m$,
- Each job i consists of a sequence of N_J operations O_{ij}, $i = 1, 2, ... n$; $j = 1, 2, ... , N_J$,
- Execution of each operation O_{ij} requires a resource from a set of available machines.

- Each machine can perform only one operation at a time,
- Assigning an operation O_{ij} with a machine M_k causes the occupation of this machine throughout the execution time of the operation, denoted by $p_{i, j, k}$,
- Operation preemption is not allowed.

3   Chemical Reaction Optimization

Chemical Reaction Optimization is a meta-heuristic developed by [Lam et al, 2010] for optimization inspired by the nature of chemical reactions. Chemical modification of a molecule is started by a collision. There are two types of collision: uni-molecular and inter-molecular collision. The first type describes the situation when the molecule hits on some external substances; the second type is where the molecule collides with other molecules. The corresponding reaction change is called an elementary reaction; we consider four kinds of elementary reactions: On-wall collision ineffective, Decomposition, Inter-molecular collision ineffective and Synthesis. The two collisions ineffective implement local search (intensification) whereas decomposition and synthesis provide the diversification effect. The mixture between intensification and diversification make an effective search of the global minimum in the solution space. Each molecule has several attributes:

- The molecular structure (ω) captures a solution of the problem; it may be a number, a vector or a matrix.
- PE: Potential energy is defined as the objective function $PE_{\omega} = f(\omega)$.
- KE: Kinetic energy is the positive number and it quantifies the system tolerance to accept a worse solution than the existing one.
- NumHit: is a record of the total number of hits (collisions) a molecule has taken.
- MinStruct: the molecular structure with the minimum PE.
- Min PE: when a molecule achieved its MinStruct.
- MinHit: is the number of hits when a molecule performs MinStruct.

There are four basic types of reactions, each of which occurs in each iteration of CRO, they work to manipulate solutions (explore the solution space) and redistribute the energy among the molecules and the buffer, see algorithm 1.

On-wall ineffective collision represents the situation when the molecule collides with a wall of the container.

Decomposition refers to the situation in which a molecule hits a wall and breaks into several parts. The idea of decomposition is to allow the system to explore other areas of the solution space after enough local searches by the ineffective collisions.

Inter-molecular ineffective collision represents the situation when the molecules collide with each other.

Synthesis happens when two molecules hit against each other and fuse together. One molecule is produced. This function is the opposite of decomposition. The idea behind Synthesis function is the diversification of solutions.

Algorithm 1 CRO

1: **Input:** Objective function f and the parameter values
2: \\ Initialization
3: Set *PopSize, KELossRate, MoleColl, buffer, InitialKE,* α, and β
4: Create *PopSize* number of molecules
5: \\ Iterations
6: **while** the stopping criteria not met **do**
7:   Generate $b \in [0, 1]$
8:   **if** $b > MoleColl$ **then**
9:     Randomly select one molecule M_ω
10:     **if** Decomposition criterion met **then**
11:       Trigger *Decomposition*
12:     **else**
13:       Trigger *OnwallIneffectiveCollision*
14:     **end if**
15:   **else**
16:     Randomly select two molecules M_{ω_1} and M_{ω_2}
17:     **if** Synthesis criterion met **then**
18:       Trigger *Synthesis*
19:     **else**
20:       Trigger *IntermolecularIneffectiveCollision*
21:     **end if**
22:   **end if**
23:   Check for any new minimum solution
24: **end while**
25: \\ The final stage
26: Output the best solution found and its objective function value

---

4   Multi Agent Model Based on CRO Proposed for FJSP

We propose a multi-agent model based on the Chemical Reaction Optimization to solve the Flexible Job Shop Problem named MACRO—FJSP "Multi-Agent model based on CRO for Flexible Job Shop Problem" in order to minimize the makespan or the maximum completion time (CMAX). The architecture of our model consists of two classes of agents: an Interface agent and n Scheduler agents where n is the number of jobs in the problem. The objective of our model is to solve the Flexible Job Shop Problem by satisfying the various constraints such as the temporal constraints (precedence constraints) and the resource constraints. In the following, we describe each agent class.

4.1   Interface Agent

This agent is considered as an interface between the user and the program. It launches the program and creates the n Scheduler agents where n corresponds to the number of jobs, and then it generates initial solutions according to a predefined number of populations. It supplies the necessary information for every Scheduler agent to begin the global optimization phase; which is based on the Chemical Reaction Optimization (CRO) meta-heuristic; then it posts the final result.

The Interface agent has two types of knowledge. The static knowledge such as:

— The jobs and the operations to realize so its durations in the various resources.
— The initials solutions created by the Interface agent itself.
— The stopping criterion for the global optimization phase defined as "nb_iteration".

— The necessary variables for the execution of our program.
— The solutions found after the phase of optimization.
— The best solution found after the phase of optimization.
— β the threshold of diversification.
— α the maximal number of iteration allowed between two successive improve-
  ments
  And the dynamic knowledge as:
— The current initial solution and its makespan (CMAX)
— NumHit: the number of iteration for the on wall function
— MinHit: the value of the iteration which offers us an improved solution
— KE: the kinetic energy.

4.2   Scheduler Agent

This class of agent aims to make the phase of optimization based on the CRO meta-
heuristic.

The static knowledge of the Scheduler agent are cited as following:

— The random solution chosen by the Interface agent
— The job's operations associated to the agent and its various durations.

In the other hand, the dynamic knowledge of the Scheduler agent are:

— The current solution and its makespan.
— The inactive time interval on every resource.

4.3   Optimization Process

4.3.1   Global Optimization Process

The Interface agent sends one/two initial solutions to Scheduler agents for making the
global optimization phase based on the CRO meta-heuristic, see algorithm 2.

Algorithm 2 Main Algorithm

1.Input :n,m,nbop[],dure[][],α, β .MinHit,NumHit,popsize,KElossrate,initialKE,Molecoll
2.create(interface agent,scheduler agent,n)
3.send(interface agent,scheduler agent,duration)
4.generate(interface agent, initials-solutions)
5. While the stopping criterion not met {
6.generate(interface agent, b)
7. if (b>Molecoll) then send(interface agent,scheduler agent, s1)
else send(interface agent,scheduler agent, s1, s2)
8.if (b>Molecoll) and (NumHit-MinHit> α) then { Trigger Decomposition
send(scheduler agent, interface agent,s'1,s'2)   NumHit =0; MinHit=0; }
else if (b>Molecoll) and (NumHit-MinHit ≤ α) then {
9. trigger On wall ineffective collision
10. send(scheduler agent, interface agent,s'1); NumHit++; if (f(s'1) is the
best) then MinHit++ }
11.if (b ≤ Molecoll) and (Ke ≤β) then { Trigger Synthesis
send(Last scheduler agent, interface agent,s'1) } else if (b ≤ Molecoll) and
(Ke >β) {
12. Trigger inter-molecular ; KE=KE-KElossrate
13. send(Last scheduler agent, interface agent,s'1,s'2) }
14.}

4.3.2   Determination of the Initial Solution

The Interface agent is in charge of the creation of the initial solution by randomly generating it, see algorithm 3.

Algorithm 3 Generate initial solution
---
1. begin-date=0
2. end-date = 0
3. Foreach i in jobs do
4. Foreach j in operations do
5. Choose a random machine r such that the duration $(i,j,r) \neq 0$
6.end-date = begin-date+duration
7.begin-date = end-date
8.EndFor
9.EndFor

4.3.3   The Diversification Techniques

To guarantee better resolution of flexible job shop problem, we need to explore more search space. That's why diversification phase should be performed. When a better solution cannot be found after a certain threshold "α", the diversification phase is triggered. When the number of iterations is increased without a better solution after a certain threshold, it means that the best solution found was not replaced by one of its neighbors for a certain time, which is a sign that our method is probably trapped in a local optimum, In this situation, the "decomposition" must be performed with another initial solution and the values of MinHit and NumHit are reset. The other way to set the diversification phase is to execute synthesis function; after each execution of the "On Wall Function", the value of KE is reduced by the value of KElossrate, so after some iteration KE becomes lower than β so the Synthesis function to be executed.

The Decomposition. The aim of this function is to find two new solutions from the initial solution. This treatment is repeated twice and for each iteration we obtain a new solution, see algorithm 4.

Algorithm 4 Decomposition
---
1. for (int i = 1 ; i<2 ;i++) {
2. determine inactive time interval (scheduler agent, s1)
3. While (all operations not moved) and (one of operations cannot being moved)do
4. Foreach i in operations
5. Foreach rj in inactive time interval
6. if (the duration of the operation is included in an interval) and (i $\neq$ 1) and (precedence constraint= True) then
7. place(i,rj)
8. if (the duration of the operation is included in an interval) and (i = 1) then place(i,rj)
9. endwhile
10. if (not-affected-operations) then {
11. if (i = 1) then place (i,r) with r have a minimum cmax
12. else determine ressource r with minimum cmax and cmax > end-date (i-1)
13. place(i, r) }
14. Foreach not-affected-operations
15. place (i,r) with satisfying precedence constraint
16. place (i,r)/ i $\notin$ job which corresponds to the current scheduler agent
17. }

Synthesis. The principle of this function is to find a solution from both random solutions, this function is the opposite of the decomposition function, see algorithm 5.

Algorithm 5 Synthesis

---
1. If (the scheduler agent belongs to the first half of the solution) then {
2. aff=same-Affectations (scheduler agent,s1)
3. send (scheduler agent, aff, following scheduler agent) }
4. If (the scheduler agent belongs to the second half of the solution) then {
5. Foreach i in operations
6. if (i = 1) then
7. attribute1(i,r1,aff)
8. if (i ≠ 1) then
9. attribute2(i,r2,aff)
10.send (scheduler agent, aff, following scheduler agent) }

4.3.4 The Intensification Techniques

The intensification phase is one of the most important phases to better exploit the search space by executing the "On Wall Function" and "Inter Ineffective Collision"

On-wall Ineffective Collision. The idea of this function is to find one new solution from the initial solution. Every Scheduler agent receives the random solution from the Interface agent and tries to move the operations of the job associated to it. The treatment of this function is the same for the "decomposition".

Inter-molecular Ineffective Collision. The idea of this function is to find two new solutions from both random solutions.

The first new solution is composed of the first part of the first solution and the second part of the second solution. However, the second new solution is composed of the first part of the second solution and the second part of the first solution. The treatment of this function is the same for the "synthesis".

5   Computational Results

5.1   The Program Settings

The parameters of the CRO are set as follows, see table 1.

Table 1. The parameters of the CRO

Molecoll	b	a	NumHit	MinHit	buffer	InitialKE	KElossrate	beta	popsize	nb_iteration
0.5	[0..1]	3	0	0	0	200	12	25	1000	10

5.2   Results on Kacem Benchmark

We tested our model MACRO—FJSP on some instances of benchmark of Kacem [12] and we compared the results obtained by our model with the exact solutions of these instances provided by [12]. The experiment shows that our model reach the optimum for 25% on the Kacem benchmark, see table2.

Table 2. Results on instances of [kacem et al, 2002]

Name	n	m	UB	MACRO—FJSP
Instance 1	4	5	16	**16***
Instance 2	10	7	15	25
Instance 3	8	8	14	38
Instance 4	10	10	7	21

5.3   Comparisons on Fattahi Benchmark

We also tested our model MACRO—FJSP on some instances of benchmark of Fattahi [8] and we compared the results obtained by our model with those obtained by integrated approaches [8] called integrated approach with tabu search heuristic (ITS) and integrated approach with simulated annealing heuristic (ISA) and the approach of [2] based on natural immune system, see table 3. The experiment shows that our model reaches the optimum for 9 instances of 17 with a percentage equal to 53% of the instances that we have tested.

Table 3. Results on instances of [Fattahi et al, 2007]

Name	n	m	LB	UB	AIA	ISA	ITS	MACRO—FJSP
SFJS1	2	2	66	66	66	66	66	**66***
SFJS2	2	2	107	107	107	107	107	**107***
SFJS3	3	2	221	221	221	221	221	**221***
SFJS5	3	2	119	119	119	119	137	**119***
SFJS6	3	3	320	320	320	320	320	**320***
SFJS7	3	5	397	397	397	397	397	**397***
SFJS9	3	3	210	210	210	215	215	**210***
MFJS1	5	6	396	470	468	488	548	477
MFJS2	5	7	396	484	448	478	457	**464**
MFJS3	6	7	396	564	468	599	606	578
MFJS4	7	7	496	684	554	703	870	889
MFJS5	7	7	414	696	527	674	729	729
MFJS6	8	7	469	786	635	856	816	856

6   Conclusion and Perspectives

In this paper, we study the flexible job shop problem and we propose, to solve this problem, a multi-agent model named MACRO—FJSP "Multi-Agent model based on CRO for Flexible Job Shop Problem" which consists of a society agent and based on the meta-heuristic CRO for the optimization phase. Our model is composed of an Interface agent and Scheduler agents, where the Interface agent is considered as an interface between the user and the program and the Scheduler agents aim to make the phase of optimization by using the meta-heuristic CRO. The results obtained in our work now encourage us to continue the study of certain research lines such as the multi-objective FJSP, we can consider larger benchmarks and adapt other methods to generate initial solutions such as tabu search, Particle Swarm Optimization...

References

1. Azzouz, A., Ennigrou, M., Jlifi, B., Ghédira, K.: Combining tabu search and genetic algorithm in a multi-agent system for solving flexible job shop problem. In: Mexican International Conference on Artificial Intelligence (2012)
2. Bagheri, A., Zandieh, M., Mahdavi, I., Yazdani, M.: An artificial immune algorithm for the flexible job-shop scheduling problem. Future Generation Computer Systems 26(4), 533–541 (2010)
3. Brandimarte, P.: Routing and scheduling in a flexible job shop by tabu search. Annals of Operations Research 4(1–4), 157–183 (1993)
4. Bruker, P., Schlie, R.: Job-shop scheduling with multipurpose machines. Computing 45, 369–375 (1990)
5. Chambers, J.B.: Classical and Flexible Job Shop Scheduling by Tabu Search. PhD thesis, University of Texas at Austin, Austin, U.S.A (1996)
6. Dauzère-Pérès, S., Paulli, J.: An Integrated Approach for Modeling and Solving the General Multiprocessor Job Shop Scheduling Problem using Tabu Search. Annals of Operations Research 70(3), 281–306 (1997)
7. Ennigrou, M., Ghédira, K.: Solving flexible job shop scheduling with a multi-agent system and tabu search. In: Procs. 6th Int. J. Conf. on International Conference on Enterprise Information Systems, Porto, Portugal, vo. l2, pp. 22–28 (2004)
8. Fattahi, P., Saidi Mehrabad, M., Jolai, F.: Mathematical modeling and heuristic approaches to flexible job shop scheduling problems. Journal of Intelligent Manufacturing 18(331–342), 2007 (2007)
9. Garey, E.L., Johnson, D.S., Sethi, R.: The Complexity of Flow-shop and Job-shop scheduling. Mathematics of Operations Research 1, 117–129 (1976)
10. Jin, Xu, Lam, A., li, V.: Chemical Reaction Optimization for the Grid Scheduling Problem. IEEE ICC (2010)
11. Jun, L., Quan-ke, P.: Chemical-reaction optimization for flexible job-shop scheduling problems with maintenance activity. Applied Soft Computing 12, 2896–2912 (2012)
12. Kacem, I., Hammadi, S., Borne, P.: Approach by Localization and Multi-Objective Evolutionary Optimization for Flexible Job Shop Scheduling Problems. IEEE Trans. Systems, Man and Cybernetics 32(1), 1–13 (2002)
13. Lam, A., Li, V.: Chemical-reaction-inspired metaheuristic for optimization. IEEE Transactions on Evolutionary Computation 14 (2010)

14. Mastrolilli, M., Gambardella, L.M.: Effective neighborhood functions for the flexible job shop problem. Journal of Scheduling (2000)
15. Motaghedi-larijani, A., Sabri-laghaie, K., Heydari, M.: Solving Flexible Job Shop Scheduling with Multi Objective Approach. International Journal of Industrial Engineering and Production Research 21(4), 197–209 (2010)
16. Nouri, H.E., Belkahla Driss, O., Ghedira, K.: Genetic algorithm combined with tabu search in a holonic multiagent model for flexible job shop scheduling problem. In: 17th International Conference on Enterprise Information Systems (ICEIS) (2015)
17. Ponnambalam, S.G., Aravindan, P., Rajesh, S.V.: A Tabu Search Algorithm for Job Shop Scheduling. International Journal of Advanced Manufacturing Technology (2000)
18. Saad, I., Boukef, H., Borne, P.: The comparison of criteria aggregative approaches for the multi-objective optimization of flexible job-shop scheduling problems. In: Fourth Conference on Management and Control of Production and Logistics, MCPL 2007, pp. 603–608, 2007
19. Wannaporn, T., Thammano, A.: Modified Genetic Algorithm for Flexible Job-Shop Scheduling Problems. Procedia Computer Science 12, 122–128 (2012)
20. Zhang, H., Gen, M.: Multi stage based genetic algorithm for flexible job shop scheduling problem. Journal of Complexity International 11, 223–232 (2005)

MATS–JSTL: A Multi-Agent Model Based on Tabu Search for Job Shop Problem with Time Lags

Madiha Harrabi[2(✉)] and Olfa Belkahla Driss[1,2]

[1] Stratégies d'Optimisation et Informatique IntelligentE, High Institute of Management, University of Tunis, 41, Street of Liberty Bouchoucha-City CP-2000-Bardo, Tunis, Tunisia
olfa.belkahla@isg.rnu.tn
[2] Higher Business School of Tunis, University of Manouba, Manouba, Tunisia
madiha.harrabi@gmail.com

Abstract. The Job Shop problem with Time Lags (JSTL) is an important extension of the classical job shop scheduling problem, in that additional constraints of minimum and maximum time lags existing between two successive operations of the same job are added. The objective of this work is to present a distributed approach based on cooperative behaviour and a tabu search metaheuristic to finding the scheduling giving a minimum makespan. The proposed model is composed of two classes of agents: a Supervisor Agent, responsible for generating the initial solution and containing the Tabu Search core, and Resource_Scheduler Agents, which are responsible for moving several operations and satisfaction of some constraints. Good performances of our model are shown through experimental comparisons on benchmarks of the literature.

Keywords: Job shop scheduling · Time lags · Tabu search · Multi-agent system

1 Introduction

Different works on scheduling were concerned with the analysis of single-machine systems, parallel-machines systems, and shop problems. Among all these problems, we focus in this paper to optimize the total duration of the scheduling often called makespan for the Job Shop problem with time lags. The job shop problem with time lags is defined by a set of n jobs which have to be sequenced on a set of m machines. The constraints of time lags is a generalization of precedence constraints which concern the minimum and maximum waiting time existing between two successive operations of the same job. In this paper, we propose a Multi-Agent model based on Tabu Search for the Job Shop problem with Time Lags (MATS–JSTL) in order to minimize the makespan. This paper is organized as follows. In Section 2, we present related works. In Section 3, we present the problem formulation. Then, we describe the tabu search method on which our model is based. In the following sections, the Multi-Agent system, and the adaptation of the tabu search parameters are provided. Then, we discuss experimental results of our model. Finally, section 8 contains our conclusions and remarks with some future directions.

© Springer International Publishing Switzerland 2015
M. Núñez et al. (Eds.): ICCCI 2015, Part I, LNAI 9329, pp. 39–46, 2015.
DOI: 10.1007/978-3-319-24069-5_4

2   Related Works

2.1   Previous Published Works on Scheduling Problems with Time-Lags

The time lags constraints were introduced for the first time by [14] in a problem with parallel machines. In the literature, there are two kinds of time lags constraints: we found time lags between successive operations of the same job and generic time lags between whatever pairs of operations introduced by [13] in the Job Shop problem. The first kind of time lags was introduced by [15] in the single-machine problems by including minimum and maximum times lags. [10] introduced both minimum and maximum time lags in the single machine problem. [6] proposed different methods for solving Flow Shop problem with minimum and maximum time lags. In [9] an exact procedure for a general RCPSP with minimum and maximum time lags is proposed.

2.2   Previous Published Works on Job Shop Problem with Time-Lags

The job shop problem with time lags constraints is very little studied in the literature and the number of research and solution methods is limited. [1] proposed a branch and bound method. In [3] an adaptation of the tabu search method was presented. [5] investigated a memetic algorithm. In [4], a method of priority rules had been proposed. A construction heuristics had been proposed by [6]. In [11] a CDS (Climbing Discrepancy Search) was proposed. [2] used a cooperative Multi-Agent system, composed of Interface Agent, Job Agent, and Resource Agent.

3   Problem Formulation

The Job Shop Scheduling problem with Time Lags is formulated as follows.

M	set of machines
O	set of operations to schedule
Z	completion time of the last operation
E_k	set of operations processed on machine k
t_i	starting time of operation i
p_i	processing time of operation i
s_i	next operation of operation i depending on job
TL_i^{min}	minimal time-lags between i and si
TL_i^{max}	maximal time-lags between i and si

$$a_{ij} = \begin{cases} 1 & \text{if operation } j \text{ is processed before } I \\ 0 & \text{otherwise} \end{cases}$$

H	a positive large number

Minimize Z,
Subject to

$$(H + p_j) \, a_{ij} + (t_j - t_i)\_\geq p_i \qquad \forall \, (i, j) \in E_k, \forall k \in M \qquad (c1)$$

$$(H + p_j) \, (1 - a_{ij}) + (t_i - t_j) \geq p_j \quad \forall \, (i, j) \in E_k, \forall k \in M \qquad (c2)$$

$$t_i + p_i + TL_i^{min} \leq t_{si} \qquad \forall i \in O \qquad (c3)$$

$$t_{si} \leq t_i + p_i + TL_i^{max} \qquad \forall i \in O \qquad (c4)$$

$$Z \geq t_i + p_i \qquad \forall i \in O \qquad (c5)$$

$$t_i \geq 0 \qquad \forall i \in O \qquad (c6)$$

$$a_{ij} \in \{0; 1\} \qquad \forall \, (i, j) \in E_k, \forall k \in M \qquad (c7)$$

Constraints (c1) and (c2) represent the machine disjunctions, which mean that two operations i and j use the same resource. These constraints express either that operation i precedes the operation j or the reverse. Constraint (c3) relates to the minimum time lags, and the constraint (c4) represents the maximum time lags. Constraint (c5) determines the completion time of the last operation, which represents the optimization criteria (makespan or Cmax). Constraints (c6) and (c7) give the domain of definition of variables a_{ij} and t_i.

4   Tabu Search

The Tabu Search (TS) algorithm was first proposed by [8] which has been successfully used for solving a large number of combinatorial optimization problems specially scheduling problems. Tabu search uses a local neighborhood search procedure to iteratively move from one potential solution S to an improved solution S' in the neighborhood of S, until some stopping criterion has been satisfied. A particularity of TS is that it explicitly employs the history of the search, and storing the selected move in a structure named tabu list for a certain number of iterations called Tabu list size.

5   The Multi-Agent Model

5.1   Global Dynamic

A multi-agent system is a computerized system composed of multiple interacting intelligent agents who cooperate, communicate, and coordinate with each other to reach common objectives. We propose a multi-agent model based on tabu search for job shop problem with time lags. It is composed of two classes of agents: Supervisor Agent and a set of m Resource_Scheduler Agents with m refers to the number of machines, see figure 1.

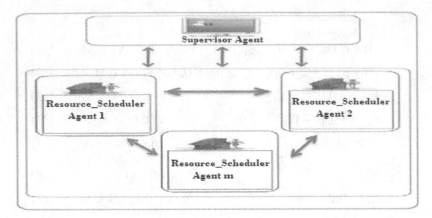

Fig. 1. The Multi-Agent model based on Tabu Search for Job Shop problem with Time Lags

5.2   Supervisor Agent Description

This agent contains the core of the tabu search algorithm. It aims to launch the program, build an initial solution, create Resource_Scheduler agents, and assign for each agent its workspace defined by all operations belonging to the same resource. It subsequently provides the necessary information for each Resource_Scheduler Agent. This agent is unsatisfied as long as the maximal number of iteration is not reached, and it aims to detect that the problem has been resolved. Static knowledge of this agent is composed of the initial solution S0 from which begins the optimization process, the used stopping criterion, the size of the Tabu List, the used diversification criterion. Its dynamic knowledge consists of: the current solution and its makespan, neighbor solutions and their makespans, the best found schedule and its makespan, the tabu list elements, the performed number of iterations, the number of iterations after the last improvement. The Supervisor agent is satisfied when the stopping criterion is reached; in this case it provides the found-solution to the user.

5.3   Resource_Scheduler Agent Description

The Resource_Scheduler Agents are created by the Supervisor Agent according to resource number. The decisions of scheduling are progressively negotiated between the different Resource_Scheduler Agents involved in the system to satisfy different constraints of precedence, disjunctive and time lags. Static knowledge of this agent is composed of the execution time of each job on all machines and time lags between operations. Dynamic knowledge as current solution and the job to move sent by the Supervisor Agent. Communication of Resource_Scheduler Agents handles the sending and receiving messages between each other containing the new locations of operations. This agent is satisfied when the constraints of succession and disjunction on his workspace operations are met, otherwise it is dissatisfied and he tries to solve the problem. The algorithm describes the moving operation procedure. cf. algo1.

Moving operation algorithm

1. : *Start*
2. : *move__operation ← true*;
3. : *msg.receiver("job""op")*;
4. : *compute__inactivity__intervals*;
5. : *insert__op ← satisfaction__verification(c3, c4, c5)*;
6. : if *constraints__satisfaction ← false*;
7. : search new location
8. : else
9. : *insert__op ← true*;
10. : *msg.setContent("job""op", start__date, end__date)*;
11. : Update inactivity__intervals ;
12. : *End*

Algo. 1. Moving operation algorithm

6 Adaptation of Tabu Search Elements

6.1 Initial Solution

The initial solution is the starting embedding used for the algorithm to begin the search of better configurations in the search space. In this implementation the starting solution is randomly generated. Then it proceeds iteratively to visit a series of locally best configurations following a neighborhood function.

6.2 Neighborhood Function

In our approach, we choose to use a neighborhood structure based on insertion move. In fact, the move strategy is to move various operations in the inactivity intervals previously calculated by the Resource_Scheduler Agent. Respecting the precedence constraints and time lags constraints, the Resource_Scheduler Agents work together to find the new location for each operation that meets all these requirements. After moving all operations, each Resource_Scheduler Agent sends a message to the Supervisor Agent containing new locations of operations moved, so the Supervisor Agent generates the neighbor solution and launches the evaluation phase.

6.3 Evaluation of the Current Solution

The best non-tabu neighbor among the neighborhoods of the current solution will be selected for the next iteration, for this it would be necessary to evaluate all neighbors. The Supervisor Agent, after receiving all new locations of the operations of different jobs, from Resource_Scheduler Agents, it builds neighbor solution and calculates its makespan, chooses the best non-tabu neighbor and start the next iteration.

6.4   Tabu-List

A fundamental element of tabu search is the use of a flexible memory, short-term, which keeps track of some recent past movements. The size of the tabu list used is a static size "LT". If the size of the tabu list exceeds the maximum allowed size, we remove from this list the oldest item (FIFO strategy: First In First Out: The first arrival element is the first to be deleted).

6.5   Diversification

When the number of iterations continuously increases without any improvement in the current solution, this means that the best solution found has not been replaced by one of these neighbors for some time, which is a sign that tabu search was probably trapped in a local optimum. In this situation a new schedule is built by the Supervisor Agent using a new insertion order of jobs in order to explore a new region of the search space. And it restarts again the resolution process. After that, the number of iterations diversification, i.e. the number of iterations after the last improvement, is reset and the research process continuous by considering the solution obtained by diversification phase as a new current solution until reaching the stopping criterion.

6.6   Stopping Criterion

In our model, we adopted as a stopping criterion, the maximum number of iterations. The tabu search process attempts to improve the solution after a maximum number of iterations. Then, the best found scheduling during the search and its makespan are returned.

7   Experimentation Results

Several experiments were conducted on a set of benchmarks for Job Shop problems with minimum and maximum time lags constraints existing in the literature, to test the effectiveness and the performance of our model using the JADE platform. These instances correspond to classical problems of Job Shop, and were constructed by [3]. Instances la01 to la05 (10 jobs, 5 machines) of [12] and instances ft06 (6 jobs, 6 machines) of [7] modified to introduce the constraints of time lags. The number of iterations is 1000.

Finally we choose to compare our distributed model MATS–JSTL with two centralized approaches, the approach of [4] which is based on a operation insertion heuristic OI, and the approach of [1] that is based on a job insertion heuristic JI. Then, we choose to compare our approach with the only distributed approach MAJSPTL of [2] which is a multi-agent approach. The results of comparison between several approaches are given in table 1 showing results of different instances. The obtained results show that our model MATS–JSTL provides good results for most existing benchmarks in the literature in terms of makespan. For all used instances, we find a

Table 1. Results for JI Heuristic/OI Heuristic/MAJSPTL/MATS–JSTL

| Instances | Makespan | | | | CPU time (sec) |
	JI Heuristic	OI Heuristic	MAJSPTL	MATS–JSTL	MATS–JSTL
ft06_0_0	96	83	80	**80***	7
ft06_0_0.5	72	109	96	**69***	4
ft06_0_1	72	58	67	64	5
ft06_0_2	70	55	72	60	2
la01_0_0	1258	1504	1341	**1020***	1117
la01_0_0.5	1063	1474	1231	**980***	94
la01_0_1	928	1114	879	907	62
la01_0_2	967	948	980	**894***	37
la02_0_0	1082	1416	1264	1085	213
la02_0_0.5	1011	1207	1099	**973***	154
la02_0_1	935	1136	923	**694***	121
la02_0_2	928	895	978	**874***	473
la03_0_0	1081	1192	1023	1178	198
la03_0_0.5	930	1085	898	946	246
la03_0_1	886	931	904	894	168
la03_0_2	808	787	854	**761***	34
la04_0_0	1207	1346	1387	1393	218
la04_0_0.5	870	1156	921	1234	21
la04_0_1	1010	857	1053	915	78
la04_0_2	892	838	870	**827***	31
la05_0_0	1080	1224	1219	1168	178
la05_0_0.5	935	1208	1094	**962***	87
la05_0_1	814	964	952	**803***	61
la05_0_2	749	683	751	763	44

makespan value better than Artigues's method [1] (JI heuristic) in 62% of instances, better than Caumand's method [4] (OI heuristic) in 75% of instances, and better than Ben Yahia and Belkahla's method [2] (MAJSPTL) in 75%. CPU times (sec) of the solutions found by our model given in table 1, are reasonable.

8   Conclusion and Future Researches

In this paper, we have proposed a multi-agent model based on Tabu Search meta-heuristic. It consists of Supervisor Agent and Resource_Scheduler Agents in interaction, trying to find the best scheduling. Experiments made on already existing benchmarks, show the effectiveness of our model in some cases in terms of makespan.

The results were encouraging; it would be interesting to develop other aspects considered in our future researches. We extend our model by introducing a heuristic in the construction of initial solution and diversification phase. We can study another extension of the Job Shop problem with time lags introducing constraints time lags between generalized pair of operations that is very little studied in the literature. We can also propose hybridization of metaheuristics via the Multi-Agent dynamics, such as simulated annealing, ant colonies or the genetic algorithm to better explore the search space to improve the solution.

References

1. Artigues, C., Huguet, M., Lopez, P.: Generalized disjunctive constraint propagation for solving the job shop problem with time lags. Engineering Applications of Artificial Intelligence **24**, 220–231 (2011)
2. Ben Yahia, A., Belkahla Driss, O.: A multi-agent approach for the job shop problem with time lags. In: International Conference on Artificial Intelligence, ICAI, Tunisie (2013)
3. Caumond, A., Gourgand, M., Lacomme, P., Tchernev, N.: Métaheuristiques pour le problème de jobshop avec time lags : Jm|li,s j(i)|Cmax. In: 5ème Confêrence Francophone de Modélisation et SIMulation (MOSIM 2004). Modélisation et simulation pour l'analyse et l'optimisation des systèmes industriels et logistiques, Nantes, France, pp. 939–946 (2004)
4. Caumond, A., Lacomme, P., Tchernev, N.: Feasible schedule generation with extension of the Giffler and Thompson's heuristic for the job shop problem with time lags. In: International Conference of Industrial Engineering and Systems Management, pp. 489–499 (2005)
5. Caumond, A., Lacomme, P., Tchernev, N.: A memetic algorithm for the job-shop with time-lags. Computers Operations Research **35**, 2331–2356 (2008)
6. Deppner, F.: Ordonnancement d'atelier avec contraintes temporelles entre opérations. PhD thesis, Institut National Polytechnique de Lorraine (2004)
7. Fisher, H., Thompson, G.L.: Probabilistic learing combination of local jobshop scheduling rules. In: Industrial Scheduling, pp. 225–251. Prentice Hall (1963)
8. Glover, F.: Future paths for Integer Programming and Links to Artificial Intelligence. Computers and Operations Research **5**, 533–549 (1986)
9. Heilman, R.: A branch-and-bound procedure for the multi-mode resource-constrained project scheduling problem with minimum and maximum time lags. European Journal of Operational Research **144**(2), 348–365 (2003)
10. Hurink, J., Keuchel, J.: Local search algorithms for a single machine scheduling problem with positive and negative time-lags. Discrete Applied Mathematics **112**, 179–197 (2001)
11. Karoui, W., Huguet, M.-J., Lopez, P., Haouari, M.: Méthode de recherche à divergence limitée pour les problèmes d'ordonnancement avec contraintes de délais [Limited discrepancy search for scheduling problems with time-lags]. In: 8ème ENIM IFAC Confêrence Internationale de Modélisation et Simulation (MOSIM 2010), Hammamet, Tunisie, May 10–12, 2010
12. Lawrence, S.: Supplement to Resource Constrained Project Scheduling : An experimental investigation of Heuristic Scheduling Techniques. Graduate School of Industrial Administration, Carnegie Mellon University, Pittsburgh, PA (1984)
13. Lacomme, P., Huguet, M.J., Tchernev, N.: Dedicated constraint propagation for Job-Shop problem with generic time-lags. In: 16th IEEE Conference on Emerging Technologies and Factory Automation IEEE Catalog Number: CFP11ETF-USB, Toulouse, France (2011). ISBN 978-1- 4577-0016-3
14. Mitten, L.G.: Sequencing n jobs on two mahines with arbitrary timlags. Management Science 5(3) (1958)
15. Wikum, E.D., Llewellyn, D.C., Nemhauser, G.L.: One-machine generalized precedence constrained scheduling problems. Operations Research Letters **16**, 87–99 (1994)

A Model of a Multiagent Early Warning System for Crisis Situations in Economy

Marcin Hernes[1], Marcin Maleszka[2(✉)], Ngoc Thanh Nguyen[2], and Andrzej Bytniewski[1]

[1] Wrocław University of Economics, ul. Komandorska 118/120, 53-345 Wrocław, Poland
{marcin.hernes,andrzej.bytniewski}@ue.wroc.pl
[2] Wroclaw University of Technology, Wybrzeże Wyspiańskiego 27, 50-370 Wrocław, Poland
{marcin.maleszka,ngoc-thanh.nguyen}@pwr.edu.pl

Abstract. During last decades, the world has experienced a large number of economic crises, which were not confined to an individual economy, but affected directly, or in-directly almost every country all over the world. As a result, a number of international organizations, governments, and private sector institutions have begun to develop an early warning system as a monitoring system to detect the possibility of occurrence of an economic crisis in advance and to alert its users to take preventive actions. However, each of the systems addresses just one selected branch of economy, or particular spheres of company operation, and only a limited and small group of users apply them. Therefore this paper focuses on developing a conception of a model of a multiagent early warning system for crisis situations in economy. The system will include all branches of economy, and it may be used by any group of users. It will have the ability to predict unfavorable economic situations at a local, scattered, and integral level.

Keywords: Multiagent systems · Crisis in economy · Multiagent prediction · Prediction system · Knowledge management

1 Introduction

Early warning systems (EWS) enable people to recognize hazards and to start adequate recovery processes depending on circumstances taking place in various spheres of human activity. The types of systems are used for example to warn against imminent natural disasters or epidemics.

The systems are becoming more and more important from an economic point of view [7, 9]. There are already in use systems of early warning with respect to mainly financial markets or conditions of companies [8, 9]. However, each of the systems addresses just one selected branch of economy, or particular spheres of company operation, and only a limited and small group of users apply them [3].

The occurrence of unfavourable economic phenomena (for example bankruptcies of companies providing financial or tourist services, considerable deviations of exchange rates of some currencies, power industry problems, considerable drop in prices of agricultural products) requires, however, a complex approach to the functioning of

© Springer International Publishing Switzerland 2015
M. Núñez et al. (Eds.): ICCCI 2015, Part I, LNAI 9329, pp. 47–56, 2015.
DOI: 10.1007/978-3-319-24069-5_5

early warning systems before a crisis in economy appears. On the one hand, functioning of such systems shall focus on all branches of economy. On the other hand, however, various groups of users shall make use of the systems, i.e. entrepreneurs, government administration clerks, civil servants, as well as individual users. The system should also be capable of understanding phenomena taking place in economy, determining their context and automatically reacting to possible crisis events (for example by suggesting solutions to problems, or even taking action aimed at avoiding threats). Realization of early warning processes using the multiagent system involves the need for permanent cooperation between a human (people) and an agent (agents). The cooperation may take place on various levels, for example it may concern sources of information on the basis of which agents function, this may be information from electronic sources, or opinions of experts (people). The cooperation may also concern interpretation of phenomena taking place in a given economic environment or actions connected with reacting to the possibility of crisis events. In order to obtain general knowledge of forecasted crisis events, knowledge of individual members of a collective needs to be integrated.

The aim of this paper is to develop a model of a multiagent early warning system for crisis situations in economy. The system will include all branches of economy, and it may be used by any group of users.

This paper is organized as follows: the first part shortly presents the state-of-the-art in the considered field; next, the conception of a model of multiagent early warning system is developed; the method for verification of the system is discussed in the last part of paper.

2   Related Works

There have been a lot of theoretical and empirical works to understand the economic crises and build EWS that provides signal for possible economic crisis. Based on the surveys of the literature [2], [7], [8],[9], two methods are widely used to build EWS: the Probability model and the Signal approach. The probability model is based on regression model using limited dependent variables to estimate the possibility of a financial crisis [2]. The signal approaches are the most commonly used method which monitors the evolution of several indicators. If any of signal indicators exceeds a given threshold during the period preceding a crisis, the EWS is activated to warn about a possible crisis [9]. Meanwhile, some research works considered EWS as a pattern recognition problem since the distinctive feature of crisis makes it possible to distinguish critical and normal economic situations using a pattern classifier [7]. The tools used in EWS can be based on mathematics, statistics, economics, or artificial intelligence [8], [10].

The overall model of existing EWSs consists of the following modules [15]:

1. **Monitoring and warning module.** The module is responsible for continuous monitoring of the environment (by gathering from different sources information concerning values of parameters, attributes of objects, actions, events) and for predicting and forecasting values of parameters, attributes, operations and events in an economy. In order to perform these actions, various methods are used. The module

cooperates with a risk knowledge module. As a result of module's operation we are able to assess economic situation, generate warnings and raise alarms. The results are then sent to the dissemination and communication module.

2. **Risk knowledge module.** The module includes patterns, created by experts, which enable assessment of the risk of possible future crisis events. Additionally, in the module takes place a process of learning on the basis of experience, i.e. economic situations which took place in the past and have now been connected with causes of a present crisis event. The module closely cooperates with the monitoring and warning module, which results in two-way information exchange.

3. **Dissemination and communication module.** The module is responsible for communication with users. It sends to them information concerning a current economic situation, level of risk, warnings and alarms. The information is personalized depending on the needs of users (e.g. entrepreneurs, natural persons, clerks).

4. **Response module.** The module contains patterns of reactions to the possibility of crisis events and suggests decisions a user shall take. The module can also react automatically (take decisions, actions) to a possibility of crisis events.

The model relates to early warning systems applied in various areas such as natural disasters, epidemics, as well as economy. The model is used also in relation to individual branches of economy. Various early warning systems (e.g. against financial crises, bankruptcies) are created on the basis of this model.

The IT tools applied today for early warning systems enable only an analysis of the form of information, relationships between economic values, yet do not support the process of analyzing its meaning. However, the need to make analysis on the basis of not only information and knowledge but also experience, which has been treated as the human domain to date, arises increasingly more frequently. Therefore, it is becoming reasonable to employ such tools as cognitive agents [13], which often cooperate with each other as part of a multiagent system in order to achieve the set goal.

Taking into consideration the human-agent collectives issues, the work [6] states, that human-agent collectives are a new class of socio-technical systems in which humans and smart software (agents) engage in flexible relationships in order to achieve both their individual and collective goals. Sometimes humans take the lead, sometimes the computer does and this relationship can vary dynamically. The authors conclude, that key research challenges in human-agent collectives include achieving flexible autonomy between humans and software, and constructing agile teams that conform and coordinate their activities.

Integration of collective knowledge is considered in different ways and on different levels. Previous attempts tended to predict group performance based on some statistics involving members' performances. For example, [1], [14], [16] reported that group performance is an average of individual performance. In parallel to the mentioned aspects, research has showed that team knowledge is greater than the collection of knowledge of individual team members and it emerges as a result of interactions among team members. Author of [5] argue that the interaction of a task and a team are important. In works [4],[11], [12], a formal mathematical model for knowledge integration is presented, in which the consensus-based knowledge functions for generating integration of knowledge have been defined. It was also showed that a collective is more intelligent than one single member.

Subject literature, however, lacks an approach that would make it possible to devise a model for the early warning system to detect possibility of economic crisis in advance related to all branches of economy. Note that such a system can be used by different groups of users (economic entities). Also human-agent collectives' knowledge integration issues are scarcely considered in literature of the subject.

3 Model of Multiagent Early Warning System

The role of the system is to predict unfavorable economic situations at the local (enterprises), scattered (branches) and integral (country economy or group of countries economy) level. Figure 1 presents overall conception of the functional architecture of a multiagent system of early warning before the occurrence of crises in the economy.

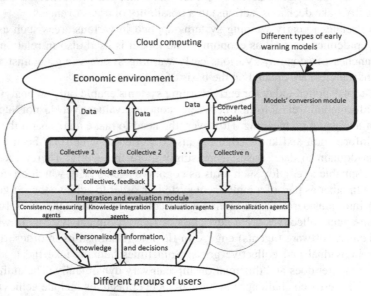

Fig. 1. Overall conception of the functional architecture of multiagent system of early warning before the occurrence of crises in the economy.

The system consist of the following components:

Human-agent collectives consist of several human and cognitive agents (based on LIDA and JADE platforms). The purpose of each collective member is to analyze a given economic environment. Each agent, running in a given collective, has been implemented with a different early warning model related to the same branch. The task of each expert is to develop an opinion about a current situation in the frame of considered branch. Each collective analyses the different branches and there are also global economy collectives. A collective member's knowledge state is represented by multi-attribute, multi-value semantic structure. Among other things, knowledge acquisition, a never-ending learning and learning by experience processes are performed by collectives.

Integration and evaluation module, consist of the following components:

1. Consistency measuring agents. These agents main task is to guarantee the consistency of the solution. According to [12] it is assumed that between elements of set X a distance function can be defined:

$$d : U \times U \to [0,1] \tag{1}$$

which is:

Nonnegative, i.e. $\forall x, y \in U : d(x, y) \geq 0$, $\tag{2}$

Reflexive, i.e. $\forall x, y \in U : d(x, y) = 0$ if $x=y$, and $\tag{3}$

Symmetrical, i.e. $\forall x, y \in U : d(x, y) = d(y, x)$, $\tag{4}$

where $[0,1]$ is the closed interval of real numbers between 0 and 1. Pair (U, d) is called a distance space. The definition of a distance function is independent the structure of elements of U. The following symbols will be used (according [11]):

- The matrix of distances between the elements of collective knowledge profile X:

$$D^X = \left[d_{ij}^x\right] = \begin{bmatrix} d(x_1, x_1) & \cdots & d(x_1, x_M) \\ \vdots & \ddots & \vdots \\ d(x_M, x_1) & \cdots & d(x_M, x_M) \end{bmatrix} \tag{5}$$

- The vector of average distances between an element to the rest (for $M > 1$)

$$W^X = (w_1^x, w_2^x, \ldots, w_M^x), \text{ where:} \tag{6}$$

$$w_i^x = \frac{1}{M-1} \sum_{j=1}^{M} d_{ij}^x, \tag{7}$$

for i = 1, 2, . . . , M. Notice that although in the above sum there are M components, the average is calculated only for $M - 1$. This is because for each i value $d_{ii}^x = 0$,

- The average distance in profile X:

$$d_{mean}(X) = \begin{cases} \dfrac{1}{M} \sum_{i=1}^{M} w_i^x & \text{for } M > 1 \\ 0 & \text{for } M = 1 \end{cases} \tag{8}$$

Additionally, for defining the consistency functions we introduce the following symbols:

- Variance of set X:

$$Variance(X) = \frac{1}{M^2} \sum_{i=1}^{M} \sum_{j=1}^{M} (d_{ij}^x - d_{mean}(X))^2 \qquad (9)$$

- Variance of vector W^x:

$$Variance(W^x) = \frac{1}{M} \sum_{i=1}^{M} (w_i^x - d_{mean}(X))^2 \qquad (10)$$

- Standard deviation of set X:

$$Stddev(X) = \sqrt{Variance(X)} \qquad (11)$$

- Standard deviation of vector W^x:

$$Stddev(W^x) = \sqrt{Variance(W^x)} \qquad (12)$$

Coefficient of variation of set X:

$$Coef(X) = \frac{stddev(X)}{d_{mean}(X)} \qquad (13)$$

Coefficient of variation of vector W^x:

$$Coef(W^x) = \frac{stddev(W^x)}{d_{mean}(X)} \qquad (14)$$

These parameters are applied for defining the following consistency functions:

$$c_1(X) = 1 - Variance(X), \qquad (15)$$

$$c_2(X) = 1 - Variance(W^x), \qquad (16)$$

$$c_3(X) = 1 - Stddev(X), \qquad (17)$$

$$c_4(X) = 1 - Stddev(W^x), \qquad (18)$$

$$c_5(X) = 1 - Coef(X), \qquad (19)$$

$$c_6(X) = 1 - Coef(W^x). \qquad (20)$$

2. Knowledge integration agents. The task of this agent is to integrate knowledge of the collective. The most basic approach is to calculate the average or median of member knowledge states. For complex knowledge structures this is often a non-trivial task. We define the integration function as follows [11]:

By an integration function in space (U, d) we call a function

$$I : \prod(U) \to 2^U \qquad (21)$$

For a collective $X \in \prod(U)$ the value $I(X)$ is called the *knowledge state* of collective X.

3. Evaluation agents. The evaluation functions will be defined on two levels:
   - consistency of knowledge; in considered system, the postulates for consistency functions and the consistency functions for the human-agent collectives knowledge takes into consideration different structures of economic knowledge representation (particular the semantic net).
   - performance (efficiency) of decisions or reaction results (the examples of evaluation functions are described in section 4 of this paper). The evaluation results are sent to collectives as feedback.

4. Personalization agents. As each person using the system may have different usage requirements, the system consist of some methods to personalize the system services to that user. The most basic approach allows users to choose the services they want to use. Our proposal is to observe user interactions with the system and based on the feedback provided by his behavior automatically personalize the services.

Finally, the agents running in **the models' conversion module** perform a conversion of existing and newly created early warning models (related to different branches of economy (or related to global economy) – the overall conception of these models is presented on fig. 1– into the universal language of model description – there are several agents in this module. On the basis of results of conversion, particular agents will be running within a given collective.

4   Method for Evaluation of the Developed Model

Generally speaking, evaluation of early warning systems relies on the direct comparison of these crisis occurrence probabilities with an original crisis dating, which constitutes the benchmark [3]. This comparison implies two inputs of different nature: a sequence of probabilities and a crisis dating that takes the form of a Boolean variable, labelled $\hat{y}_t$. By convention, it is assumed that $\hat{y}_t$ takes a value equal to one if a crisis is identified at time t, and zero otherwise. The forecasted probabilities are thus transformed into a Boolean variable, known as crisis forecast. Formally, if we denote $\hat{p}_t$ the estimated (or forecasted) crisis probability at time t issued from an EWS model, the crisis forecast variable $\hat{y}_t$ is computed as follows benchmark [3]:

$$\hat{y}_t(c) = \begin{cases} 1, \text{if } \hat{p}_t > c \\ 0, \text{othervise} \end{cases} \tag{22}$$

where $c \in [0;1]$ represents the cut-off. In this perspective, the first step of any EWS evaluation consists of determining an optimal cut-off that discriminates between predicted crisis periods ($\hat{y}_t(c) = 1$) and predicted calm periods ($\hat{y}_t(c) = 0$). Noise to Signal Ratio, a credit-scoring approach, and accuracy measures methods will be used to identify the optimal cut-off. To assess the developed model we will use such criteria, as the Quadratic Probability Score (QPS - simply a mean square error measure comparing the crisis probability with an indicator variable for the crisis) and the Log Probability Score (corresponds to a loss function that penalizes large errors more heavily than QPS), the Area under the Receiver Operating Characteristic Curve (AUC - indicates that the early warning system is getting closer to the perfect classification) [3].Verification will also concern different branches and global economy, e.g.:

1. Financial market. The collective consist of experts and agents carrying out technical and fundamental analysis. Results will be stored with the use of ontology in the form of a semantic net with node and links activation level (the "slipnet") [13]. On the basis of this knowledge state buy-sell decisions concerning securities or currencies will be taken. Next, the experiment will be repeated several times, each time with different periods. The results obtained by using the developed model will be compared with the different benchmarks' results (Buy and Hold, Random Walk). Back testing using historical data will be used. The comparison of results will be performed with the use of the following measures (ratios): rate of return (ratio x1), number of transaction, gross profit (ratio x2), gross loss (ratio x3), total profit (ratio x4), number of profitable transactions (ratio x5), number of profitable transactions in a row (ratio x6), number of unprofitable transactions in a row (ratio x7), Sharpe ratio (ratio x8),the average coefficient of variation (ratio x9), Value at Risk (ratio x10). Evaluation function is calculated as follows:

$$y = (a_1 x_1 + a_2 x_2 + a_3(1-x_3) + a_4 x_4 + a_5 x_5 + a_6 x_6 + a_7(1-x_7) + a_8 x_8 + a_9(1-x_9) + a_{10}(1-x_{10})) \quad (23)$$

where x_i denotes normalized values of the ratios from x_1 to x_{10}. It can be adopted in that coefficients a_1 to $a_{10}=1/10$ or these coefficients may be modified with the use of, for instance, an evolution method or determined by a user (investor) in accordance with his/her preference. The function is given the values from the range [0..1], and the collective's efficiency is directly proportional to the function value.

2. Analyzing customers' opinions about products offered by enterprises. Results of an automatic analysis will be compared with results of an analysis performed by a human (an expert), i.e. a manual analysis. In order to determine the correctness of results of the automatic analysis in relation to the results of the manual annotation the following measures will be applied:

- effectiveness – this measure defines the relationship of the number of opinions whose polarity (or polarity of features) has been determined automatically to the number of opinions whose it has been determined manually; this measure enables one to determine in how many cases the polarity of opinions has not been determined by an agent;
- precision – which specifies the accuracy of classification (negative, positive) within a recognized class of opinions and it is defined in the following way

$$p = \frac{opp}{opp + onp}, \quad (24)$$

where: p – precision, opp – positive opinions recognized as positive ones, onp – negative opinions recognized as positive ones;

- sensitivity – the relationship of the number of opinions recognized by an agent as positive ones against all positive opinions is defined in the following way:

$$c = \frac{opp}{opp + opn}, \tag{25}$$

where: c- sensitivity, opn – positive opinions recognized as negative ones;

- F1 measure – measure of a test's accuracy, defined in the following way:

$$F1 = \frac{2*(p*c)}{p+c}, \tag{26}$$

where: $F1$- F1 measure, p – precision, c – sensitivity.
All the presented measures have values ranging from 0 to 1.

3. Gross domestic product prediction. For example, The Dynamic stochastic general equilibrium modeling, Generalised Dynamic Factor Model [3] models for gross domestic product forecasting will be evaluated.

5   Conclusions

The main novelties of the proposed model are: integration of knowledge related to all branches of economy; a global character; never-ending learning process will be performed by all the human agent collectives, the possibility of joining under runtime new models related to a particular branch or related to global a macroeconomic situation; the wide scope of the system user's groups; cooperation of human-agent collectives will enable the increase in the level of knowledge of an economic situation. Application of consensus methods in the process of integrating knowledge of human-agent collectives will enable users to gain knowledge of forecasted crisis events which is closest to the real knowledge.

The model of the multiagent early warning system for crisis situations in economy described in this paper can allow for the economic crisis effects to be avoided by many social groups (e.g. entrepreneurs can avoid a bankruptcy, borrowers can avoid an insolvency, state officials can avoiding increase in the level of unemployment or avoid strikes due to falling prices of agricultural products). Knowledge integration methods of human-agent collectives will expand areas of use of the multiagent systems, especially in relation to collective intelligence issues.

This paper presents only general issues related to the conception of the EWS. Further researches will be required in order to put into practice the evaluation mechanism. The evaluation will be performed on data shared by various national and global organizations deal with collecting and elaborating statistical data, as well as by enterprises and public administrations. Back-testing, which consists in checking a model on historical (past) data, will be employed as the test method.

Acknowledgement. The research has been funded by a grant by Polish Ministry of Science and Higher Education.

References

1. Barrick, M.R., Stewart, G.R., Neubert, M.J., Mount, M.K.: Relating Member Ability and Personality to Work-Team Processes and Team Effectiveness. Journal of Applied. Psychology **83**(3), 377–391 (1998)
2. Berg, A., Pattillo, C.: Predicting currency crises: The indicators approach and analternative. J. Int. Money Financ. **18**, 561–586 (1999)
3. Candelon, B., Dumitrescu, E., Hurlin, C.: How to Evaluate an Early-Warning System: Toward a Unified Statistical Framework for Assessing Financial Crises Forecasting Methods. IMF Economic Review **60**(1), 75–113 (2012)
4. Duong, T.H., Nguyen, N.T., Jo, G.-S.: A method for integration of wordnet-based ontologies using distance measures. In: Lovrek, I., Howlett, R.J., Jain, L.C. (eds.) KES 2008, Part I. LNCS (LNAI), vol. 5177, pp. 210–219. Springer, Heidelberg (2008)
5. Hackman, R.J.: The interaction of task design and group performance strategies in determining group effectiveness. Organizational Behavior and Human Performance **16**(2), 350–365 (1976)
6. Jennings, N.R., Moreau, L., Nicholson, D., Ramchurn, S., Roberts, S., Rodden, T., Rogers, A.: Human-Agent Collectives. Communications of the ACM **57**(12) (2014)
7. Kim, D.H., Lee, S.J., Oh, K.J., Kim, T.Y.: An early warning system for financial crisis using a stock market instability index: Expert system. J. Knowl. Eng. **26**(3), 260–273 (2009)
8. Korol, T.: Early warning models against bankruptcy risk for Central European and Latin American enterprises. Economic Modelling **31**, 22–30 (2013)
9. Lin, C., Khan, H.A., Chang, R., Wang, Y.: A new approach to modeling early warning systems for currency crises: Can a machine-learning fuzzy expert system predict the currency crises effectively? Journal of International Money and Finance **27**(7), 1098–1121 (2008)
10. Mitra, S., Erum: Early warning prediction system for high inflation: an elitist neurogenetic network model for the Indian economy. Neural Comp. and Appli. **22**(1) (2012)
11. Nguyen N.T.: Advanced Methods for Inconsistent Knowledge Management. Springer-Verlag London (2008)
12. Sliwko, L., Nguyen, N.T.: Using Multi-agent Systems and Consensus Methods for Information Retrieval in Internet. International Journal of Intelligent Information and Database Systems **1**(2), 181–198 (2007)
13. Snaider, J., McCall, R., Franklin, S.: The LIDA framework as a general tool for AGI. In: Schmidhuber, J., Thórisson, K.R., Looks, M. (eds.) AGI 2011. LNCS, vol. 6830, pp. 133–142. Springer, Heidelberg (2011)
14. Tziner, A., Eden, D.: Effects of Crew Composition on Crew Performance: Does the Whole Equal the Sum of Its Parts? Journal of Applied Psychology **70**(1), 85–93 (1985)
15. Wiltshire, A.: Developing early warning systems: a checklist. In: Proceedings of the 3rd International Conference on Early Warning EWC III, Bonn (2006)
16. Woolley, A.W., Chabris, C.F., Pentland, A., Hashmi, N., Malone, T.: Evidence for a Collective Intelligence Factor in the Performance of Human Groups. Science **330**(6004), 686–688 (2010)

Combining Machine Learning and Multi-agent Approach for Controlling Traffic at Intersections

Mateusz Krzysztoń[1][(✉)] and Bartłomiej Śnieżyński[2]

[1] Institute of Control and Computation Engineering,
Warsaw University of Technology, Warsaw, Poland
mkrzyszt@mion.elka.pw.edu.pl
[2] AGH University of Science and Technology, Krakow, Poland
bartlomiej.sniezynski@agh.edu.pl

Abstract. Increasing volume of traffic in urban areas causes great costs and has negative effect on citizens' life and health. The main cause of decreasing traffic fluency is intersections. Many methods for increasing bandwidth of junctions exist, but they are still insufficient. At the same time intelligent, autonomous cars are being created, what opens up new possibilities for controlling traffic at intersections. In this article a new approach for crossing an isolated junction is proposed - cars are given total autonomy and to avoid collisions they have to change speed. Several methods for adjusting speed based on machine learning (ML) are described, including new methods combining different ML algorithms (hybrid methods). The approach and methods were tested using a specially designed platform MABICS. Conducted experiments revealed some deficiencies of the methods - ideas for addressing them are proposed. Results of experiments made it possible to verify the proposed idea as promising.

Keywords: Intersection · Self-driven car · Reinforcement learning · Hybrid methods

1 Introduction

With improvements in technology, multi-robot systems become more present in our life, increasing standard of our living in the vast majority of cases. To make this impact even more efficient, multi-robot systems exploit artificial intelligence (AI) very often. Robots, often identified with agents [1], can use multiple algorithms within AI. A very popular choice for implementing decision-making mechanism for robots is to use machine learning (ML) methods, which allow robots to learn from their experience [2].

One of the natural environments for multi-robot systems is a city, especially its transport infrastructure. According to EU Commission report [3], a large number of European towns and cities suffer from increasing traffic congestion, which causes not only great costs (estimated at €80 billion annually), but also has significant influence on environment and hence on citizens' life and health

© Springer International Publishing Switzerland 2015
M. Núñez et al. (Eds.): ICCCI 2015, Part I, LNAI 9329, pp. 57–66, 2015.
DOI: 10.1007/978-3-319-24069-5_6

(around 23% of all CO_2 emissions from transport in UE comes from urban areas). Unfortunately, extending existing infrastructure is limited. Hence, methods for improving fluency of traffic using existing infrastructure need to be found.

The main cause of limited traffic fluency in cities are intersections [4], because cars often have to stop before crossing them. Some existing solutions (both already implemented in real cities and these at the research stage) for increasing fluency at junctions will be discussed. The main aim of this article is to present a new approach for managing traffic at the intersection, based on giving full autonomy to self-driven, intelligent vehicles, which are slowly but successfully being introduced on real world roads [5]. Then, several decision-making methods for a vehicle at an intersection, based on ML, are proposed. They are evaluated using a specially developed platform and the results are discussed.

2   Related Work

Improving traffic at intersections in terms of safety and fluency was under study since the 19th century when the first traffic lights were introduced [6]. This is still the most popular solution for controlling traffic at junctions. Despite many advantages, like simplifying making decisions by drivers and increasing safety, traffic lights have negative influence on traffic fluency [6]. Hence, many systems were developed to adjust traffic lights' cycles on neighboring intersections to increase overall bandwidth of the infrastructure [7–9].

To optimize traffic multi-agent systems are commonly used. For example, in [10] particle swarm optimization techniques are applied to adapt traffic lights. More flexible and innovative approach is described in [11], where every car is an autonomous agent and there can be many cars crossing a junction in different directions at the same time. To cross an intersection, an agent has to send a reservation request to the intersection control system. The control system simulates the path of the car at the junction and checks if it does not contain already reserved fields in the given time. The agent sends requests until one of them does not conflict with already accepted requests. The main problem with deploying this method in real world is a restriction that every intersection needs to be equipped with a special control system. Ensuring safety for other road users (cyclist, pedestrians) also has to be considered.

In multi-agent systems two main techniques applied for learning are reinforcement learning, and evolutionary computation. However, other techniques, such as supervised learning are also applied. Good survey of learning in multi-agent systems working in various domains can be found in [12] and [13]. Learning can be applied in various environments. Predator-Prey is one of them, where several learning techniques were applied. [14] is an example of reinforcement learning application. In this work predator agents use reinforcement learning to learn a strategy minimizing time to catch a prey.

Another domain, where several learning techniques were applied is target observation. In [15] rules are evolved to control large area surveillance from the air. In [16] Parker presents cooperative observation tasks to test autonomous generation of cooperative behaviors in robot teams. Lazy learning based on reinforcement learning is used to generate strategy better than a random, but worse than a manually developed one. Results of application of reinforcement learning mixed with state space generalization method can be found in [17], where Evolutionary Nearest Neighbor Classifier - Q-learning (ENNC-QL) is proposed. It is a hybrid model, a combination of supervised function approximation and state space discretization with Q-learning. This method has similar goals to hybrid algorithm presented in this paper: reduction of state space for reinforcement learning, with minimal possible information loss, so that the Markov property can be still satisfied after applying the reduction. For ENNC-QL this works best in deterministic domains. Technically, ENNC-QL algorithm works as a very sophisticated function approximator with built-in discretization support. The main application domain of ENNC-QL approach consists of problems with possibly large, continuous state spaces. [17] gives no information about experiments with pure discrete state spaces, so the range of applications is basically somewhat different than for the hybrid model proposed here. Additionally, the ENNC-QL algorithm requires several predefined phases in order to compute discretization and state space representation, including explicit exploration phase and two learning phases, so it might be hard to apply in non-stationary, changing environments. On the other hand, it is more generic than hybrid model described here, because it can be easily applied to any continuous state space problem without making any assumptions on the problem's domain.

There are also several other works about learning in multi-agent systems that are using supervised learning. Rule induction is used in a multi-agent solution for vehicle routing problem [18]. However, in this work learning is done off-line. In [19], agents learn coordination rules, which are used in coordination planning. If there is not enough information during learning, agents can communicate additional data during learning. Singh et. al add learning capabilities into BDI model [20]. Decision tree learning is used to support plan applicability testing.

3   New Approach for Intersection Crossing

It is assumed that with an increasing number of intelligent, self-driven cars it will be possible to resign from some current rules of crossing intersections in favor of higher degree of cars' autonomy. To increase fluency of traffic in existing infrastructure, a new approach SInC (Simultaneous Intersection Crossing) for controlling traffic at a single junction is proposed.

3.1   General Idea

The SInC approach is inspired by the zipper method [21], which is used when few lanes reduce to a single one. It increases fluency during changing lane by eliminating unnecessary car stops to avoid collision. Analogously, cars at a junction do not have to stop to give way to other cars but they can cross it simultaneously, reducing or increasing speed to avoid collision. Decisions about speed change are made by fully autonomous agent that is responsible for steering car while crossing intersection. The main issue in SInC is making decisions that minimize crossing time and prevent accidents. These two goals are often contradictory. The agents make decision about velocity change basing on current situation at intersection in real time.

3.2 Model

To simplify the examination of SInC the following assumptions were made. Cars cross intersection straight only and they do not change lanes. The environment is (according to features described in [13]):

- discrete - intersection is modeled as a grid, time is divided into steps, speed and acceleration of the car are natural and integer numbers, respectively;
- fully observable - agent knows the whole state of the intersection in every time step;
- deterministic (partially) - agent's decision to change speed from s to s' always causes that in the next time step car speed is s';
- non-episodic - every decision made by the agent is independent on previous ones.

The state of the car on the intersection is defined by its' location, direction and value of speed (s). The state of intersection is composed of all cars' states.

In every step of simulation agent chooses action $a_i \in A$, which represents decision to change speed by i. The set of actions A is defined, according to maximal speed change a_max, as:

$$A := \{a_{-a\_max}, a_{-a\_max+1}, \ldots, a_{a\_max-1}, a_{a\_max}\} \ . \tag{1}$$

If the speed after change is beyond the permitted range of speed values, the resulting speed value is set to the adequate end of the range.

3.3 Methods

Decision-making methods proposed in this article are based on ML. Of course, every other algorithm may be used to implement this approach.

The agent steering the car knows the state of intersection and it uses that state to describe the state of the car with following attributes:

- distance to target - end of intersection (d_t);
- speed of the car (s_c);
- distance of car to the nearest collision point with a car coming from the crossing road (collision car) (d_{cp});
- distance from the collision car to the collision point (d_{ccp});
- speed of the collision car (s_{cc}).

Below term "state" refers to these attributes. In Fig. 1 (left) an example realization of a state is shown, where $d_t=18$, $s_c=2$, $d_{cp}=10$, $d_{ccp}=8$, $s_{cc}=1$.

To avoid collisions and minimize time for crossing the intersection, agent uses ML to create mapping of every possible processed state of the car to the best action $a_i \in A$ in that processed state.

Below methods for making decisions by agents are proposed.

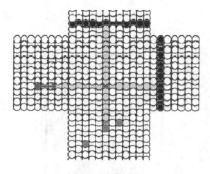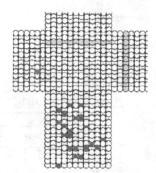

Fig. 1. On the left the example state at an intersection is shown. Place where collision between the blue car and the nearest car to it (middle green) can occur is marked with a red dot. The end of intersection is marked with black lines. On the right initial situation at intersection for experiment with consistent goals is presented. Intelligent agent is marked with green color, agents simulating real traffic with red.

3.3.1 Simple Method Based on Reinforcement Learning (RL)

In reinforcement learning, an agent experiments with the environment by taking actions that are not always optimal in the current state [2]. To teach an agent to make best decisions, agent gets reward r for every action which is sum of:

- r_c - negative reward for collision
- r_t - positive reward for reaching target
- r_s - negative reward for making move

The reinforcement learning algorithm is used by the agent to learn how to get the biggest total reward during crossing the intersection. Rewards r_t and r_s promote crossing it as quickly as possible. Reward r_c teaches the agent to avoid a collision.

3.3.2 Hybrid Method Based on Reinforcement Learning and State Reduction (RLSR)

A drawback of the method above is a fast growing number of states with the increase of the number of possible values of every part of the state (e.g. possible values of car speed). To limit the number of possible states, the part of state that describes possibility of collision (s_c, d_{cp}, d_{ccp}, s_{cc}) can be reduced to a bivalent attribute informing if in the given state a collision is possible [22, 23]. To reduce the state, a classifier is used [24]. To teach the classifier which states are dangerous, all states visited by car are classified as "collision possible" if a collision happened during crossing the intersection and as "collision impossible" if the car safely reached the end of intersection. Again the reinforcement learning is used to learn the best action in a given reduced state. Scheme of the method is presented in Fig. 2. An obvious deficiency of this method is losing considerable part of training information for reinforcement learning, which may affect the quality of agent decision. Some possible improvements are discussed in section 5.

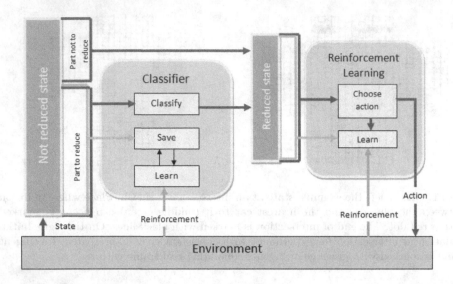

Fig. 2. General scheme of hybrid method based on reinforcement learning and state reduction.

3.3.3 Hybrid Method Based on Reduced State Meaning (RSM)

This method is similar to the previous one. Again classifier is used to reduce the state. Then, depending on the value of the reduced state various methods can be performed. For example if no collision is possible search algorithm A* [25] can be used to get to target with the shortest path. Since in this model it is assumed that the car does not change lanes, simple methods are implemented - if a collision is possible, the agent reduces the speed of the car, otherwise accelerates it. In both cases speed is changed by one.

3.3.4 Hybrid Method Based on Reinforcement Learning with Reduced State and Supervisor (RLS)

Reinforcement learning is used as in the second method - with state reduction. Then the decision (speed value change) obtained from the reinforcement algorithm together with the state of car (not reduced) is passed to the supervisor, who decides if the decision is correct or not. If not, the supervisor can change the decision and that decision is executed by the agent. The supervisor is implemented using a classifier.

To teach the classifier how to rate decisions, examples of following form are used:

$$\langle s_v, a_{s_v}, class \rangle \ . \tag{2}$$

Classifier has to store every visited state s_v (vector of attributes) combined with action a_{s_v} (one attribute) performed by the agent in that states as not labeled example. When collision occurs or car gets to the end of intersection safely, stored examples are labeled with *class* "bad" or "good" respectively. Then these

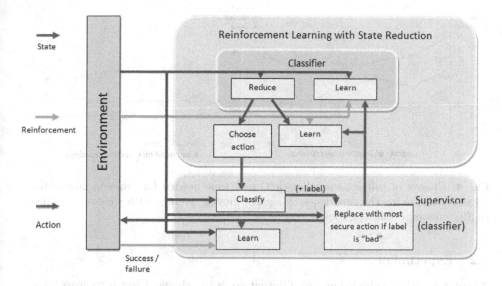

Fig. 3. General scheme of hybrid method based on reinforcement learning with reduced state and supervisor.

examples are added to training set. The classifier should either be able to learn online (increase quality of decision with every new example in scalable fashion) or be retrained with updated training set periodically. The decision, if action chosen by the reinforcement learning algorithm is correct, is based on classifying that action together with the current state. The supervisor decides that action is wrong only if he is sure to a certain, configurable, extent. The scheme of this method is shown in Fig. 3.

4 Realization and Evaluation of SInC

4.1 Platform MABICS

To verify the SInC approach and methods presented above, a special platform MABICS (Multi-Agent Based Intersection Controlling System) was created. The platform simulates real traffic on an isolated, fully configurable intersection. An external application is used to generate possible moves of vehicles during intersection crossing [26]. MABICS allows implementing different methods for controlling single vehicle. Methods discussed previously were implemented using RLPark [27] and WEKA [28] libraries for reinforcement learning and classifiers, respectively.

The platform supports also conducting experiments. During single experiment vehicles cross the intersection many times to generate knowledge how to make the best decisions. The outcome of the experiment is two graphs. They present change of the collisions number and the times of passing the intersection with gaining the experience by agent.

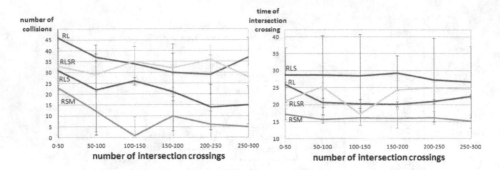

Fig. 4. Change of collisions number (left) and time needed for crossing intersection (right) with gaining the experience by agent - moving average with window size equal to fifty crossings of intersection.

4.2 Experiment

Agent controlling vehicle has two different goals to achieve. First is to cross intersection as fast as possible, second is not to collide with other vehicles. These two goals can be consistent (to avoid collision agent has to speed up) or contradictory (avoiding collision requires slowing down what causes later achieving end of junction). In experiment consistent configuration was used.

The initial situation for experiment with consistent goals is shown on Fig. 1 (right). On the left side of intersection vehicle controlled by intelligent agent is placed. On the bottom there are 21 vehicles simulating real traffic moving with different, random, constant speeds. It is easy to observe that the optimal strategy for intelligent agent to achieve both targets is to accelerate to maximal speed.

4.3 Results

The experiment was repeated three times. Obtained average results of time and collision numbers for different methods are presented on Fig. 4. All agents except the one using RSM method were improving their decisions, but could not find the optimal strategy, which occurred to be too far from the initial, random decisions' strategy. The agents learn to increase speed, which is expected behavior in problem with consistent targets, but they do accelerate only in some steps. Hence, they do not cross intersection fast enough to pass the vehicles coming from their right side. In such situation, when they "meet" other vehicles in the middle of crossroad, they try to avoid collisions by increasing and reducing speed. It means, that they look for suboptimal strategy in contradictory goals situation and they partially succeed (except RLSR) - the number of collisions is reduced with continuing learning. Different behavior is shown by the agent that uses the RSM method. This agent achieves the best results. Analysis of every serie shows that in two of them this agent finds the best strategy (no collisions in last periods of series). In one serie the agent behaves similarly to other agents and looks for suboptimal strategy in the contradictory situation, but still achieves better results (0.3 collision per crossing).

5   Conclusions and Future Work

The presented work is only an introduction to research on possibility of using ML to control autonomous vehicles at an intersection. The conducted experiments made it possible to verify new idea preliminarily as promising - some of the presented methods increased the quality of decisions with successive crossings, which may indicate that in a more realistic environment collisions will be completely eliminated.

The future work should include targeting the problem of poor exploration of the search space, for example by introducing variable parameters for reinforcement learning. Additionally proposed methods should be improved with ideas that occurred during work - adding negative reward for stopping, marking not all visited states as "bad" if a collision occurred, but only a few last states and adding new values of the reduced state in the RLSR method for different types of collisions. It is also necessary to integrate system MABICS with more realistic and more efficient application for generating possible moves of vehicles. Then experiments with different configurations of intersections and longer learning periods should be conducted.

References

1. Kaminka, G.A.: Robots are agents, too! In: 6th International Joint Conference on Autonomous Agents and Multiagent Systems, Honolulu, Hawaii, USA (2007)
2. Sutton, R.S., Barto, A.G.: Reinforcement Learning: An Introduction. MIT Press, Cambridge (1998)
3. European Commission: Urban mobility package - frequently asked questions. Brussels (2013). http://europa.eu/rapid/press-release_MEMO-13-1160_en.doc (accessed October 15, 2014)
4. Xia, X., Xu, L.: Coordination of urban intersection agents based on multi-interaction history learning method. In: Tan, Y., Shi, Y., Tan, K.C. (eds.) ICSI 2010, Part II. LNCS, vol. 6146, pp. 383–390. Springer, Heidelberg (2010)
5. Markoff, J.: Google Cars Drive Themselves, in Traffic, p. A1. The New York Times, New York (2010)
6. Datka, S., Suchorzewski, W., Tracz, M.: Traffic Engineering. Wydawnic two Komunikacji i Łaczności, Warsaw (1999). pp. 282;324;328 (in Polish)
7. Robertson, D.I.: TRANSYT: a Traffic Network Study Tool, Transport and Road Research Laboratory Report (1969)
8. Taale, H., Fransen, W.C.M., Dibbits, J.: The second assessment of the SCOOT system in Nijmegen. In: IEE Road Transport Information and Control, Conference Publication No 454 (1998)
9. Sims, A.G., Dobinson, K.W.: The Sydney coordinated adaptive traffic (SCAT) system philosophy and benefits. IEEE Transactions on Vehicular Technology 29(2), 130–137 (1980)

10. Cajias, R.H., Gonzalez-Pardo, A., Camacho, D.: A multi-agent traffic simulation framework for evaluating the impact of traffic lights. In: Proceedings of the 3rd International Conference on Agents and Artificial Intelligence, vol. 2 (2011)

11. Dresner, K., Stone, P.: Multiagent traffic management: opportunities for multiagent learning. In: Tuyls, K., 't Hoen, P.J., Verbeeck, K., Sen, S. (eds.) LAMAS 2005. LNCS (LNAI), vol. 3898, pp. 129–138. Springer, Heidelberg (2006)

12. Panait, L., Luke, S.: Cooperative multi-agent learning: The state of the art. Autonomous Agents and Multi-Agent Systems 11, 387–434 (2005)

13. Weiss, G.: Multiagent Systems: A Modern Approach to Distributed Artificial Intelligence. The MIT Press, London (1999)

14. Tan, M.: Multi-agent reinforcement learning: independent vs. cooperative agents. In: Proc. of 10th Int'l. Conference on Machine Learning (ICML 1993), pp. 330–337. Morgan Kaufmann (1993)

15. Wu, A.S., Schultz, A.C., Agah, A.: Evolving control for distributed micro air vehicles. In: Proc. of IEEE Int'l. Symp. on Computational Intelligence in Robotics and Automation (CIRA 1999), pp. 174–179. IEEE (1999)

16. Parker, L.E., Touzet, C.: Multi-robot learning in a cooperative observation task. In: Parker, L.E., Bekey, G., Barhen, J. (eds.) Distributed Autonomous Robotic Systems, vol. 4, pp. 391–401. Springer, Berlin (2000)

17. Fernandez, F., Borrajo, D., Parker, L.E.: A reinforcement learning algorithm in cooperative multirobot domains. Journal of Intelligent Robotics Systems 43, 161–174 (2005)

18. Gehrke, J.D., Wojtusiak, J.: Traffic prediction for agent route planning. In: Bubak, M., van Albada, G.D., Dongarra, J., Sloot, P.M.A. (eds.) ICCS 2008, Part III. LNCS, vol. 5103, pp. 692–701. Springer, Heidelberg (2008)

19. Sugawara, T., Lesser, V.: On-line learning of coordination plans. In: Proc. of the 12th Int'l. Workshop on Distributed Artificial Intelligence (1993)

20. Singh, D., Sardina, S., Padgham, L., Airiau, S.: Learning context conditions for bdi plan selection. In: Proceedings of the 9th International Conference on Autonomous Agents and Multiagent Systems, Richland, SC, vol. 1, pp. 325–332 (2010)

21. Minnesota Department of Transportation: Zipper Merge (2014). http://www.dot. state.mn.us/zippermerge/ (accessed November 10, 2014)

22. Wiatrak, Ł.: Hybrid Learning in agent systems. Master thesis, Cracow (in Polish) (2012)

23. Śnieżyński, B., Wójcik, W., Gehrke, J.D., Wojtusiak, J.: Combining rule induction and reinforcement learning: an agent-based vehicle routing. In: Proc. of the ICMLA 2010. Washington D.C., pp. 851–856 (2010)

24. Sammut, C., Webb, G.I.: Encyclopedia of Machine Learning, 1st edn. Springer Publishing Company, Incorporated (2011)

25. Barr, A., Feigenbaum, E.: The Handbook of Artificial Intelligence, vol. 1, pp. 64–67. HeurisTech Press, Los Altos (1981)

26. Mozgawa, J., Kaziród, M.: Steering vehicles in discrete space (2013). https://github.com/myzael/Sterowanie-pojazdami-w-przestrzeni-dyskretnej/ wiki (accessed November 10, 2014) (in Polish)

27. RLPark, Introduction to RLPark (2013). http://rlpark.github.io/ (accessed November 10, 2014)

28. Witten, I.H., Frank, E., Hall, M.A.: Data Mining: Practical Machine Learning Tools and Techniques, 3rd edn. Elsevier (2011)

A Scalable Distributed Architecture
for Web-Based Software Agents

Dejan Mitrović[1]([✉]), Mirjana Ivanović[1], Milan Vidaković[2],
and Zoran Budimac[1]

[1] Department of Mathematics and Informatics, Faculty of Sciences,
University of Novi Sad, Novi Sad, Serbia
{dejan,mira,zjb}@dmi.uns.ac.rs
[2] Faculty of Technical Sciences, University of Novi Sad, Novi Sad, Serbia
minja@uns.ac.rs

Abstract. In recent years, the web has become an important software
platform, with more and more applications becoming purely web-based.
The agent technology needs to embrace these trends in order to remain
relevant in the new era. In this paper, we present recent developments of
our web-based multiagent middleware named Siebog. Siebog employs
enterprise technologies on the server side in order to provide automatic agent load-balancing and fault-tolerance. On the client, it relies
on HTML5 and related standards in order to run on a wide variety of
hardware and software platforms. Now, with automatic clustering and
state persistence, Siebog can support thousands of external devices hosting tens of thousands of client-side agents.

1 Introduction

During the last decade, there has been an obvious paradigm shift in software
development. The *web* has evolved into an environment capable of providing
functionalities not so long ago available only in desktop applications. One of the
main reasons for the increasing popularity of web-only applications is their cross-
platform nature, which allows the end-users to access their favorite applications
in a wide variety of ways. As the overall result, the desktop-only technologies
are becoming less and less relevant.

Computer clusters play an important role in modern web and enterprise
applications. They provide *high-availability* of deployed applications [18]. This
feature is concerned with continuous, uninterrupted delivery of services, regardless of hardware and software failures, or numbers of incoming requests. The
high-availability is achieved through the so-called *horizontal scaling*, which is
the processing of adding more nodes to the cluster as the demands for processing power increase.

Siebog is our multiagent middleware designed to provides support for intelligent software agents in this new setting [24]. It efficiently combines the *HTML5*
and related web standards on the client side [20,22], and the *Enterprise edition*

M. Núñez et al. (Eds.): ICCCI 2015, Part I, LNAI 9329, pp. 67–76, 2015.
DOI: 10.1007/978-3-319-24069-5_7

of Java (Java EE) on the server side [21,23,29], in an effort to bridge the gap between the agent technology and industry applications.

By utilizing the standards and technologies readily-available in Java EE, Siebog offers "native" support for computer clusters on the server. The purpose of this paper is to discuss how this support is extended to the client side of Siebog as well. The goal is to provide the support for clusters that consist of arbitrary client devices, such as personal computers, smartphones, and tablets.

The motivation for this approach is straightforward. There can be an order of magnitude more external client devices than there can be server-side computers. Siebog could be used to distribute agents among connected clients and to support applications that require launching large populations of agents. Since the state-of-the-art smartphones have more processing powers that many laptop or desktop machines, they represent a significant computational resource.

This approach, however, does pose some technical challenges. Client-side clusters are highly dynamic, in the sense that clients are able to join and leave at any time. In order to deal with this situation, we added a highly-scalable infrastructure for agent state persistence which allows the agents to "rise above" the interruptions, and to operate regardless of their physical locations.

The rest of the paper is organized as follows. Section 2 discusses the overall motivation behind this paper, and presents relevant existing work. Section 3 shows how the support for dynamic and heterogeneous client-side clusters was added to Siebog. The overall performance evaluation is presented in Section 4. The final conclusions and future research directions are given in Section 5.

2   Background

Web application play an increasingly important role in the modern-day computing. They offer a number of advantages over traditional desktop application, such as the lack of need for installations, configurations, or upgrades. The importance of web applications is emphasized by the continuously increasing sales of mobile devices and the ability of web applications to run as native applications on these devices [30].

In order to maintain its relevance in this new era, the agent technology not only needs to move to the web, but it needs to do so in accordance to the modern standards and the end-users' expectations. Agent-based applications need to seamlessly be integrated into web and enterprise applications in order to reach the end-users more easily, and to stay relevant in this new state of affairs.

Currently, there exists a large number of both open-source and commercial agent middlewares [3,4]. However, almost none of these systems has fully exploited the advantages of web environments. Some efforts aimed at extending existing systems with web support have been made, but usually in a inefficient manner. For example, in many Java-based middlewares, such as *JADE* [2] or *JaCa-Web* [19], the extensions are based on Java *applets*. But, Java applets require a browser plug-in to run, which is unavailable on some platforms (e.g. *iOS* and *Smart TVs*). With some desktop-based browsers also starting to disable

Java support[1], the applicability of Java-based web solutions becomes limited to a narrower set of hardware and software platforms.

One of the prominent ways of migrating the agent technology to the web is to use the expanding *HTML5* and related set of standards [12]. HTML5 covers various aspects of web and enterprise applications, from audio and video playbacks, to offline application support, to more advanced features, such as multi-threaded execution and *push*-based communication. In addition, since web browser vendors keep investing significant resources into improving the overall performance of their respective JavaScript virtual machines, the HTML5 is expected to become "a mainstream enterprise application development environment" [11].

2.1 Related Work

Over the years, many practical applications of agents on the web have been proposed. One domain that appears to be the most interesting to agent researchers and practitioners is *content personalization* and various *recommender systems* [17,26]. Another possible thriving area is to use the so-called *pedagogical agents* in web-based e-learning systems [6,15], as well as in knowledge management [9].

Along with the more recent trends, there have been several proposals of using mobile agents within the so-called *Internet of Things* (IoT) concept [27], in *smart objects* [1,10] and in *smart cities* [28].

The role of Siebog in these practical applications would be in providing a standards-compliant, platform-independent, and efficient [22] multiagent middleware. In addition, with the work presented in this paper, we intend to bring the more traditional agent applications to the web. As discussed later, Siebog is suitable for distributed systems with large populations of agents. A concrete example of its possible practical application would be in the area of swarm intelligence, such as [14].

As we discussed previously in [20,22], many traditional (i.e. desktop- or server-based) multiagent middlewares have exposed their functionalities to the web through Java applets. This approach does provide many important benefits, such as the immediate availability of complex reasoning agents in web browsers [19]. However, with the lack of Java support in many popular modern platforms, this approach is no longer sufficient.

To the best of our knowledge, currently there exists one additional HTML5-based multiagent middleware [16,27]. The middleware is focused on using (primarily) mobile agents to support the IoT requirements. On the technical viewpoint, the client side of their system does not utilize the full range of HTML5 and related standards (such as *Web Workers* [22]). It is also not clear how multiple agents could be started within the same host, and how would they interact with each other without the server. The backend is conveniently based on *Node.js*[2],

---

[1] https://java.com/en/download/faq/chrome.xml, retrieved on March 12, 2015.
[2] https://nodejs.org/, retrieved on March 12, 2015.

which simplifies certain development aspects (such as mobility), but lacks several advanced features found in Siebog, namely automated agent load-balancing and fault-tolerance.

3 Clustering Client-Side Siebog Agents

In this section we describe how Siebog is extended to support automatic clustering and load-balancing of its client-side agents. More concretely, we discuss how a possibly large set of heterogeneous client-side devices can be observed as a coherent cluster. The cluster can then be used to execute resource-demanding and computationally-expensive tasks, such as launching large populations of agents.

3.1 The Existing Architecture

The overall existing architecture of Siebog is shown graphically in Fig. 1. On the server side, Siebog includes the following three main modules [21,23,24, 29]: *Agent Manager*, which acts as an agent directory and controls the agents' life-cycles, *Message Manger*, in charge of inter-agent communication, and the *WebClient Manager*, which acts as an intermediary for server-to-client (i.e. push) messaging, and also handles state persistence for client-side agents.

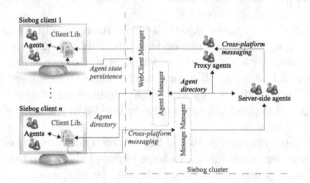

Fig. 1. The overall architecture of Siebog (adapted from [24]).

Client-side agents are executed inside web browsers [20,22,24] or possibly in dedicated JavaScript runtimes of external devices. Inside a device, agents rely on the Siebog client library for execution support and for communicating with the server. For example, the client library offers *proxy* implementations of server-side components. In order to send a message to the server-side agent, the client-side agent simply invokes the proxy implementation of the Message Manager. Underneath, the proxy then turns this invocation into a corresponding *AJAX* call to the server.

Another important feature of Siebog is that its server can (if configured) hold a proxy representation of each client-side agent. This representation simply forwards all incoming messages to the corresponding client-side counterpart (and through the WebClient Manager). This feature opens-up the possibility for transparent agent communication across different devices. An example is shown in Fig. 2. Let there be two agents, *AgA* and *AgB*, hosted by two different devices, *Device A* and *Device B*, respectively. When the AgA decides to send a message to AgB, the message will be delivered in the following way:

- AgA makes the appropriate call to the Message Manager Proxy.
- This call is transformed into an AJAX call to the server-side Message Manager.
- The Message Manager delivers the message to the AgB proxy.
- Since this is a proxy representation, it forwards the message to the WebClient Manager.
- The WebClient Manager, which is aware of all external clients, finally pushes the message to the target agent.

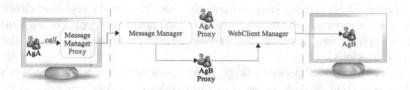

Fig. 2. Transparent communication of client-side agents across different devices.

Due to limitations imposed by certain web browsers [22], each client device can run up to a few dozens of agents. But, there can be large numbers of physical devices active simultaneously. The main idea here is to exploit this possibility in order to distribute portions of a large population of agents. This is conceptually similar to, for example, the famous *SETI@home* scientific experiment[3]. As an important advantage, the Siebog does not require any software installation: all the end-user needs to do is to visit the corresponding web page.

The main issue here is how to efficiently support these large numbers of external devices, and to deal with their inherently dynamic availability.

3.2 Managing Heterogeneous and Dynamic Clusters

A central component in a computer cluster is a *load-balancer* with the task of distributing the work across available machines. In the context of Siebog, the load-balancer continuously accepts tasks that need to be solved. For example, it

---

[3] http://setiathome.berkeley.edu/, retrieved on March 12, 2015.

can accept large maps for the *Traveling Salesman Problem*[4] and then partition each map [13] and send it (along with the corresponding set of ants) to a subset of available devices.

In the majority of existing non agent-based distributed architectures, the load-balancer selects the target device randomly. In this way, the workload is distributed "for free" and, in the longer run, equally among all available devices. However, the clusters that consist of Siebog clients are heterogeneous, in the sense that they can include devices with very different processing capabilities. Therefore, the load-balancing process is a bit more complex.

When it comes to load-balancing in heterogeneous systems, the agent-oriented research has proposed some rather complex approaches (e.g. [5,25,31]). In case of Siebog, however, we decided to follow the industry norms of keeping things as simple as possible. Once a device joins the cluster for the first time, a performance benchmark is executed. The results of this benchmark are used to assign a number of *compute units* (CUs) to the device. Now, during the load-balancing phase, the target device is selected with the probability that corresponds to its number of CUs.

From the end-user's point of view, joining the Siebog cluster is fairly simple: he/she only needs to visit the appropriate web page that hosts the worker agents. Unfortunately, it is also very easy to leave the cluster; once the end-user closes the web browser or switches off the device, the hosted agents are lost. For meaningful practical applications, however, the agents need to be able to run regardless of these interruptions.

In order to support possibly large numbers of agents, Siebog needs a scalable datastore, one capable of serving multitudes of requests per second. The datastore should also be fault-tolerant – capable of surviving server crashes. More formally, these requirements can be described as principles of the so-called *Dynamo systems* [8]. Currently, there exist several concrete Dynamo realizations. After a careful evaluation of these solutions, we determined that the open-source *Apache Cassandra*[5] datastore fulfills the needs of the Siebog client-side clusters. The client-side Siebog library has been extended to allow the agents the interact with the datastore directly, and over the *WebSocket* protocol [20,22]. The performance evaluation of the new architecture is discussed in the following section.

4   Performance Evaluation

The newly proposed architecture of Siebog needs to be able to serve large numbers of running agents, which are concurrently, and at high frequencies, storing and retrieving their respective internal states. In order to evaluate this feature, we used the open-source *Yahoo! Cloud Serving Benchmark* (YCSB) [7] tool. YCSB is designed for load-testing of (primarily) NoSQL databases, and can be

---

[4] http://www.math.uwaterloo.ca/tsp/data/, retrieved on March 12, 2015.

[5] http://cassandra.apache.org/, retrieved on March 12, 2015.

configured through a range of parameters, including the desired number of operations per second (throughput), the number of concurrent threads, maximum execution time, etc.

The experiments were performed using two machines, each with 8 virtual CPUs and 28 GB of RAM, running 64-bit version of *Ubuntu 14.04 LTS*. One machine was hosting the Apache Cassandra datastore and the Siebog server, while the other one was used to launch YCSB-simulated external devices. Each external device was represented by a separate WebSocket that the server needed to maintain.

In a realistic use-case, there will be many more writes to the store than reads. That is, the internal state of an agent will usually be read only once: when the web page is loaded and the agent is started. On the other hand, the state can be stored multiple times during the agent's execution, e.g. after each processed message or after each computational sub-step. Therefore, the YCSB workload was set up as *write-heavy*, so that 90% of all operations are *writes*.

The goal of the experiment was to determine the maximum number of external devices as well as client-side agents that our system can support. For this goal, a number of test-cases was executed. With each successive test-case, the total number of connected devices was increased. Then we would try to find the maximum throughput (i.e. the number of operations per second) that the Siebog server can support. A test-case was executed for one hour, and the maximum throughput value that was stable during this period was taken as the end-result.

We started with 100 external devices, increased the number for each test-cased, and finally reached the limit of approximately 16,000 devices. This number actually represents the maximum number of open connections that the operating system could support. Nonetheless, being able to support 16,000 external devices using a single-node Siebog cluster is an excellent result, given the fact that the cluster can easily be extended with more nodes as the demands grow. The results of this test-case are shown in Fig. 3. More concretely, the figure shows average read and write latencies during the one hour period, calculated at one minute intervals. The latencies are very low (expressed in microseconds), due to the WebSocket protocol's support for asynchronous I/O.

For each test-case, through trial-and-error, we determined that the value of approximately 6,000 operations per second is the maximum throughput that remained stable during the one hour period. Our systems is capable of serving much larger numbers than this (i.e. up to 100,000 operations per second), but only in "short bursts," after which the backend datastore needed some time to manage all the write operations.

Although the 6,000 operations per second might not seem as a large number at first, it is worth noting that an agent is not supposed to store its internal state at every second. So even if agents store their respective states at every 10 seconds, we reach the conclusion that our Siebog multiagent middleware can manage 60,000 agents distributed across 16,000 devices, using only one server-side node.

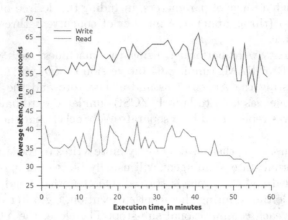

Fig. 3. Average read and write latencies of the test-case simulating 16,000 external devices and 6,000 operations per second, during the one hour period. The latencies are calculated at one minute intervals.

5   Conclusions and Future Work

Siebog is a web-based, enterprise-scale multiagent middleware. It combines the enterprise technologies on the server with HTML5 and related standards on the client in order to support multiagent solutions whose functionalities meet the expectations of modern software systems.

In this paper, we have presented how Siebog was recently updated to support dynamic clusters of heterogeneous client-side devices. The two new components of our system are the load-balancer, which is in charge of distributing agents across the connected devices, and a highly-scalable backend datastore used for persisting the internal states of client-side agents.

For any meaningful application of Siebog, its client-side agents need to become "detached" from their host environments (e.g. web pages). As shown in the paper, thanks to the use of a Dynamo architecture and the WebSocket protocol, on just one server node the state persistence system in Siebog can support thousands of external devices hosting tens of thousands of client-side agents, which is an excellent result.

Future developments of Siebog will be focused on an even tighter integration of client-side and server-side agents. Also, the system will be extended with an interoperability module, allowing it to interact with third-party multiagent solutions. Although Siebog already supports BDI agents on the server, the work is underway to develop a unique architecture for intelligent agents.

Acknowledgments. This work was partially supported by the Ministry of Education, Science and Technological Development of the Republic of Serbia, Grant III-44010, Title: Intelligent Systems for Software Product Development and Business Support based on Models.

References

1. Aiello, F., Fortino, G., Gravina, R., Guerrieri, A.: A java-based agent platform for programming wireless sensor networks. Computer Journal 54(3) (2011)
2. Bellifemine, F., Caire, G., Greenwood, D.: Developing Multi-Agent Systems with JADE. John Wiley & Sons (2007)
3. Bordini, R.H., Braubach, L., Dastani, M., El, A., Seghrouchni, F., Gomez-sanz, J.J., Leite, J., Pokahr, A., Ricci, A.: A survey of programming languages and platforms for multi-agent systems. Informatica 30, 33–44 (2006)
4. Bădică, C., Budimac, Z., Burkhard, H.D., Ivanović, M.: Software agents: languages, tools, platforms. Computer Science and Information Systems, ComSIS 8(2), 255–298 (2011)
5. Cao, J., Spooner, D.P., Jarvis, S.A., Nudd, G.R.: Grid load balancing using intelligent agents. Future Generation Computer Systems 21(1), 135–149 (2005)
6. Cheng, Y.M., Chen, L.S., Huang, H.C., Weng, S.F., Chen, Y.G., Lin, C.H.: Building a general purpose pedagogical agent in a web-based multimedia clinical simulation system for medical education. IEEE Transactions on Learning Technologies 2(3), 216–225 (2009)
7. Cooper, B.F., Silberstein, A., Tam, E., Ramakrishnan, R., Sears, R.: Benchmarking cloud serving systems with YCSB. In: Proceedings of the 1st ACM Symposium on Cloud Computing, SoCC 2010, pp. 143–154. ACM, New York (2010)
8. DeCandia, G., Hastorun, D., Jampani, M., Kakulapati, G., Lakshman, A., Pilchin, A., Sivasubramanian, S., Vosshall, P., Vogels, W.: Dynamo: Amazon's highly available key-value store. In: Proceedings of Twenty-First ACM SIGOPS Symposium on Operating Systems Principles, SOSP 2007, pp. 205–220 (2007)
9. Dignum, V.: An overview of agents in knowledge management. In: Umeda, M., Wolf, A., Bartenstein, O., Geske, U., Seipel, D., Takata, O. (eds.) INAP 2005. LNCS (LNAI), vol. 4369, pp. 175–189. Springer, Heidelberg (2006)
10. Fortino, G., Guerrieri, A., Russo, W.: Agent-oriented smart objects development. In: 16th International Conference on Computer Supported Cooperative Work in Design (CSCWD), pp. 907–912, May 2012
11. Gartner identifies the top 10 strategic technology trends for 2014, October 2013. http://www.gartner.com/newsroom/id/2603623 (retrieved on March 12, 2015)
12. HTML5: a vocabulary and associated APIs for HTML and XHTML, October 2014. http://www.w3.org/TR/html5/ (retrieved on March 12, 2015)
13. Ilie, S., Bădică, A., Bădică, C.: Distributed agent-based ant colony optimization for solving traveling salesman problem on a partitioned map. In: Proceedings of the International Conference on Web Intelligence, Mining and Semantics, WIMS 2011, pp. 23:1–23:9. ACM (2011)
14. Ilie, S., Bădică, C.: Multi-agent approach to distributed ant colony optimization. Science of Computer Programming 78(6), 762–774 (2013)
15. Ivanović, M., Mitrović, D., Budimac, Z., Jerinić, L., Bădică, C.: HAPA: Harvester and pedagogical agents in e-learning environments. International Journal of Computers Communications and Control 10(2), 200–210 (2015)
16. Jarvenpaa, L., Lintinen, M., Mattila, A.L., Mikkonen, T., Systa, K., Voutilainen, J.P.: Mobile agents for the internet of things. In: 17th International Conference on System Theory, Control and Computing (ICSTCC), pp. 763–767, October 2013
17. Lops, P., Gemmis, M., Semeraro, G.: Content-based recommender systems: state of the art and trends. In: Recommender Systems Handbook, pp. 73–105 (2011)

18. Michael, M., Moreira, J.E., Shiloach, D., Wisniewski, R.W.: Scale-up x scale-out: a case study using Nutch/Lucene. In: IEEE International Parallel and Distributed Processing Symposium, pp. 1–8, March 2007
19. Minotti, M., Santi, A., Ricci, A.: Developing web client applications with JaCa-Web. In: Omicini, A., Viroli, M. (eds.) Proceedings of the 11th WOA 2010 Workshop, Dagli Oggetti Agli Agenti, Rimini, Italy, September 5–7, 2010. CEUR Workshop Proceedings, vol. 621. CEUR-WS.org (2010)
20. Mitrović, D., Ivanović, M., Bădică, C.: Delivering the multiagent technology to end-users through the web. In: Proceedings of the 4th International Conference on Web Intelligence, Mining and Semantics, WIMS 2014, pp. 54:1–54:6. ACM (2014)
21. Mitrović, D., Ivanović, M., Budimac, Z., Vidaković, M.: Supporting heterogeneous agent mobility with ALAS. Computer Science and Information Systems 9(3), 1203–1229 (2012)
22. Mitrović, D., Ivanović, M., Budimac, Z., Vidaković, M.: Radigost: Interoperable web-based multi-agent platform. Journal of Systems and Software 90, 167–178 (2014)
23. Mitrović, D., Ivanović, M., Vidaković, M., Budimac, Z.: Extensible Java EE-based agent framework in clustered environments. In: Müller, J.P., Weyrich, M., Bazzan, A.L.C. (eds.) MATES 2014. LNCS, vol. 8732, pp. 202–215. Springer, Heidelberg (2014)
24. Mitrović, D., Ivanović, M., Vidaković, M., Budimac, Z., Bădică, C.: An enterprise-scale multiagent middleware based on HTML5 and Java EE technologies. Advances in Electrical and Computer Engineering (in print)
25. Nehra, N., Patel, R.: Towards dynamic load balancing in heterogeneous cluster using mobile agent. In: International Conference on Conference on Computational Intelligence and Multimedia Applications, vol. 1, pp. 15–21, December 2007
26. Swezey, R.M.E., Shiramatsu, S., Ozono, T., Shintani, T.: Intelligent page recommender agents: real-time content delivery for articles and pages related to similar topics. In: Mehrotra, K.G., Mohan, C.K., Oh, J.C., Varshney, P.K., Ali, M. (eds.) IEA/AIE 2011, Part II. LNCS, vol. 6704, pp. 173–182. Springer, Heidelberg (2011)
27. Systä, K., Mikkonen, T., Järvenpää, L.: HTML5 agents: mobile agents for the web. In: Krempels, K.-H., Stocker, A. (eds.) WEBIST 2013. LNBIP, vol. 189, pp. 53–67. Springer, Heidelberg (2014)
28. Verma, P., Gupta, M., Bhattacharya, T., Das, P.K.: Improving services using mobile agents-based iot in a smart city. In: International Conference on Contemporary Computing and Informatics (IC3I), pp. 107–111 (2014)
29. Vidaković, M., Ivanović, M., Mitrović, D., Budimac, Z.: Extensible Java EE-based agent framework – past, present, future. In: Ganzha, M., Jain, L.C. (eds.) Multiagent Systems and Applications. Intelligent Systems Reference Library, vol. 45, pp. 55–88. Springer, Heidelberg (2013)
30. Xanthopoulos, S., Xinogalos, S.: A comparative analysis of cross-platform development approaches for mobile applications. In: Proceedings of the 6th Balkan Conference in Informatics, BCI 2013, pp. 213–220. ACM, New York (2013)
31. Zhang, Z., Zhang, X.: A load balancing mechanism based on ant colony and complex network theory in open cloud computing federation. In: 2nd International Conference on Industrial Mechatronics and Automation (ICIMA), vol. 2, pp. 240–243, May 2010

Mapping BPMN Processes to Organization Centered Multi-Agent Systems to Help Assess Crisis Models

Nguyen Tuan Thanh Le[1,2](✉), Chihab Hanachi[3], Serge Stinckwich[4,5,6], and Tuong Vinh Ho[4,5,7]

[1] IRIT Laboratory, University Paul Sabatier Toulouse III, Toulouse, France
[2] University of Science and Technology of Hanoi, Hanoi, Vietnam
Nguyen.Le@irit.fr
[3] IRIT Laboratory, University of Toulouse I, Toulouse, France
hanachi@univ-tlse1.fr
[4] IRD, UMI 209, UMMISCO, IRD France Nord, 93143 Bondy, France
serge.stinckwich@ird.fr
[5] Sorbonne Universités, Univ. Paris 06, UMI 209, UMMISCO, 75005 Paris, France
[6] Université de Caen Basse-Normandie, Caen, France
[7] Institute Francophone International, Vietnam National University, Hanoi, Vietnam
ho.tuong.vinh@ifi.edu.vn

Abstract. Coordination is one of the most important issues in order to reduce the damage caused by a crisis. To analyze the efficiency of a coordination plan, a BPMN (Business Process Modeling Notation version 2.0) model is usually used to capture the processes of activities and messages exchanged between the actors involved in a crisis, while an OCMAS (Organization Centered Multi-Agent System) model is used to represent the roles, their interactions and the organizational structures. In this paper, we describe a proposal that allows to perform an automatic transformation between BPMN and OCMAS models of the same coordination plan. The proposal is illustrated through a coordination plan of a tsunami evacuation.

Keywords: Coordination representation · Process mapping · Process assessment · Organization centered multi-agent system · Crisis management

1 Introduction

Crisis situations such as natural disasters with environmental consequences impose the coordination of various stakeholders: firemen, medical organizations, police, etc ... In the context of crisis resolution, coordination plans could be examined under different representations. The mostly used representation in reality

N.T. Thanh Le—The work presented in this paper has been funded by the ANR Genepi project.

© Springer International Publishing Switzerland 2015
M. Núñez et al. (Eds.): ICCCI 2015, Part I, LNAI 9329, pp. 77–88, 2015.
DOI: 10.1007/978-3-319-24069-5_8

is the textual format. This latter has several drawbacks [1]. Its ambiguity makes the coordination among stakeholders difficult. Moreover, it cannot support the direct and autonomous analysis or simulation. Another possible representation of coordination is the process model as shown in [1], with a BPMN diagram. This diagram, built by analyzing an official textual plan, can support process simulation and analysis [2]: process complexity, end-to-end process time, resources costs.

For some time, we have witnessed an increasing interest in research aiming at modeling and simulating complex systems (such as Crisis Management) with the multi-agent paradigm. However, one drawback of this approach is the lack of means allowing to design and visualize the whole system behavior before programming it. In software development, we always perform the design phase before the implementation phase. Thus, we argue that multi-agent system (MAS) should also follow this way.

The idea of combining business process techniques with multi-agent system has appeared in order to improve agent-based design [3]. One reason is that process models are complementary to agent ones, since they provide means to represent an aggregate view of a MAS behavior. In addition, process models share several concepts with MAS. Therefore, we believe that the combined used of Agent and Process models is suitable for designing coordination in multi-agent system [4] and especially in crisis management: while stakeholders and their behaviors may be described by agents, the crisis resolution plan is amenable to a process.

Regarding notations, there are few efforts using UML or BPMN to design agents' interactions or activities [4], [5]. In our approach, we choose BPMN, instead of UML. Firstly, in crisis management, the stakeholders have sometimes non-IT background, thus it is difficult for them to understand a UML diagram. In comparison with UML, BPMN is easier to understand and to validate. Secondly, BPMN can support most of MAS' concepts e.g. *participant* (corresponding to an agent), *message flow* (corresponding to communication and interaction among agents) [4], [3], [7].

The work reported in this paper follows a life cycle shown in Fig. 1. Combining BPMN and OCMAS models helps us to perform several analysis based on the strong sides of both representations (such as control-flow complexity metric [11] for BPMN, organizational structure metrics [10] for OCMAS) in order to improve the quality of the coordination. More precisely, we use the Agent Group Role model of Ferber [8] as an OCMAS representation.

In this paper, our contribution consists in the definition of a framework for coordination models mapping. We provide four complementary views and a transformation procedure: a BPMN view, a BPD-graph view as pivot representation, a Role view and an AGR view. Even if our work examines a concrete case (i.e. the Ho Chi Minh City tsunami response plan), our approach can be applied to any coordination plan.

This paper is organized as follows. We first recall the process representation of our previous tsunami response plan. The transformation procedure is defined

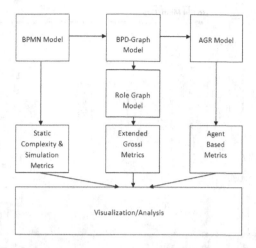

Fig. 1. Mapping Life-cycle from BPMN Model to OCMAS Model

in section 3. We then represent related works about the process-agent transformation and organizational structure assessment in section 4. Finally, we conclude our work with some perspectives.

2   BPMN for Crisis Plan Representation

The coordination among stakeholders, in general, can be represented by several formats such as: text, process diagram, Petri net diagram, conventional graph etc. Each of them expresses some aspects of a coordination plan but they differ from one another in term of abstraction level, precision and expressive power. Combining various representations helps us to have an overall view of the crisis management. In our previous work [1], we have introduced a process-based model, visualized by a BPMN diagram (Fig. 2), with coordination information extracted from a textual rescue plan.

We can identify in the process model eight participants represented by rectangular boxes, called Swimlane Objects (aka: Swimlanes). Besides, in order to visualize coordination process, we use the activity notation (like *T1: Detect tsunami risk*), represented by a rounded-corner rectangle. These activities are connected by the Connectors such as *Sequence Flow* and *Message Flow*, and the Flow Objects like *Start Event, Intermediate Event, End Event*. Moreover, the control structures help to coordinate the different activities, such as parallelism (diamond including " + ") or alternatives (diamond with " × ").

3   Mapping from BPMN to OCMAS Description

The advantages of each representation is given in Table 1.

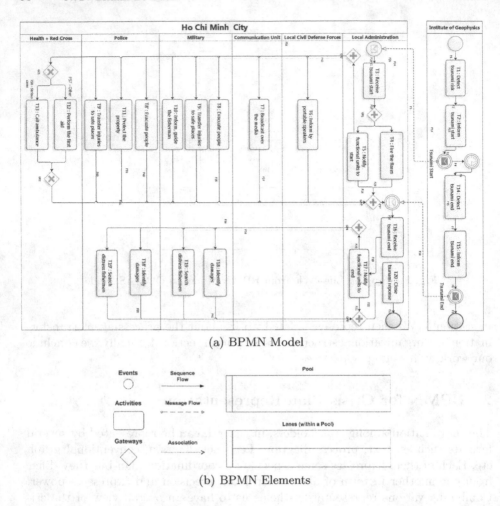

(a) BPMN Model

(b) BPMN Elements

Fig. 2. BPMN Representation for Tsunami Response Plan

3.1 BPD-Graph Representation

In this section, we explain how to transform BPMN to BPD-Graph. In our context, BPD-Graph constitutes a pivot language towards the others representations and also it is in a formal format that enables graph analysis (control flow complexity, termination ...). BPMN has four basic categories of elements: *Swimlanes, Flow Objects, Connectors and Artifacts* (Fig. 2). In [7], the authors defined a graph-based representation corresponding to each BPMN diagram, called BPD-Graph (Business Process Diagram Graph) in which BPMN elements are represented as nodes and edges of a graph. Following the notation presented in [7], we formulate three categories of mapping rules:

- *Swimlanes* (pool and lane) are represented by nodes.

Table 1. Advantages of Different Coordination Plan Representations

Notation	Views and Advantages
BPMN	An understandable and aggregate representation of a MAS behavior and possibility of analysis and macro-simulation
BPD Graph	Formal pivot representation on top of which formal analysis are possible
Role Graph	Focusing on dependency between the roles and enables analysis robustness, flexibility and efficiency of organization structure [2]
AGR Model	A complete OCMAS representation and possible macro and micro simulation

Table 2. Mapping from BPMN Elements to Graph Notations

Swimlanes	Notation	Flow Objects	Notation	Connectors	Notation
Pool	O^P	Start Event	O_S^E	Sequence Flow	F^S
Lane	O^L	Intermediate Event	O_I^E	Message Flow	F^M
		End Event	O_E^E	Association	F^A
		Atomic Activity	O_{At}^A		
		Sub-Process Activity	O_{Sub}^A		
		Exclusive Gateway	O_X^G		
		Inclusive Gateway	O_O^G		
		Parallel Gateway	O_A^G		

- *Flow Objects* (event, activity and gateway) are represented by nodes.
- *Connectors* are represented by edges.

We have defined in Table 2 all the possibility of mapping where O corresponds to nodes and F to edges. O and F could be indexed if we have several occurrences of the same object (event, activity ...). We do not count on the *Artifact* elements (or Properties) such as: *Group* (of activities), *Annotation, Image, Header or Formatted Text*. However, these elements might be useful during the mapping.

Using these notations, Endert et al. in [7] defined BPD-Graph as a tuple (O, F, src, tar) where O is the set of nodes (objects), F is the set of edges (message and sequence flows), src, tar are two functions: $F \rightarrow O$, identify the source and target nodes of each edge. In this paper, we use the notation of [7] but we refine it to give the context (swimlanes) of each node O with two attributes: *name* and (swimlane) *parent*. For example, we formulate the pool *Institute of Geophysics* by $O^P(1)$, where $O^P(1).name =$ "*Institute of Geophysics*" and $O^P(1).parent = \{\emptyset\}$; for the first activity of $O^P(1)$, we present it by $O_{At}^A(1)$, where $O_{At}^A(1).name =$ "*T1: Detect tsunami risk*" and $O_{At}^A(1).parent = \{O^P(1)\}$.

Table 3. Part of the Mapping from the BPMN Model to the BPD-Graph

NODES (O)	EDGES (F)	SOURCE	TARGET
$O^P(1)$ $O^P(1).name =$ "*Institute of Geophysics*" $O^P(1).parent = \emptyset$	$F^S(1)$	$O^E_S(1)$	$O^A_{At}(1)$
$O^P(2)$ $O^P(2).name =$ "*Ho Chi Minh City*" $O^P(2).parent = \emptyset$	$F^S(2)$	$O^A_{At}(1)$	$O^A_{At}(2)$

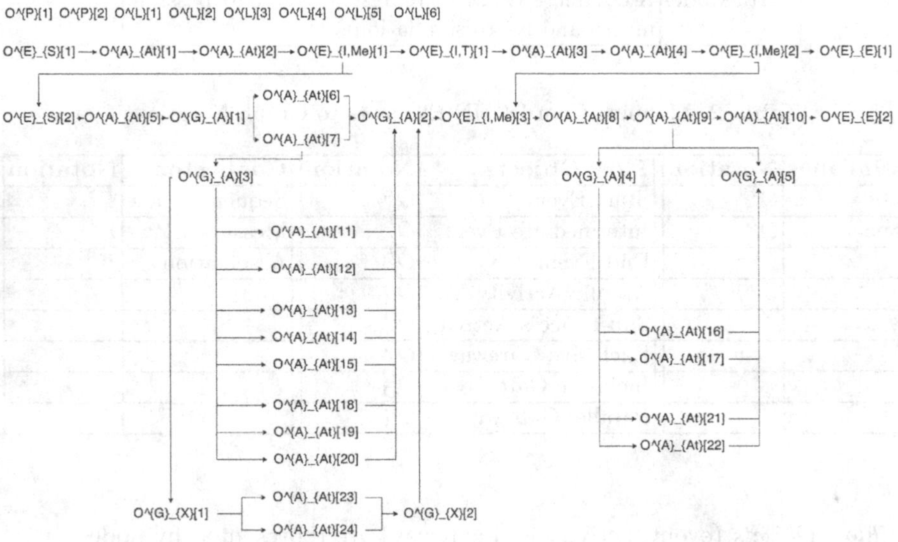

Fig. 3. BPD-Graph corresponding to BPMN Diagram

In Table 3, we just show part of the mapping from our previous formal process model (Fig. 2) to BPD-Graph, performed in two main pools.

In [7], they do not use the *Lane* concept, which is important in our approach. The BPD-Graph corresponding to the BPMN model is shown in Fig. 3.

3.2   Deriving Role Graph

The Role Graph aims at analyzing the properties of the organization involved in crisis plan, notably its robustness, flexibility and efficiency as done in [2]. The BPD-Graph thus can be used to build a Role Graph corresponding to our tsunami respond plan. This type of representation describes the roles and the relationships between them. Following the typology introduced by Grossi et al.

[10], we can distinguish three types of relations: *power* which corresponds to task delegation; *coordination* which represents flow of information among actors; and *control* relation between actors: actor A controls actor B if A monitors agent B activities. Regarding the mapping rules, the roles correspond to the name of lanes (or pools without lane) in a process. For the relationships between roles, we met a difficult problem. Because the lanes' relationships are not defined clearly in a BPMN diagram. Hence, we propose three patterns to detect three types of relation by analyzing the semantics of BPMN *Connectors* (Sequence Flow, Message Flow), as follows:

- *Power* relation: if we detect a lane/pool A with *only one-direction* message or sequence flows with another lane/pool B, we can assume that there is a power relation from A to B. For example, as depicted in Fig. 4 (messages underlined in bold), the pool *Institute of Geophysics* sends two messages to the lane *Local Administration* and there is no message in the opposite direction. Thus we conclude that the *Institute of Geophysics* role has a power relation with the *Local Administration* role.
- *Coordination* relation: if we identify a lane/pool A with *bidirectional* message or sequence flows with another lane/pool B, we can assume that there is a coordination relation between A and B, as illustrated in Fig. 6 (see Appendix).
- *Control* relation: if we detect a lane/pool A with *bidirectional* message or sequence flows *with all tasks* of another lane/pool B, we can assume that A controls B, as illustrated in Fig. 6 (see Appendix).

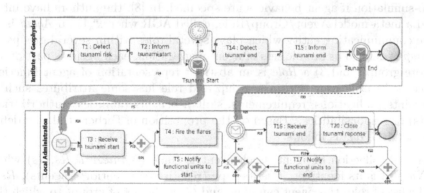

Fig. 4. Illustration of POWER Relation between Two Actors

We provide a unique semantic to this communication BPMN patterns. We could have used the Artifact elements to describe directly with a text the relationships between roles. However, we cannot assume that when designing a process model, the users will be able to supply the relation information.

As the result, the extracted Role Graph based on three above patterns is depicted in Fig. 5, each circle corresponds to an actor (IG - Institute of Geophysics, LA - Local Administration, LCDF - Local Civil Defense Forces, CU - Communication Unit, M - Military, P - Police, HR - Health & Red Cross).

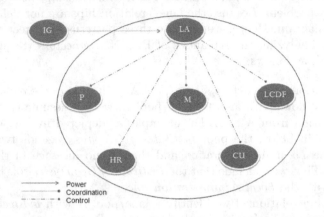

Fig. 5. Role Graph corresponding to the BPMN Diagram

3.3   Deriving AGR Model

An organization centered multi-agent system (OCMAS) view, as proposed by Ferber et al. [8], eases macro-simulation regarding the organization and also micro-simulation if agent behaviors are specified. In [8], the authors have introduced a meta-model Agent/Group/Role, called AGR where:" 1) an *Agent* is an active communicating entity which plays several roles within several groups; 2) a *Group* is defined as atomic sets of agent aggregation, each agent is part of one or more groups; and 3) a *Role* is an abstract representation of agent function, service or identification within a group and role has some attributes such as constraints (obligations, requirements, skills), benefits (abilities, authorization, profits) or responsibilities". Based on the proposition of Ferber et al., we define $AGR = (A, G, R)$, as follows:

- A is a collection of agent. Each agent is tuple $(NameA, T, Rs, Gs)$ where $NameA$ is its identifier; T is its type (reactive or intentional agents); Rs is the list of roles this agent can play; and Gs is the list of groups to which this agent may belong.
- G is a collection of groups. Each group is couple $(NameG, Rs)$ where $NameG$ is its identifier; and Rs is the list of roles involved in this group.
- R is set of roles where a role is tuple $(NameR, C, B, D, Pc, I)$ where $NameR$ is its identifier; C is the list of constraints (obligations, requirements, skills); B is the list of benefits (abilities, authorization, profits); D is the list of duties or responsibilities; Pc is the pattern of communication or interaction; I is the list of useful information.

In order to complete the process-agent mapping, we also define some notations, as follows: $x.send(y, m)$ means agent x sends message m to agent y; $x.Start$ means agent x initiates his state and/or work; $x.Do(act)$ means agent x performs the activity act; $x.Wait(time)$ means agent x has to wait for a time; $x.End$ means agent x terminates his work.

We consider a lane or a pool without lane as a role. A group constitutes a context of interaction for agents. Hence, we consider two cases: 1) each pool with more than one lane becomes a group; 2) for each message flow between two pools A and B, we create also a new group where the role A and B can be played. Regarding agents, they are not given by the BPMN diagram but by some additional information (comments) giving the number of occurrences of each roles. Thus, we just have to create as much agent by role as indicated in the additional document.

Our mapping process from BPD-Graph to AGR consists in five steps, as shown in Table 4 (see Appendix). **In step 1**, we identify the roles and groups extracted from BPMN diagram. Let us illustrate this step through our tsunami response case study. We have the first pool $O^P(1)$ where $O^P(1).name =$ "*Institute of Geophysics*" and it has no lane, therefore we consider it as a role $R(1)$. In contrary, for the second pool $O^P(2)$ where $O^P(2).name =$ "*Ho Chi Minh City*", it has six lanes: $O^L(1)$ ($O^L(1).name =$ "*Local Administration*"), $O^L(2)$ ($O^L(2).name =$ "*Local Civil Defense Forces*"), $O^L(3)$ ($O^L(3).name =$ "*Communication Unit*"), $O^L(4)$ ($O^L(4).name =$ "*Military*"), $O^L(5)$ ($O^L(5).name =$ "*Police*") and $O^L(6)$ ($O^L(6).name =$ "*Health & Red Cross*"). Thus we transfer them respectively to six roles $R(2)$, $R(3)$, $R(4)$, $R(5)$, $R(6)$ and $R(7)$. All these six roles belong to group $G(1)$. **In step 2**, we obtain the information extracted from the Artifact elements to identify roles' properties. **In step 3**, by analyzing additional data, we identify the agents' attributes such as their type, the number of agents playing a role, the number of agents belonging to a group etc. **In step 4**, we identify the communication or interaction protocols between groups and create new possible groups by analyzing message flows. In our case study, we create a new group $G(2)$ based on the message flows between two roles $R(1)$ and $R(2)$. Finally, **in step 5**, we identify the roles' activities by following sequence flows.

As the result, we have two groups and seven roles with their attributes and interactions. $R(2)$ is the only role which belongs to two groups.

4   Related Works

In [8], authors highlight the software engineering benefits of differentiating agent aspect (the Agent Centered Multi-Agent System or ACMAS) from social aspect (the Organization Centered Multi-Agent System or OCMAS). They presented the essential drawbacks which cannot be solved with ACMAS. Instead of using ACMAS, Ferber et al. attempt to view complex system under the eye of organizational structure. They propose the Agent/Group/Role meta-model (or AGR) as a means to combine efficiently these two aspects in a uniform framework.

In our work, we use this meta-model as the destination of the process-agent mapping.

In [7], Endern et al. described the mapping from BPMN to agents using an agent-centered approach (ACMAS). The agents are represented according to the *Believe-Desire-Intention (BDI)* type, which is in our opinion not fully compliant with BPMN model where the notion of goal and intention are not given. Authors consider that sub-processes determine goals which in our point of view is a strong assumption. They also omit to consider the lane concept during the mapping. In our approach, we do not use an ACMAS approach since we believe that an organizational view (OCMAS) [8] is more compliant with BPMN.

In [4] and [3], the authors presented a model to text transformation (M2T), from BPMN model to a specific agent-oriented language (JADE). This work is too much specific so that we cannot extend it to other languages. In contrary, we follow a model to model approach (M2M). Thus, the destination model (AGR) can be implemented by any agent-based language.

In [5] and [6], Onggo introduced a BPMN pattern used to represent agent-based models. He developed specific BPMN diagrams to describe activities of agents according to an ACMAS approach. In our work, we can use an arbitrary BPMN diagram and map it to the corresponding agent model, based on an OCMAS approach.

In [9], the authors described the mapping from BPMN models to Alvis language in order to formally verify these models. However, the transformation which focuses on the activities forgets the organizational structure.

Concerning the organization assessment, Grossi et al. in [10] proposed a set of equations in order to evaluate organizational structure based on the role graph with three dimensions: power, coordination and control. Comparing the results with standard values, we can determine the robustness, flexibility and efficiency of our organization. All these metrics can be performed in our Graph Role as we have demonstrated it in [2].

In [11], the author presented a metric to measure control-flow complexity of a workflow or a process. He also suggested other metrics such as: *Activity Complexity, Data-Flow Complexity, Resource Complexity*. These metrics, combined with the equations of Role Graph [10] [2], can help us to determine the quality of a coordination plan according to two points of view: process and organization.

5   Conclusion

In this paper, we have introduced an approach mapping a process model to several different graphical models: BPG-Graph, Role Graph and AGR model (following OCMAS paradigm). Combining several views of the same plan enables the authorities to benefit from the advantages of each representation. This work is the first step towards a visualization and assessment platform for Crisis Management. In a future research, we will continue on developing the assessment part of the different models, illustrated in this paper, using a set of static and dynamic metrics.

Appendix: Mapping Table & Role Relations

Table 4. Mapping BPD-Graph to AGR Model for Tsunami Response Plan

	BPD-Graph	AGR
Step 1 Identify Roles & Groups	$O^P(1)$ has no lane	$R(1) = (O^P(1).name, C\{\}, B\{\}, D\{\}, Pc\{\}, I\{\})$
		$R(2) = (O^L(1).name, C\{\}, B\{\}, D\{\}, Pc\{\}, I\{\})$
		$R(3) = (O^L(2).name, C\{\}, B\{\}, D\{\}, Pc\{\}, I\{\})$
		$R(4) = (O^L(3).name, C\{\}, B\{\}, D\{\}, Pc\{\}, I\{\})$
	$O^P(2)$ has 6 lanes	$R(5) = (O^L(4).name, C\{\}, B\{\}, D\{\}, Pc\{\}, I\{\})$
		$R(6) = (O^L(5).name, C\{\}, B\{\}, D\{\}, Pc\{\}, I\{\})$
		$R(7) = (O^L(6).name, C\{\}, B\{\}, D\{\}, Pc\{\}, I\{\})$
		$G(1) = (O^P(2).name, \{R(2), R(3), R(4), R(5), R(6), R(7)\}$
Step 2 Identify Roles' Properties by examining Artifacts		
Step 3 Identify Agents by reading additional data		$A(i) = (Name, Type, Rs(i), Gs(i))$
		$Rs(i) = \{R_k : Nb_k, R_{k+1} : Nb_{k+1}, ...\}$
		$Gs(i) = \{G_j, G_{j+1}, ...\}$
Step 4 Identify communication between Groups & Create new group based on Message Flow	$F^M(1)$ $F^M(2)$	$R(1).Pc \leftarrow \{send(R(2), F^M(1).msg)\}$
		$R(2).Pc \leftarrow \{receive(R(1), F^M(1).msg)\}$
		$R(1).Pc \leftarrow \{send(R(2), F^M(2).msg)\}$
		$R(2).Pc \leftarrow \{receive(R(1), F^M(2).msg)\}$
		$G(2) = (F^M(1, 2).msg, \{R(1), R(2)\})$
Step 5 Identify Roles' activities based on Sequence Flow	$F^S(1), F^S(2)$ $F^S(4), F^S(5)$ $F^S(6), F^S(8) ...$	$R(1).D \leftarrow \{Start, Do(O^A_{At}(1))\}, \{Do(O^A_{At}(2))\}$
		$R(1).D \leftarrow \{Wait(O^E_{I,T}(1))\}, \{Do(O^A_{At}(3))\}$
		$R(1).D \leftarrow \{Do(O^A_{At}(4))\}, \{End\} ...$

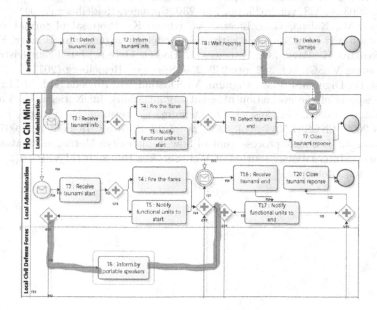

Fig. 6. Illustration of COORDINATION & CONTROL Relation between Two Actors

References

1. Thanh, L.N.T., Hanachi, C., Stinckwich, S., Vinh, H.T.: Representing, simulating and analysing Ho Chi Minh City Tsunami plan by means of process models. In: ISCRAM Vietnam 2013. Information Systems for Crisis Response and Management (2013)
2. Thanh, L.N.T., Hanachi, C., Stinckwich, S., Vinh, H.T.: Combining process simulation and agent organizational structure evaluation in order to analyze disaster response plans. In: 9th International KES Conference on Agents and Multi-Agent Systems: Technologies and Applications (2015)
3. Küster, T., Lützenberger, M., Heßler, A., Hirsch, B.: Integrating process modelling into multi-agent system engineering. Multiagent and Grid Systems **8**(1), 105–124 (2012). IOS Press
4. Küster, T., Heßler, A., Albayrak, S.: Towards process-oriented modelling and creation of multi-agent systems. In: Dalpiaz, F., Dix, J., van Riemsdijk, M.B. (eds.) EMAS 2014. LNCS, vol. 8758, pp. 163–180. Springer, Heidelberg (2014)
5. Onggo, B.S.S.: BPMN pattern for agent-based simulation model representation. In: Winter Simulation Conference (WSC), pp. 1–10. IEEE (2012)
6. Onggo, B.S.S.: Agent-based simulation model representation using BPMN. In: Formal Languages for Computer Simulation: Transdisciplinary Models and Applications, pp. 378–399 (2013)
7. Endert, H., Küster, T., Hirsch, B., Albayrak, S.: Mapping BPMN to agents: an analysis. In: Agents, Web-Services, and Ontologies Integrated Methodologies, pp. 43–58 (2007)
8. Ferber, J., Gutknecht, O., Michel, F.: From agents to organizations: an organizational view of multi-agent systems. In: Giorgini, P., Müller, J.P., Odell, J.J. (eds.) AOSE 2003. LNCS, vol. 2935, pp. 214–230. Springer, Heidelberg (2004)
9. Szpyrka, M., Nalepa, G.J., Ligeza, A., Kluza, K.: Proposal of formal verification of selected BPMN models with alvis modeling language. In: Brazier, F.M.T., Nieuwenhuis, K., Pavlin, G., Warnier, M., Badica, C. (eds.) Intelligent Distributed Computing V. SCI, vol. 382, pp. 249–255. Springer, Heidelberg (2011)
10. Grossi, D., Dignum, F.P.M., Dignum, V., Dastani, M., Royakkers, L.M.M.: Structural aspects of the evaluation of agent organizations. In: Noriega, P., Vázquez-Salceda, J., Boella, G., Boissier, O., Dignum, V., Fornara, N., Matson, E. (eds.) COIN 2006. LNCS (LNAI), vol. 4386, pp. 3–18. Springer, Heidelberg (2007)
11. Cardoso, J.: Business process control-flow complexity: Metric, evaluation, and validation. International Journal of Web Services Research (IJWSR) **5**(2), 49–76 (2008). IGI Global

Agent Based Quality Management in Lean Manufacturing

Rafal Cupek[1](✉), Huseyin Erdogan[2], Lukasz Huczala[1], Udo Wozar[2], and Adam Ziebinski[1]

[1] Institute of Informatics, Silesian University of Technology, Gliwice, Poland
{Rafal.Cupek,Lukasz.Huczala,Adam.Ziebinski}@polsl.pl
[2] Conti Temic Microelectronic GmbH, Ingolstadt, Germany
{Hueseyin.Erdogan,Udo.Wozar}@continental-corporation.com

Abstract. Quality Management (QM) issues are together with production costs and delivery time one of the three main pillars of Lean Manufacturing. Although, Quality Operations Management should be supported by IT Manufacturing Execution Systems (MES), in practice it is very difficult to automate QM support on MES level because of its heterarchical and unpredictable nature. There is a lack of practical models that bind QM and MES. Authors try to fill this gap by proposed agent based MES architecture for QM support. This paper shows both concept of proposed architecture and its practical realisation on the example of automotive electronics device manufacturing.

Keywords: Multi agent systems (MAS) · Manufacturing execution system (MES) · Quality management(QM) · Lean manufacturing · Signalr library

1 Introduction

Nowadays, producers have to respond quickly to varying customer needs and rapid market changes while still controlling production costs and quality. As was defined in [1], an Advanced Manufacturing Production System is capable of furnishing a mix of products in small or large volumes with both the efficiency of mass production and the flexibility of custom manufacturing in order to respond rapidly to customer demands and the desired quality. These trends have also been stressed as critical paths in a recent foresight study on the 2025 perspective. In [2] it was highlighted that "the production of goods and services will therefore have to address mass customisation and become localised and networked to be closer to customers, to respond to local demand and to decrease costs".

Quality Management (QM) together with production costs and delivery time is one of the three main pillars of a Lean manufacturing strategy. Early detection of quality issues and quick reactions to them during the production process appear to have become a significant source of manufacturing cost reduction. The efforts that are necessary to solve quality problems in the early stages of production are many times lower than the costs related to fixing the quality problems that are reported by end users.

© Springer International Publishing Switzerland 2015
M. Núñez et al. (Eds.): ICCCI 2015, Part I, LNAI 9329, pp. 89–100, 2015.
DOI: 10.1007/978-3-319-24069-5_9

This means that QM is one of the essential components of a Lean strategy in all types of production and on that is particularly important in the case of a short series production. The QM support for mass customised production and short series production has become the most pressing problem that has to be solved.

The change of the production mode from mass manufacturing to customised manufacturing and short series production requires a change in Quality Management (QM) philosophy. Among the reasons behind the requirement for rapid reaction on quality issues, two aspects seem to be very important in the case of short series production. In the case of mass production, there is enough time to stabilise the production process both in terms of device parameters and also by the properties of the materials that are used for the production. Also, the manual operations that are performed by operators are more stable and the final results are more repeatable. The loss that is incurred in the initial phase of production is relatively small and often negligible in comparison to the total costs of mass production. QM issues in mass production can be easily detected and removed during the startup phase of production. In the case of the mass customised production and short series manufacturing, the configuration and parameterisation of a production line must be performed very rapidly in order to reduce the loss from non-productive time. Moreover, any quality problems must be detected immediately and corrected quickly to allow the target product parameters to be reached in a very short time. Otherwise, the losses that are caused by quality problems may outweigh the profits that are made from the sale of short series of product.

On the other hand, mass production makes it easy to fill the product gaps that are caused by quality problems. Production scheduling software at the MES (Manufacturing Execution System) level takes into account any differences between the expected and produced quantity of items and compensates for any losses by planning the missing products in the next production cycle. Such an approach is not possible in the case of short series production, which in many cases is composed of only one production cycle. Any additional production volume may force successive production profile changes and may be the cause of consecutive quality issues that will force the next repetition of a production cycle.

In this paper, the authors propose a new IT architecture at the MES level that is based on three-level multi-agent environments that are realised under the MVC 4.0 model through the application of the SignalR library. It replaces the classical QM model that is used in MES, which is based on the analysis of post-production reports. The execution of orders is attended to on line by the Product Supervisory Agents that are responsible for the management of the production process including quality issues. Quality problems are detected and recognised by Holons that operate at the level of quality test devices and quality management support is given by the QM Agents. Thanks to this approach, the response to quality problems can be implemented instantaneously without it being necessary to wait for the final realisation of the order. The link to the actual production process is established by means of Holons, which makes the system more flexible and which is also highly important in the case of short series production. However, this paper focuses on the agent-based IT architecture that is necessary for the quality information management and the production execution issues are not within the scope of the presented work.

The paper is organised as follows: the second chapter highlights the most important aspects that are necessary to realise the Lean management philosophy. The literature review on multi-agent systems that are used in discrete manufacturing is presented in the third section. The fourth section presents the use case of QM data flow on the example of the Prototyping Department of Continental Ingolstadt. The quality management process is illustrated on the example of a test line that is used to produce automotive electronic devices. The fifth chapter presents some details of the implementation of the system. It shows the main assumptions of a multi-agent application that was created based on the MVC 4.0 environment. The proposed mechanisms of event-based information exchange are presented. The implementation of a communication framework based on the SignalR library is also described. The conclusions are presented in chapter six.

2   Quality Management Aspects in Lean Manufacturing

Lean Manufacturing (LM) is "lean" because it uses less of everything when compared to mass production – less manufacturing space, less investment in tools and less engineering hours to develop a new product in shorter time. It also requires keeping far less inventory on site, results in many fewer defects and produces a greater and ever growing variety of products [3].

Although LM is characterised [4] as creating value, it is also about customer service, about the revolution and the evolution of an enterprise's systems and processes and it is also an endless journey towards success. LM techniques are crucial for surviving in today competitive global economy. A company that applies LM has big advantages compared to a company that does not use LM because it has less of everything that is associated with waste but generates high quality and a greater variety of products while using less inventory.

Making the value flow in response to the needs of the customer prevents and eliminates the previously mentioned waste in the processes and also within the whole company. Waste is categorised as part of the seven Muda wastes overproduction, inventory, waiting, motion, transportation, production rejects and processing [5]. All of these wastes have a direct impact on costs because they are non-value adding operations.

LM does not only eliminate non value-added steps, it eliminates the quality problems that stem from those cost, time and resource drains. Even as the terminology becomes more complex by implanting LM tools, companies that learn these methods and incorporate them into production notice that the manufacturing becomes simpler, products become better and lead times become shorter. Lean manufacturing is about eliminating Muda waste, while increasing mistake proofing by Poka-yoke and striving for continuous improvement through the use of Kaizen. Poka-yoke is a quality technique that proofs mistakes. Kaizen is focussed on activities where a team attempts to identify and implement a significant improvement in an existing process. Some of addressable quality problems are transportation damage or breakdowns as opposed to preventive maintenance and setup adjustments. LM targets overproduction, inventory, transportation, work-in-progress, defect reworking and underutilisation of the employees' talents and knowledge.

According to MEP studies (Manufacturing Extension Partnership) [6], typical LM benefits include productivity improvements, a reduction in work-in-progress, an increase in space utilisation, an improvement in quality and a reduction in lead times The quality improvement can be achieved by three main ways:

1. Evaluating the waste in the system. Wasted time and raw materials are useless costs. The root causes should be identified and eliminated.
2. Early and close quality inspection after each step of production offers a perfect result of production. Errors should be corrected or sorted before they cause damage.
3. If employees are shown all of the processes in manufacturing, it would be advantageous for their knowledge and understanding of the whole process. Therefore, they will have a better opportunity to make good suggestions.

This paper focuses on an IT system that supports quality improvement through early inspections, which can be achieved by an agent-based quality management system that can be applied for mass customised and short series production. This system helps to recognise weaknesses in quality as early as possible and to initiate counter measures to compensate for these weaknesses.

3   Holon and Agent-Based Architectures for Manufacturing

Holonic manufacturing is based on an autonomous and cooperative entity called a Holon, which combines the advantages of hierarchical and heterarchical organisational structures [7]. It can provide the flexibility and adaptability of heterarchical control by reacting to changes while also allowing the stability of hierarchical control to be maintained [8]. Therefore, holonic manufacturing can meet the requirements for adaptability.

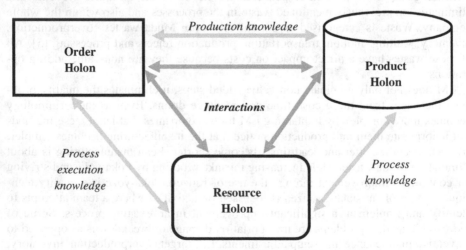

Fig. 1. Basic Holons and types of associated knowledge

PROSA (product, resource, order, staff, architecture) is one of the reference architectures for a Holonic Manufacturing System (HMS) [9]. Class models for PROSA are used to implement Holon-based dynamic control mechanisms. The example Holon types and associated knowledge are presented in Figure 1. The HMS is centred on the whole life-cycle activities of order Holons. The typology of resource Holons is based on the characteristics of the tasks that they perform – supply, transformation, assembly or disassembly. For each type of resource Holon, specific heuristics are used to solve the problem of scheduling tasks. These mechanisms and algorithms are implemented within a multi-agent system, which supports the development of the manufacturing control system.

In [10] we can find another example of the development of a robot control system for intelligent manufacturing in which the Conceptual Holonic Model was used for the control software. A networked robotised job shop assembly structure is composed of a number of robotic resources that are linked by a closed-loop transportation system was described in [11]. The example of an MES application as an HMS and multi-agent system (MAS) is presented in [12]. A manufacturing system that produces laminated bullet-proof security glasses that are ready-to-assemble on vehicles was described. Software agents work collaboratively to assist in the production, planning and sales departments for the generation of the production plans.

Agents that are organised into a heterarchical structure with a high-level of autonomy and co-operation based on the client–server structure with no fixed relations are presented in [13]. Such a concept allows for high performance against disturbances through global optimisation. The decision making is local and autonomous, without a global view of the system. For the expansibility of the system, the functioning of some agents can be modified or new agents can be added to the control system. The agent-based production planning approach can be applied for the optimisation of time scheduling and/or resource allocation in various domains of the production [14]. The planning mechanism and collaboration of the autonomous agents allows efficient resource management during the production of goods in many manufacturing processes. Based on the coordination between the planning and production agents, the efficiency of the production devices can be increased by calculating the workload and directing the most appropriate device in a specific product line.

Product Data Management (PDM) [15] considers the product development stages and evolved into Product Lifecycle Management (PLM) by including information about resources, performance, risk management and all of the stages along the complete product lifecycle as well as the PLM that integrates business process models and information exploitation. Product Lifecycle Management is essential to achieve efficiency and maintain consistency during the lifecycle of a product from the early stages of the development process of new products to the end of production cycle. Software agents can perform their tasks for sales, production and planning [16]. An autonomous system is capable of deriving materials and production plans for energy cost savings by taking into account the electricity time zones for a plant and the workload of the factory. Sales agents allow alternative plans including their costs to be introduced to the sales representatives during negotiations with customers. Such a system allows for significant energy cost savings.

4   Quality Assurance Process for Short Series Production

There are two main kinds of production that are carried out by the Prototyping Department of Continental Ingolstadt. The first is focussed on prototypes of new automotive electronics devices that have to be produced in order to allow the testing of new Advanced Driver Assistance Systems (ADAS) before starting their mass production. This kind of production has an obviously low product volume. However, product quality must be ensured at the same level as in the case of mass production. The second kind of production is short series production. In this case advanced car electronic equipment is produced in a short series often with many variants.

All orders are collected by an electronic order acquisition system. Each order is supervised by a dedicated product manager. This process is partially manual since the product manager has to contact the customer and agree on both the technical issues and logistic aspects of production. Afterwards, orders are qualified for execution. The logistics team is responsible for preparing all of the necessary materials together with the technical documentation for the production process. Production is split between SMD production, which is focussed on the electronics components and the backend production that is related to assembling the housing and connectors. In this paper we focus on the quality test operations that are performed between SMD production and backend production.

In the Prototyping Department, the test section consists of tree positions – AOI (Automatic Optical Inspection), Flying Probe and Rework station. AOI tests allow any irregularities on the SMD surface to be detected throug the automatic visual inspection of particular electronic components. The tests are carried out by automated stations that use vision recognition algorithms. In the case of error detection, additional verification and error validation is carried out manually by an operator. Error information is processed by the MES. Depending on the type of production and the type of error, products can be scrapped or directed to a Rework Station. Reworking can also be used for additional product modifications in the case of prototype production. Additional devices and connectors that are not used in the final products can be added at a Rework Station if they are required in the specification of given order. Next, products are examined for the correctness of any electrical connections. Electrical tests are performed by an automatic Flying Probe station. In the case of an error, the product can be classified as scrap or a rework. The product flows on to the Test section as illustrated in Fig.2. The maximum possible number of repetitions of each production phase is limited by the product description. This means that the actual production cycle is not predetermined by the product definition but emerges during the production realisation. Therefore, classical MES cannot be used in such a variable production and product supervisory operations have to be executed manually.

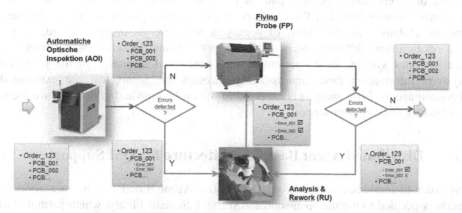

Fig. 2. Production flow in the test section (Prototyping Department in Continnental Ingolstadt)

In order to overcame this limitation, authors propose an agent-based IT system that works at the MES system level and supports the required functionality in dynamic and variable short series production. The proposed architecture consists of three levels as depicted in Fig.3. There are Holons on the bottom that are directly connected to the devices that are used for production. Three types of holons were realised – AOI holons that are directly linked to the Automatic Optical Inspection, FP Holons with the Flying Probe stations and RU Holons with the Rework Units. Since final error information is verified and entered by operators, the Holons are equipped with a user interface that is accessible through Internet Explorer. Error information is stored in an Oracle database. Business logic and related operations are excluded due to the stored procedures that are implemented in Oracle.

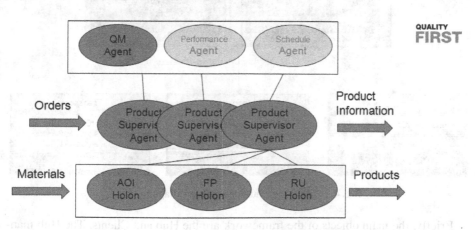

Fig. 3. Proposed IT architecture for agent QM based support

The Product Supervisory Agents are placed above the Holons level and perform operations that are related to product supervisor activities. Because of the flexible mode of production, a product path is not fixed before production but its actual implementation is traced by Product Supervisor Agents based on the information sent by the Holons. The third level of agents is necessary in order to aggregate the information from different orders and to support finding the origin of any errors. In the previous system. these operations were performed manually. In the proposed system, information flows automatically through the events that are exchanged between agents. Error analysis is executed in the on-line mode. In this paper we focus on Quality Management support.

5   Distributed Agent Based Architecture for QM Support

In order to implement and test our Distributed Agent Based Architecture concept, authors decided to use the open-source ASP.NET SignalR library, which permitted to build a flexible application layer based on a simple network communication framework [17] (Fig. 4). The popular Java-based agent frameworks (e.g., JADE, WADE) could not be used in our solution due to the development policy in that corporation, which excludes Java applications from productive solutions.

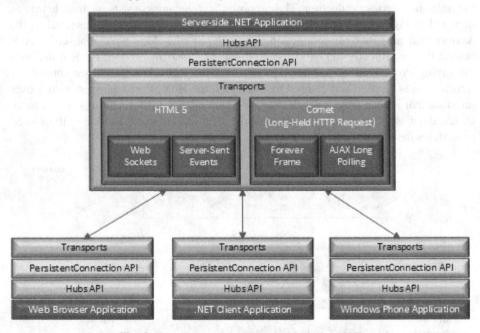

Fig. 4. SignalR communication layer diagram [17].

Briefly, the main objects of the framework are the Hub and Clients. The Hub manages the connections and message calls that are invoked by the Clients. The connection between the client and server is persistent, unlike a classic HTTP connection, which is re-established for each communication. A SignalR connection starts as an HTTP, and is then promoted to a WebSocket connection if one is available. If the client application does not support the HTML 5 standard, older transports can be used (Server Sent Events, Forever Frame, Ajax long polling). The server and client implementation can be made in many different constellations as is shown in Fig. 5. As for our solution application, the server is created as a Windows .NET Service and the clients are ASP.Net MVC4 server and Windows Console application.

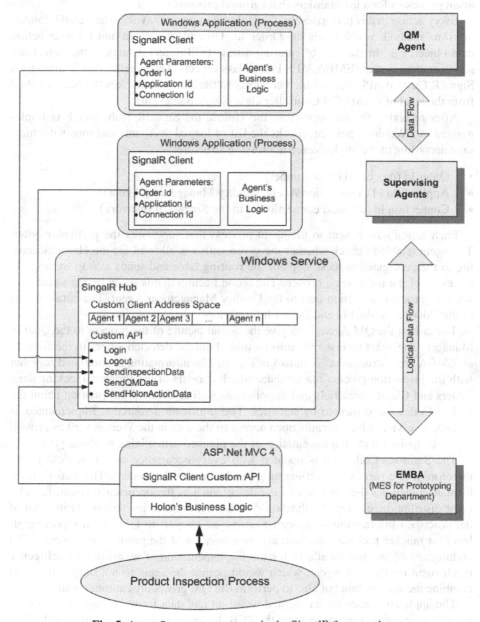

Fig. 5. Agent System components in the SignalR framework

In presented application the Agents are clients that log into the Hub, which routes the messages and provides the communication API. The Address Space of the agents is created during the Login action to the Hub that is invoked by every client on the connection. The user has the possibility to create the agents manually for single orders (console process) or to use the Hub user interface to automatically create a set of agent processes for a list of orders (background processes).

Every action that is performed by the Holon component invokes the SendHolonActionData method, which sends the Order Id, User Id, Action Id and relevant action data object, e.g., in the case of entering a new PCB defect detection, the defect description data is sent (EMBA.MES.Error class object). Technically, this is done by a SignalR Custom API object, which connects to the Hub and invokes the Send method from the level of the MVC 4 Controller class (OnAction event).

After receiving the message from the Holon, the SignalR Hub, which is implemented as a Windows Service, checks the list of logged in Agents and routes the message according to the implemented addressing structure:

- Order Id (processed order number)
- Application Id (Supervision Agent, Quality Management Agent)
- Connection Id (physical connection id in the SingalR framework)

Each action data is sent to the agent process that supervises the particular order. The agent is able to check whether the given action is allowed for the Holon according to the configured process steps of the routing table and sends a Stop message in the event of the route being broken. The second action that is made by the agent is to send the product inspection data to the Quality Management Agent. The data consists of the Oder Id, Product Id and Defect Information.

The task of the QM Agent is to give the actual picture of the process to the Quality Manager. The agent receives the information about the detected defect in the form of an EMBA error structure as is shown in Fig. 6. The information is combined together with the inspection process master data, which consists of the Process Tracking data, Orders and Clients, Materials and Suppliers as well as detailed information about the Defect Codes supplemented by statistics. The information source is implemented as an OData service, which permits open access to the data in the Web as well as providing an endpoint for the implementation of the planned data mining methods [18].

The Statistics Collection is updated with every occurrence of a new defect, and therefore it is a part of the self-learning mechanism of the system. The statistical information includes the attributes of the defect such as the occurrence count, fail element distribution, element localisation distribution or detection source station. All of the described information is presented online as a report to QM. Such reporting allows for quicker problem detection and optimisation of the prototyping process. The architecture of the system allows for the implementation of an artificial intelligence mechanism for the QM Agent, which would permit the agent to not only collect and combine the information but also to perform the QM procedures automatically.

The application tests were executed based on real data from the inspection department. The inspection operators enter the PCB defects that are detected into the EMBA Web client, which sends the data to the ASP.Net MVC 4 application server. There, the inspection data is transferred from the Controller object to our new Agent-Based system. The current tests are concentrated on gathering the Quality Management information for the product quality. Thanks to the Agent-Based data flow, the inspection information and quality report for the product is accessible online. The information changes after every new entry appears in the inspection systems.

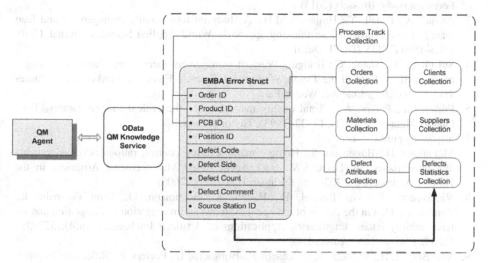

Fig. 6. OData QM Knowledge Service

6   Conclusions

In this paper the authors have described how a Multi-Agent System can increase the flexibility of the Manufacturing Execution Systems that are used in discrete production. Special attention was paid to Quality Management, which is one of pillars of Lean production. Because of the field of application, it was impossible to use classical MAS frameworks. Instead, the authors decided to prepare a dedicated environment that supports Holon and Agent-based architectures.

The biggest advantage of the proposed approach is the very short path between error detection and the reaction of product supervisor. A previous business model was based on a pull mechanism of information flow and therefore did not allow for a quick reaction to QM issues. The proposed architecture is based on the push information flow model, which has a shorter reaction time and allows losses to be reduced. Agents support distributed information processing and an appropriate reaction in the event of quality problems.

Acknowledgment. This work was supported by the European Union from the FP7-PEOPLE-2013-IAPP AutoUniMo project "Automotive Production Engineering Unified Perspective based on Data Mining Methods and Virtual Factory Model" (grant agreement no: 612207).

References

1. http://www.wilsoncenter.org/sites/default/files/Emerging_Global_Trends_in_Advanced_Manufacturing.pdf
2. A Manufacturing Industry Vision 2025, European Commission (Joint Research Centre) Foresight study, Brussels (2013)
3. Anvari, A., Ismail, Y., Hojjati, S.M.H.: A study on total quality management and lean manufacturing: through lean thinking approach. World Applied Sciences Journal 12(9), 1585–1596 (2011). IDOSI, Dubai
4. Sui Pheng, L., Shang, G.: Bridging Western management theories and Japanese management practices: Case of the Toyota Way model. Emerald Emerging Markets Case Studies 1(1), 1–20 (2011). Emerald, Wagon Lane
5. Porter, L.J., Parker, A.J.: Total quality management—the critical success factors. Total Quality Management 4(1), 13–22 (1993). Taylor & Francis
6. http://mep.purdue.edu/
7. McFarlane, D., Bussmann, S.: Holonic manufacturing control: rationales, developments and open issues. In: Deen, S.M. (ed.) Agent Based Manufacturing Advances in the Holonic Approach, pp. 303–326. Springer, Heidelberg (2003)
8. Valckenaers, P., Van Brussel, H., Hadeli, K., Bochmann, O., Saint Germain, B., Zamfirescu, C.: On the design of emergent systems: an investigation of integration and interoperability issues. Engineering Applications of Artificial Intelligence 16(4), 377–393 (2003). Elsevier B.V, Amsterdam
9. Van Brussel, H., Wyns, J., Valckenaers, P., Bongaerts, L., Peeters, P.: Reference architecture for Holonic manufacturing systems: PROSA. Computers in Industry 37(3), 255–274 (1998). Elsevier B.V, Amsterdam
10. Rannanjärvi, L., Heikkilä, T.: Software development for Holonic manufacturing systems. Computers in Industry 37(3), 233–253 (1998). Elsevier B.V, Amsterdam
11. Borangiu, T., Gilbert, P., Ivanescu, N.A., Rosu, A.: An implementing framework for Holonic manufacturing control with multiple robot-vision stations. Engineering Applications of Artificial Intelligence 22(4), 505–521 (2009). Elsevier B.V, Amsterdam
12. Blanc, P., Demongodin, I., Castagna, P.: A Holonic approach for manufacturing execution system design. An Industrial Application Engineering Applications of Artificial Intelligence 21(3), 315–330 (2008). Elsevier B.V, Amsterdam
13. Diltis, D., Boyd, N., Whorms, H.: The evolution of control architectures for automated manufacturing systems. Journal of Manufacturing Systems 10(1), 63–79 (1991). Elsevier B.V, Amsterdam
14. Aissani, N., Beldjilali, B., Trentesaux, D.: Dynamic scheduling of maintenance tasks in the petroleum industry: A reinforcement approach. Engineering Applications of Artificial Intelligence 22(7), 1089–1103 (2009). Elsevier B.V, Amsterdam
15. Marchetta, M.G., Mayer, F., Forradellas, R.Q.: A reference framework following a proactive approach for product lifecycle management. Computers in Industry 62, 672–683 (2011). Elsevier B.V, Amsterdam
16. Yildirim, O., Kardas, G.: A multi-agent system for minimizing energy costs in cement production. Computers in Industry 65(7), 1076–1084 (2014). Elsevier B.V, Amsterdam
17. Microsoft. http://www.asp.net/signalr/
18. Cupek, R., Huczala, L.: OData for service-oriented business applications. In: 2015 IEEE International Conference on Industrial Technology, IEEE Xplore on line digital library (2015)

Towards a Framework for Hierarchical Multi-agent Plans Diagnosis and Recovery

Said Brahimi[1,2]($\boxtimes$), Ramdane Maamri[2], and Sahnoun Zaidi[2]

[1] Univertity 8 Mai 1945 of Guelma, Guelma, Algeria
[2] Lire Laboratory, University of Constantine 2, Constantine, Algeria
{brahimi.said,rmaamri,sahnounz}@yahoo.fr

Abstract. This paper aim to outline an abstract template of theoretical framework for diagnosis and recovery of hicrarchical multi-agent plans to deal with the plans execution exceptions. We propose a recovery policy based on the runtime diagnosis of resources and task failure. To this end, we take advantage of the ability to reason (using the formalism Hierarchical-Plan-Net-with-Synchronization, or *SHPlNet*) on the abstract tasks interaction and interference in order to localize the recovering region in the global plan.

Keywords: Multi-agent planning · Execution recovery · Plans diagnosis · Partial global plan repairing · Petri net

1 Introduction

In complex multi-agent systems evolving in dynamic and unpredictable environments, the planning agents must be able to monitor the execution of their plans and the evolution of their environment in order to react adequately to eventual exceptions. These exceptions are caused by the failure of tasks execution, the failure of required resources, the fault in execution platform, or the errors of the designer or programmer. In this paper, we are interested only of the exception related to the plans execution and resource failure.

The works that have been proposed to deal with this issue may be grouped in two main approaches: fault-tolerant and recovery based approaches. The first approach is based on the tasks or plans replication techniques. The critical tasks or plans, which the execution may fail, are replicated. In this manner, if the execution of one fails the other continues its execution normally. This approach is not appropriate to cope with the different types of failure and is not practical in the case of complex systems, because it is too expensive (require too many resources). The second approach is based on the use of techniques that allow the agents to detect and to handle the eventual plans execution errors. Techniques proposed in the context of this second approach are multiples. Some classical techniques are based on the re-planning from scratch. These techniques are not practices because they are too expensive and are not suitable to handle the dynamic aspect of environment. To cope with this issue, others works proposed in [5–7] have used plans repairing techniques, which may be considered

© Springer International Publishing Switzerland 2015
M. Núñez et al. (Eds.): ICCCI 2015, Part I, LNAI 9329, pp. 101–110, 2015.
DOI: 10.1007/978-3-319-24069-5_10

as enhancement of the re-planning techniques, where the recovering involves the repairing of the failed plans' segments only.

In this paper, we propose an alternative technique based on hierarchical plan repairing. In previous work, we proposed a formalism, called Hierarchical-Plan-Net-with-Synchronization or SHPlNet [1,3], for representing agent and multi-agent plans as a model taking into account the synchronization and the flexible execution of concurrent plans. In another works [3], we used this formalism to propose a hierarchical coordination scheme. Viewed as a hierarchical-like structure of coordination cells (CC), this scheme allows the merge, in partially centralized way, the agents' plans. Based on the ability to localize the interference between tasks, supported by SHPlNet, the global plan is dynamically decomposed to set of partial-global-plan that may be analyzed in decentralized manner. This hybridization, between centralized and distributed coordination, can favor the interleaving of planning and execution. This ability aids the individual agents to deal with some execution exceptions by adaption, during the execution phase, the refinement of their local plans (that are partially elaborated) according to the new circumstances.

This paper aim to complement these previous works by proposing a recovery policy based on the runtime diagnosis of resources and task failure. We take advantage of the ability to reason on the abstract tasks interaction in order to localize the recovering region in the global plan. The recovery policy can reinforce the ability of agents to collectively deal with a lost set of execution exception.

The remaining of this paper is organized as follows. In section 2, we outline the formalism used to represent agents' plans and the coordination process. In section 3, we explain a technique for diagnosis of agent and multi-agent plans. In section 4, we outline the proposed strategies for recovering from failure. Finally, we conclude the paper.

2   Preliminaries

we advocate an approach where a multi-agent system is defined by three-tuple $\langle As, Ps, R \rangle$ where $As = \{A_i/i = 1..n\} \cup \{mA_i/i = 1..m\}$ is a set of domain-dependent agents and meta-level agents; $Ps = \{P_i/i = 1..n\}$ is a set of concurrent hierarchical plans, each P_i is assigned to A_i); R is a set of resources related to (to produce or to consume) agents' plans. These resources are shared between agents. Each plan P_i includes elementary (or primitive) tasks (or actions), abstract (or component) tasks (or goal), and decomposition (or refinement) methods for abstract tasks. The execution of tasks may require certain quantity of some resources and may produce others. Tasks in local or global plan may be interrelated due to the functional relationships or to the sharing of common resources. With the participation of domain-dependent agents, the meta-level agents must ensure that the concurrent execution of plans must be safe and lead to accomplish the global goal of the entire system.

In this section, we firstly present the formalism, that we call SHPlNet (Hierarchical Plan Net with Synchronization), to use for representing the agents and multi-agent plans. In second time, we formalize the status of agent and multi-agent plans. Finally, we present the principal actions that can be applied in order to reorganize the tasks in plans.

2.1   SHPlNet Formalism

SHPlNet formalism has been proposed to represent hierarchical plans with multiple level of abstraction and with explicit features to handle tasks and plan synchronization. The foundation of this formalism is based on three concepts, hierarchy, abstraction, and synchronization concepts. These concepts are featured in two components: Hierarchical-Plan-Net (*HPlNet*) and synchronization net (S). The figure 1 shows an example of a hierarchical plan, $\mathcal{P}$, represented by a SHPlNet where HPlNet and S are separated by a vertical line.

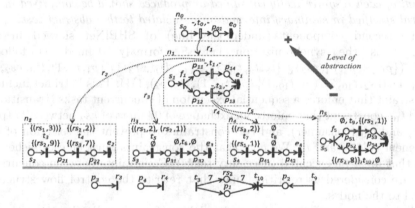

Fig. 1. Graphical representation of an example of SHPlNet.

The HPlNet component (upper the line) is represented as a tree-based structure that includes the root n_0 (that must include always one task, t_0), leaves nodes n_2, n_3, n_4, n_5, and an intermediate node n_1. Each node is a special Petri net that we call the Task Petri Net (TaskPN). The formal definition of HPlNet component of $\mathcal{P}$ is: $(n_0, Res, \{r_1\})$, where $r_1 = (t_0, pl_1)$ and $pl_1 = (n_1, Res, \{r_2, r_3, r_4, r_5\})$ such that $r_2 = (t_1, pl_2)$, $r_3 = (t_1, pl_3)$, $r_4 = (t_2, pl_4)$, $r_5 = (t_2, pl_5)$ where $pl_2 = (n_2, Res, \{\})$, $pl_3 = (n_3, Res, \{\})$, $pl_4 = (n_4, Res, \{\})$, and $pl_5 = (n_5, Res, \{\})$. The nodes n_0, n_2, n_3, and n_4 are called Sequential-TaskPN representing total-ordered tasks network. The nodes n_1 and n_5 are called Parallel-TaskPN representing partial-ordered task network. Tasks in HPlNet can be abstract (that represent goals), e.g. t_0, t_1, t_2, or elementary, e.g. t_3, t_4, t_5, t_6, t_7, t_8, t_9, t_{10}. Each abstract task in one level is connected to one or more nodes of a lower level by refinement rules, which describes how this task can be decomposed (accomplished). Leave nodes must include elementary tasks only. Each transition, t, in HPlNet is annotated, by means of the function $Res(t)$, with information summarizing the set the potential resources to consume and to produce by the task represented by this transition. The summary information associated to an abstract transitions are derived according to particular computation schemes [2]. In the example of the figure 1, the set of resources to consume (resp. to produce) of task t_4 (as is specified by the designer) is $Res(t_4).cons = \{(rs_2, 2)\}$ (resp. $Res(t_4).prod = \{(rs_3, 7)\}$) and the set of resources to consume (resp. to produce) of task t_1 (computed from lowest-level tasks and represented in terms of lower and upper values) is $Res(t_1).cons = \{(rs_1, (2, 3)), (rs_2, (0, 1))\}$ (resp. $Res(t_1).cons = \{(rs_3, (0, 7))\}$). If we are not interested in resources set, we replace it with'-'. We note that $Res(t_0)$ summarize the consumption and the production of the entire plan regardless the used refinement rules or the execution order of concurrent tasks. Overall, the summary information of an abstract task is computed according the following semantic: *Regardless of the sequence of elementary tasks that may be carried out for performing an abstract task, the amount of each resource really consumed or produced should be comprised in the interval specified in summary information associated to this abstract task.*

The second component (under the line) of SHPlNet showed in the figure 1 is the synchronization net S, formaly defined as follows: $S = (\{p_1, p_2, p_3, p_4, rs_2\}, \{t_3, t_9, t_{10}, r_3, r_4\}, \{(p_3, r_3, 1), (p_4, r_4, 1), (t_3, rs_2, 7), (t_3, p_1, 1), (rs_2, t_{10}, 7), (p_1, t_{10}, 1), (p_2, t_{10}, 1), (t_9, p_2), 1)\})$. This Petri net includes a constraint that enforce a sequential execution of concurrent tasks (transitions) t_9 and t_{10}, constraint that enforce an exchange of 07 units of rs_2 between t_3 (producer) and t_{10} (consumer), and two constraints to prevent the selection of the refinement rule r_3 and r_4. We note that the constraints must be established on tasks that must be executed (necessary tasks). Constraints between concurrent tasks are considered as rewriting rules that rewrite the control flow structure related to the nodes.

All formal representations are omitted here, for more detail related to, the reader can refer to our previous works [1,3].

2.2   Status of Agent and Multi-agent Plans

The *SHPlNet* formalism deals with structural and behavioral aspects of both agents and multi-agent plans. We mean by the structural aspect of plan the organization of its tasks as they are specified by the designer or by the planning and coordination process. SHPlNet allows especially to represent concurrent and hierarchical tasks of both agents and multi-agent plans. The link between agent and multi-agent plans is such as between subjective and objective view [4]. The Synchronization net (S) represents intra-agent and inter-agents synchronization component. The plan represented by the *SHPlNet* showed in figure 1 may be considered as global (multi-agent) plan, $\mathcal{GP}$, for two planning (domain-dependent) agents $\mathcal{A}_1$ and $\mathcal{A}_2$, having respectively the task t_1 and t_2. In this case, the individual plan of $\mathcal{A}_1$ and $\mathcal{A}_2$ are respectively $\mathcal{P}_1$ and $\mathcal{P}_2$ (figure 2), which are interrelated by resources exchange (production/consumption) relationship,

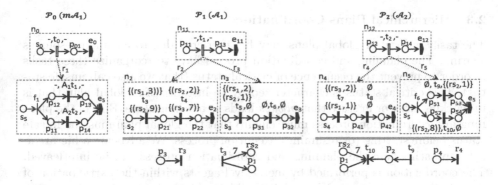

Fig. 2. Subjective Individual agents' plans

between t_3 (producer) and t_{10} (consumer). The plan $\mathcal{P}_0$ in this figure concerns a Meta-level agent that monitor the coordination of planning agents.

The behavioral aspect of plans denotes their evolution and their interfering with the environment. It described in terms of execution states and steps. In this perspective, agent and multi-agent plans (or simply individual-agent and global-system) status are featured by their structure state and execution state.

Formally, global system status is defined by $\langle \mathcal{GP}, Tr, Cxt, Ms \rangle$ where (Tr, Cxt, Ms) is the current execution state of the current instance of $\mathcal{GP}$. As it is detailed in [3], the state of $SHPlNet$ is defined by (Tr, Cxt, Ms) where (Tr, Cxt) is the state of $HPlNet$ and Ms is the marking of S. Tr is tree of nodes indicating the activated nodes (defined in $HPlNet$) and the marking of places and abstract transitions in these nodes. Cxt is the execution context of tasks, which describing the amount of available resources defined in $\mathcal{R}$. The initial state of $\mathcal{GP}$ is $(Tr_0, Cxt_0, 0)$ such that Tr_0 is formed by the highest-level node n_0 only. however, final state (or goal state) is $(Tr_f, Cxt_f, 0)$ such that Tr_f is an empty tree (that has no node), noted $Tr_f = \bot$. A step sp between two states (Tr, Cxt, Ms) and (Tr', Cxt', Ms'), denoted by $(Tr, Cxt, Ms) \xrightarrow{sp} (Tr', Cxt', Ms')$, may concern the execution (or firing) of elementary task (or transition) or a decomposition of an abstract task (selection on a refinement-rule). These steps can be performed according to the tokens in activated nodes places, the availability of required resources, and the fact that these steps are enabled in S.

In the same manner, the status of individual agent $\mathcal{A}_i$ is defined by $\langle \mathcal{P}_i, Tr_i, Ms_i, Cxt_i \rangle$ that is considered as a projection of the global status on its individual tasks. We note that in initial execution state $(Tr_{i_0}, Ms_{i_0}, Cxt_{i_0})$ (resp. in final execution state $(\bot, Ms_{i_f}, Cxt_{i_f})$) of agent plan $\mathcal{P}_i$, it is possible that $Ms_{i_f} \neq 0$ (resp. $Ms_{i_f} \neq 0$).

The synchronization in execution phase can be implemented based on the shared memory principle or on the messages exchange. In the last case, if the tokens are delivered in shared place (connecting multiples agents' tasks) by one agent task, the communication module will proceed to inform the concerned agents. In this paper, we are not interested in the concrete implementation of the synchronization technique.

2.3   Hierarchical Plans Coordination

The tasks in local or global plans may be interfered due to common resources sharing. The planning and coordination process have to reorganize these tasks in order to prevent the occurrence of critical status (that we also call unsafe status). This critical status is characterized by an instance of a global plan that is in blocking state or in state that *always* leads to blocking state. Hence, the planning and coordination process must ensure that the global plan must be in safe status before starting (or resuming) execution process. We advocate a hierarchical approach where the planning and coordination process can be interleaved. The coordination is performed by meta-level agents within the participation of domain-dependant agents. We exploit the hierarchical coordination scheme that we proposed in [3]. Viewed as a hierarchical-like structure of coordination cells (CC), this scheme allows the merge, level by level, the agents' plans in a global plan, $\mathcal{GP}$. In this process, $\mathcal{GP}$ that is represented by a SHPlNet is analyzed from top level of abstraction to ground level. At each step, the synchronization net is modified according to the strategy of consumption minimization and production maximization of all critical resources. The decisions that can be taken are [3]: (i) selection of one refinement rule (from those that can be used to refine an abstract task) and blocking the others; (ii) imposing a new sequential execution between concurrent tasks (transitions); (iii) Addition of producer/consumer link between two tasks (who that provide or releases resources and who that uses or consumes these resources). Based on the ability to localize the interference between abstract tasks (supported by SHPlNet), the global plan is dynamically decomposed to set of partial-global-plan that may be analyzed in decentralized manner. In this paper, we are interested in the use of this coordination scheme to cope with run time plans adaptation and repairing in order to deal with unexpected perturbations.

3   Diagnosis of Plans

The diagnosis is the process that enables the agents to estimate the impact of unexpected perturbations on their plans. The perturbation sources can be summarized on a failed execution of a task, lack of resources and absence or death of an agent that is committed to providing a service to others. The diagnosis consist of identifying if a status is still safe or become unsafe regarding the execution state. In the case where the plan becomes unsafe, the recovery region must be recognized.

To be able to recognize the occurrence of perturbations, we assume that each agent can (i) evaluate the availability of required resources in the execution phase, (ii) verify the success or the failure of the execution of its tasks by checking the amount of produced resources or by other information that is not explicitly represented in the plans, (ii) communicate with others agents and identify their absence, dead, or inability to provide the committed services.

According to the sematic associated to the summary information, we propose a diagnosis method that consist principally of comparing the summary information concerning the resources required by the highest-level task $(Res(t_0).cons)$ of a plan $\mathcal{P}_i$ with the available resources specified in Cxt_i as follows. Let $St = \langle \mathcal{P}_i, Tr_i, Ms_i, Cxt_i \rangle$ be the current status of the agent $\mathcal{A}_i$:

- St is *Safe*: iff $\forall (rs, q) \in Res(t_{0i}).cons, qt_{Cxt_i}(rs) \geq q^+$. It means that if no perturbations, St will lead always to a safe final state.
- St is *Critical*: iff $\exists (rs, q) \in Res(t_{0i}).cons, qt_{Cxt_i}(rs) < q^-$. It means that the missing of resources in Cxt_i prevent St to lead to a safe final state.
- St is *Ambiguous*: iff $\forall (rs, q) \in Res(t_{0i}).cons, qt_{Cxt_i}(rs) > q^-$ and $\exists (rs, q) \in Res(t_{0i}).cons, qt_{Cxt_i}(rs) \in [q^-, q^+[$. It means that, According to an execution order of concurrent tasks, St may (but not sure) lead to a safe state.

For this end, this summary information must be updated when the plan execution is evolved so that it does not take into account the required resources of the tasks performed successfully and the expected resources that are not really provided. In the case when certain resource is defined as goal (to be delivered when the execution is finished), the safeness property must take into account the verification of availability of these resources in the final state. We can deal with this issue by defining the goal resources as resources to must be consumed by the terminal transition.

In the case of critical or ambiguous status, the agent must identify the tasks that are not yet executed (or Open-Tasks), necessary missing resources (or Critical Resources), and the tasks requiring missing resources (or Critical Tasks) in order to be able to recover from failure. In the definition 1 we formally define these concepts.

Definition 1 (Open-Task, Critical-Task, and Critical-Resource) *Let* $\langle \mathcal{P}_i, Tr_i, Ms_i, Cxt \rangle$ *be a status of an agent* $\mathcal{A}_i$. *(i) A task* t *in* $\mathcal{P}_i$ *is Open-Task (OTsk) iff* $(Tr_i, Ms_i, Cxt) \xrightarrow{*} (Tr'_i, Ms'_i, Cxt')$ *and* $(Tr'_i, Ms'_i, Cxt') \xrightarrow{t}$; *(ii) A resource* rs *is Critical-Resource (CR) for the execution of* $\mathcal{P}_i$ *in* Cxt *iff* $qt_{Cxt}(rs) < q^+$ *such that* $(rs, q) \in Res(t_{0i}).cons$; *(iii) An open-task* t *in* $\mathcal{P}_i$ *is Critical-Task iff* $(rs, q) \in Res(t_{0i}).cons \cup Res(t_{0i}).prod$ *and* rs *is Critical-Resource.*

We advocate in this paper an approach where the diagnosis process is initiated by an individual agent with the possibility to be spread to others agents (where the list of critical task can be expanded to include the tasks of other agents).

4 Plan Adaptation and Recovery from Failure

The SHPlNet formalism is proposed to be able to represent hierarchical plans and to reason on their abstract level. If individual plans are coordinated in a high level of abstraction, they may be refined in several ways. The agents may therefore interleave local planning and execution while ensuring some execution adaptation. To cope with unexpected disruptions, agents can act individually or collectively through appropriate recovery strategies based on the impact of these disturbances on the plans.

4.1   Local Strategy for Recovery from Failure

The local strategy is applied by single agent without implying other. It is applied when perturbations' effects can be absorbed (treated) locally without involving other agents. For implementing the local recovering strategy, we propose three policies: Further-Refinement based Policy, Local-Execution Backtracking based Policy, and Local-planning Backtracking based Policy.

Further-Refinement Based Policy: This policy is applied when the agent status is *Ambiguous* (not safe and not critical) and the perturbations effects are minor. It is based on the refinement of local plan by imposing additional constraints in order to push its execution towards a goal state. The application of this policy is considered as resuming of local planning where only the open tasks can be considered. As the planning process, the further-refinement process consists of applying of the three decisions presented in the section 2.3. It must also take into account the keeping of causality relationship with the other agents. we note that before applying this policy, the individual plans must be corrected by depositing the missed tokens (only the control flow token) in the SHPlNet representing these plans.

The Further-Refinement decision can be formally defined as follows (definition 2):

Definition 2 (Further-Refinement). *Let $St_1 = \langle \mathcal{P}_i, Tr_i, M_i, Cxt_i \rangle$ and $St_2 = \langle \mathcal{P}'_i, Tr'_i, M'_i, Cxt'_i \rangle$ be two status of an agent $\mathcal{A}_i$, St_2 is: (i) a Further-Refinement of St_1 iff $\mathcal{P}'_i = (P(\mathcal{P}_i), T(\mathcal{P}_i), F(\mathcal{P}_i), W(\mathcal{P}_i), s(\mathcal{P}_i), e(\mathcal{P}_i), Res(\mathcal{P}_i), R(\mathcal{P}_i), S'_i)$ such that $S'_i \supseteq S(\mathcal{P}_i)$, $Tr'_i = Tr_i$, $M'_i \supseteq M_i$, $Cxt'_i = Cxt_i$; (ii) a safe Further-Refinement of St_1 iff St_2 is safe status .*

As ambiguous status can not ensure a successful execution, a policy based Further-Refinement can not always produce a final safe status.

Local-Execution Backtracking Based Policy: Local-Execution Backtracking based Policy is applied when also the agent status is *Ambiguous* or *Critical*. It consist in making a backtracking of the execution process to an anterior safe status. The execution will be resumed by the questioning of the effects of the tasks performed. Local-Execution Backtracking decision is formally defined as follows (definition 3):

Definition 3 (Local-Execution Backtracking). *Let $St_1 = \langle \mathcal{P}_i, Tr_i, M_i, Cxt_i \rangle$ and $St_2 = \langle \mathcal{P}'_i, Tr'_i, M'_i, Cxt'_i \rangle$ be two status of an agent $\mathcal{A}_i$, St_2 is: (i) a Local-Execution Backtracking of St_1 iff $\mathcal{P}'_i = \mathcal{P}_i$, $Tr'_i \xrightarrow{*} Tr_i$ or Tr'_i is a correction of Tr_i, $M'_i \xrightarrow{*} M_i$ or M'_i is a correction of M_i, and $Cxt'_i = Cxt_i$ or Cxt'_i is a correction (by releasing of reusable resources) of Cxt_i; (ii) is a safe Local-Execution Backtracking of St_1 if St_2 is safe status.*

This policy may be only applied if the agents have other recovery plans. These plans enable the mapping between harmful execution state and the partial description of plan status from which the execution will be resumed. They may be defined as domain-dependant knowledge defined by the designer, or control knowledge automatically inferred by the system. For example, If a task t, belonging to refinement node n of an abstract task t', fails to be correctly execution, the execution may be resumed from a state where the execution of t' will be enabled. In this target state, this task may be refined by another node.

Local-Planning Backtracking Based Policy: This policy can be applied when the agent status is *Ambiguous* and especially when it is *Critical*. It consists of replanning (some fragments in) the local plan according to the impact of unexpected perturbations on this plan. This policy enable the agent to both impose *additional* constraints and to question some local planning decision. It is involved generally when the other previous policies fail to find a way that can lead to a safe status (if the ambiguous status is deemed critical). It can also be applied together with the previous policies according to a particular strategy.

Is a policy that allows to resume the local planning process from an anterior planning-state. It is described in terms of a corresponding between a Critical-Tasks and a state from which the local planning will be resumed. A Local-planning Backtracking decision is formally defined as follows (definition 4):

Definition 4 (Local-planning Backtracking). *Let* $St_1 = \langle P_i, Tr_i, M_i, Cxt_i \rangle$ *and* $St_2 = \langle P_i', Tr_i', M_i', Cxt_i' \rangle$ *be two status of an agent* A_i, St_2 *is: (i) a Local-planning Backtracking of* St_1 *if and only if* $P_i' = (P(\mathcal{P}_i), T(\mathcal{P}_i), F(\mathcal{P}_i), F(\mathcal{P}_i), s(\mathcal{P}_i), e(\mathcal{P}_i), Res(\mathcal{P}_i), R(\mathcal{P}_i), S_i')$ *such that* $S_i' \subseteq S$, $Tr_i' = Tr_i$, $M_i' = M_i$, $Cxt_i' = Cxt_i$; *(ii) is a safe Local-planning Backtracking of* St_1 *if* St_2 *is safe status.*

Backtracking states can be generated during the planning process progression, some planning states (selected according to a particular heuristic that we do not detail in this paper) are kept as historical trace of planning process. We note that this policy must not question the tasks that are in causal link with other agent. If these tasks are canceled, multi-agent plan recovery policies must be applied.

4.2   Global or (Partial global) Strategy for Recovery from Failure

The global or partially-global strategy for plan recovery is required when local strategy is non applicable or not appropriate. This may concern the case where it may not succeed to lead to a safe (local) status or it implies questioning of causal link between different agents. To avoid the recovery of global plan, and hence to avoid the interruption of complete multi-agent plan execution, the recovery must be applied to the concerned tasks of the concerned agents. In this paper, we advocate an approach where recovery is applied to repair only limited regions.

We propose to implement the global strategy for recovering from multi-agent plan failure as partial-global planning strategy. For this end we can refer to our previous work proposed in [3]. We propose to resume the coordination process in the execution phase. In this case, multi-agent plan recovery can be governed by the meta-level agent with the participation of domain-dependent agents. If a domain-dependant agent esteems that local recovery is not possible, it call all CC mediators related to the critical-tasks. The recovery region will then be determined with the participation of the CC mediators according the resuming states kept in the historical trace.

5 Conclusion

In this paper, we proposed some ideas of theoretical framework for diagnosis and recovery of hierarchical multi-agent plans to deal with the plans execution exceptions. We proposed several recovery policies based on the runtime diagnosis of resources and task failure. The diagnosis is based on the runtime comparing of summary information on the required resources with the amount of resources available in each execution state. For this end, we are based on the SHPlNet formalism and the underlying reasoning techniques. The SHPlNet formalism enable the agents to have an abstract control structure that aids to monitor their plans, tasks, and resources evolution. the recovery strategies are formed on the basis of these abilities.

The future work will be focused on the expansion of the proposed recovery technique and on its use in a framework for interleaving multi-agent planning and execution.

References

1. Brahimi, S., Maamri, R., Sahnoun, Z.: Model of petri net for representing an agent plan in dynamic environment. In: O'Shea, J., Nguyen, N.T., Crockett, K., Howlett, R.J., Jain, L.C. (eds.) KES-AMSTA 2011. LNCS, vol. 6682, pp. 113–122. Springer, Heidelberg (2011)
2. Brahimi, S., Maamri, R., Sahnoun, Z.: Hierarchical Multi-Agent Plans Using Model-Based Petri Net. International Journal of Agent Technologies and Systems (IJATS) **5**(2), 1–30 (2013). IGI Global
3. Brahimi, S., Maamri, R., Sahnoun, Z.: Partially centralized approach for dynamic hierarchical-plans merging. Neurocomputing **146**, 187–198 (2014)
4. Decker, K., Lesser, V.: Task environment centered design of organizations. In: Proceedings of the AAAI Spring Symposium on Computational Organization Design, vol. 127 (1994)
5. de Jonge, F., Roos, N.: Plan-execution health repair in a multi-agent system. In: PlanSIG (2004)
6. Komenda, A., Novak, P., Pechoucek, M.: Decentralized multi-agent plan repair in dynamic environments. In: Proceedings of the 11th International Conference on Autonomous Agents and Multiagent Systems. International Foundation for Autonomous Agents and Multiagent Systems, vol. 3, pp. 1239–1240 (2012)
7. Musliner, D.J., Durfee, E.H., Shin, K.G.: Execution monitoring and recovery planning with time. In: Proceedings of the Seventh IEEE Conference on Artiïñ̈Acial Intelligence Applications, 1991, vol. 1, pp. 385–388. IEEE (1991)

Social Networks and NLP

A Survey of Twitter Rumor Spreading Simulations

Emilio Serrano[✉], Carlos A. Iglesias, and Mercedes Garijo

Universidad Politécnica de Madrid, Madrid, Spain
emilioserra@fi.upm.es, {cif,mga}@dit.upm.es

Abstract. Viral marketing, marketing techniques that use pre-existing social networks, has experienced a significant encouragement in the last years. In this scope, Twitter is the most studied social network in viral marketing and the rumor spread is a widely researched problem. This paper contributes with a survey of research works which study rumor diffusion in Twitter. Moreover, the most useful aspects of these works to build new multi-agent based simulations dealing with this interesting and complex problem are discussed. The main four research lines in rumor dissemination found and discussed in this paper are: exploratory data analysis, rumor detection, epidemiological modeling, and multi-agent based social simulation. The survey shows that the reproducibility in the specialized literature has to be considerably improved. Finally, a free and open-source simulation tool implementing several of the models considered in this survey is presented.

Keywords: Agent-based social simulation · Agent theory and application · Rumor spreading model · Data mining for social networks · Information diffusion model · Social networks · Twitter · Review

1 Introduction

Viral marketing, marketing techniques that use pre-existing social networking services, has experienced a significant encouragement over the past few years because a number of reasons. Among others: the low cost of these campaigns; traditional marketing techniques do no longer cause the desired effect; and, people influence each other's decisions considerably [12]. Rumors are the basis for viral marketing [18] and, therefore, rumors diffusion is a topic widely studied. Besides, Twitter is the most studied social network in viral marketing. Twitter allows researchers to study global phenomena from a quantitative point of view for the first time in humanity's history [3]. The main reason for this is that, unlike the leading social network Facebook[1], users' messages in Twitter are public by default.

Multi Agent Based Simulation (MABS) combines computer simulation and agent theory by using a simple version of the agent metaphor to specify single

---

[1] Leading social networks: http://goo.gl/bjFfWC

© Springer International Publishing Switzerland 2015
M. Núñez et al. (Eds.): ICCCI 2015, Part I, LNAI 9329, pp. 113–122, 2015.
DOI: 10.1007/978-3-319-24069-5_11

components and interactions among them [23][2]. MABS has become one of the most popular technologies to model and study complex adaptive systems such as: emergency management [24], intelligent environments [6], e-commerce [25], economy [4], trust and reputation [26], and marketing [20]. In the rumor case, MABS allows researchers to understand how a piece of information spreads on a network and evaluate strategies to control its diffusion; maximizing it in the case of advertisement or minimizing in the case of malicious rumors.

This paper contributes with a survey of research works which study rumor diffusion in Twitter. Moreover, the most useful aspects of these works to build new MABS dealing with this interesting and complex problem are discussed. After revising the review questions in section 2, the main research lines found in the specialized literature are discussed. Section 3 covers works which address an exploratory data analysis of gossips. Section 4 details works which attempt to detect Twitter rumors by several techniques. Section 5 introduces the epidemiological models for rumors dissemination. Section 6 details research works which study Twitter hearsay under the multi-agent based simulation paradigm. Finally, section 7 concludes and gives future works.

2   Review Questions

In the spirit of the systematic review methods [28], several review questions were formulated before locating and selecting relevant studies. These questions are the following:

- Q1. Does the work deals with rumors dissemination?
- Q2. Does it include the Twitter case?
- Q3. Real data is employed in the study?
- Q4. Does the paper simulate the information spread?
- Q5. Is there multi-agent based simulation?
- Q6. Are there what-if scenarios?
- Q7. A general methodology is presented to evaluate and use simulations?
- Q8. Is the data provided?
- Q9. Is the implementation given?
- Q10. Is it free and open source software?

Note that these questions fall in three main categories: (1) type of target studied (Q1-Q3); (2) method employed (Q4-Q7); (3) reproducibility of the research (Q8-Q10). Moreover, the questions are not disjoint, e.g. if no real data is employed (Q3), data cannot be provided (Q8).

Table 1 summarizes the works reviewed and answers for these review questions. A quick glance at the reproducibility fields show that there is great room for improvement in this matter.

---

[2] With some significant differences, MABS can also be referred as agent-based models (ABM), agent-based social simulations (ABSS), or social simulation (SocSim) [15].

Table 1. Review questions for survey. Check mark: yes, empty space: No, UR: under request.

Ref.	Target system			Method				Reproducibility		
	Q1	Q2	Q3	Q4	Q5	Q6	Q7	Q8	Q9	Q10
Valecha et al. [32]	✓	✓	✓					UR		
Mendoza et al. [17]	✓	✓	✓							
Starbird et al. [29]	✓	✓	✓							
Cha et al. [2]	✓	✓	✓							
Weng et al. [33]		✓	✓	✓	✓					
Gupta et al. [8,9]	✓	✓	✓							
Kwon et al. [13,14]	✓	✓	✓					UR		
Qazvinian et al. [19]	✓	✓	✓					UR		
Nekovee et al. [18]	✓			✓						
Zhao et al. [35]	✓			✓						
Shah and Zaman [27]	✓			✓						
Domenico et al. [3]	✓	✓	✓	✓						
Jin et al. [11]	✓	✓	✓	✓						
Tripathy et al. [31]	✓	✓	✓	✓	✓	✓				
Liu and Chen [16]	✓	✓		✓	✓					
Seo et al. [22]	✓	✓	✓	✓	✓	✓				
Yang et al. [34]	✓	✓	✓	✓	✓	✓				
Gatti et al. [7]		✓	✓	✓	✓	✓				

3   Exploratory Data Analysis Studies

This section deals with works which, without simulating the rumor propagation, conduct a exploratory data analysis of rumor data to gain insights into this problem.

Valecha et al. [32] analyze Twitter data of the Haiti earthquake in 2010[3]. The authors categorize seven different communication modes for four time stages at this occurrence. The paper concludes that information with credible sources contributes to suppress the level of anxiety in Twitter community, which leads to rumor controlling and high information quality.

In this vein, Mendoza et al. [17] explore the behavior of Twitter users in the 2010 earthquake in Chile. The authors classify the tweets manually in affirms, denies, or unknown. They also conclude that rumors tend to be questioned more than news by the Twitter community.

Starbird et al. [29] present another exploratory work which deals with the 2013 Boston Marathon Bombing[4] and conclude that corrections to the misinformation emerge but are muted compared with the propagation of the misinformation.

---

[3] On January 12, 2010, a devastating earthquake with a magnitude of 7.3 struck Haiti. More than 220,000 people were killed and over 300,000 injured.

[4] The Boston Marathon bombings were a series of attacks and incidents which began on April 15, 2013, when two pressure cooker bombs exploded during the Boston Marathon, killing 3 people and injuring an estimated 264 others.

Cha et al. [2] use Twitter data to gain insights into viral marketing and, more specifically, to compare three measures of influence: indegree, retweets, and mentions. These authors conclude that popular users who have high indegree are not necessarily influential in terms of retweets or mentions, while influence is gained limiting tweets to a single and specialized topic.

These works hint at the potential of understanding rumors spread and having strategies to control them. Nevertheless, they do not cope with these strategies or their evaluation by simulation techniques.

4   Rumor Detection Studies

Another important research line in rumor diffusion is the rumor detection, specially with machine learning techniques but also with social network analysis methods.

Weng et al. [33], without dealing with rumors specifically, address meme propagation in Twitter. Memes are parts of cultural tradition, e.g. thoughts, cultural techniques, behaviors, etcetera [5]. In Weng et al.'s work, memes are identified with a Twitter hashtag, i.e. a metadata tag used in Twitter and which consists of a word or an unspaced phrase prefixed with '#'. The authors, based on real data, compare memes spread with four simple simulated models: random, cascade, social reinforcement, and homophily. Finally, the authors present a method to detect if a meme will go viral depending on the meme first 50 tweets and machine learning techniques. Although this is a very significant work which gives sound results to support the hypothesis presented, it does not intend to give realistic simulated models or use them for designing and testing any what-if scenario. Moreover, as displayed in table 1, data and implementations are not given.

Other works also propose machine learning models after an exploratory data analysis of Twitter. Gupta et al. [8,9] study tweets of the Boston marathon blasts and propose a regression prediction model. This model allows calculating the number of nodes which will be infected in a network assuming that fake content is published by a specific user.

In this vein, Kwon et al. [14] identify a large number of characteristics in rumors under three main categories: temporal, structural, and linguistic. Then these features are used in several machine learning algorithms to classify a Tweet as rumor or non-rumor.

Qazvinian et al. [19] also deal with rumor detection and explore the effectiveness of three categories of features: content-based, network-based, and specific memes.

These machine learning models are important contributions for viral marketing, but they do not allow researchers to test marketing strategies with them. Moreover, as pointed out in some works [19], identifying new emergent rumors directly from the Twitter data is more challenging than the classification of a dataset previously retrieved.

In a sense, the research line presented in these works is complementary of the use of rumor spreading MABSs. On the one hand, machine learning approaches may employ features taken from simulated models [14]. On the other hand, the strategies tested with simulation can be undertaken when detected rumors by these machine learning approaches.

5   Epidemiological Modeling

The epidemiological modeling is the hegemonic research line to model rumor spread. In this line, the population is divided into several classes such as susceptible (S), infected (I), and recovered (R) individuals. These analytical models are usually formulated using differential equations since the transition rates from one class to another are mathematically expressed as derivatives. The standard model in this line is the *SIR* model [10] (susceptible, infected, recovered). Moreover, the *SI* (susceptible, infected) and *SIS* (susceptible, infected, susceptible) models are also very used.

Nekovee et al. study the SIR model applied to rumor spread in complex social network [18]. In this vein, Zhao et al. [35] extends the SIR model with forgetting mechanisms. Shah and Zaman [27] use a SI model to study algorithms to find a rumor source in a network. Domenico et al. [3] study Twitter rumors about the Higgs boson discovery and reproduce the global behavior using the SI model and extending it. Jin et al. [11] employees the *SEIZ* model (which considers exposed individuals, E, and sceptics, Z) for capturing diffusion of rumors and news in Twitter.

The main appealing of these works is the accuracy they achieve by adjusting automatically the model parameters, e.g. population size, with fourth generation programming languages such as MATLAB. On the other hand, comparing these model to real-world data is difficult and they often require overly simplistic assumptions [20].

These works employ social simulation (a society is modeled), but they are not MABS works (equations describe the society instead of agents). Furthermore, unlike MABS, they do not allow the exploration of individual-level theories of behavior which can be used to examine larger scale phenomenon [20]. For example, if a single Twitter user gives extensive information for an event while the remaining users post just one tweet (as in Mendoza et al.'s [17] work); MABS allows this special user to be modeled.

6   Multi-agent Based Simulations

Works studied above do not use MABSs except for Weng et al. paper [33], i.e. question five has "no" as an answer in table 1. However, there are a few works in this line.

Tripathy et al. [31] present a study and an evaluation of rumor-like methods for combating the spread of rumors on social networks. They use variants of the independent cascade model [33] for rumor spread. Besides, the authors criticise epidemic spread models such as SIS and SIR because, among others, anti-rumors can be spread from person to person unlike vaccines for viruses which can only be administered to individuals. Tripathy et al. also propose an anti-rumor strategy which consists of embedding agents called *beacons* in the network which detect rumors and spread anti-rumors.

Liu and Chen [16] build an agent-based rumor spread model using SIR as baseline and implemented in NetLogo [30], a popular MABS framework. Although the authors find out interesting conclusions with regard to the Twitter case using the simulation model, this model is not founded on real data.

Seo et al. [22] present a simple MABS based on gathering retweets (not necessarily rumors), getting the largest connected component in the network, and calculating the retweet probability of each edge $x \to y$ with the number of retweets given in that edge. More than the simulation, the contribution rests on the use of this model to evaluate a method to identify rumors and their sources by injecting special nodes called *monitors*.

Yang et al. [34] employ MABS to analyze the 2013 Associated Press hoax incident[5]. The authors give three profiles for twitter users (broadcaster, acquaintances, and odd users); probability density functions for each profile; and a study of the effects of removing relevant nodes of the network in the information spread. The authors conclude that removing the node of the highest *betweenness centrality* [21] has the optimal effect in reducing the spread of the malicious messages.

Gatti et al. [7] address the general information diffusion modeling instead of the rumor spread. These authors explore President Obama's Twitter network as an egocentric network and present an MABS approach where each agent behavior is determined by the Markov Chain Monte Carlo simulation method. As in other works revised [34], simulation is employed to find users with more impact on the information flow.

The last works revised present significant contributions in the use of MABS to study information dissemination in Twitter. Nonetheless, as shown in table 1, the efforts in reproducibility are quite questionable. None of them give: the data the results are based on, the simulation implementation, or the source code (three last questions in the table). This hinders researchers from verifying the results or reusing these works in their research or developments.

7   Conclusion and Future Works

Although creating virtual populations to test viral marketing strategies is considered an effective and useful approach [1,4,20], there are a number of shortcomings in the specialized literature which hinder researchers from learning and reusing

---

[5] On April 23 2013, the Associated Press Twitter account was hacked and a malicious message was sent stating that the White house had been attacked and President Obama was injured.

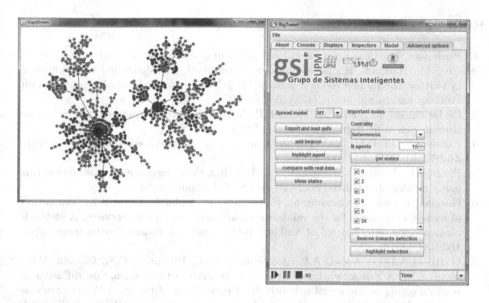

Fig. 1. BigTweet, a rumor spread simulator for evaluating viral marketing strategies.

these works for new cases. More specifically, for the Twitter rumor spreading case, the authors have found a lack of: (1) general methods to conduct such research, (2) data to validate the realism of the proposed models, and (3) tools (specially free and open-source code) to deploy these simulations. As in many other problems in computer sciences, without these three elements researchers are condemned to reinvent the wheel for each case. Besides, the Big Data technologies, which provide researchers with a great deal of information about prolific users, make the transition from analytical models to multi-agent based simulation models a must because the latter modeling paradigm allow the exploration of individual-level theories.

Under the Big Market research project ("Big Data platform to simulate and evaluate marketing techniques in realistic environments"), the authors have developed free and open-source simulation tool whose interface is shown in figure 1. This simulator called "Big Tweet"[6] implements several of the models considered in this survey to evaluate viral marketing strategies.

Acknowledgments. This research work is supported by the Spanish Ministry of Economy and Competitiveness under the R&D project CALISTA (TEC2012-32457); by the Spanish Ministry of Industry, Energy and Tourism under the R&D project Big-Market (TSI-100102-2013-80); and, by the Autonomous Region of Madrid through the program MOSI-AGIL-CM (grant P2013/ICE-3019, co-funded by EU Structural Funds FSE and FEDER).

---

[6] GitHub repository https://github.com/gsi-upm/BigTweet, presentation video https://www.youtube.com/watch?v=rGROCQllNxo

References

1. Buchanan, M.: Economics: Meltdown modelling. Nature **460**(7256), 680 (2009)
2. Cha, M., Haddadi, H., Benevenuto, F., Gummadi, K.: Measuring user influence in twitter: the million follower fallacy. In: 4th International AAAI Conference on Weblogs and Social Media (ICWSM) (2010)
3. De Domenico, M., Lima, A., Mougel, P., Musolesi, M.: The Anatomy of a Scientific Rumor. Scientific Reports **3**, October 2013
4. Farmer, J.D., Foley, D.: The economy needs agent-based modelling. Nature **460**(7256), 685–686 (2009)
5. Flentge, F., Polani, D., Uthmann, T.: Modelling the emergence of possession norms using memes. J. Artificial Societies and Social Simulation **4**
6. Garcia-Valverde, T., Campuzano, F., Serrano, E., Villa, A., Botia, J.A.: Simulation of human behaviours for the validation of ambient intelligence services: A methodological approach. Journal of Ambient Intelligence and Smart Environments **4**(3), 163–181 (2012)
7. Gatti, M.A.D.C., Appel, A.P., dos Santos, C.N., Pinhanez, C.S., Cavalin, P.R., Neto, S.B.: A simulation-based approach to analyze the information diffusion in microblogging online social network. In: Proceedings of the 2013 Winter Simulation Conference: Simulation: Making Decisions in a Complex World, WSC 2013, pp. 1685–1696. IEEE Press, Piscataway (2013)
8. Gupta, A., Lamba, H., Kumaraguru, P.: $1.00 per RT #BostonMarathon #PrayForBoston: Analyzing fake content on twitter, San Francisco, CA, September 2013
9. Gupta, A., Lamba, H., Kumaraguru, P., Joshi, A.: Faking sandy: characterizing and identifying fake images on twitter during hurricane sandy. In: Proceedings of the 22Nd International Conference on World Wide Web Companion, WWW 2013 Companion, pp. 729–736. International World Wide Web Conferences Steering Committee, Republic and Canton of Geneva (2013)
10. Hethcote, H.W.: The mathematics of infectious diseases. SIAM Review **42**, 599–653 (2000)
11. Jin, F., Dougherty, E., Saraf, P., Cao, Y., Ramakrishnan, N.: Epidemiological modeling of news and rumors on twitter. In: Proceedings of the 7th Workshop on Social Network Mining and Analysis, SNAKDD 2013, pp. 8:1–8:9. ACM, New York (2013)
12. Kostka, J., Oswald, Y.A., Wattenhofer, R.: Word of mouth: rumor dissemination in social networks. In: Shvartsman, A.A., Felber, P. (eds.) SIROCCO 2008. LNCS, vol. 5058, pp. 185–196. Springer, Heidelberg (2008)
13. Kwon, S., Cha, M., Jung, K., Chen, W., Wang, Y.: Aspects of rumor spreading on a microblog network. In: Jatowt, A., Lim, E.-P., Ding, Y., Miura, A., Tezuka, T., Dias, G., Tanaka, K., Flanagin, A., Dai, B.T. (eds.) SocInfo 2013. LNCS, vol. 8238, pp. 299–308. Springer, Heidelberg (2013)
14. Kwon, S., Cha, M., Jung, K., Chen, W., Wang, Y.: Prominent features of rumor propagation in online social media. In: Xiong, H., Karypis, G., Thuraisingham, B.M., Cook, D.J., Wu, X. (eds.) 2013 IEEE 13th International Conference on Data Mining, Dallas, TX, USA, December 7–10, 2013, pp. 1103–1108. IEEE Computer Society (2013)
15. Li, X., Mao, W., Zeng, D., Wang, F.-Y.: Agent-based social simulation and modeling in social computing. In: Yang, C.C., Chen, H., Chau, M., Chang, K., Lang, S.-D., Chen, P.S., Hsieh, R., Zeng, D., Wang, F.-Y., Carley, K.M., Mao, W., Zhan, J. (eds.) ISI Workshops 2008. LNCS, vol. 5075, pp. 401–412. Springer, Heidelberg (2008)

16. Liu, D., Chen, X.: Rumor propagation in online social networks like twitter - a simulation study. In: Proceedings of the 2011 Third International Conference on Multimedia Information Networking and Security, MINES 2011, pp. 278–282. IEEE Computer Society, Washington, DC (2011)

17. Mendoza, M., Poblete, B., Castillo, C.: Twitter under crisis: Can we trust what we rt? In: Proceedings of the First Workshop on Social Media Analytics, SOMA 2010, pp. 71–79. ACM, New York (2010)

18. Nekovee, M., Moreno, Y., Bianconi, G., Marsili, M.: Theory of rumour spreading in complex social networks. Physica A: Statistical Mechanics and its Applications 374(1), 457–470 (2007)

19. Qazvinian, V., Rosengren, E., Radev, D.R., Mei, Q.: Rumor has it: identifying misinformation in microblogs. In: Proceedings of the Conference on Empirical Methods in Natural Language Processing, EMNLP 2011, pp. 1589–1599. Association for Computational Linguistics, Stroudsburg (2011)

20. Rand, W., Rust, R.T.: Agent-based modeling in marketing: Guidelines for rigor. International Journal of Research in Marketing 28(3), 181–193 (2011)

21. Rolla, V.G., Curado, M.: A reinforcement learning-based routing for delay tolerant networks. Engineering Applications of Artificial Intelligence 26(10), 2243–2250 (2013)

22. Seo, E., Mohapatra, P., Abdelzaher, T.: Identifying rumors and their sources in social networks (2012)

23. Serrano, E., Moncada, P., Garijo, M., Iglesias, C.A.: Evaluating social choice techniques into intelligent environments by agent based social simulation. Information Sciences 286, 102–124 (2014)

24. Serrano, E., Poveda, G., Garijo, M.: Towards a holistic framework for the evaluation of emergency plans in indoor environments. Sensors 14(3), 4513–4535 (2014)

25. Serrano, E., Rovatsos, M., Bota, J.A.: Data mining agent conversations: A qualitative approach to multiagent systems analysis. Information Sciences 230, 132–146 (2013)

26. Serrano, E., Rovatsos, M., Botia, J.: A qualitative reputation system for multiagent systems with protocol-based communication. In: Proceedings of the 11th International Conference on Autonomous Agents and Multiagent Systems, AAMAS 2012, vol. 1, pp. 307–314. International Foundation for Autonomous Agents and Multiagent Systems, Richland (2012)

27. Shah, D., Zaman, T.: Rumors in a network: Who's the culprit? IEEE Transactions on Information Theory 57(8), 5163–5181 (2011)

28. Shamshirband, S., Anuar, N.B., Kiah, M.L.M., Patel, A.: An appraisal and design of a multi-agent system based cooperative wireless intrusion detection computational intelligence technique. Engineering Applications of Artificial Intelligence 26(9), 2105–2127 (2013)

29. Starbird, K., Maddock, J., Orand, M., Achterman, P., Mason, R.M.: Rumors, false flags, and digital vigilantes: Misinformation on twitter after the 2013 boston marathon bombing. In: iConference 2014 Proceedings, pp. 654–662 (2014)

30. Tisue, S., Wilensky, U.: NetLogo: A Simple Environment for Modeling Complexity (2004)

31. Tripathy, R.M., Bagchi, A., Mehta, S.: A study of rumor control strategies on social networks. In: Proceedings of the 19th ACM International Conference on Information and Knowledge Management, CIKM 2010, pp. 1817–1820. ACM, New York (2010)

32. Valecha, R., Oh, O., Rao, H.R.: An exploration of collaboration over time in collective crisis response during the haiti 2010 earthquake. In: Baskerville, R., Chau, M. (eds.) Proceedings of the International Conference on Information Systems, ICIS 2013, Milano, Italy, December 15–18, 2013. Association for Information Systems (2013)
33. Weng, L., Menczer, F., Ahn, Y.-Y.: Virality prediction and community structure in social networks. Scientific Reports **3**, August 2013
34. Yang, S.Y., Liu, A., Mo, S.Y.K.: Twitter financial community modeling using agent based simulation. SSRN scholarly paper, Rochester, NY. IEEE Computational Intelligence in Financial Engineering and Economics, London (2013)
35. Zhao, L., Cui, H., Qiu, X., Wang, X., Wang, J.: {SIR} rumor spreading model in the new media age. Physica A: Statistical Mechanics and its Applications **392**(4), 995–1003 (2013)

Mining Interesting Topics
in Twitter Communities

Eleni Vathi[(✉)], Georgios Siolas, and Andreas Stafylopatis

Intelligent Systems Laboratory, School of Electrical and Computer Engineering,
National Technical University of Athens, Athens, Greece
{elvathi,gsiolas}@islab.ntua.gr, andreas@cs.ntua.gr

Abstract. We present a methodology for identifying user communities
on Twitter, by defining a number of similarity metrics based on their
shared content, following relationships and interactions. We then introduce a novel method based on latent Dirichlet allocation to extract user
clusters discussing interesting local topics and propose a methodology to
eliminate trivial topics. In order to evaluate the methodology, we experiment with a real-world dataset created using the Twitter Searching API.

Keywords: Twitter · Communities · Topics

1   Introduction

Twitter[1] is an online social network that enables its users to send and read
messages of up to 140 characters, known as Tweets. Since its launch in 2006 it
has gained worldwide popularity and currently has 288 million monthly active
users, with 500 million Tweets sent per day. The content shared on Twitter
represents the thoughts and activities of its users. Additionally, the "following"
relationships between users may be an indicator of their interests, as people tend
to follow other accounts based on how interesting they consider their Tweets.
Therefore, analysis of the shared content and the graph formed by a set of
Twitter users can provide insight on their interests, opinions and behavior.

An important task in social network analysis is to identify the underlying
communities. Communities can be defined as groups of users that are more
densely connected to each other than to the rest of the network, interact
more between them and share common interests. In [1], the authors address
two kinds of community detection approaches, topology-based and topic-based.
Since communities detected by the topology-based approaches tend to contain
different topics within each community, while meaningful topology-based sub-
communities exist inside each topic-based community, the authors suggest that
community detection should consider both the graph structure and textual information of the networks.

There exist a number of approaches for detection of communities in large
social networks that combine the topology of social connections and the topic

---

[1] https://twitter.com/

© Springer International Publishing Switzerland 2015
M. Núñez et al. (Eds.): ICCCI 2015, Part I, LNAI 9329, pp. 123–132, 2015.
DOI: 10.1007/978-3-319-24069-5_12

features [2–4]. In [5], the authors use a clustering algorithm to divide the members of their dataset into topical clusters and then perform link analysis on each topical cluster to detect the communities. In [6], the authors present a machine learning-based framework which is capable of discovering the top similar users for each user on Twitter based on the similarity of the content produced by the users. In [7], the notion of user similarity from both textual contents and social structure is applied and classical clustering algorithms are used to discover Twitter communities based on users interests. Other directions include overlapping community detection [8], or unifying approaches such as the edge structure and node attributes model of [9].

In this paper, we present a novel user distance metric that can be combined with a distance-based clustering algorithm to extract user communities on Twitter. Our work follows an approach which is roughly analogous to [7] in the overall perspective, yet we propose different techniques for tackling most issues addressed in the paper. We first define a number of similarity metrics based on Twitter network properties, such as following relationships, hashtags, etc. These metrics are then used to compute the distance (or inversely the similarity) between each pair of users. The result of this process is a similarity matrix, which is used to group the users into communities by applying Affinity Propagation, a similarity-based clustering algorithm. We then propose a novel methodology to automatically extract cluster topics using the latent Dirichlet allocation (LDA) model, remove the trivial topics and detect the most interesting Twitter communities. We evaluate our approach on a group of users interested in programming, using a well-defined clustering quality criterion.

2   Community Detection and User Clustering Based on Twitter Attributes

Our methodology consists of two steps. Initially, the concept of user similarity between Twitter users is defined and the distance between each pair of users is computed. The second step involves the process of clustering the users into meaningful communities based on their distances.

The similarity between a pair of Twitter users is derived from their properties and the interactions recorded in their tweeting history. The similarity measures based on Twitter attributes will be presented in the next paragraphs.

Following Relationship Similarity. The most evident way of determining how similar two users are, is by examining their following relationship. It is known that the Twitter graph is directed, meaning that a user can follow another user without being followed in return. Let u_i and u_j be two Twitter users. Then, their user similarity based on the following relationship is computed as:

$$S_1(u_i, u_j) = \begin{cases} 1, & \text{if } u_i \text{ follows } u_j \text{ and } u_j \text{ follows } u_i \\ 0.5, & \text{if only one of the users } u_i, u_j \text{ follows the other} \\ 0, & \text{otherwise} \end{cases} \quad (1)$$

Common Followers Similarity. A user's home timeline on Twitter displays a stream of Tweets from accounts the user has chosen to follow, so, usually, people tend to follow other accounts based on their interests. Consequently, we can calculate the user similarity based on common followers as shown in Equation 2, where $followers_i$ is the set of u_i's followers and n is the number of users. We normalize the number of common followers, so as to ensure that all values range between 0 and 1.

$$S_2(u_i, u_j) = \frac{|followers_i \cap followers_j|}{\max_{1 \leq l \leq n} followers_l} \tag{2}$$

Common Friends Similarity. In Twitter, the set of users one follows are called his friends. As previously stated, we can assume that similar users would have common friends. Considering that, we define user similarity as follows:

$$S_3(u_i, u_j) = \frac{|friends_i \cap friends_j|}{\max_{1 \leq l \leq n} friends_l} \tag{3}$$

Hashtag Similarity. The # symbol preceding a keyword or a phrase without spaces is called a hashtag and its purpose is to categorise Tweets. To compute the hashtag similarity, we must first determine the importance of each keyword to a specific user. This can be accomplished by computing the tf-idf weights of the vector space model [10] representation of the hashtags, considering that all hashtags in the tweets of a user form a single document. Therefore, if h_i is the tf-idf vector of the hashtags used by u_i, the hashtag similarity is defined as follows:

$$S_4(u_i, u_j) = cos(h_i, h_j) \tag{4}$$

Reply Similarity. The frequency of replies between two users, as well as the number of users that both users reply to, are two indicators of the users' reply similarity, which is defined in Equation 5. R_i and R_j are the sets of users that u_i and u_j have replied to, respectively, so $|R_i \cap R_j|$ is the number of users they have both replied to. nr_{ij} is the number of times u_i replied to u_j, and NR_i is the number of times u_i replied to another user's Tweet.

$$S_5(u_i, u_j) = \frac{|R_i \cap R_j|}{\sqrt{|R_i|}\sqrt{|R_j|}} + \frac{nr_{ij} + nr_{ji}}{NR_i + NR_j} \tag{5}$$

User Mention Similarity. By placing the @ symbol in front of a username, a user can tag another user in a Tweet. User mention similarity is computed as follows:

$$S_6(u_i, u_j) = \frac{|M_i \cap M_j|}{\sqrt{|M_i|}\sqrt{|M_j|}} + \frac{nm_{ij} + nm_{ji}}{NM_i + NM_j} \tag{6}$$

M_i and M_j are the sets of users that u_i and u_j have mentioned in their Tweets, respectively, nm_{ij} is the number of times u_i mentioned u_j, and NM_i is the number of times u_i mentioned another user. The latter two measures are adapted from a relevant measure defined in [7].

Total User Similarity. The user similarity measure is expected to be a linear combination of the individual similarity measures, as shown in the following equation:

$$S(u_i, u_j) = \sum_{m=1}^{6} a_m S_m(u_i, u_j) \tag{7}$$

The values assigned to the a_m parameters are between 0 and 1 and $a_1 + a_2 + a_3 + a_4 + a_5 + a_6 = 1$. The values are obtained through validation, as described in Subsection 4.2.

Given the similarity measures for each pair of users, the final step of community detection is the clustering of users for the formation of communities. It is well known that there is a wide range of clustering algorithms. In this case, the chosen algorithm must not require as input the absolute positions of the data points, since they are not available, but should take as input the measures of similarity between pairs of data points.

Affinity Propagation [11] is an algorithm that identifies exemplars among data points and forms clusters of data points around these exemplars. It simultaneously considers all data points as potential exemplars and iteratively exchanges messages between data points until a good set of exemplars and clusters emerges. Affinity propagation takes as input a collection of real-valued similarities between data points, where the similarity $s(i, k)$ indicates how well the data point with index k is suited to be the exemplar for data point i. The resulting number of clusters is not predefined. The algorithm terminates if decisions for the exemplars and the cluster boundaries remain unchanged for a number of iterations, or if the maximum number of iterations is attained.

An alternative would be to apply hierarchical clustering using partition density to automatically extract the clusters [12].

3   Local Topics Extraction and Trivial Topic Elimination

In order to determine whether the users who belong in the same cluster tend to tweet about the same topics, their shared content must be studied. In this section, we describe the process of extracting the topics discussed by the users and propose a method for eliminating the trivial topics.

In our method, the words from all the Tweets published by the same user are aggregated into a document. This process is repeated for all the users, therefore we can consider that our entire dataset is a collection of documents. The extraction of topics can be achieved with latent Dirichlet allocation (LDA) [13], a generative probabilistic model of a corpus, which is based on the idea that documents are represented as random mixtures over latent topics, where each topic is a probability distribution over words.

By using the LDA algorithm on the users' collection of documents we obtain a set of N topics (and distribution of words per topic) and a corresponding topic distribution for each user's document, the *User Topic Distribution* (UTD):

$$UTD_i = [topic_{1,u_i}, topic_{2,u_i}, ..., topic_{N,u_i}] \tag{8}$$

where $topic_{p,u_i}$ is the probability value of topic p in the topic distribution of u_i.

Since our goal is to find the topic distribution for each cluster, and by extension the interests of the users, we can consider that all users in a cluster are forming a document, aggregate the cluster's documents and finally get a cluster specific document collection. By calculating the topic distribution of this cluster collection, while keeping the same topics as with the users' collection, we obtain the topic distribution for each cluster r, the *Local Topic Distribution* (LTD):

$$LTD_r = [topic_{1,C_r}, topic_{2,C_r}, ..., topic_{N,C_r}] \tag{9}$$

where $topic_{p,C_r}$ is the probability value of topic p in the document corresponding to cluster C_r. In a similar manner, we can compute the total topic distribution, or *Collection Topic Distribution* (CTD), by aggregating all documents:

$$CTD = [topic_{1,C}, topic_{2,C}, ..., topic_{N,C}] \tag{10}$$

where $topic_{p,C}$ is the probability value of topic p for the whole collection $C = \bigcup_{r=1}^{k} C_r$ and k is the number of clusters. Then, we define a novel *Local Topic Interestingness* measure (LTI) for cluster r as follows:

$$LTI_r = \|CTD - LTD_r\| \tag{11}$$

This measure gives an estimate on how much the discussion in a community deviates from the general discussion and focuses on specific subjects.

While extracting topics using LDA is very useful, not all topics are equally interesting from a mining perspective. Some topics consist of general, everyday words, while others represent common interests for the majority of the examined users. For example, keywords such as "autism, spectrum, support, asperger", "paper, research, student, phd" and "race, vettel, alonso, hamilton" represent meaningful and important topics, while the topic "year, week, time, day, today" contains general words. In addition, the topic "build, develop, base, create, release" seems to be representative for the users' interests. However, in a collection of documents about programming, it would be considered trivial.

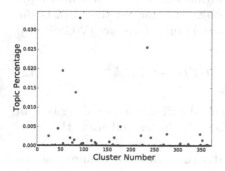

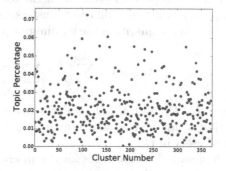

Fig. 1. Percentage of topic "race, vettel, alonso" per cluster.

Fig. 2. Percentage of topic "year, week, time" per cluster.

For this reason, we developed a novel method for eliminating trivial topics. For each cluster we normalize the topic distributions and calculate each topic's in-cluster percentage. A plot of cluster percentages for the interesting topic "race, vettel, alonso" is illustrated in Figure 1. The percentage of this topic is very high for very few clusters, while it is nearly zero in all the other clusters. As a result, the Mean Cluster Percentage (MCP) is also low, near zero.

On the contrary, for a generic topic such as "year, week, time" (Figure 2) we note high percentages in almost all clusters and a similarly high MCP value. So, it becomes clear that there is an MCP threshold over which the topics are trivial and below which the topics are interesting. In Subsection 4.3 we show a method for calculating the MCP threshold value.

4   Experiments

The dataset utilized in this work was collected using the Twitter Searching API. Since our aim was to include users with common interests (i.e. who tweet about similar topics) in the dataset, we selected the followers of @isocpp as our collection of users. @isocpp is the Twitter account of the ISO C++ standards committee, so it is expected that the users following this account would be interested to some extent and tweet about programming. This resulted in a set of 5077 users. For each of these users we crawled all of their published tweets and we retrieved a list of their followers' IDs and a list of their friends' IDs.

From this original set of users, we excluded those who had less than twenty published Tweets. The result set consisted from 2728 users. It should also be noted that one of the similarity measures described earlier, the following similarity measure, has a value of zero for every pair of users in the dataset.

4.1   Clustering Quality Criterion

As explained earlier, the parameter validation of our model will be achieved using a clustering quality criterion. Typical functions for clustering evaluation focus on high intra-cluster similarity and low inter-cluster similarity, so the goal is to have similar topic compositions for the users within a cluster and dissimilar topic compositions between different clusters. An objective function that depicts the intra-cluster similarity is the within-group (cluster) sum of squares (WGSS) [14]:

$$WGSS = \sum_{r=1}^{k} \frac{1}{2n_r} \sum_{u_i \in C_r} \sum_{u_j \in C_r} \|UTD_i - UTD_j\|^2 \qquad (12)$$

where UTD_i and UTD_j are the user topic distributions, as described in Section 3, for user u_i and user u_j, C_r is the r-th cluster and n_r is the number of users in C_r.

A similar formula exists for the inter-cluster similarity, but requires the use of the distances between the centroids of the clusters. Since we do not know the absolute poisitions of the centroids, we define the distance between two clusters as the euclidean distance of their local topic distributions. Therefore, between group (cluster) sum of squares can be calculated as follows:

$$BGSS = \frac{1}{2} \sum_{r=1}^{k} \sum_{s=1}^{k} \|LTD_r - LTD_s\|^2 \tag{13}$$

A criterion applicable in cluster analysis, which combines the WGSS and BGSS functions is the Calinski-Harabasz criterion [14], which is sometimes called the variance ratio criterion (VRC). The Calinski-Harabasz index is defined as:

$$VRC = \frac{BGSS}{k-1} \Big/ \frac{WGSS}{n-k} \tag{14}$$

where n, k denote the number of users and the number of clusters, respectively. It seems natural that the optimal clustering of users maximizes the VRC index.

4.2 Similarity Measure Parameters

As mentioned in Section 2, we use six parameters to determine the way individual similarity measures affect the final similarity measure. Initially, we compute the similarity defined in Equation 7 for all possible combinations of the parameters, with step 0.1, where each parameter is between 0 and 1 and their sum equals to 1. Subsequently, the affinity propagation algorithm is executed with each different similarity metric as input. The result of this process is a number of possible partitions, or clusterings, of the users, which must be evaluated.

The VRC index is applied to all the clusterings produced by every execution of the affinity propagation algorithm. The combinations of the parameters which result in a normalized value of the index that is less than 0.9 are discarded. For the remaining combinations we compute the mean value for each parameter. This process results in the optimal values which are $a_2 = 0.15$ (followers), $a_3 = 0.11$ (friends), $a_4 = 0.41$ (hastags), $a_5 = 0.15$ (replies) and $a_6 = 0.15$ (user mentions). We observe that the hashtags parameter has a greater value than the rest of the parameters. This is expected, considering that the optimal values were calculated using the topic distributions of the Tweets' text and hashtags are labels (or tags) for Tweets with specific subjects, so they are indicators of the topics discussed in a Tweet.

4.3 Number of Topics and Trivial Topics Removal

When executing the LDA algorithm, one of the most common questions is how to determine the optimal number of topics. This depends on the size of the documents' collection and the desired outcome: a small number of topics will provide broader topics, while a large number will make topics harder to understand.

In order to identify the ideal number of topics for our dataset, we compute the Calinski-Harabasz index for the clustering derived from the optimal values of the a_m parameters and for different numbers of topics. The result is depicted in Figure 3. As can be seen, the optimal value for VRC occurs at 450 topics.

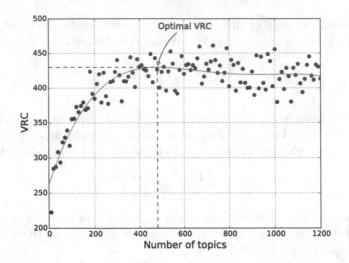

Fig. 3. The value of the clustering criterion for the different numbers of topics.

As explained in Section 3, the MCP threshold is used to distinguish interesting topics from trivial ones. In order to determine the MCP threshold for a given number of topics, we first compute the MCP values for each topic. The MCP values are sorted in increasing order (Figure 4) and the differences of adjacent MCP values are compared to a small quantity Δ. If the difference is greater than Δ, we set the MCP threshold equal to the average of the two MCP values. Topics above this threshold are considered trivial and are discarded, while we consider that the topics corresponding to MCP values below the threshold are interesting. The value of Δ depends on the collection of documents and the number of topics. For a given number of topics, we observe jumps in the MCP values as we approach the trivial topics, which can help define the value of Δ. For our dataset, the number of trivial topics is approximately 20% of the total number of topics.

4.4   Interesting Cluster Extraction

In a final step we use all the methods introduced previously to perform the final clustering and extract the most interesting topics. We use the optimal similarity measure parameters given in Subsection 4.2, then perform LDA with the optimal number of topics and remove the trivial topics as described in Subsection 4.3. Finally, we recalculate LTI for all clusters, this time with the optimal parameters and having removed the trivial topics. We can then define the *Local Twitter Community Interestingness* (LTCI) as follows:

$$LTCI_k = n_k \cdot LTI_k \tag{15}$$

where n_k is the number of users in cluster k. High LTCI values signify large communities, with interesting/non-trivial topics, which discuss more specific subjects, since they differentiate from the general discussion. Table 1 shows the

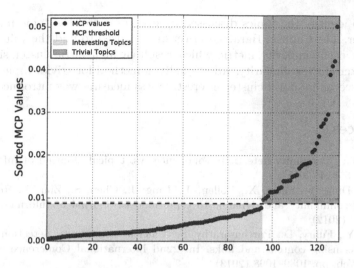

Fig. 4. Sorted MCP values

Table 1. Example of LTCI values

LTCI	Topic Percentage	Topic Keywords	
0.8	50.15%	delphi, rad, studio, develop, firemonkey	
0.75	42.40%	diary, vampire, men, big, bang, half	
	23.11%	codec, fastpictureviewer, image, file, raw	
0.51	55%	firebird src database sql git	

normalized value of LTCI for three representative clusters and the corresponding percentages and keywords for the most popular topics in these clusters. The extracted topics shown in Table 1 are interesting, in the sense that on the top LTCI clusters two are not so much related to C++, but discuss databases ("firebird"), image viewers ("fastpictureviewer") and TV series ("the big bang theory", "two and a half men").

5 Conclusion

In this paper, we proposed a novel methodology for identifying Twitter communities and extracting interesting user clusters. This approach uses the content shared by users, the graph structure and the interactions between users to define a number of similarity metrics. Affinity propagation, a clustering algorithm which requires the similarity between data points and not their absolute positions, was used for grouping the users into communities. The LDA algorithm was executed in order to identify the topics discussed by each user and the topic mixture of each cluster. The Calinski-Harabasz criterion was used to find the values of the weights for each similarity metric which result in high intra-cluster similarity and low inter-cluster similarity. Additionally, a method for trivial topic removal was proposed and a novel cluster interestingness measure was introduced.

References

1. Ding, Y.: Community detection: Topological vs. topical. Journal of Informetrics **5**(4), 498–514 (2011)
2. Li, D., Ding, Y., Shuai, X., Bollen, J., Tang, J., Chen, S., Zhu, J., Rocha, G.: Adding community and dynamic to topic models. Journal of Informetrics **6**(2), 237–253 (2012)
3. Ruan, Y., Fuhry, D., Parthasarathy, S.: Efficient community detection in large networks using content and links. In: 22nd International Conference on World Wide Web, pp. 1089–1098 (2013)
4. Lim, K.H., Datta, A.: Tweets Beget Propinquity: Detecting Highly Interactive Communities on Twitter Using Tweeting Links. IEEE/WIC/ACM International Conference on Web Intelligence **1**, 214–221 (2012)
5. Zhao, Z., Feng, S., Wang, Q., Huang, J.Z., Williams, G.J., Fan, J.: Topic oriented community detection through social objects and link analysis in social networks. Knowledge-Based Systems **26**, 164–173 (2012)
6. Goel, A., Sharma, A., Wang, D., Yin, Z.: Discovering similar users on twitter. In: 11th Workshop on Mining and Learning with Graphs, Chicago, Illinois (2013)
7. Zhang, Y., Wu, Y., Yang, Q.: Community Discovery in Twitter Based on User Interests. Journal of Computational Information Systems **8**(3), 991–1000 (2012)
8. Xie, J., Kelley, S., Szymanski, B.K.: Overlapping Community Detection in Networks: The State-of-the-art and Comparative Study. ACM Computing Surveys **45**(4), 1–35 (2013)
9. Yang, J., McAuley, J.J., Leskovec, J.: Community detection in networks with node attributes. In: IEEE 13th International Conference on Data Mining, pp. 1151–1156 (2013)
10. Salton, G., Wong, A., Yang, C.S.: A Vector Space Model for Automatic Indexing. Communications of the ACM **18**(11), 613–620 (1975)
11. Frey, B.J., Dueck, D.: Clustering by Passing Messages Between Data Points. Science **315**, 972–976 (2007)
12. Sander, J., Qin, X., Lu, Z., Niu, N., Kovarsky, A.: Automatic extraction of clusters from hierarchical clustering representations. 7th Pacific-Asia Conference on Advances in Knowledge Discovery and Data Mining. LNCS, vol. 2637, pp. 75–87. Springer, Heidelberg (2003)
13. Blei, D.M., Ng, A.Y., Jordan, M.I.: Latent Dirichlet Allocation. Journal of Machine Learning Research **3**, 993–1022 (2003)
14. Caliński, T., Harabasz, J.: A dendrite method for cluster analysis. Communications in Statistics-Simulation and Computation **3**(1), 1–27 (1974)

Community Division of Bipartite Network Based on Information Transfer Probability

Chunlong Fan[✉], Hengchao Wu, and Chi Zhang

College of Computer, Shenyang Aerospace University, Shenyang 110136, China
754713151@qq.com

Abstract. Bipartite network is a performance of complex networks,The divided of unilateral node of bipartite network has important practical significance for the study of complex networks of community division. Based on the diffusion probability of information and modules ideas in the network,this paper presents a community divided clustering algorithm (IPS algorithm) for bipartite network unilateral nodes.The algorithm simulates the probability of information transfer in the network,through mutual support value between the nodes in network,selecting the max value as the basis for merger different communities.Follow the module of the definition for division after mapping the bipartite network nodes as a single department unilateral network.Finally,we use actual network test the performance of the algorithm.Experimental results show that,the algorithm can not only accurate divided the unilateral node of bipartite network,But also can get high quality community division.

Keywords: Modularity · Bipartite network · Support value · Community division

1 Introduction

Complex networks is a rise in recent years involving physics, biology, mathematics and computer science and other fields of interdisciplinary research community structure for roughly classified into three types: Community found that evolutionary analysis community, and the community structure and network dynamics and network compression indicate other relations between functional features. Where the research community found mostly concentrated in a single unit network, the relative lack of research bipartite network of community division. Community research division of unilateral node bipartite network, focusing only on the use of two sub-networks to build a single unit network, using sophisticated algorithms have been divided communities. We can use the bipartite network partitioning algorithm community as a class of two types of processing nodes, perform community is divided and then extracted unilateral community nodes, such as the reality of bipartite network Scientist - Research Collaboration Network [2-3] of scientists collaborative research, movie - actor cooperative research actors network [4] actor, disease - gene networks [5] disease relevance, audience - songs network [6] audience community and computer terminals - P2P data network terminal groups [7].

© Springer International Publishing Switzerland 2015
M. Núñez et al. (Eds.): ICCCI 2015, Part I, LNAI 9329, pp. 133–143, 2015.
DOI: 10.1007/978-3-319-24069-5_13

2   Current Research Presentation

The bipartite networks are mapped into a single network using relatively convention-al, relatively mature single network of community division algorithm is divided into bipartite networks of community common treatment. But whether it is entitled to a weighted projection or projection, each node can only be divided unilateral and pro-jected results will lead to lack of bipartite network information. The division directly on the original two points of the network, often dichotomous types of nodes in the network as a class for processing, so far no one can recognize bipartite network com-munity divided evaluation criteria. Therefore there is a need for further study two points online community divided.

According to the characteristics bipartite network, such as clustering coefficient, edge betweenness, degree and degree distribution, the number of overlapping topo-logical potential, all kinds of bipartite network partitioning algorithm communities have been proposed. Yongcheng Xu et al. [8] proposed a bipartite network communi-ty mining transformed into an ant search graph model for the optimization problem, the algorithm based on the definition of the vertex topology heuristic information, construct a community divided result; but ants swarm algorithm uses random assign-ment process vertex belongs to the community with a large degree of randomness, so the whole algorithm run time overhead is relatively large. Chenbo Lun et al. [9] through the bipartite network diagram that corresponds to the matrix recursively split application matrix decomposition methods were divided community, also known as MP algorithm; but every time you want to split the result matrix remained intact Community Information very difficult, so the results may vary depending on the divided matrix decomposition process is different.Yajing Wu et al. [10] using cluster-ing method original bipartite network, the bipartite network resource distribution matrix and fuzzy clustering method for vector combining clustering proposed determi-nation F statistic most clustering methods, the algorithm The disadvantage is the need to know in advance the number of associations to be subdivided; do not know the number of associations in the case, this method is difficult to get an accurate division of the community. Gao et al. [11] Newman algorithm based on the idea of   BRIM fast algorithm is proposed to improve the degree of aggregation algorithm module (MAB), although compared to the BRIM algorithm does not require additional number given community, but the algorithm complexity than BRIM Many high algorithms. Foreign studies have associated Raghavan et al. [12] Reference numeral propagation method (LPA), the algorithm first to each vertex is assigned a unique label, after selection of each vertex adjacent to them as labels, and finally with the same iterative vertex labels form a single community, but the algorithm is only suitable for small networks. Murata et al. [13] The LPA algorithm has been improved to the expense of accuracy obtained LPA algorithm suitable for large bipartite network and a large network of parallel real dichotomy community analysis. Italian scholars Dorigo et al. [14] pro-posed ant colony optimization algorithm has been successfully applied to multi com-binatorial optimization problem, the experimental results show that the algorithm is universal, global, distributed computing, and robustness advantages; the same because the algorithm the combination is more, the algorithm complexity is still very high.

Newman Barber expand the definition of a single sub-network modularity presents a dichotomy modularity [15], combined with Adaptive maximization algorithm BRIM dichotomy modularity to get to divide the community, the algorithm can achieve better classification results. To get the optimal solution to try to go through the exchange community nodes defined in advance the number of community bipartite network. Therefore, the higher complexity of the algorithm, application limitations, does not apply to large-scale networks.

Based on information diffusion probabilistic thinking, combined with the definition of Newman et al modularity [16] and two points mapped network nodes unilateral proposed community bipartite network partitioning algorithm a unilateral nodes, the algorithm first two points of a network side of each vertex seen as a community, on both sides of the node distribution information to initialize. Six times the probability of information in accordance with the small world theory of six degrees of separation to get a rear diffuser exchange information resource matrix. Then the matrix values as a basis to judge the merger communities. After the last node will be unilateral projection mapping, combined with Newman's definition of modularity, as community standards division. Disconnect the combined maximum value at the Q to get the final result is divided community. Experimental results show that the algorithm can not only get an accurate number of communities, but also can get high-quality results divided communities.

3   Information Diffusion Probability (IPS) Model

Bipartite network can be represented as a bipartite graph G=(U,V,E), G, the nodes are divided into two parts, U and V, E is the edge of graph G . No edge is connected between the set U (or V) nodes, any edge of set E (u_i, v_j) , there must be $u_i \in U, v_j \in V$. Set up a collection of nodes in U is m, the number of nodes in the set V is n, two adjacency matrix can be expressed as:

$$\widetilde{A} = \begin{bmatrix} 0_{m \times m} & A_{m \times n} \\ A_{n \times m}^T & 0_{n \times n} \end{bmatrix} \tag{1}$$

Where matrix A is the sub-matrix of bipartite adjacency matrix , We call that matrices A is the   relationship matrixof bipartite graph G , the relationship matrix A are:

$$A_{ij} = \begin{cases} 1 & if \ i \, connect \, j \ \ i \in U, j \in V \\ 0 & else \end{cases} \tag{2}$$

Each row of the matrix A is the connected case of a node of Set U, each row is the connected case of a node of Set V. Initialize set U in each node an information unit, the amount of information, given the set V of units of information in each node 0. To obtain set U of an initial unit information matrix $I_{m \times m}^U$:

$$I_{ij}^{U} = \begin{cases} 1 & if \ \ i = j \\ 0 & else \end{cases} \quad (3)$$

Set U of the unit amount of information obtained as above matrix equation (3), each row of the unit matrix or the amount of information in each column represents a node in the set U,U has set the amount of information of the number of other nodes, initially only with its own node one unit of information, at this time it can be represented by a vector $\overline{x_i}$ owned by the node set U which is the amount of information to other nodes, the value of the information element value vector i matrix row or column i. Each node has a set amount of information $\overline{x_i}$, in accordance with rules such as the diffusion equation in equation (4).

$$R_i = \begin{cases} \dfrac{1}{k_i}\overline{x_i} & if \ \ i \ connect \ j \\ 0 & else \end{cases} \quad (4)$$

Where k_i is the set of nodes of degree U, if there is an edge node u_i, v_j connecting, in this case the information of node u_i diffuse into the node v_j, each node vector v_j information obtained which are summed, the amount of information $R_i = \sum_{i=1}^{m} R_i^{'}$ obtained from set U in each node diffuse over , the combination of R_i of diffusion can be left matrix $R_{m \times n}$, in the node information set U diffusion matrix set V of nodes is expressed as:

$$S_{m \times n} = I_{m \times m}^{U} \bullet R_{m \times n} \quad (5)$$

Each row where $R_{m \times n}$ represents a node in set U diffusion to the collection of information for each node V, each column represents a set V of set U in node which receives information from each node. In this case each node in the set V is obtained from the amount of information for each node in the set U.

Finally, the collection information vector V on each node, according to the formula 6 and then spread to the set U in each node.

$$T_j = \begin{cases} \dfrac{1}{k_j}\overline{y_j} & if \ \ i \ connect \ j \\ 0 & else \end{cases} \quad (6)$$

Node $\overline{y_j}$ which has vector v_j information from each node in the set U, k_j is the degree of the node v_j. Combination of vector T_j information to get a n*m right information diffusion matrix $T_{n \times m}$, at this time, each node of set U itself diffuse to the collection of information of matrix U in the amount of information to other nodes that can be expressed as:

$$S_{m \times m}^{1} = I_{m \times m}^{U} \bullet R_{m \times n} \bullet T_{n \times m} \quad (7)$$

the elements R_{ij} of the matrix R_{mon} can be obtained by corresponding elements of A_{ij} divided by relation matrix A where the degree of node i gived, $T_{n\times m}$ divided by the transposed matrix A^T may be the relationship of each element of the matrix A corresponding to A_{ij}^T degree of node i to give and therefore, the above-described steps n times the cycle information obtained after diffusion:

$$S_{m\times m}^n = (I_{mon}^U \bullet R_{m\times n} \bullet T_{n\times m})^n \tag{8}$$

Where S_{ij}^n represents the diffusion of information after the loop n times, node u_i to node u_j of the amount of information collected.

4   Select Determination Principle

4.1   Merge Determination Principle

Defines the degree of mutual support for the information size value of the diffusion matrix, when the information in the network for circulating diffusion times, In matrix S^n corresponding to the size of the value S_{ij} which represents the link between the two nodes tightness. Six Degrees principle states that the average distance between people to 6, after the information in the network will be able to meet the six diffusion determination result, therefore, the information diffusion cycle number n is 6.

Information Diffusion probabilistic algorithms of this paper is clustering algorithm, initialization set U of each node in a community, the diffusion of information on the diagonal matrix elements in addition to the traverse, select the elements in each row or column of the matrix S^6 of the maximum corresponding to merge the two communities. When communities merge, update the corresponding matrix S^6 corresponding i-th row and j-th row, the updated strategy for the community i or j is the largest value of other community support as a community updated external support, and delete information diffusion matrix j rows. Continue to traverse the updated next row until all communities merged into one community, then get the result of the merger of tree tree.

4.2   Principles of Defining Choice

The results of Single network and the bipartite networks division is determined by the merits of the community that have a lot of evaluation criteria, such as the clustering coefficient, modules, and so on. But there are two sub-types of the demarcation of the online community will be seen as the same type of node node treat, so this method is not applicable in this paper unilateral dichotomy network nodes in the network community is divided on the clustering. Bipartite network entitled to draw projection mapping and evaluation criteria divided community of single points on the network, so the community is divided on this paper, the results of Newman quality evaluation standard modules of the Q value [16] is an effective evaluation methods.

Physical meaning of modularity are: using a network connection belongs to the same community proportion edge node minus the proportion of community structures in the same random edges connecting these two nodes expectations. For obvious network community Q value is between 0.3 and 0.7. Module is defined as:

$$Q = \sum_i (e_{ii} - a_i^2) = Tre - \| e^2 \|$$ (9)

Where e_{ii} represents the internal side of the community share in the proportion of all edges, which represents the ratio of the i-th community connected to the side edges of the proportion of all, $\| x \|$ represents all elements of the matrix and $Tre = \sum_i e_{ii}$ represents the matrix of and each element of the corner line.

4.3   Information Diffusion Example

For bipartite network shown in Figure 1, we study the situation on the community side of the divide nodes, for example, first get the dichotomous relationship matrix network A and A^T, according to the formula wherein R is a matrix of four changes from the information diffusion left matrix, the matrix T is changes come right information diffusion matrix A^T. After the diffusion node information to the upper edge of the network, the information diffusion matrix at this time becomes $I_{6\times6} \bullet R_{6\times5}$. After the current side node receives the information from each node, then the information on average each node after returning to the side connected to it, then get the information diffusion matrix $I_{6\times6} \bullet R_{6\times5} \bullet T_{5\times6}$ expression of which is $R_{6\times5}, T_{5\times6}$:

$$R_{6\times5} = \begin{bmatrix} 1/2 & 1/2 & 0 & 0 & 0 \\ 1/2 & 1/2 & 0 & 0 & 0 \\ 1/3 & 1/3 & 1/3 & 0 & 0 \\ 0 & 0 & 1/3 & 1/3 & 1/3 \\ 0 & 0 & 0 & 1/2 & 1/2 \\ 0 & 0 & 0 & 1/2 & 1/2 \end{bmatrix}$$ (10)

$$T_{5\times6} = \begin{bmatrix} 1/3 & 1/3 & 1/3 & 0 & 0 & 0 \\ 1/3 & 1/3 & 1/3 & 0 & 0 & 0 \\ 0 & 0 & 1/2 & 1/2 & 0 & 0 \\ 0 & 0 & 0 & 1/3 & 1/3 & 1/3 \\ 0 & 0 & 0 & 1/3 & 1/3 & 1/3 \end{bmatrix}$$ (11)

After the information diffusion six times to get a matrix, the matrix is a matrix of information diffusion between nodes on the network side, the resulting matrix diffusion six:

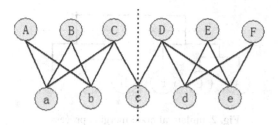

Fig. 1. bipartite network

Table 1. six iterative diffusion matrix

	A	B	C	D	E	F
A	0.2470	0.2470	0.3182	0.1101	0.0389	0.0389
B	0.2470	0.2470	0.3182	0.1101	0.0389	0.0389
C	0.2121	0.2121	0.2838	0.1451	0.0734	0.0734
D	0.0734	0.0734	0.1451	0.2838	0.2121	0.2121
E	0.0389	0.0389	0.1101	0.3182	0.2470	0.2470
F	0.0389	0.0389	0.1101	0.3182	0.2470	0.2470

Table is obtained after six iterations upper node information spread to other nodes on the side of the case. Column in the table indicates the information of other nodes similar to the volume diffusion node corresponding to the column, in the row corresponding to the line information indicating which node of the node spread to his information, the value of the diagonal line indicates after after six times the number of information diffusion itself still remaining information. For example, the amount of information of the node A diffusion node C to a value of 0.3182 on the amount of information diffusion node C to node A is 0.2121, the node A itself is 0.2470 in the remaining amount of information has experienced six times diffused.

A community of C community support is greater than the degree of support to other nodes, so the community A and C are combined into a single AC community, then updates the A and C corresponding line in the column, after updating the new community support for the selection of external a or C external support node greatest values, mutual support matrix updated as shown in Table 2, when the update is mutual support matrix, AC community's support for community-B maximum, perform a merge operation and updates matrix, and so on, until finally the entire unilateral nodes in the network are combined into a community until, as shown in Fig 2.

Table 2. AC merge support updated

	AC	B	D	E	F
AC	0.3182	0.2470	0.1451	0.0734	0.0734
B	0.3182	0.2470	0.1101	0.0389	0.0389
D	0.1451	0.0734	0.2838	0.2121	0.2121
E	0.1101	0.0389	0.3182	0.2470	0.2470
F	0.1101	0.0389	0.3182	0.2470	0.2470

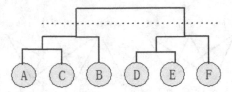

Fig. 2. unilateral node merging process

Figure 4 is a change in the value of Q in Fig. 3 node merging process, the Q value in the fourth step after completion of the merger maximum, where C and D after merging stops merge. To make the community is divided on the lower side in Figure 1 node, the same steps above. The results can be seen through, there is a loss of information, although after the mapping, but here as a community is divided on the basis of judgment is an effective method.

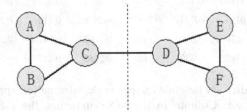

Fig. 3. unilateral node mapping

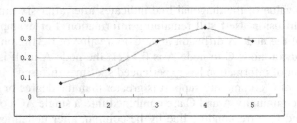

Fig. 4. Changes in the merger of Q value

4.4   Algorithm Complexity Analysis

The algorithm consists of: energy diffusion calculation, unilateral node mapping and Q value is calculated in three parts. Performed in the matrix multiplication energy dispersal process, this step algorithm complexity is O (n2 + m2), where m is the number of nodes in the bipartite graph node in the set U, n represents the number of nodes in the set V. After making six matrix multiplication algorithm complexity $O(6(m^2+n^2))$, to traverse the projection matrix unilateral node, this step of the algorithm complexity $O(mn)$, computational complexity Q value is approximately O (m2) so the final complexity of the algorithm is $O(k_1 n^2 + k_2 m^2)$ where k1 and k2 are constants.

5   Experimental Results and Analysis

5.1   Tested on Southern Women Datasets

South African Women's Network (southern women) dataset [18] is the southern United States women's participation in the activities of the line into a real network, proposed by Stephen, the network consists of 18 women and 14 activities, numbered 1-18 points for the Women's Day, 19 -32 for the active node. As shown below:

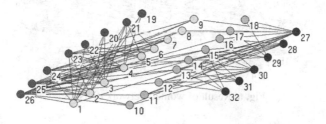

Fig. 5. South African women - Event Network

Experimental results show that the network is divided into four communities, including women 1-9 and 10-18 were each divided into a community, 19-26 and 27-32 are classified as another community, which is consistent with the original observations Davis . Women's network and the network event networks are projected, and dividing the result with several other algorithms divide the result of this algorithm compared with the module as a measure of the standard, to obtain results as shown in Table 1:

Table 3. Women - Event Network division results were compared with the Q value

Algorithm	Women's Network	Q Value	Event Network	Q Value
IPS	{1-9}{10-18}	0.3705	{19-26}{27-32}	0.3864
BRIM	{1-6}{7,9,10} {11-15}{8,16-18}	0.3455	{19-26}{27-32}	0.3864
LPA	{1-7,9}{8、 10-18}	0.3692	{19-24}{25-26} {27,29}{28,30-32}	0.3572
MP	{1-6}{7-10}{11-18}	0.3364	{19-24}{25-28} {29-32}	0.3487
ACODC	{1-9}{10-18}	0.3705	{19-26}{27-32}	0.3864

The figure shows the results of comparison of various algorithms division, the division of the results of the IPS module unilateral network algorithm to get the maximum value. Division results based on ant colony optimization algorithm for bipartite network of community division (ACODC algorithm) identical; the same division results Davis1 Davis did in the original literature. After mapping division of the women's network, obtain the following diagram shown in Figure 6, and FIG two are completely connected within the community, between the community is not completely connected. If there is no connection between whole numbers 2,4,5,6,7 and left women, although women and another one on the 1st of all women in the community who are now out of 21 in the same community, but if you put a woman on the 1st in another community, the module of the Q value of the entire network will be reduced, therefore the 1st women on the right side of the community can be better divided quality.

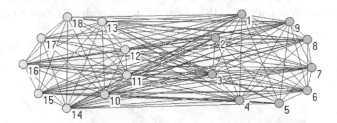

Fig. 6. Result of women network division

6   Conclusion

In order to divide the community bipartite network, we propose a bipartite network of community division algorithm based on information diffusion. The algorithm does not require additional parameters in the case, one side of the bipartite network node precise community division respectively.

The algorithm also exists insufficient; vertex mapping during unilateral calculate the Q value, the mapping information is missing presence. Ministry of network complexity and single community partitioning algorithm is also compared to the algorithm needs to be further improved. Therefore, the research division of bipartite network community needs to be further explored in solving practical problems.

References

1. Wang, X., Li, X., Chen, G.: Complex network theory and its applications. Tsinghua University Press, p. 9 (2006)
2. Newman, M.E.J.: Scientific collaboration networks. I. network construction and fundamental results. Physical Review E **64**, 016131 (2001)
3. Newman, M.E.J.: Scientific collaboration networks. II. network construction and fundamental results. Physical Review E **64**, 016132 (2001)
4. Liu, A., Chunhua, F., Zhang, Z.: Empirical statistical study Chinese mainland movie network. Complex Systems and Complexity Science **4**(3), 10–16 (2007)
5. Chen, W., Junan, L., Liang, J.: Disease gene network analysis bipartite graph projection. Complex Systems and Complexity Science **6**(1), 13–19 (2009)
6. Lambiotte, R., Ausloos, M.: Uncovering collective listening habits and music genres in bipartite network. Physical Review E **72**, 066107 (2005)
7. Le Blond, S., Guillaume, J.-L., Latapy, M.: Clustering in P2P exchanges and consequences on performances. In: van Renesse, R. (ed.) IPTPS 2005. LNCS, vol. 3640, pp. 193–204. Springer, Heidelberg (2005)

8. Xu, Y., Chen, J.: ACO based bipartite network community mining. Computer Science and Exploration **8**(3), 286–304 (2014)

9. Chen, B., Chen, J., Zou, S.: Mining algorithm based on matrix decomposition bipartite network community. Computer Science **41**(2), 55–101 (2014)

10. Zou, Y., Di, Z., Pan, Y.: Based on the distribution of resources bipartite network Fabric Matrix class methods. Beijing Normal University **46**(5), 643–646 (2010)

11. Gao, M.: Bipartite network community discovery algorithm. Hefei University of Technology, p. 04 (2012)

12. Raghavan, U.N., Albert, R., Kumara, S.: Near linear time algorithm to detect community structures in large-scale networks. Physical Review E **76**, 036106 (2007)

13. Liu, X., Murata, T.: Community detection in large-scale bipartite newworks. In: Proceedings of the 2009IEEE/WIC/ACM International Joint conference on Web Intelligence and Intelligent Agent Technology(WI-IAT 2009), pp. 5–8. IEEE Computer Society, Washington, DC (2009)

14. Dorigo, M., Maniezzo, V., Colorni, A.: Ant system:optimization by a colony of cooperating agents. IEEE Transactions on Systems, Man, and Cybernetics: Part B **26**(1), 29–41 (1996)

15. Barber, M.J.: Modularity and community detection in bipartite networks. Physical Review E **76**, 066102 (2007)

16. Newman, M.E.J.: Modularity and community structure in networks. Proceedings of the National academy of sciences of the United states of America **103**(23), 8577–8582 (2006)

17. Milgram, S.: The small world problem. Psychology Today, 60–67, May 1967

18. Davis, A., Gardner, B.B., Gardner, M.R.: Deep south. University of Chicago Press, Chicago (1941)

19. Scott, J., Hughes, M.: The anatomy of Scottish capital: Scottish companies and Scottish capital. Croom Helm, London (1980). 1990–1979

20. Fan, C.L., Yu, L.: 2013 2nd International Conference on Mechatronic Sciences, Electric Engineering and Computer, (MEC2013) ShenYang, liaoning, China (2013)

User-Tweet Interaction Model and Social Users Interactions for Tweet Contextualization

Rami Belkaroui[1](✉), Rim Faiz[2](✉), and Pascale Kuntz[3](✉)

[1] LARODEC, ISG Tunis, University of Tunis, Bardo, Tunisia
rami.belkaroui@gmail.com
[2] LARODEC, IHEC Carthage, University of Carthage, Carthage Presidency, Tunisia
rim.faiz@ihec.rnu.tn
[3] LINA, la Chantrerie, BP 50609, 44360 Nantes Cedex, France
pascale.kuntz@univ-nantes.fr

Abstract. In the current era, microblogging sites have completely changed the manner in which people communicate and share information. They give users the ability to communicate, interact, create conversations with each other and share information in real time about events, natural disasters, news, etc. On Twitter, users post messages called tweets. Tweets are short messages that do not exceed 140 characters. Due to this limitation, an individual tweet it's rarely self-content. However, users cannot effectively understand or consume information.

In order, to make tweets understandable to a reader, it is therefore necessary to know their context. In fact, on Twitter, context can be derived from users interactions, content streams and friendship. Given that there are rich user interactions on Twitter. In this paper, we propose an approach for tweet contextualization task which combines different types of signals from social users interactions to provide automatically information that explains the tweet. To evaluate our approach, we construct a reference summary by asking assessors to manually select the most informative tweets as a summary. Our experimental results based on this editorial data set offers interesting results and ensure that context summaries contain adequate correlating information with the given tweet.

Keywords: Tweet contextualization · Tweet influence · Conversation aspects · User interactions

1 Introduction

Recent years have revealed the exponential growth of microblogging platforms, offering to users an easy way to share different kinds of information like common knowledge, opinions, emotions under the form of short text messages. Twitter, the microblogging service addressed in our work, is a communication mean and a collaboration system that allows users to share short text messages (tweets), which doesn't exceed 140 characters with a defined group of users called followers. Twitter's data flow is examined in order to measure public sentiment, trends

© Springer International Publishing Switzerland 2015
M. Núñez et al. (Eds.): ICCCI 2015, Part I, LNAI 9329, pp. 144–157, 2015.
DOI: 10.1007/978-3-319-24069-5_14

monitoring, reputation management, follow political activity and news. However, tweets in its raw form can be less informative, but also overwhelming. For both end-users and data analysts, it is a nightmare to plow through millions of tweets which contain a lot of noise and redundancy. Furthermore, an individual tweet is short and without sufficient contextual information, it is often hard to capture the associated information. For example, a tweet posted by Darrell (one of the most popular twitter user) just contains a single hashtag "#PrayForTunisia" during the terrorist museum attacks in Tunisia in 2015. When reading this tweet, without knowing related news, it would be very difficult to understand this tweet topic (what is this tweet about? what happened?). Furthermore, tweets may contain information that is not understandable to user without some context. All these obstacles impede users from effectively understanding or consuming information, which can either make users less engaged or even unfastened from using Twitter.

Traditional contextualization techniques only consider text information which is insufficient for tweet contextualization task, since text information on twitter is very sparse. In addition, tweets are short and not always written maintaining a formal grammar and proper spelling. Given that there are rich user interactions on Twitter called social conversations [3], this paper describes an approach that exploits social conversations to provide some context for a given tweet, in order to help users effectively understand the tweet context. Typically, we focus, in our case on exploiting multiple different types of signals such as social signals, user-tweet influence signals, temporal signals and text based signals, which can be potentially useful to improve tweet contextualization task.

The remainder of this paper is organized as follows: we begin by describing some related works presented in INEX 2012, 2013 and 2014. In section 3, we present our approach that exploits social conversations to provide some context for a given tweet, and introduce tweet-user influence model in section 4. In section 5, we explain different types of signals used in our work. Our experimental results are presented in Section 6. Finally, we conclude and present some future works.

2 Related Work

In this section, we report related work exploiting tweet contextualization task. Moreover, there have been some studies done for this task.

In [9], the authors proposed a new method based on the local Wikipedia dump. They used TF-IDF cosine similarity measure enriched by smoothing from local context, named entity recognition and part-of-speech weighting presented at INEX 2011. Recently, [10] modified the method presented at INEX 2011, 2012 and 2013 [12,13] by adding the influence of topic-comment relationship on contextualization. The approach proposed in [6] described a hybrid tweet contextualization system using Information Retreival (IR) and Automatic Summarization (AS). They used nutch architecture and TF-IDF based sentence ranking and sentence extracting techniques for Automatic Summarization. While, in the same way [1] described a pipeline system where first extracted phrases from tweets by using ArkTweet toolkit and some heuristics; then retrieved relevant documents for these phrases from Wikipedia before summarizing those with MEAD toolkit.

In [21], the authors developed and tested a statistical word stemmer which used by the CORTEX to preprocess input texts and generate readable summary. Recently, they presented three statistical summarizer systems to build the tweet context applied to CLEF-INEX 2014 task [20]. The first one is Cortex summarizer based on the fusion process of several sentence selection metrics and an optimal decision module to score sentences from a document source. The second one is Artex summarizer uses a simple inner product among the topic-vector and the pseudo-word vector and the third is a performant graph-based summarizer. While, in [8], the authors used a method that allows to automatically contextualize tweets by using information coming from Wikipedia. They treat the problem of tweet contextualization as an Automatic Summarization task, where the text to resume is composed of Wikipedia articles that discuss the various pieces of information appearing in a tweet, whereas, in [16] the authors combined Information Retrieval, Automatic Summarization and Topic Modeling techniques to provide the context of each tweet. They took advantage of a larger use of hashtags in the topics and used them to enhance the retrieval of relevant Wikipedia articles. In [2], the authors have simply treated contextualization as a passage retrieval task. They used the textual tweet content as a query to retrieve paragraphs or sentences from Wikipedia corpus. Another approach proposed by [14] used latent Dirichlet analysis (LDA) to obtain a tweet representation in a thematic space. This representation allows finding a set of latent topics covered by the tweet. Lately, [24] described a new method for tweet contextualization based on association rules between sets of terms. This approach allows the extension of tweet's vocabulary by a set of thematically related words.

These works presented in INEX Tweet Contextualization track are based on the assumption that it is possible to overcome the tweets' lack of knowledge by providing a bunch of sentences that give some context or additional information extracted from Wikipedia about a given tweet. However, sometimes after news events, the Wikipedia information is not immediately available. For example: After, the terrorist attack against newspaper Charlie Hebdo, there were no articles on Wikipedia describing the topic #jesuischarlie. Indeed, the first article that explains this topic was available 7 hours after this attack. While, the first tweet was launched at 11:52h (less than an hour after the attack). At the same time, this event demonstrates a scenario where users urgently need information, especially if they are directly affected by the event. Unexpected news events represent an information access problem where the approaches using Wikipedia to contextualize a tweet fail.

3   Our Proposed Approach

On Twitter, users are posting millions of tweets in order to express what they are thinking about natural disasters, political debates and sporting events followed by some comments, retweet or favorite. The users' interactions essentially

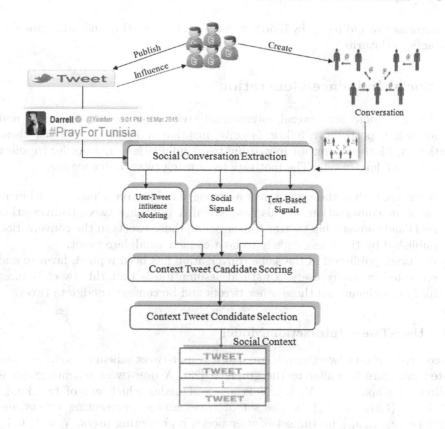

Fig. 1. Overview of our proposed approach

reflect the importance of different tweets and can be used to improve the tweet contextualization quality.

Compared to traditional contexts that are defined based on textual information, we modeled social tweet context using various dynamic social relationships such as following relationships between users, retweeting and replying relationships between tweets. In this paper, our proposed approach (Figure 1) consists of extracting social tweet context by means of social Twitter conversations. In addition, we defined a social tweet context and social Twitter conversations as follows:

Social Tweet Context
Given a tweet t its social context C_t is defined as $<I_t, U_t>$ where I_t is a set of interactions (comment, retweet,...) on t written by users U_t in a social network.

Social Twitter Conversations[5]
There is a set of short text messages posted by a user at specific timestamp on the same topic. These messages can be directly replied to other users by using

"@username" or indirectly by liking, retweeting, commenting and other possible interactions (favorite).

4   Social Influence Generation

On Twitter, there are several interactional relationships between users and tweets such as post, reply, follow, favorite, mention and retweet. We take these relationships into account for measuring tweet influence score in order to select context candidate tweets. The motivations of using tweet influence are:

- If we know that the user A has a strong influence on a user B within in the same conversation, in this case, when A publish a tweet (conversation root) and causes a big twitter conversation, those tweets in the conversation published by B are more likely to be a context candidate tweet.
- If a tweet published in the same conversation has been replied, favorite and retweeted by many users, a natural assumption is that this tweet is most likely to influence all those other tweets and be context candidate tweet.

4.1   User-Tweet Interaction Model

To construct a user-tweet graph, we define a user-tweet schema graph, as illustrated in Figure 2 similar to the graph in [23]. A user-tweet schema graph is a directed graph G = (V, E). V is a set of nodes which are of two kinds. Let $V_t = \{t_1, t_2, \ldots, t_m\}$ be the set of tweet nodes representing tweets and $V_u = \{u_1, u_2, \ldots, u_n\}$ be the set of user nodes representing users. V = $V_t \cup V_u$ nodes. E is the edge set consisting of post, reply, follow, retweet, mention and favorite edges.

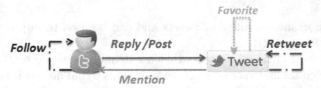

Fig. 2. user-tweet schema graph

A reply edge is from a user u to a tweet t posted by u. A follow edge is from a user u to another user who follows u. A retweet edge is from a tweet t to another tweet which retweets t. A mention edge is from a tweet t to a user u who comments t. A favorite edge is from a tweet t to another tweet which favorite's t. A user-tweet interaction model is shown in Figure 3. In the case of user-tweet interaction model, it consists of user nodes, tweets nodes and six kinds of edges.

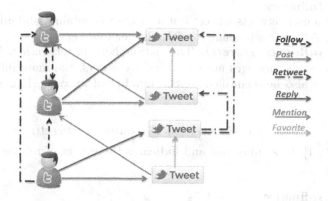

Fig. 3. user-tweet interaction model

4.2 Influence Measuring Based on User-Tweet Interaction Model

In Twitter microblog, the tweet of a user who has more followers always draw more attention, so they are evidently exists a correlation between tweet characteristics influence and tweet's author influence. We exploit two types of score for tweet influence measuring:

- Measuring tweet influence score refers to those features which represent the particular characteristics of tweet such as reply influence.
- Measuring tweet's author influence score refers to those features which represent the influence of tweet's author such as follow influence.

Measuring of Tweet Influence. The tweet influence is calculated from reply influence, retweet influence and favorite influence.

- **Reply Influence**
  When a user replied to tweet, it means she/he has taken time to react to the posted content. She/he is reacting to what this user tweeted and is most likely sharing her personal opinion in published content.
  Reply influence score(t): The action here is replying. The more replies a tweet receives, the more influential it is. This influence can be quantified by the number of replies the tweet receives. The reply influence is defined as follow:

$$Reply\_influence(t) = \alpha \times number\_reply(t). \tag{1}$$

  $\alpha \in (0, 1]$. It is adjustable and indicates the weight of reply edge.

– **Retweet Influence**

Generally, a user retweets a tweet if it appears to contain useful information, because he/she wants to share it with his/her followers.

Retweet influence score(t): The action here is retweeting. The more frequently user's messages are retweeted by others, the more influential it is. This can also be quantified by the number of retweets. It is defined as follow:

$$Retweet\_influence(t) = \beta \times number\_retweet(t). \tag{2}$$

$\beta \in (0, 1]$. It is adjustable and indicates the weight of retweet edge.

– **Favorite Influence**

Favorites are described as indicators that a tweet is well-liked or popular among online users. A tweet can be identified as a favorite by the small star icon seen beside the post. When a user mark tweets as favorites, she/he can easily find useful and relevant information. In addition, she/he can also spark the interest of other online users to start a conversation or comment on the tweet.

Favorite influence score(t): The action here is favoriting. The more favorites a tweet receives, the more influential it is. This influence can be quantified by the number of favorite the tweet receives. It is defined as follows:

$$Favorite\_influence(t) = \gamma \times num\_favorite(t). \tag{3}$$

$\gamma \in (0, 1]$. It is adjustable indicates the weight of favorite edge.

According to the experience, α is bigger than β and γ, it means that the users who reply on tweet t are more interested in it than others who only retweet or favorite it.

It is obvious that microblogging users mainly focus on the current tweets. However, temporal aspect can also provide valuable information for tweet contextualization due to the real-time characteristics of Twitter. Therefore, the tweet timestamp plays an important role on the tweet influence .i.e. a recent tweet has larger chance to have bigger influences compared to old published tweet. So to cope with, we use Gaussian Kernel [15] to estimate a distance Δt between tweet conversation root time d and other tweet time d' within the same conversation, i.e., $\Delta t = |d' - d|$. It is defined as follow:

$$\Gamma(\Delta t) = \exp\left[\frac{-\Delta t^2}{2\sigma^2}\right] with\sigma \in \mathbb{R}+. \tag{4}$$

Finally, the tweet influence score is defined as follows:

$$Tweet\_influence(t) = \Gamma(\Delta t) \times Reply\_influence(t) + Retweet\_influence(t)$$
$$+ Favorite\_influence(t). \tag{5}$$

Measuring of Tweet's Author Influence. In Twitter, many celebrities post relevant messages but got many followers simply because of their influence in real life. Thus, considering only the number of followers can not show real influence in microblogs. It has been proved that attributes expressing engaging audience links such as mention relationship are better to represent user influence. Furthermore, in our work we consider both the follow relationship and the mention relationship.

- **Mention Influence** whch we measure through the number of mentions containing one's name, indicates the ability of that user to engage others in a conversation. The mention influence score is defined as follows:

$$Mention\_influence(u) = \delta \times number\_Mention(u). \qquad (6)$$

$\delta \in (0, 1]$. It indicates the weight of mention edge.

- **Follow Influence**

A user followed by many users is likely to be an authoritative user and their post is also likely to be useful. In addition, the followers number of a user directly indicates the audience size for that user. The follow influence score is defined as follows:

$$Follow\_influence(u) = \omega \times number\_follow(u). \qquad (7)$$

$\omega \in (0, 1]$. It indicates the weight of follow edge.

Finally, the tweet's author influence score is defined as follows:

$$Tweetauthor\_influence(u) = Mention\_influence(u) + Follow\_influence(u) \qquad (8)$$

5   Context Candidates Tweets Extraction

Besides user-tweet influence model we also included text based signals and social signals.

5.1   Text-Based Signals

In this section, we assign score to a candidate tweet based on the similarity between different tweets in the whole conversation. Therefore, From each tweet t in a conversation C, we derive a vector V using the vector space model [17]. Thus, the set of conversation is viewed as a set of vector.

- Similarity to tweet root

We used cosine similarity to measure the similarity between tweet root vector $V_{t_{root}}$ and other tweets vector V_t within the same conversation. In addition, we aim to measure how much a tweet would be related to tweet root's content.

$$cosine(V_t, V_{t_{root}}) = \frac{V_t . V_{t_{root}}}{||V_t|| . ||V_{t_{root}}||} \qquad (9)$$

– Similarity of content [7]

In our case, it measures how many tweet of the whole conversation C are similar in content with current tweet $t_{current}$. We calculate cosine similarity score for every pair of tweets. The similarity is calculated using Lucene similarity function [1]. We denote current tweet modeled as a vector:

$$cosine(t_{current}, C) = \frac{\sum_{t_{current} \neq t'} sim(t_{current}, t')}{|C| - 1} \qquad (10)$$

5.2   Social Signals

– **Context Candidates Tweets Regarding the URLs**
  By sharing an URL, an author would enrichment the information published in his tweet. When a URL is present in the tweet root, we download the page and extract its title as well as the body content. For each candidate tweet t we computed:
  - The word overlap between a candidate tweet t and the web page title, and between t and the body content of the web page.
  - The cosine similarity between t and the web page title, and between t and the body content of the web page.

– **Context Candidates Tweets Regarding the Hashtags**
  The # symbol, called hashtags, is a very important pieces of information in tweet, since they are tags that were generated by user. Hashtag is used to mark a topic in a tweet or to follow conversation. In addition, publishers can use hashtag to provide implicit tweet context. We used this feature to collect candidates tweets that share the same tweet root hashtags.

$$F1(t, t_{root}) = \begin{cases} 1 \text{ if t contains the same hashtag.} \\ 0 \text{ otherwise.} \end{cases} \qquad (11)$$

5.3   Supervised Learning Framework

Given the above signals, we could convert them as features, then cast the Twitter context summarization task into a supervised learning problem. After training a model, we could predict a few tweets as its summary for all tweets in a new context tree. In this paper, we choose Gradient Boosted Decision Tree (GBDT) algorithm [11] to learn a non-linear model. GBDT is an additive regression algorithm consisting of an ensemble of trees, fitted to current residuals, gradients of the loss function, in a forward step-wise manner.

---

[1] http://lucene.apache.org/core/3_6_1/scoring.html

6 Experiments and Results

6.1 Twitter Conversations Data Set

As celebrities are highly influential in Twitter [22], celebrities initiated tweets would lead to large context trees. We extract 50 Twitter context trees from January 7th to March 22th, 2015, using our conversations trees detection system [4] to construct a data set in our work. These 50 context trees are initiated by many celbrities as like as Lady Gaga, who is the most popular elite user on Twitter, Manuel Valls, Olivia Wilde, J. k. Rowling, Norman Thavaud, Fran?ois Hollande. From another perspective, 26 out of 50 context trees are about the terrorist attack on charlie hebdo, another 16 context trees are related to the Tunisian museum terror attack, while the remaining 8 are about different topics.

6.2 Reference Summary

To the best of our knowledge, there is no data set available to evaluate social Twitter contextualization. Thus, we conduct a pilot study to construct an editorial data set in our work. The pilot study goal is to construct a reference summary generating by humans which can be useful to evaluate our results. Thus, we only focus on 15 context trees about three different topics, but ask 10 assessors for judgments for every context tree.The assessors selected among students and colleagues of the authors (with backgrounds in computing and social sciences). In addition, we ask each assessor to first read the tweet root and open any URL inside to have a sense of what the tweet root is about. Then, the assessor reads through all contexts candidates tweets to get a sense of the overall set of data. Thus, for each context tree, we will have 10 independent judgments. Finally, the assessor selects 5 to 10 tweets ordered sequentially as the summary, which respond or extend the original tweets by providing extra information about it.

6.3 Evaluation Metrics

Tweet contextualization is evaluated on both informativeness and readability [18]. Informativeness aims at measuring how well the summary explains tweet or how well the summary helps user to understand tweet content. On the other hand, readability aims at measuring how clear and easy is to undersatnad the summary.

- **Informativeness:** The objective of this metric is to evaluate relevant tweets selection. Informativeness aims at measuring how well summary helps user to understand tweet content. Therefore, for each tweet root, each candidate tweet will be evaluated independently from the others, even in the same summary. The 10 best tweets summary for each tweet root are selected for evaluation. This choice is made based on the score assigned by the automatic system tweets contextualization (high scores). The dissimilarity between a human selected summary (constructed using a pilot study) and the proposed summary (using our approach) is given by:

$$Dis(T,S) = \sum_{t\in T}(P-1) \times \left(1 - \frac{min(log(P),log(Q))}{max(log(P),log(Q))}\right) \tag{12}$$

where $P = \frac{f_T(t)}{f_T} + 1$ and $Q = \frac{f_S(t)}{f_S} + 1$.

T is the set of terms presented in reference summary. For each term $t \in T$, $f_T(t)$ represents the frequency of occurrence of t in reference summary and $f_S(t)$ its frequency of occurrence in the proposed summary. More Dis (T,S) is low, more the proposed summarry is similar to the reference. T may take three distinct forms:

- Unigrams made of single lemmas.
- Bigrams made of pairs of consecutive lemmas (in the same sentence).
- Bigrams with 2-gaps as well as the bigram, but can be separated by two lemmas.

Our results in the informativeness evaluation presented in Table 1 .

Table 1. Table of Informativeness Results

	Unigrams	Bigrams	Skipgrams
Topic1			
Human Summary	0.7263	0.8534	0.9213
Our Proposed Summary	**0.7909**	**0.8865**	**0.9355**
Topic2			
Human Summary	0.7932	0.9137	0.9361
Our Proposed Summary	**0.8105**	**0.9408**	**0.9592**
Topic3			
Human Summary	0.7786	0.9172	0.9426
Our Proposed Summary	**0.8272**	**0.9438**	**0.9617**

– **Readability:** readability aims at measuring how clear and easy it is to understand summary. By contrast, readability is evaluated manually (cf.Table 2). Each summary has been evaluated by considering the following two parameters [19]:

- Relevance: judge if the tweet make sense in their context (i.e. after reading the other tweets in the same context). Each assessor had to evaluate relevance with three levels, namely highly relevant (value equal to 2), relevant (value equal to 1) or irrelevant (value equal to 0).
- Non-Redundancy: evaluates the ability of context does not contain too much redundant information, i.e. information that has already been given in a previous tweet. Each assessor had to evaluate redundancy with three levels, namely not redundancy (value equal to 2), redundancy (value equal to 1) or highly redundancy (value equal to 0).

6.4   Experimental Outcomes and Interpretation Results

A good summary should have good quality but with less redundancy. The obtained informativeness (cf.Table 1) evaluation results shed light that our proposed approach offers interesting results and ensure that context summaries contain adequate correlating information with the tweet root. In addition, based on editorial data set, our experimental results show that user influence information is very helpful to generate a high quality summary for each Twitter context tree. Furthermore, our tweet contextualization approach based on social Twitter conversations leads to the context informativeness improvement ; we note also that the tweets selection impacts the context quality. The contexts are less readable; it may be that they contain some noises which need to be cleaner.

Table 2. Table of Readability Results

	Relevance	Non Redundancy	AVG
Topic1			
Human Summary	88.65%	66.33%	77.49%
Our Proposed Summary	**89.72%**	**69.78%**	**79.75%**
Topic2			
Human Summary	90.72%	65.82%	78.27%
Our Proposed Summary	**91.03%**	**67.49%**	**79.26%**
Topic3			
Human Summary	90.23%	69.06%	79.64%
Our Proposed Summary	**92.24%**	**69.72%**	**80.98%**

7   Conclusion

we explored in this paper the tweet contextualization problem. We proposed an approach that combined different types of signals from social user interactions and exploited a set of conversational features, which help users to get more context information when using Twitter. Traditional contextualization methods only consider text information and we focused on exploiting multiple types of signals such as social signals, user-tweet influence signals and text based signals. All signals are converted into features, and we throw tweet contextualization into a supervised learning problem. Our approach was evaluated by using an editorial data set in which 10 assessors are employed to generate a reference summary for each context tree.

Future work will further research the conversational aspects by including human communication aspects, like the degree of interest in the conversation by gathering data from multiple sources such as comments on news articles or comments on Facebook pages.

References

1. Ansary, K.H., Tran, A.T., Tran, N.K.: A pipeline tweet contextualization system at INEX 2013. In: Working Notes for CLEF 2013 Conference, Valencia, Spain, September 23–26, 2013
2. Bandyopadhyay, A., Pal, S., Mitra, M., Majumder, P., Ghosh, K.: Passage retrieval for tweet contextualization at INEX 2012. In: CLEF 2012 Evaluation Labs and Workshop, Online Working Notes, Rome, Italy, September 17–20, 2012
3. Belkaroui, R., Faiz, R., Elkhlifi, A.: Conversation analysis on social networking sites. In: Tenth International Conference on Signal-Image Technology and Internet-Based Systems, SITIS 2014, Marrakech, Morocco, November 23–27, 2014, pp. 172–178 (2014)
4. Belkaroui, R., Faiz, R., Elkhlifi, A.: Social users interactions detection based on conversational aspects. In: Barbucha, D., Nguyen, N.T., Batubara, J. (eds.) New Trends in Intelligent Information and Database Systems. SCI, vol. 598, pp. 161–170. Springer, Heidelberg (2015)
5. Belkaroui, R., Faiz, R., Elkhlifi, A.: Using social conversational context for detecting users interactions on microblogging sites. Revue des Nouvelles Technologies de l'Information Extraction et Gestion des Connaissances **RNTI–E–28**, 389–394 (2015)
6. Bhaskar, P., Banerjee, S., Bandyopadhyay, S.: A hybrid tweet contextualization system using IR and summarization. In: CLEF 2012 Evaluation Labs and Workshop, Online Working Notes, Rome, Italy, September 17–20, 2012
7. Damak, F., Pinel-Sauvagnat, K., Boughanem, M., Cabanac, G.: Effectiveness of state-of-the-art features for microblog search. In: Proceedings of the 28th Annual ACM Symposium on Applied Computing, SAC 2013, pp. 914–919. ACM, New York (2013)
8. Deveaud, R., Boudin, F.: Contextualisation automatique de tweets à partir de wikipédia. In: CORIA 2013 - Conférence en Recherche d'Infomations et Applications - 10th French Information Retrieval Conference, Neuchâtel, Suisse, April 3–5, 2013, pp. 125–140 (2013)
9. Ermakova, L., Mothe, J.: IRIT at INEX: question answering task. In: Geva, S., Kamps, J., Schenkel, R. (eds.) INEX 2011. LNCS, vol. 7424, pp. 219–226. Springer, Heidelberg (2012)
10. Ermakova, L., Mothe, J.: IRIT at INEX 2014: tweet contextualization track. In: Conference on Multilingual and Multimodal Information Access Evaluation (CLEF), Sheffield, UK, September 15–18, 2014, pp. 557–564 (2014)
11. Friedman, J.H.: Greedy function approximation: A gradient boosting machine. The Annals of Statistics **29**, 1189–1232 (2000). The Institute of Mathematical Statistics
12. Liana, E., Josiane, M.: IRIT at INEX2012: tweet contextualization. In: Conference on Multilingual and Multimodal Information Access Evaluation (CLEF), Rome, Italy, 17/09/2012-20/09/2012 (2012)
13. Liana, E., Josiane, M.: IRIT at INEX 2013: tweet contextualization track. In: Conference on Multilingual and Multimodal Information Access Evaluation (CLEF), Valencia, Spain, September 23–26, 2013
14. Morchid, M., Linarès, G.: INEX 2012 benchmark a semantic space for tweets contextualization. In: CLEF 2012 Evaluation Labs and Workshop, Online Working Notes, Rome, Italy, September 17–20, 2012
15. Phillips, J.M., Venkatasubramanian, S.: A gentle introduction to the kernel distance. Computing Research Repository abs/1103.1625 (2011)

16. Romain, D., Florian, B.: Effective tweet contextualization with hashtags performance prediction and multi-document summarization. In: Working Notes for CLEF 2013 Conference, Valencia, Spain, September 23–26, 2013
17. Salton, G., Wong, A., Yang, C.S.: A vector space model for automatic indexing. In: Communications of the ACM, No. 11, pp. 613–620. ACM, New York, November 1975
18. SanJuan, E., Bellot, P., Moriceau, V., Tannier, X.: Overview of the INEX 2010 question answering track (QA@INEX). In: Geva, S., Kamps, J., Schenkel, R., Trotman, A. (eds.) INEX 2010. LNCS, vol. 6932, pp. 269–281. Springer, Heidelberg (2011)
19. San Juan, E., Moriceau, V., Tannier, X., Bellot, P., Mothe, J.: Overview of the inex 2012 tweet contextualization track. In: Forner, P., Karlgren, J., Womser-Hacker, C. (eds.) CLEF (Online Working Notes/Labs/Workshop) (2012)
20. Torres-Moreno, J.: Three statistical summarizers at CLEF-INEX 2013 tweet contextualization track. In: Working Notes for CLEF 2014 Conference, Sheffield, UK, September 15–18, 2014, pp. 565–573 (2014)
21. Torres-Moreno, J., Velázquez-Morales, P.: Two statistical summarizers at INEX 2012 tweet contextualization track. In: CLEF 2012 Evaluation Labs and Workshop, Online Working Notes, Rome, Italy, September 17–20, 2012
22. Wu, S., Hofman, J.M., Mason, W.A., Watts, D.J.: Who says what to whom on twitter. In: Proceedings of the 20th International Conference on World Wide Web, WWW 2011, pp. 705–714. ACM, New York (2011)
23. Yamaguchi, Y., Takahashi, T., Amagasa, T., Kitagawa, H.: TURank: twitter user ranking based on user-tweet graph analysis. In: Chen, L., Triantafillou, P., Suel, T. (eds.) WISE 2010. LNCS, vol. 6488, pp. 240–253. Springer, Heidelberg (2010)
24. Zingla, M.A., Ettaleb, M., Latiri, C.C., Slimani, Y.: INEX2014: tweet contextualization using association rules between terms. In: Working Notes for CLEF 2014 Conference, Sheffield, UK, September 15–18, 2014, pp. 574–584 (2014)

Supervised Learning to Measure the Semantic Similarity Between Arabic Sentences

Wafa Wali[1](✉), Bilel Gargouri[1], and Abdelmajid Ben hamadou[2]

[1] MIR@CL Laboratory, FSEGS, Sfax, Tunisia
{wafa.wali,bilel.gargouri}@fsegs.rnu.tn
[2] MIR@CL Laboratory, ISIMS, Sfax, Tunisia
abdelmajid.benhamadou@isimsf.rnu.tn

Abstract. Many methods for measuring the semantic similarity between sentences have been proposed, particularly for English. These methods are considered restrictive as they usually do not take into account some semantic and syntactic-semantic knowledge like semantic predicate, thematic role and semantic class. Measuring the semantic similarity between sentences in Arabic is particularly a challenging task because of the complex linguistic structure of the Arabic language and given the lack of electronic resources such as syntactic-semantic knowledge and annotated corpora.

In this paper, we proposed a method for measuring Arabic sentences' similarity based on automatic learning taking advantage of LMF standardized Arabic dictionaries, notably the syntactic-semantic knowledge that they contain. Furthermore, we evaluated our proposal with the cross validation method by using 690 pairs of sentences taken from old Arabic dictionaries designed for human use like Al-Wassit and Lissan-Al-Arab. The obtained results are very encouraging and show a good performance that approximates to human intuition.

Keywords: Sentence similarity · Automatic learning · Arabic language · Syntactico-semantic knowledge · LMF-ISO 24613 standardized dictionaries

1   Introduction

Today, people are surrounded by a huge amount of information due to the rapid development of the Internet and its associated technologies. Also, increasingly, the techniques related to information retrieval, knowledge management, Natural Language Processing (NLP), and so on, are becoming increasingly important and are being developed to help people manage and process information. However, one of the key problems of these themes is sentence similarity, which has a close relationship with psychology and cognitive science. There are numerous studies that have been previously developed with the aim of computing sentence similarity. The problem was formally brought to attention and the first solutions were proposed in 2006 with the works that are reported in [1] and that takes into account the syntactic information via the word order and the semantic one via the semantic similarity of words using

© Springer International Publishing Switzerland 2015
M. Núñez et al. (Eds.): ICCCI 2015, Part I, LNAI 9329, pp. 158–167, 2015.
DOI: 10.1007/978-3-319-24069-5_15

knowledge-based and corpus-based methods. Several methods have adapted the proposition of Li et al. [1] and improved by adding to it other features such as Longest Common subsequence (LCS) and Word Sense Disambiguation (WSD).

Recently, some methods have been suggested such as [2] and [3] which take into consideration in their computation the sentence similarity, the syntactic dependencies, and the semantic similarity between words using WordNet.

Nevertheless, one of the main problems of existing sentence similarity methods is that most of them have neglected some elements of semantic knowledge such as semantic class and thematic role. Additionally, the syntactic-semantic knowledge that can be extracted from the sentence and that is highly relevant in the computation of sentence similarity is ignored. Indeed, these elements of knowledge, notably the semantic predicate, the thematic role and the semantic class, supply a mechanism of interaction between the syntactic processor and the discourse model, provide information about the relationships between words, and perform an important role in conveying the meaning of a sentence.

Concerning the research works on semantic similarity for the Arabic language, they are focused on words similarity as [4] whereas to the best of our knowledge, there is no sentence similarity measure developed specifically for Arabic compared to the works in the English language, which has already benefited from extensive research in this field. There are some aspects that slow down progress in Arabic NLP compared to the accomplishments in English and other European languages such as the absence of diacritics (which represent most vowels) in the written text, which creates ambiguity, agglutination and complexity in grammar.

In addition to the linguistic issues, there is also a lack of Arabic corpora, lexicons that treat the syntactic-semantic knowledge, which are essential to any advanced research in different areas. In fact, the standardization committee ISO TC37/SC4 has validated the Lexical Markup Framework norm (LMF) project under the standard ISO 24 613 [5] which allows for the encoding of rich linguistic information, including among others morphological, syntactic, and semantic aspects covering several languages. The Arabic has taken advantage of this standard and an LMF standardized Arabic dictionary [6] that incorporates multi-data knowledge has been developed within the MIR@CL research team.

The focus of this paper is on the production of the first sentence similarity benchmark dataset for Modern Standard Arabic (MSA) that we will use in evaluating the content of Arabic dictionaries [7]. Indeed, the similarity function is devoted to the measurement of the similarity degree between the definitions of lexical entries or the examples of one definition in order to avoid redundancy [8].

In this paper we propose a novel method, to compute the similarity between Arabic sentences by taking into account semantic, syntactic and syntactic-semantic knowledge and taking advantage of the LMF standardized Arabic dictionary. The proposal measures the semantic similarity via the synonymy relations between words in sentences. Besides, the syntactic similarity is measured based on the co-occurrences of dependent structures between two sentences after a dependency parsing. The syntactic-semantic similarity is measured on the basis of the common semantic arguments, associated with the semantic predicate in terms of thematic role and semantic class, between the pair of sentences.

An experiment was carried out on 1380 Arabic sentences, taken from various definitions found in old Arabic dictionaries designed for human use like Alwasit and Lissan Al Arab. The experiment is based on supervised learning which shows a good accuracy that approximates to human judgment. Therefore, our method proves the importance of semantic and syntactic-semantic knowledge in the computation of sentence similarity.

The next section presents a brief review of the approaches used to compute sentence similarity. Section 3 presents the main features of the Arabic language. Section 4 gives the details of the proposed method of measuring sentence similarity. Section 5 covers the experiments and the obtained results. Lastly, Section 6 sums up the work, draws some conclusions and announces prospective future works.

2   Related Works

During the last decade, several methods to measure sentence similarity were established based on semantic and/or syntactic knowledge. In this section, we examine some related works, called hybrids, which are based on both syntactic and semantic knowledge and have the same contribution as our proposal, in order to explore the advantages and limitations of the previous methods.

Stefanescu et al. [3] introduced a method for measuring the semantic similarity between sentences, which is based on the assumption that the meaning of a sentence is captured by its constituents and the dependencies between them. The method considers that every chunk has a different importance with respect to the overall meaning of a sentence that is computed according to the information content of the words in the chunk. The disadvantage of this method is that the semantic measurement is isolated from the syntactic measurement. In this method, the semantic similarity is calculated based on words semantic similarity, while syntactic dependency is counted to compute the syntactic similarity.

Lee et al. [2] proposed a sentence similarity algorithm that takes advantage of a corpus-based ontology via Wu Palmer's measure [9] and grammatical rules. Nevertheless, the Wu and Palmer measure presents the following drawback: in some situations, the similarity of two elements of an IS-A ontology contained in the neighborhood exceeds the similarity value of two elements contained in the same hierarchy. This situation is inadequate within the measuring sentence similarity.

The Semantic Textual Similarity task (STS) organized as part of the Semantic Evaluation Exercises (see [10] for a description of STS 2013) provides a common platform for the evaluation of such systems through a comparison with human annotated similarity scores over a large dataset. The authors introduced a benchmark for measuring Semantic Textual Similarity (STS) between similar sentences. They used the similarity between words using Knowledge-based Similarity, Syntactic Analysis, Nnamed Entity Recognition, Semantic Role Labeling, String Similarity, and Word Sense Disambiguation. The main drawback of this method is that it computes the similarity of words from different features, which is not computationally efficient.

In STS 2012, Saric et al. [11] proposed a hybrid method that derives sentence similarity from semantic information using WordNet (similarity between words) and corpus (the information content) and syntactic information that is based on the syntactic roles, the overlap syntactic dependencies and the named entities. However, the judgment of similarity is situational and depends on time. Indeed, the information collected in the corpus may not be relevant to the present.

Furthermore, we can see that all of the hybrid methods presented above exploit insufficiently the sentence information. However, the major disadvantage of these hybrid methods is that some elements of the semantic knowledge such as the semantic class and the thematic role of the sentence's words are not considered in calculating sentence similarity. Also, the relationship between the syntactic and semantic level such as semantic predicate is not taken into account. We discuss how this knowledge enhances the sentence similarity in the following section.

3   Arabic Language Background

Arabic is one of the world's major languages. It is the fifth most widely spoken language in the world and is the second in terms of the number of speakers with over than 250 million Arabic speakers, of whom roughly 195 million are first language speakers and 55 million are second language speakers. Arabic is characterized by a complex morphology and a rich vocabulary. It is a derivational and flexional language. Indeed, an Arabic word may be composed of a stem plus affixes (to refer to tense, gender, and/or number) and clitics (including some prepositions, conjunctions, determiners, and pronouns). For instance, the word "الكُتُب", transliterated al-kutubu and meaning books, is derived from the stem "كِتَاب", transliterated kitAb and meaning book, which is derived from the root "كَتَب", transliterated katab and meaning to write. Moreover, Agglutination in Arabic is another specific phenomenon. In fact, in Arabic articles, prepositions, pronouns, etc. can be affixed to the adjectives, nouns, verbs and particles to which they are related. The derivational, flexional and agglutinative aspects of the Arabic language yield significant challenges in the NLP. Thus, many morphological ambiguities have to be solved when dealing with the Arabic language.

Moreover, many Arabic words are homographic: they have the same orthographic form, though their pronunciation is different. In most cases, these homographs are due to the non-vocalization of words. This means that a full vocalization of words can solve these ambiguities but most of the Arabic texts are not vocalized such as the word «كتب-ktb» has 16 vocalizations and which represent 9 different grammatical categories like "كَتَبَ – kataba- write", "كُتِبَ – kutiba – was written" and "كُتُب- kutub- books". In addition, Arabic grammar is a very complex subject of study; even Arabic-speaking people nowadays are not fully familiar with the grammar of their own language. Thus, Arabic grammatical checking is a difficult task. The difficulty comes from several reasons: the first is the length of the sentence and the complex Arabic syntax; the second is the flexible word order nature of the Arabic sentence and the third is the presence of an elliptic personal pronoun "alDamiirAlmustatir".

4   The Proposed Method

This section is devoted to the presentation of our suggested method. The suggested method for measuring semantic similarities between Arabic sentences has two phases: the learning phase and the test phase.

4.1   Method Overview

The first phase of our proposal requires a training corpus, the features extracted from the learning corpus and an LMF standardized Arabic dictionary [9]. Indeed, the features are lexical, semantic and syntactico-semantic. This phase includes two processes: the first is the pre-processing that aims to have an annotated corpus and the second is the training used to get a hyper plane equation via the learning algorithm. The second phase implements the learning results from the first phase to achieve the similarity score of the Arabic sentence in order to classify the sentences as similar and not similar. The phases of our approach are illustrated in the following figure.

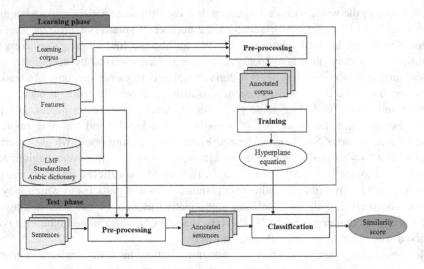

Fig. 1. The suggested method

4.2   The Learning Phase

The learning phase involves the use of a training corpus, a set of features extracted from the learning corpus analysis and an LMF standardized Arabic dictionary in order to train the learning algorithm. It is composed of the following two processes:

Pre-processing: In the pre-processing process, we are applying the defined features on the learning corpus taking advantage of the LMF standardized Arabic dictionary [9] in order to have an annotated corpus. Indeed, the features are classified into three classes' namely the lexical, the semantic and the syntactic-semantic features. The lexical feature specifies the common words between the pair of sentences. The semantic feature analyzes the synonymous words among the words of the sentences. And the syntactic-semantic feature detects the common semantic arguments between sentences in terms of semantic class and thematic role [12].

Also, in this step we aim to annotate each word of a sentence in the learning corpus according to the different extraction features presented above. Each pair of sentences is described by a vector called extraction vector.

The value of the lexical feature SL (S1,S2) corresponds to determining the similar stems between the pair of sentences. In this step, we use the Jaccard coefficient [13] to compute the lexical feature. The following formula shows how to calculate the lexical feature.

$$SL_{(S1,S2)} = M_C / (M_{S1} + M_{S2} - M_C) \tag{1}$$

Where:
MC: the number of common stems between the two sentences
M_{S1}: the number of stems contained in the sentence S_1
M_{S2}: the number of stems contained in the sentence S_2

This extraction vector is completed by the semantic features selected from the semantic annotations in the corpus. Indeed, the semantic feature is derived from the LMF standardized Arabic dictionary [9].The procedure to compute the semantic feature is to form a joint word set only by using the distinct stem in the pairs of sentences. For each sentence, a raw semantic feature is derived with the assistance of the semantic annotation. Indeed, each sentence is readily represented by the use of the joint word set as follows: The vector derived from the joint word set denoted by Š. Each entry of the semantic vector corresponds to a stem in the joint word set, so the dimension equals the number of stems in the joint word set. The value of an entry of the lexical semantic vector, $Š_i$ (i=1, 2, …, m) is determined by the semantic similarity of the word corresponding to a word in the sentence. Take $S_1$is made up by W_1 $W_2...W_m$ as an example:

Case 1: if W_i appears in the sentence, $Š_i$ is set to 1.

Case 2: If W_i is not contained in S_1, a semantic similarity score is computed between W_i and each word in the sentence S_1, using the semantic annotation. Thus, the most similar word in T_1 to W_i is the one with the highest similarity score δ. If δ exceeds a preset threshold, then $Š_i$ =δ; otherwise, $Š_i$=0. Once the two sets of synonyms for each stem are collected, we calculate the degree of similarity between them using the Jaccard coefficient [13].

$$Sim_{(W1,W2)} = M_C / (M_{W1} + M_{W2} - M_C) \tag{2}$$

Where:
M_C: the number of common words between the two synonym sets
M_{W1}: the number of words contained in the w_1 synonym set
M_{W2}: the number of words contained in the w_2 synonym set

From the semantic vectors generated as described above, we compute the semantic feature between the pair of sentences, which we call SM(S_1, S_2) using the Cosine similarity.

$$SM_{(S1,S2)}= V1.V2/(\|V1\|*\|V2\|) \tag{3}$$

Where:

V_1: the semantic vector of sentence S_1

V_2: the semantic vector of sentence S_2

Also, this extraction vector is completed by the syntactic-semantic feature selected from the syntactico-semantic annotations in the corpus. The syntactic-semantic feature is derived from the LMF standardized Arabic dictionary [9]. Indeed, on the one hand each sentence is syntactically parsed using a syntactic analyzer and on the other hand it is semantically analyzed by an expert giving its semantic predicate. The correspondence between syntactic and semantic analysis is then defined in the LMF standardized Arabic dictionary [9] in order to extract the semantic arguments such as semantic class and thematic role. Then, the value of syntactico-semantic feature corresponds to extraction the similar semantic arguments in terms of semantic class and thematic role between the pair of sentences. To compute the syntactico-semantic feature, which we call $SSM(S_1,S_2)$, using the Jaccard coefficient [13]. The following formula shows how to compute the syntactico-semantic feature.

$$SSM_{(S1,S2)}= AS_C/(AS_{S1}+AS_{S2}-AS_C) \tag{4}$$

Where:

AS_C: the number of common semantic arguments between the two sentences

AS_{S1}: the number of semantic arguments contained in the sentence S_1

AS_{S2}: the number of semantic arguments contained in the sentence S_2

This extraction vector is completed by the appropriate similarity decision (similar, not similar) that is provided by an expert. The set of extraction vectors forms an input file for the learning stage. At the end of this process, the learning corpus is converted from its original format into a vector format and we obtain a tabular corpus which consists of a set of vectors separated by a return as shown in the following example:

Vector1: 0.3, 0.4, 1, not similar

Vector2 :0.8, 0.7, 1, similar

Vector3 :0, 0, 0, not similar

In fact, the values of lexical, semantic, and syntactic-semantic features are included between 0 and 1.

Training: This stage uses the previously generated extraction vectors in order to produce an equation known as hyperplane equation. There are many learning algorithms proposed in the literature such as the SVM, the Naïve Bayes and the J48 decision tree algorithms. In fact, these algorithms generate the equation which is used to compute a similarity score in order to classify the sentences (similar, not similar). The training stage generates a hyperplane equation. It is not worthy that the learning stage is done only once and is only repeated in case we increase the size of the corpus, or change the type of corpus. This step is performed using the data set of sentences and the Weka library [14]. This tool takes as input extraction vectors in the form of an ".arff" file and an output hyperplane equation.

4.3   The Test Phase

This phase implements the results of the learning phase in order to measure the semantic similarity between the Arabic sentences. The user must provide segmented sentences as input to our system. This phase proceeds in two steps as follows: Firstly, a pre-processing is applied to the input pair of sentences. Indeed, we use the features and the LMF standardized Arabic dictionary to process the sentences in order to have the vector format as presented in the learning stage. This pre-processing generates extraction vectors like those generated as input for the learning stage. The only difference is that these vectors do not contain the similarity decision (similar or not similar). This information will be calculated by the learning algorithms. Then, the extraction vectors generated in the first step and the hyperplane equation generated in the learning stage are provided as input to the classification module. Indeed, for each vector, we calculate a score using the hyperplane equation. Each equation discriminates between two similarity decision classes. So every vector will have a score according to the coefficients of three features as lexical, semantic and syntactic-semantic. The score and its sign are used to identify the similarity decision class for the test vector. At the end of this stage we obtain a similarity decision (similar, not similar) between the pair of sentences.

5   Experimentation

5.1   Learning Corpus

There are currently no suitable Arabic data sets (or even standard text sets) annotated with syntactic-semantic knowledge, notably the semantic predicate, the thematic role and the semantic class. Building such a data set is not a trivial task due to subjectivity in the interpretation of language, which is in part due to the lack of deeper contextual information. In our work, we collected a set of 1380 sentences that consist of the dictionary definitions and the examples of definitions of words, taken from Arabic dictionaries like Lissan Al-Arab and Al-Wassit. The data were annotated with the features described above. In fact, the lexical knowledge was derived from the MADAMIRA analyzer [18] while the semantic and the syntactic-semantic knowledge were taken from the LMF standardized Arabic dictionary [9].

5.2   Results

The evaluation of our method is achieved following the cross-validation method using the Weka tool [15]. To realize that, we divided the training corpus into two distinct parts, one for learning (80%) and one for the test (20%). The classification constitutes the most important stage of our proposal that is why we did some experiments to evaluate its impact in the whole solution. We chose five classifiers from the java API weka. The choice covers different algorithms of different classifiers types; namely: the rule-based, the probabilistic, the case-based, the functions and the decision trees methods. The results are listed in the table below.

Table 1. Evaluation results

	Probabilistic (NaiveBayes)	Decision tree (J48)	Function (SMO)	Empiric (KStar)	Rule-based (DecisionTable)
Precision	0.968	0.987	0.979	0.987	0.985
Recall	0.962	0.988	0.978	0.987	0.985
F-measure	0.964	0.988	0.977	0.987	0.985

The results obtained are encouraging and represent a good start for the implementation of automatic learning for measuring semantic similarity between Arabic sentences. We noticed that the analysis of short sentences (<=10 words) presents the highest measures of recall and precision. As the sentence gets longer, there will be a more complex calculation, which reduces the system's performance. We believe that these results can be improved. In fact, we think that we can improve the learning stage by adding other features besides the lexical, semantic and syntactico-semantic features.

6   Conclusion and Perspectives

Measuring the semantic similarity between the Arabic sentences is not a trivial task given the major problems of the Arabic language. In this paper, we presented, in our knowledge, the first work of computation of the semantic similarity between Arabic sentences based on a supervised learning. Indeed, our proposal uses three linguistic features. The first is lexical and consists of extracting the similar stems between the pair of sentences; the second is semantic and allows extracting the common synonyms of the words of a sentence; and the third is syntactic-semantic and consisting to extract the common semantic arguments between the sentences to be compared. Further, the proposal is specific to the Arabic language and not applicable to other languages since it treats the specificities of the Arabic language such as the stem, which is absent in others languages. The experiments conducted on a set of Arabic sentence pairs demonstrate that the proposed method provides a significant accuracy that approximates to human intuition. Additionally, the developed system can be used and integrated in several applications of NLP such as automatic summarization, question answering and machine translation.

As a perspective, we plan to integrate a more efficient stemmer and another syntactic parser in our system in order to process the complex sentences. However, the incompleteness issue in databases in the used tools still remains a major problem in the Arabic research works. We want to enrich the corpus training with complex sentences covering anaphora and ellipses phenomenon.

References

1. Yuhua, L., McLean, D., Bandar, Z.A., O'Shea, J.D., Crockett, K.: Sentence similarity based on semantic nets and corpus statistics. In: IEEE Transactions on Knowledge and Data Engineering, vol. 18, pp. 1138–1150 (2006)

2. Lee, M., Chang, J., Hsieh, T.: A Grammar-Based Semantic Similarity Algorithm for Natural Language Sentences. The Scientific World Journal **2014**, Article ID 437162, 17 (2014). http://dx.doi.org/10.1155/2014/43716
3. Ştefănescu, D., Banjade, R., Rus, V.: A sentence similarity method based on chunking and information content. In: Gelbukh, A. (ed.) CICLing 2014, Part I. LNCS, vol. 8403, pp. 442–453. Springer, Heidelberg (2014)
4. Almarsoomi, F., O'Shea, J., Bandar, Z, Crockett, K.: Arabic word semantic similarity. In: World Academy of Science, Engineering and Technology (WASET 2012), vol. 6, pp. 81–89 (2012)
5. Francopoulo, G., George, M.: Language Resource Management. 2008. Lexical Markup Framework (LMF). Technical report, ISO/TC 37/SC 4 N453 (N330 Rev.16) (2008)
6. Khemakhem, A., Gargouri, B., Ben Hamadou, A.: LMF standardized dictionary for arabic language. In: International Conference on Computing and Information Technology (2012)
7. Wali, W., Gargouri, B., Ben Hamadou, A.: Towards detecting anomalies in the content of standardized LMF dictionaries. In: Recent Advances in Natural Language Processing (RANLP 2013), pp 719–726 (2013)
8. Wali, W., Gargouri, B., Ben Hamadou, A.: LMF-based approach for detecting semantic anomalies in electronic dictionaries. In: The Asialex International Conference (ASIALEX 2013), pp 242–252 (2013)
9. Wu, Z., Palmer, M.: Verb semantics and lexical selection. In: The Proceedings of the Annual Meeting of the Association for Computational Linguistics (1994)
10. Agirre, E., Cer, D., Diab, M., Gonzalez-Agirre, A., Guo, W.: Shared task: semantic textual similarity, including a pilot on typed similarity. In: Proceedings of the Second Joint Conference on Lexical and Computational Semantics (*SEM 2013), Atlanta, Georgia (2013)
11. Saric, F., Glavas, G., Karan, M., Snajder, J., Basi, B.: TakeLab: systems for measuring semantic text similarity. In: First Joint Conference on Lexical and Computational Semantics (*SEM 2012), Montreal, Canada, pp 441–448, June 7-8, 2012. @2012 Association for Computational Linguistics (2012)
12. Wali, W., Gargouri, B., Ben Hamadou, A.: Using standardized lexical semantic knowledge to measure similarity. In: 7th International Conference on Knowledge Science, Engineering and Management (KSEM 2014), pp. 93–104 (2014)
13. Jaccard, P.: Etude comparative de la distribution florale dans une portion des Alpes et des Jura. In : Bulletin de la Société Vaudoise des Sciences Naturelles, vol. 37, pp. 547–579 (1901)
14. Frank, E., Witten, I.H.: Practical machine learning tools and techniques. In: 2nd Morgan Kaufmann series in data management systems (2005)
15. Pasha, A., Al-Badrashiny, M., Diab, M., El Kholy, A., Eskander, R., Habash, N., Pooleery, M., Rambow, O., Roth, R.M.: MADAMIRA: a fast, comprehensive tool for morphological analysis and disambiguation of arabic. In: The Proceedings of 9th Edition of the Language Resources and Evaluation Conference (LREC2014), Reykjavik, Iceland, pp. 26–31, May 2014

Sentiment Analysis

How Impressionable Are You? - Grey Knowledge, Groups and Strategies in OSN

Camelia Delcea[1(✉)], Ioana Bradea[1], Ramona Paun[2], and Emil Scarlat[1]

[1] Bucharest University of Economic Studies, Bucharest, Romania
camelia.delcea@csie.ase.ro, alexbradea1304@gmail.com
[2] Webster University, Bangkok, Thailand
paunr@webster.ac.th

Abstract. Today's leading businesses have understood the role of "social" into their everyday activity. Online social networks (OSN) and social media have melt and become an essential part of every firm's concerns. Brand advocates are the new leading triggers for company's success in online social networks and are responsible for the long term engagement between a firm and its customers. But what can it be said about this impressive crowd of customers that are gravitating around a certain brand advocacy or a certain community? Are they as responsive to a certain message as one might think? Are they really impressed by the advertising campaigns? Are they equally reacting to a certain comment or news? How they process the everyday grey knowledge that is circulating in OSN? In fact, how impressionable they are and which are the best ways a company can get to them?

Keywords: OSN · Grey knowledge · Grey clustering analysis · Strategies in OSN

1   Introduction on Online Social Networks (OSN)

A short and easy to understand definition of OSN is that they "form online communities among people with common interest, activities, backgrounds, and/or friendships. Most OSN are web-based and allows users to upload profiles (text, images, and videos) and interact with others in numerous ways" [1]. As structure, the OSN are usually perceived as a set of nodes – represented by its users, and a set of directed (e.g. "following" activity on Facebook) of undirected (e.g. friendship relationships) edges connecting the various pairs of nodes [2].

Apart from the research and development role kept by the OSN mostly through the usage of the customers' feedback to improve and solve problems or even to create and diversify different categories of products / services, the word-of-mouth marketing and targeted advertising are also some of the benefits brought by these networks. In a recent paper about social intelligence and customer experience, Synthesio [3] votes for learning who your customers really are in order to address them a proper campaign. Also, by identifying the "army" of product advocates among different communities and getting

© Springer International Publishing Switzerland 2015
M. Núñez et al. (Eds.): ICCCI 2015, Part I, LNAI 9329, pp. 171–180, 2015.
DOI: 10.1007/978-3-319-24069-5_16

them closer to a specific brand, will give that firm a tremendous opportunity to gain more customers and to extend a real relationship on an individual basis [3].

But before this, it still remains unsolved the question of identifying those particulars advocates among the "big crowd" of customers on the OSN. For this, in the following we are going to use some elements extracted form grey systems theory for spreading the customer's crowd into groups based on how impressionable they are by commercials, adds, videos, comments in OSN.

2   Grey Knowledge

Grey systems theory is one of the newest theories in the field of artificial intelligence and starts from a definition stipulated in the control theory in which an object was considered black when nobody knows anything about its inner structure and white when this structure was completely known. Therefore, a grey object is that particular entity whose structure is just partially known [4].

By developing its own methods and techniques, the grey systems theory succeeds to extract and bring some new knowledge about a specific object, process or phenomenon. Along with the grey relational analysis, one of the most known and used methods from the grey systems theory, the grey clustering is also one of the techniques that is bringing some new knowledge regarding a specific community.

In OSN more than in other type of communities and networks, grey systems theory finds its applicability due to the nature of the relationships between its main actors. Forrest [5] identifies two main types of relationships in a system: the generative and the non-generative ones. While the generative relationships are due to the interactions among different elements of a system, the non-generative are represented by the inner characteristics of these elements [6].

Even though most researches focus mainly on one or another type of these relationships, some advances have been made recently to include both of these aspects, as they are giving more substance about what is really happening at each network's level [5].

Due to these different approaches related to a system, the amount of knowledge extracted it is limited and can be easily regarded as grey [7, 8]. Even more, by adding the human component, through the consumers' demands and needs, strictly related to preferences, self-awareness, self-conscience, free-will, etc. the study of the knowledge that can be extracted through OSN is becoming more complicated [9].

For this, it can be identified a new type of knowledge: the grey knowledge, which is lying between the two well-known types of knowledge: the tacit and the explicit one and is continually circulating and transforming within the network. It can be encountered in the internalized and externalized feedback loops that are formed between different network users and it accompanies the external (chatting, e-mails sending, etc.) and internal (listening, watching a commercial, reading a comment, evaluation, observing, etc.) processes.

Considering the everyday activities, is can easily be seen that the grey knowledge is the most predominant type of knowledge that can be encountered and, therefore, the study of it can reveal new information that can be used in understanding the OSN's complexity.

3 Grey Clustering Analysis

Assume that there are n objects to be clustered according to m cluster criteria into s different grey classes [4, 10].

A function noted $f_j^k(\cdot)$ is called the "whitenization" weight function of the k-th subclass of the j criterion, with: $i = 1, 2, \ldots, n$; $j = 1, 2, \ldots, m$; $1 \leq k \leq s$. [11]

Considering a typical whitenization function as described by [4, 12, 13] with four turning points noted as: $x_j^k(1), x_j^k(2), x_j^k(3)$ and $x_j^k(4)$:

$$f_j^k(x) = \begin{cases} 0, & x \notin [x_j^k(1), x_j^k(4)] \\[2mm] \dfrac{x - x_j^k(1)}{x_j^k(2) - x_j^k(1)}, & x \in [x_j^k(1), x_j^k(2)] \\[2mm] 1, & x \in [x_j^k(2), x_j^k(3)] \\[2mm] \dfrac{x_j^k(4) - x}{x_j^k(4) - x_j^k(3)}, & x \in [x_j^k(3), x_j^k(4)] \end{cases} \tag{1}$$

Or the whitenization weight function of lower measure (a particular case of the typical whitenization function presented above, where the first and the second turning points $x_j^k(1), x_j^k(2)$ are missing):

$$f_j^k(x) = \begin{cases} 0, & x \notin [0, x_j^k(4)] \\[2mm] 1, & x \in [0, x_j^k(3)] \\[2mm] \dfrac{x_j^k(4) - x}{x_j^k(4) - x_j^k(3)}, & x \in [x_j^k(3), x_j^k(4)] \end{cases} \tag{2}$$

Or the whitenization function of moderate measure (also a particular form of the whitenization function, where the second and the third turning points $x_j^k(2), x_j^k(3)$ coincide):

$$f_j^k(x) = \begin{cases} 0, & x \notin [x_j^k(1), x_j^k(4)] \\[2mm] \dfrac{x - x_j^k(1)}{x_j^k(2) - x_j^k(1)}, & x \in [x_j^k(1), x_j^k(2)] \\[2mm] 1, & x = x_j^k(2) \\[2mm] \dfrac{x_j^k(4) - x}{x_j^k(4) - x_j^k(2)}, & x \in [x_j^k(2), x_j^k(4)] \end{cases} \tag{3}$$

Or the whitenization weight function of upper measure (another particular form of the whitenization function where the final third and fourth points $x_j^k(3), x_j^k(4)$ are missing):

$$f_j^k(x) = \begin{cases} 0, & x < x_j^k(1) \\ \dfrac{x - x_j^k(1)}{x_j^k(2) - x_j^k(1)}, & x \in [x_j^k(1), x_j^k(2)] \\ 1, & x \geq x_j^k(2) \end{cases} \qquad (4)$$

The grey clustering analysis can be performed by following the next steps [4]:

Step 1: Determining the form of the whitenization function $f_j^k(\cdot)$, for $j = 1, 2, \ldots, m$; $1 \leq k \leq s$;

Step 2: Attributing a cluster weight η_j to each criterion based on external information such as prior experience or qualitative analysis, with $j = 1, 2, \ldots, m$;

Step 3: Calculating all fixed weight cluster coefficients from the whitenization function $f_j^k(\cdot)$ determined at step 1, cluster weights η_j at step 2 and observational values x_{ij} of the object i for the j criterion, with $= 1, 2, \ldots, n$; $j = 1, 2, \ldots, m$; $1 \leq k \leq s$:

$$\sigma_i^k = \sum_{j=1}^{m} f_j^k(x_{ij}) * \eta_j \qquad (5)$$

Step 4: If $\sigma_i^{*k} = \max_{1 \leq k \leq s}\{\sigma_i^k\}$, then the object i is belonging to the $k*$th grey class.

Let us perform the grey clustering analysis in the case study on the OSN users in order to determine which of them are being impressed by the marketing campaigns, comments, articles, videos, etc. in the online environment and what conclusions can be drawn from studying their personal and cluster characteristics.

4   Case Study on OSN Users

For conducting the cluster analysis, a questionnaire was applied to the online social networks' users, 211 persons answering to all the addressed questions. Having the answers, a confirmatory factor analysis was accomplished in order to validate the construct, validity and reliability of the questionnaire. After proceeding this, the selected factors were passed through the grey clustering method, obtaining three relevant used categories as it will be showed in the following sections.

4.1   Questionnaire and Data

The 211 questionnaire's respondents can be divided into five age categories 104 having between 18-25 years old, 76 having between 26-35 years old, 20 having between 36-45 years old, 7 between 46-55 years old and 4 having between 56-65 years old;, 61.61% of them being female and 38.39% male. Along with the questions regarding the personal characteristics, the respondents were asked to answer to the following questions, evaluated through a Likert scale taking values between 1 and 5:

- When I want to buy a product: (DM_1)
  - I buy it immediately without hesitation;
  - I am thinking a while on this opportunity and in a couple of days I decide whether to buy it or not ;
  - I am asking for my close friends' advice;
  - I am asking for my friends and family's advice;
  - I am asking for advice from friends and family, I am searching other buyers' comments on internet and on social websites.
- In general: (DM_2)
  - I make my own decision and I stick to it no matter what happens;
  - I make a set of possible decisions, I analyse them a couple of days and after that I take my decision;
  - I have a set of possible decisions and for validation, sometimes, I ask someone else's opinion;
  - I discuss the possible decisions set with close friends/ co-workers/ family;
  - I always discuss the decisions with other people, read news and comments.
- When I cannot identify the product I am looking for: (PL_1)
  - I buy another product from the same producer, but from a different assortment;
  - I am looking for another store where I can buy my product;
  - I buy another product from a similar producer;
  - I am looking for new information that can help me in finding what I need;
  - I cannot evaluate this situation;
- When choosing a particular product/service, I am taking into consideration the following aspects: (please select among: strongly disagree; disagree; undecided; agree; strongly agree)
  - The product and availability term: (P_1);
  - Product's inner characteristics: (P_2);
  - Package characteristics: (P_3);
  - Brand awareness: (P_4);
  - Information received recently about that product: (P_5).
- Which of the following actions have you made on friends' recommendation: (please select de appropriate answer between: never, sometimes, often, usually, always):
  - I have watched a commercial: (INT_1);
  - I have looked for a product promotion campaign: (INT_2);
  - I have informed about an event of a certain company: (INT_3);
  - I have participated on a contest organized by a firm: (INT_4);
  - I have followed that company's activity on social media: (INT_5).

These questions have been divided into three categories, as it can also be observed from the labels attached to them: Decision making and product placement (DM), Product (P) and Interaction with friends on social networks (INT).

4.2   Model Fit Through a Confirmatory Factor Analysis

Having the answers to the questionnaire above, a confirmatory factor analysis was conducted in order to validate its main constructions.

The starting construction contained 13 latent factors, but due to the poor values obtained for main confirmatory factor analysis's indices such as: CMIN/DF of 4.612, GFI of 0.836, AGFI of 0.760, CFI of 0.778, NFI of 0.737, RFI of 0.670, IFI of 0.782, RMSEA of 0.131, etc. the construction from Fig. 1 has been validated and used in the next section.

Goodness of Fit (GOF)

The goodness of fit indicates how well the specified model reproduces the covariance matrix among the indicator variables, establishing whether there is similarity between the observed and estimated covariance matrices.

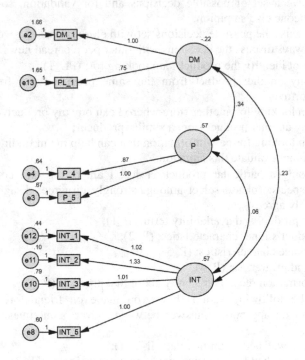

Fig. 1. Latent Construct and the Measured Variables

Table 1. Result Table (AMOS 22 Output)

Minimum was achieved
Chi-square = 25.212
Degrees of freedom = 17
Probability level = .090

Table 2. CMIN (AMOS 22 Output)

Model	NPAR	CMIN	DF	P	CMIN /DF
Default model	19	25.212	17	.090	1.483
Saturated model	36	.000	0		
Independence model	8	514.628	28	.000	18.380

One of the first measure of GOF is Chi-squares statistic through which is tested the null hypothesis that no difference is between the two covariance matrices, with an acceptance value for the null hypothesis of > 0.050. As Table 1 indicates, this value is exceeded. The improved model has a CMIN/DF of 1.483 less than the threshold value 2.000 (Table 2).

Moreover, the values of GFI and AGFI are above the limit of 0.900, recording a 0.972, respectively a 0.940 value, while CFI is exceeding 0.900 (being 0.983 – see Table 3) the imposed value for a model of such complexity and sample size. As for the other three incremental fit indices, namely NFI, RFI and IFI, the obtained values are above the threshold value 0.900 for NFI and closely to 1.000 for RFI and IFI.

Table 3. Baseline Comparisons (AMOS 22 Output)

Model	NFI Delta1	RFI rho1	IFI Delta2	TLI rho2	CFI
Default model	.951	.919	.983	.972	.983
Saturated model	1.000		1.000		1.000
Independence model	.000	.000	.000	.000	.000

Table 4. RMSEA (AMOS 22 Output)

Model	RMSEA	LO 90	HI 90	PCLOSE
Default model	.048	.000	.085	.494
Independence model	.288	.266	.310	.000

As Table 4 shows, the root mean squared error approximation (RMSEA) has a value below 0.100 for the default model, showing that there is a little degree to which the lack of fit is due to misspecification of the model tested versus being due to sampling error. The 90 percent confidence interval for the RMSEA is between LO90 of 0.000 and HI90 of 0.085, the upper bound being close to 0.080, indicating a good model fit.

Validity and Reliability

For testing the construct's validity and reliability, first of all, the standardized loadings should be analyzed and should be higher than 0.500, ideally 0.700 or higher. In this case these values are between 0.634 and 0.953, confirming the validity.

The convergent validity is given by two additional measures: the average variance extracted (AVE) and construct reliability (CR). As these two measures are not computed by AMOS 22, they have been determined by using the equations presented in the literature [14].

The following values have been obtained for P, DM and INT: AVE: 0.431, 0.527 and 0.600 and CR: 0.714, 0.793 and 0.909. An AVE of 0.500 indicates an adequate convergent validity, while a CR of 0.700 or above is suggest a good reliability. Having the obtained values, it can be concluded that the overall construct validity and reliability is good and that the considered measures are consistently representing the reality.

4.3   Grey Clustering

Using the GSTM 6.0 software, the grey cluster analysis was performed and the results are showed in the following:

*********************************Start*********************************

(1) Compute coefficients of clustering
(2) Comparison clustering coefficients, and determine the object belongs to which grey class
Objects of belonging to the grey class [1-th]: 2, 3, 7, 8, 9, 10, 12, 13, 26, 34, 38, 39, 42, 44, 45, 46, 47, 54, 55, 56, 57, 60, 73, 75, 78, 79, 80, 81, 88, 96, 97, 100, 101, 102, 103, 104, 107, 119, 122, 123, 124, 128, 129, 137, 141, 142, 143, 145, 146, 156, 159, 160, 161, 173, 174, 179, 180, 181, 188, 190, 191, 198, 199, 204,
Objects of belonging to the grey class [2-th]: 5, 23, 24, 27, 28, 29, 30, 52, 66, 69, 70, 82, 83, 91, 92, 93, 110, 113, 114, 132, 135, 136, 144, 151, 154, 163, 164, 165, 166, 171, 186, 193, 201, 202, 206, 207, 209,
Objects of belonging to the grey class [3-th]: 1, 4, 6, 11, 14, 15, 16, 17, 18, 19, 20, 21, 22, 25, 31, 32, 33, 35, 36, 37, 40, 41, 43, 48, 49, 50, 51, 53, 58, 59, 61, 62, 63, 64, 65, 67, 68, 71, 72, 74, 76, 77, 84, 85, 86, 87, 89, 90, 94, 95, 98, 99, 105, 106, 108, 109, 111, 112, 115, 116, 117, 118, 120, 121, 125, 126, 127, 130, 131, 133, 134, 138, 139, 140, 147, 148, 149, 150, 152, 153, 155, 157, 158, 162, 167, 168, 169, 170, 172, 175, 176, 177, 178, 182, 183, 184, 185, 186, 187, 189, 192, 194, 195, 196, 197, 200, 203, 205, 208, 210, 211,
*********************************End*********************************

Based on the answers received, it can be concluded that the second grey cluster is formed mostly by impressionable persons which are positively reacting to promotion campaigns in online environment and which are taking into account other's opinions when making a decision.

Considering the members of the second grey cluster, it has been established that they have an average age of 24,1 years old, their majority being formed by women, with an average number of friends on OSN of 516, who are accessing the OSN more than once a day. Also, the persons in this category are spending more than four hours per day in online social networks and actively participating in forms and discussions in online environment.

Having these information about the most impressionable members in OSN, the companies can adapt their strategies in order to deliver the new pieces of information directly to these users. From here, a specific analysis can be done for each new user in order to determine which group he belongs to.

Also, another further research direction can be the identification of the most important nodes among the ones that can easily be impressed using a grey approach similar to the one proposed by Wu et al. [15]. In this way, by knowing both the nodes that are easily to be impressed and the ones that have great influence in each network, the companies' strategies can be adapted in order to better target the OSN audience.

5   Conclusions

OSN are becoming more and more a now-a-days reality. In this context, companies have adapted their strategies in order to meet the target audience. This paper presents a method for selecting the most impressionable members of a network. For this, a questionnaire has been deployed, applied and validated for better extracting the most impressionable members. Grey clustering was used as the information flowing within the feedback loops in OSN is a grey one.

As further research, a grey relational analysis will be used for identifying the most important and influential node among the most impressionable modes within an OSN. Having this information, each company can adapt or create a specific strategy that will target this persons in order to increase and strengthen competitive position on the market.

Acknowledgments. This paper was co-financed from the European Social Fund, through the Sectorial Operational Programme Human Resources Development 2007-2013, project number POSDRU/159/1.5/S/138907 "Excellence in scientific interdisciplinary research, doctoral and postdoctoral, in the economic, social and medical fields -EXCELIS", coordinator The Bucharest University of Economic Studies. Also, the authors gratefully acknowledge partial support of this research by Webster University Thailand and by Leverhulme Trust International Network research project "IN-2014-020".

References

1. Schneider, F., Feldmann, A., Krishnamurthy, B., Willinger, W.: Understanding online social network usage from a network perspective. In: Proceedings of the ACM SIGCOMM Conference on Internet Measurement, pp. 35–48 (2009)

2. Heidemann, J., Klier, M., Probst, F.: Online social networks: A survey of a global phenomenon. Computer Networks **56**, 3866–3878 (2012)
3. Synthesio: Drive Customer Experience with Social Intelligence (e-book) (2014). http://synthesio.com/corporate/en/2014/uncategorized/drive-customer-experience-social-intelligence-ebook/
4. Liu, S., Lin, Y.: Grey Systems – Theory and Applications. Understanding Complex Systems Series. Springer, Heidelberg (2010)
5. Forrest, J.: A Systemic Perspective on Cognition and Mathematics. CRC Press (2013)
6. Delcea, C.: Not black. Not even white. Definitively grey economic systems. Journal of Grey System **26**(1), 11–25 (2014)
7. Delcea, C., Cotfas, L.-A., Paun, R.: Grey social networks. In: Hwang, D., Jung, J.J., Nguyen, N.-T. (eds.) ICCCI 2014. LNCS, vol. 8733, pp. 125–134. Springer, Heidelberg (2014)
8. Delcea, C., Cotfas, L.-A., Paun, R.: Understanding Online Social Networks' Users – A Twitter Approach. In: Hwang, D., Jung, J.J., Nguyen, N.-T. (eds.) ICCCI 2014. LNCS, vol. 8733, pp. 145–153. Springer, Heidelberg (2014)
9. Cotfas, L.-A., Delcea, C., Roxin, I., Paun, R.: Twitter ontology-driven sentiment analysis. In: Barbucha, D., Nguyen, N.T., Batubara, J. (eds.) New Trends in Intelligent Information and Database Systems. SCI, vol. 598, pp. 131–140. Springer, Heidelberg (2015)
10. Liu, S., Scarlat, E., Delcea, C.: Grey systems in economics. Theory and applications. ASE Printing house (2014) (in Romanian)
11. Ke, L., Xiaoliu, S., Zhongfy, T., Wenyan, G.: Grey clustering analysis method for overseas energy project investment risk decision. Systems Engineering Procedia **3**, 55–62 (2012)
12. Wen, K.L.: A Matlab toolbox for grey clustering and fuzzy comprehensive evaluation. Advances in Engineering Software **39**, 137–145 (2008)
13. Lin, C.H., Wu, C.H., Huang, P.Z.: Grey clustering analysis for incipient fault diagnosis in oil-immersed transformers. Expert Systems with Applications **39**, 1371–1379 (2009)
14. Spanos, Y.E., Lioukas, S.: An examination into the causal logic of rent generation: contrasting Porter's competitive strategy framework and the resource-based perspective. Strategic Management Journal **22**, 907–934 (2001)
15. Wu, J., Liu, X., Li, Y., Shu, J., Liu, K.: Node Importance Evaluation in Complex Networks Based on Grey Theory. Journal of Information & Computational Science **10**(17), 5629–5635 (2013)

Semi-supervised Multi-view Sentiment Analysis

Gergana Lazarova[✉] and Ivan Koychev

Sofia University "St. Kliment Ohridski", Sofia, Bulgaria
{gerganal,koychev}@fmi.uni-sofia.bg

Abstract. Semi-supervised learning combines labeled and unlabeled examples in order to find better future predictions. Multi-view learning is another way to improve the prediction by combining training examples from more than one sources of data. In this paper, a semi-supervised multi-view learning approach is proposed for sentiment analysis in the Bulgarian language. Because there is little labeled data in Bulgarian, a second English view is also used. A genetic algorithm is applied for regression function learning. Based on the labeled examples and the agreement among the views on the unlabeled examples the error of the algorithm is optimized, striving after minimal regularized risk. The performance of the algorithm is compared to its supervised equivalent and shows an improvement of the prediction performance.

Keywords: Sentiment analysis · Semi-supervised learning · Genetic algorithms

1   Introduction

Humans can feel emotions, express their feelings and understand each other relying on hidden patterns of sentiment. We can formulate a simple sentence as a neutral statement, but we can also impose some emotions in it, which make it an opinion.

Sentiment analysis is a modern discipline, which is responsible for extracting subjective opinions, understanding the emotions that are expressed in a text. It is a challenging interdisciplinary subject, combining algorithms from artificial intelligence, natural language processing, information retrieval and statistics.

Social media (facebook, twitter, google+, linkedin) contains large amounts of comments, articles - all expressing personal opinions. For examples, it may be of interest to know what the opinion on some topic in social media is, is it positive or negative? Is a product positively accepted by the customers? Is a football couch popular among the fans? Those all are important questions, whose answers are valuable for many contemporary businesses. To answer these questions, we need to develop tools for automatic extraction of sentiment from texts.

There has been considerable research in the field of sentiment analysis (a.k.a. opinion mining). Most of the works were done for the English language. Bing Liu [21] overviews the state of the art in different types of sentiment analysis and discusses a wide variety of practical applications. Kennedy and Inkpen [22] present two methods for extracting the sentiment from movie reviews in the English language. Vohra and Teraiya [28] present a detailed comparative study of sentiment analysis techniques.

© Springer International Publishing Switzerland 2015
M. Núñez et al. (Eds.): ICCCI 2015, Part I, LNAI 9329, pp. 181–190, 2015.
DOI: 10.1007/978-3-319-24069-5_17

Sentiment analysis in other languages, including the Bulgarian language suffers from labeled examples shortage. There is little data, which we can use in the learning process. In such cases, semi-supervised multi-view learning turns out to be helpful. Semi-supervised learning has been applied to various problems. Blum and Mitchel [1] use a two-view co-training algorithm for faculty web-page classification. The first view contains the words on the web-pages and the second – the links that point to the web pages. They use only 12 labeled examples out of the 1051 web pages and achieve an error rate of 5%. Multi-view semi-supervised learning has also been applied to image segmentation [3], object and scene recognition [11], and statistical parsing [12]. Vikas Sindhwani, Partha Niyogi, Mikhail Belkin [2] propose a co-regularization approach to semi-supervised learning. The presented in this paper approach is similar to the last one, it also uses a co-regularization framework, but the loss function is optimized via a genetic algorithm.

Other languages like Chinese also suffer from insufficient number of labeled examples. Xiaojun Wan [5] uses a co-training bilingual approach for improving the prediction on movie reviews. Similarly, in this paper English movie reviews are translated to Bulgarian so that two views are created: X = (genuine_English, translated_to_Bulgarian). These reviews are rich in content, therefore we use them as labeled examples. Because automatic machine translation to Bulgarian is not that good, a second source of information has to be used, genuine Bulgarian reviews. These reviews are translated to English and added to the pool of examples, but they form an unlabeled set of examples: X = (translated_to_English, genuine _Bulgarian). Based on the two views and both the labeled and unlabeled examples, the loss function is optimized so that a better predictor can be found.

2   Semi-supervised Learning

Recently, there has been significant interest in semi-supervised learning. It requires less human effort and achieves higher accuracy. Sometimes, labeling an example turns out to be a time-consuming process, requiring extra knowledge and skilled experts in the field. Furthermore, there are areas of research, where new labeled examples are hard to collect and in some cases it is not possible at all.

Semi-supervised learning uses both labeled and unlabeled examples. It falls between unsupervised learning and supervised learning. A "teacher" has already labeled some of the instances - D_1, for these examples the regression function has already been defined. In semi-supervised learning unlabeled instances D_2 are also used and added to the pool of training examples. The final training data contains both the examples of D_1 and D_2 $(D = D_1 \cup D_2)$. Let the number of labeled examples be l and the number of the unlabeled examples - u.

$$D_1 = \{(x_i, y_i)\}_{i=1}^{l} \quad, \quad D_2 = \{x_j\}_{j=l+1}^{l+u},$$

2.1 Semi-supervised Multi-view Learning

When there is scarce information defining the objects, new sources of information (views) are sought. Each view consists of characteristics of the object, projected onto its data source.

Let each instance X consists of multiple views: $X = (X_1, ..., X_k)$. $X_1, ..., X_k$ represent feature sets. Each feature set defines the object based on its data source. Let $f_1,...,f_k$ are the regression functions, corresponding to the views $X_1, .., X_k$. The goal is to find such a combination of learners $f_1^*,...,f_k^*$ (for views $X_1, .., X_k$) so that they can agree with one another and a final combined loss function is minimal. Multiple hypotheses are trained from the same labeled data set, and they are expected to make similar predictions on future unlabeled instances.

2.2 Error Estimation

- *Loss function* - A loss function $c(x, y, f(x)) \in [0, \infty)$ measures the amount of loss of the prediction. It quantifies the amount by which the prediction deviates from the actual values. In regression we can define the squared loss function as:

$$c(x, y, f(x)) = (y - f(x))^2$$

- *Risk* - the risk associated with f is defined as the expectation of the loss function:

$$R(f) = E[c(x, y, f(x))]$$

- *Emperical Risk* - in general, the risk $R(f)$ cannot be computed because the distribution $P(x, y)$ is unknown to the learning algorithm. However, we can compute an approximation, called empirical risk. The empirical risk of a function is the average loss incurred by f on a labeled training sample.

$$R_{emp}(f) = \frac{1}{l}\sum_{i=1}^{l} c(x_i, y_i, f(x_i))$$

- *Regularized Risk.* A regularizer $\Omega(f)$ is a non-negative function, which has a non-negative real-valued output. If f is "smooth", $\Omega(f)$ will be close to zero. If f is too zigzagged and overfits the training data, $\Omega(f)$ will be large.

$$Err(f) = R_{emp}(f) + \lambda\Omega(f)$$

In supervised linear regression: $f(x) = w^T x; \Omega_{SL}(f) = \|w\|^2 = \sqrt{\sum_{s=1}^{d} w_s^2}$

In semi-supervised learning we can define $\Omega(f)$ as the sum of the supervised regularizer and the disagreement between the learners on the unlabeled instances, multiplied by λ_2.

$$\Omega(f) = \Omega_{SL}(f) + \lambda_2\Omega_{SSL}(f) \qquad \Omega_{SSL}(f) = \sum_{u,v=1}^{k}\sum_{i=l+1}^{l+u} c(x_i, f_u(x_i), f_v(x_i))$$

- *Minimizing the regularized risk.*

For the multi-view learning framework, where we have k regression functions, we can define $Err (f_1,... f_k)$ as the sum of the regularized risk of each f_j on $view_j$, based on the labeled examples of D_1, plus the disagreement between the pairs f_u, f_v , based on the unlabeled examples of D_2.

$$Err(f_1,...,f_k) = \sum_{v=1}^{k} (\frac{1}{l}\sum_{i=1}^{l} c(x_i, y_i, f_v(x_i)) + \lambda\Omega_{SL}(f_v)) + \lambda_2 \sum_{u,v=1}^{k} \sum_{i=l+1}^{l+u} c(x_i, f_u(x_i), f_v(x_i))$$

Let $f_1^*,..., f_k^*$ be the learners for which $Err (f_1,... f_k)$ is minimal:

$$f_1^*,...,f_k^* = \arg\min_{f_1...f_k} Err(f_1,...,f_k)$$

Example: Two-views, $x=(x_1, x_2)$, semi-supervised learning, ridge regression [15]:

$$f_1(x) = w^T x_1, f_2(x) = v^T x_2,$$

$$view\_1: \Omega_{SL}(f_1) = \| w \|^2 = \sqrt{\sum_{s=1}^{d} w_s^2}; view\_2: \Omega_{SL}(f_2) = \| v \|^2 = \sqrt{\sum_{s=1}^{b} v_s^2}$$

$$(f_1^*, f_2^*) = \arg\min_{f_1, f_2} \frac{1}{l}\sum_{i=1}^{l}((y_i - w^T x_{1,i})^2 + (y_i - v^T x_{2,i})^2) +$$

$$\lambda_1 \| w \|^2 + \lambda_1 \| v \|^2 + \lambda_2 \sum_{i=l+1}^{l+u} (w^T x_{1,i} - v^T x_{2,i}) \tag{1}$$

3   Semi-supervised Multi-view Genetic Algorithm (SSMVGA)

Darwin's concept of evolution [18] have been carried over into genetic algorithms for solving some of the most demanding problems for computer optimization. Genetic algorithms are based on natural evolution and use heuristic search techniques to find decent solutions. [17].

To find the parameters of (1), we can also solve a system of linear equations. Not always, such a solution exists. In such cases other iterative procedures like the gradient descend are preferred. In order to minimize the regularized risk, in this paper a semi-supervised multi-view genetic algorithm (SSMVGA) is proposed and applied. This genetic algorithm uses two views and a common fitness function is optimized.

3.1   Fitness Function, Crossover and Mutation

The individuals used in the semi-supervised multi-view learning genetic algorithm contain chromosomes (features) from multiple views. The features of the two views are concatenated and a common chromosome is formed.

For the example in (1) we can define individual j as:

Individual$_j$:

w_{j1}	...	w_{js}	v_{j1}	...	v_{jp}

As we want to optimize the regularized risk and find its minimum a more fitted individual will have a smaller value for $Err (f_1, \dots f_k)$.

$$fitness(p) = -Err(f_1, \dots, f_k)$$

The fitter the chromosome (based on the fitness function), the more times it is likely to be selected to reproduce. Therefore, individuals with smaller regularized risk will survive, evolve and produce offspring.

Crossover of two individuals is the process of reproduction. We used uniform crossover – with probability of 0.5 newborn children have 50% of the genes of each parent with randomly chosen crossover points.

In genetics, a mutation is a change of the genome of an organism. As the individuals consist of weights, with small probability one of the weights is replaced by a new one, generated as a real-valued small number $\in[-0.5, 0.5]$.

4   Sentiment Analysis

Sentiment analysis determines the attitude of a speaker on some topic. It predicts the contextual polarity of a document. Usually, two classes are used – positive and negative examples, but rarely, we can also use a neutral class.

There are two main approaches to opinion mining [21]. The first one is based on a sentiment lexicon [26] and relies on the polarity of the words in the documents. Usually, we sum the weights of the negative and positive words and a final decision is made based on the sum. The second approach uses a classification algorithm in order to differentiate between the classes.

Modern directions in sentiment analysis are: *positive/negative opinions*; *personality traits* - (nervous, anxious, reckless, morose, hostile, jealous); *emotions* - angry, sad, joyful, fearful, ashamed, proud, elated; *detecting irony, rumor credibility* in documents; *plagiarism, author recognition*.

It should also be noted that an opinion is generally subjective. Two people might not agree on the sentiment of a sentence. Furthermore, some sentiment analysis competitions (Semantic Evaluation – "SemEval") remove from the data set the reviews the judges don't agree on.

5   Experimental Results

5.1   Dataset

The burden in sentiment analysis in Bulgarian lies in the absence of Bulgarian training data. Bulgarian classified instances are difficult to obtain. A "teacher" should by hand label all training examples.

For the experimental results movie reviews are used. Each review consists of sentences, expressing the user's attitude towards a movie. Furthermore, the user chooses the quality of the film, based on a 1-5 scale, a rating varying from 1 to 5 (*, **, ***, ****, *****). Usually, an instance in sentiment analysis is treated either "positive" or "negative". In this publication, the rating of the movie is treated as a real-valued function.

The Internet abounds in English labeled examples (Amazon, TripAdvisor, etc.). An English source of classified instances is easy to find and use. Therefore, such resources can be used for extracting labeled examples. This paper uses 992 Amazon reviews [23] which were translated with google translate to Bulgarian. Each review has two views:

$$x_i = (original\_English\_view, translated\_to\_Bulgarian\_view)\,,\ y_i = Rating$$

$$D_1 = \{(x_i, y_i)\}_{i=1}^{l}$$

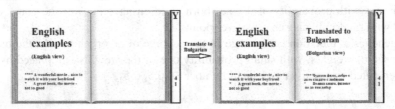

Fig. 1. Labeled examples

Unlike labeled reviews in genuine Bulgarian, a set of unlabeled examples in Bulgarian is easier to find. A set of movie reviews in Bulgarian was extracted from www.cinexio.com. It consists of 100 labeled examples, but the labels were hidden during the training process, the labels were used only for the evaluation process. This labeled set can also be added to the pool of labeled examples but this is not the purpose of this paper, usually we don't have labeled instances in Bulgarian. These Bulgarian reviews were translated to English, so that a second view is formed.

$$x_j = (translated\_to\_English\_view, original\_Bulgarian\_view)$$

$$D_2 = \{x_j\}_{j=l+1}^{l+u}$$

Fig. 2. Unlabeled examples

Construction of the training and test sets:

- The training set contains both the labeled and unlabeled examples $D = D_1 \cup D_2$.
- The test set contains all the examples in D_2. We will compare the algorithm to its supervised equivalents based on these examples (using the original known labels). Because D_2 contains only unlabeled examples, we evaluated the performance of the algorithm based on D_2, without dividing the set of unlabeled examples into bins and performing a cross-validation process. Cross-validation will not affect the supervised algorithms at all, they use only the labeled examples.

5.2   Gate and Big Data (Apache Hadoop and Apache Spark)

Gate [27] is a NLP framework, solving text processing problems. In order to extract features from the Bulgarian and English views the following pipeline was created: *Document Reset* – it enables the document to be reset to its original state, by removing all the annotation sets and their contents; *ANNIE Tokeniser* – splits the text into very simple tokens; *Snowball stemmer /BulStem – English/ Bulgarian stemmer* - reduces derived words to their word stem or root form; *Groovy Script* – self-written script for biword *tf-idf* feature extraction.

- Term frequency - the number of times that term t occurs in a document d:
  $$tf(t,d)$$
- Inverse document frequency - it measures whether the term is common or rare across all documents (N is the number of documents in D):

$$idf(t,D) = \log \frac{N}{|\{d \in D, t \in d\}|}$$

- Weights: $w_t = tfidf(t,D) = tf(t,d) * idf(t,D)$

17099 features were extracted from the Bulgarian view and 12391 – from the English.

Big data implies large amount of data, which can't be processed by a single machine and its resources. Not only can't it be stored physically, but there is no execution power to afford the learning and prediction processes.

Recently, Apache Hadoop [24] has gained popularity and turned out to be applicable in such areas as NLP. The hadoop distributed file system (*hdfs*) as well as its cluster management – yarn makes it really shine, penetrating important business fields worldwide.

Furthermore, for the experimental results a modern Apache Spark [25] framework was used - running on *hdfs* and using *yarn*. Spark supports three languages – scala, python and java. For the application implementation python was chosen.

A cluster of 2 machines was connected and used for the evaluation of the algorithms, both running under LINUX Mint.

5.3   Root Mean Squared Error (RMSE)

After minimizing (1) with the SSMVGA, we find the weights of the two views. The final prediction of a new example is done based on the errors of the k views (k is the number of views):

$$f(x) = \arg \min_{y \in Y} \sum_{i=1}^{k} c(x, y, f_i^*(x))$$

For each potential class we sum the errors of the two learners – f_1^* and f_2^* and return the rating for which the sum is minimal. This is the prediction based on the two views. Therefore, we can compare this algorithm to its supervised equivalents using a standard statistics measure – RMSE.

RMSE measures the differences between the predicted values and the observed ones with respect to all examples in the test set, averaged.

$$RMSE = \sqrt{\frac{\sum_{i=1}^{n} c(x_i, y_i, f(x_i))}{n}} = \sqrt{\frac{\sum_{i=1}^{n} (y_i - f(x_i))^2}{n}}$$

5.4   Comparison

We performed 5 experiments. Our SSMVGA is compared to four supervised genetic algorithms. All genetic algorithms use the same set of labeled examples, the same parameters – number of iterations and mutation rate. Each genetic algorithm is executed 100 times and the best performing is chosen for final comparison.

- *Supervised – English* (English supervised single-view genetic algorithm - *ESGA*): only the English view is used. For learning the original English reviews are directly used and for prediction the Bulgarian test set is translated to English and tested. No unlabeled examples are used.
- *Supervised – Bulgarian* (Bulgarian supervised single-view genetic algorithm - *BSGA*) – only the Bulgarian view is used. For learning the original English reviews are translated to Bulgarian and used, for prediction the Bulgarian test set is directly tested. No unlabeled examples are used.
- *Supervised 1-view* (supervised single-view genetic algorithm - *SGA*) – the features of the two views are concatenated in one view and used for learning and prediction. No unlabeled examples are used.
- *Supervised 2-view* (supervised multi-view genetic algorithm - *SMVGA*) – both the English and the Bulgarian views are used for learning and prediction, no unlabeled examples used.
- *Semi-supervised 2-view* (semi-supervised multi-view genetic algorithm - *SSMVGA*) – both the English and the Bulgarian views are used for learning and prediction, unlabeled examples are used as well. This is the proposed in this paper algorithm.

Table 1. Experiments: description

Algorithm	Bulgarian view	English view	Labeled examples	Unlabeled examples
Supervised 1-view (*SGA*)	V	V	V	
Supervised 2-view (*SMVGA*)	V	V	V	
Supervised – Bulgarian (*BSGA*)	V		V	
Supervised – English (*ESGA*)		V	V	
Semi-supervised 2-view (*SSMVGA*)	V	V	V	V

It can be seen from Table 2. that the English supervised GA is the weakest, the reason might lie in poor translation from Bulgarian to English. It can be seen that it worsens the prediction of both the Supervised 1-view and Supervised 2-view GAs. The multi-view supervised 2-view GA (RMSE = 2.84) performs better than its

1-view counterpart (RMSE=3.09). They are both better than the single English GA (RMSE=3.12).

The 2-view semi-supervised GA (RMSE = 2.16) manages to outperform the single Bulgarian GA, because it also relies on unlabeled examples.

The parameters of all GAs are as follows:

MAX_ITER defines the stopping criterion of the algorithm.

$MAX\_ITER = 10000, \lambda = 0.5, \lambda_2 = 0.5, N = 100,$ Mutation rate : 5% .

Table 2. Comparison of the algorithms

Algorithm	RMSE
Supervised 1-view (*SGA*)	3.09
Supervised 2-view (*SMVGA*)	2.84
Supervised – Bulgarian (*BSGA*)	2.20
Supervised – English (*ESGA*)	3.12
Semi-supervised 2-view (*SSMVGA*)	**2.16**

6   Conclusions

The Bulgarian language and sentiment analysis applied to it is a little researched area, where insufficient information has been a burning problem. In this area, there are not enough labeled examples. To overcome these difficulties, an innovative semi-supervised multi-view genetic algorithm for movie review sentiment analysis was proposed. Because, the World Wide Web abounds in English labeled examples, a second view in English was also used. The labeled instances were automatically translated to Bulgarian with google translate. Furthermore, unlabeled examples in genuine Bulgarian were also used for the learning process. These unlabeled reviews were, on the other hand, translated to English. The results from the previous section show good results. The semi-supervised multi-view algorithm outperforms its supervised competitors.

For the error minimization process a genetic algorithm was used. Because genetic algorithms are iterative procedures, taking into consideration the large number of features and number of examples, modern technologies like Hadoop and Spark played a major role in lowering the execution time.

For future research more domains should be explored. Another future application of the algorithm will be using labeled and unlabeled examples from different domains.

References

1. Blum, A., Mitchell, T.: Combining labeled and unlabeled data with co-training. In: Proceedings of the Eleventh Annual Conference on Computational Learning Theory, COLT 1998, New York, NY, USA, pp 92–100 (1998)
2. Belkin, M., Niyogi, P., Sindhwani, V.: Manifold Regularization: A Geometric Framework for Learning from Labeled and Unlabeled Examples. Journal of Machine Learning Research **7**, 2399–2434 (2006)

3. Lazarova, G.A.: Semi-supervised image segmentation. In: Agre, G., Hitzler, P., Krisnadhi, A.A., Kuznetsov, S.O. (eds.) AIMSA 2014. LNCS, vol. 8722, pp. 59–68. Springer, Heidelberg (2014)
4. Mitchell, M.: An Introduction to Genetic Algorithms. MIT Press, Cambridge (1996)
5. Xiaojun, W.: Bilingual Co-training for Sentiment Classification of Chinese Product Reviews. Computational Linguistics 37(3), 587–616 (2011)
6. Banea, C., Mihalcea, R., Wiebe, J.: A bootstrapping method for building subjectivity lexicons for languages with scarce resources. In: Proceedings of the International Conference on Language Resources and Evaluations (LREC 2008), Marrakech, Morocco (2008)
7. Zhu, X., Goldberg, A.: Introduction to Semi-Supervised Learning. Synthesis Lectures on Artificial Intelligence and Machine Learning. Morgan & Claypool Publishers (2009)
8. Chapelle, O., Scholkopf, B., Zien, A.: Semi-supervised Learning. MIT Press (2006)
9. Zhu, X., Ghahramani, Z., Lafferty, J.: Semi-supervised learning using Gaussian fields and harmonic functions. In: The 20th International Conference on Machine Learning (2003)
10. Balcan, M., Blum, A., Yang, K.: Co-training and expansion. Towards bridging theory and practice. In: Saul, L.K., Weiss, Y., Bottou, L. (eds.) Advances in Neural Information Processing Systems, vol. 17, Cambridge, MA (2005)
11. Han, X., Chen, Y., Ruan, X.: Multi-class Co-training Learning for Object and Scene Recognition. MVA, 67-70 (2011)
12. Sarkar, A.: Applying co-training methods to statistical parsing. In: Proceedings of the 2nd Meeting of the North American Association for Computational Linguistics, Pittsburgh, PA, pp. 175-182 (2001)
13. Belkin, M., Niyogi, P.: Semi-supervised Learning on Riemannian Manifolds. Machine Learning 56, 209–239 (2004)
14. Nigam, K., McCallum, A., Thrun, S., Mitchell, T.: Text classification from labeled and unlabeled data. Machine Learning 39(2/3) (2000)
15. Tikhonov, A.: Solution of incorrectly formulated problems and the regularization method. Translated in Soviet Mathematics 4, 1035–1038 (1963)
16. Whitley, D.: Applying Genetic Algorithms to Neural Network Problems, p. 230. International Neural Network Society (1988)
17. Goldberg, D.: Genetic Algorithms in Search, Optimization, and Machine Learning (1989)
18. Darwin, C.: On the Origin of Species (1859)
19. Forbes, N.: Imitation of life (2004)
20. Lazarova, G., Koychev, I.: A Semi-supervised multi-view genetic algorithm. In: Proc. of the International Conference on Artificial Intelligence, Modelling & Simulation (AIMS), Madrid (2014)
21. Bing, L.: Sentiment Anlaysis and Subjectivity, 2nd edn. Invited Chapter for the Handbook of Natural Language Processing (2010)
22. Kennedy, A., Inkpen, D.: Sentiment classification of movie reviews using contextual valence shifters. Computational Intelligence 22, 110–125 (2006)
23. McAuley, J., Leskovec, J.: Hidden factors and hidden topics: understanding rating dimensions with review text. RecSys (2013)
24. Holmes, A.: Hadoop in practice (2014)
25. Pentreath, N.: Machine Learning with Spark (2015)
26. Taboada, M., Brooke, J., Tofiloski, M., Voll, K., Stede, M: Lexicon-based Methods for Sentiment Analysis. Journal of Computational Linguists (2010)
27. Cunningham, H., Maynard, D., Bontcheva, K.: Text Processing with GATE (2011)
28. Vohra, M., Teraiya, J.: A comparative study of sentiment analysis techniques. Journal JIKRCE 2(2), 313–317 (2013)

Modelling and Simulating Collaborative Scenarios for Designing an Assistant Ambient System that Supports Daily Activities

Sameh Triki[1,4(✉)], Chihab Hanachi[2,4], Marie-Pierre Gleizes[1,4], Pierre Glize[1,4], and Alice Rouyer[3,5]

[1] Université Paul Sabatier, Toulouse III, Toulouse, France
{Sameh.Triki,Gleizes,Pierre.Glize}@irit.fr
[2] Université Toulouse Capitole, Toulouse I, Toulouse, France
Chihab.Hanachi@irit.fr
[3] Université Toulouse Jean Jaurès, Toulouse II, Toulouse, France
Rouyer@univ-tlse2.fr
[4] Institut de Recherche En Informatique de Toulouse (IRIT), Toulouse, France
[5] Laboratoire Interdisciplinaire Solidarités, Sociétés, Territoires (LISST), Toulouse, France

Abstract. This paper presents the design phase of a work aiming at designing and developing a smart living device for seniors to assist them in their daily outdoor activities. We follow a participative design approach based on scenarios in order to design a socially-adapted device that will be useful to improve seniors' life. To specify our system, we first provide an UML scenario metamodel to abstract all the concepts involved in our system collaborative functioning (interactions with stakeholders, environment ...). This metamodel is used to generate different scenarios in order to better define future users 'needs and the system requirements and notably its behavior (represented with BPMN and Petri Nets). A scenario generator has been implemented for that purpose. Finally, we show how to simulate and analyze those generated scenarios using process mining techniques.

Keywords: Scenario structure model · Daily activity scenarios · Scenarios generator · Process mining

1 Introduction

The increasing population of elder people requires that more actions should be done in order to improve their quality of life [1, 2]. One possible means in this way is to provide them with tools that assist them. Nowadays, the technological advances have led to an explosion of the number and functionalities of electronic devices. So far, a considerable amount of progress is made to help and assist seniors in their life such as home monitoring [3], fall detection[4], helpline, geo-location gadgets but all of them are either designed for indoor care or limited to a defined zone [3,4,5]. Moreover, we can notice two main insufficiencies. Firstly, the devices targeting outdoor are not yet

© Springer International Publishing Switzerland 2015
M. Núñez et al. (Eds.): ICCCI 2015, Part I, LNAI 9329, pp. 191–202, 2015.
DOI: 10.1007/978-3-319-24069-5_18

on the market and still on the experimental stage. Second, results of technological devices are often very unsatisfactory as well as socially inappropriate since they are designed in an ad-hoc way to help a senior with a specific level of frailty and in a particular environment. Furthermore, the study of their acceptance by researchers in Humanities and Social Sciences (HSS) is only performed after the device is operational.

This paper presents the design phase of a work aiming at designing and developing a usable and acceptable living device for elderly to assist them in their daily outdoor activities. This technical device, called Sadikikoi, should assist cognitive disabled people in their daily activities outdoors either through notifications and suggestions to reorder their schedule or to send alerts if troubled situations (lost, fall…) are detected depending on the context. We argue that the building of such a "well thought" technical device, should involve users and their surroundings, HSS and medical experts during all the design and development process. We also believe as demonstrated by Mitzner and Rogers that involving seniors in the design process would increase the system acceptance rate [6]. These observations lead us to follow a user-centered approach. Among the different user-centered approach, we can mention: participative design, contextual design and emphatic design [7]. The work presented in this paper, part of the Compagnon project[1], follows a participative and multidisciplinary design approach based on scenarios. Scenarios express multi-point of view use cases to capture requirements [8]. More precisely, in our context, a scenario describes the succession of the system actions to deal with possible events and activities of a user in his/her daily life. In our project, users, computer scientists and sociologists are involved.

Indeed, the users want to be in the heart of the process and like to be involved at every stage of the project to validate it in compliance with their needs to increase self-esteem and limit their dependencies to the device. The choice of scenario is justified by two reasons. It first helps to take into account several points of view and therefore builds a common language to increase understanding and sharing between the stakeholders. Also, scenarios can be used for different steps of the life-cycle of the system to be built. The paper describes an UML scenario metamodel and shows how it can be used to generate new and more complex scenarios. The generated scenarios are then used to simulate, test and validate different system behaviors, represented in Petri Nets derived from process-mining techniques.

The paper is organized as follows. Section 2 details our scenario based approach: its life-cycle and a guiding example of a scenario. Section 3 shows the proposed scenario metamodel that represents the scenario structure. It is be used to gather user needs through participatory design and also to create a scenario generator. Section 4 shows how to derive the system behavior (represented with Petri Nets) from generated scenarios using process mining techniques. Section 5 illustrates the implemented scenarios generator and its usefulness. Section 6 briefly compares our proposition to related works and concludes the paper.

---

[1]   The Compagnon project is funded by the Midi-Pyrénées Region. The system to be build is called Sadikikoi

2 A Scenario Based Approach

2.1 The Approach Life-Cycle

Now that we clearly choose to follow a participatory design, we should define how we should proceed to design and develop our system.

The life-cycle approach, given in Fig. 1, describes only the system requirement and the design phases concerned in this paper.

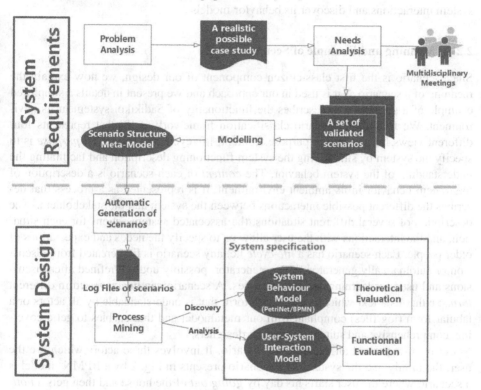

Fig. 1. The proposed approach

System Requirement Phase: we give a first realistic case study in a textual form, as a start point presentation of our system functioning, to stakeholders. Then, we start the needs analysis step with stakeholders (HSS and medical experts) through several multidisciplinary meetings in order to define and validate several possible scenarios. The main focus of this first phase is not to provide occurrences of scenarios but rather their structure. Moreover, through those collaborative brainstorming, we create simple (i.e. can be understood by any end user) and complete (i.e. involves all the elements needed to design and develop our system) scenarios. From those scenarios, we derive a scenario metamodel also discussed and validated by the same multidisciplinary study group. These scenarios represent a succession of user daily possible outdoors activities and system actions. As we chose to use a user-centered design based on scenarios, different scenarios representing different situations are required to better gather users' needs and define our system functional requirements.

Design Phase: we start by generating and producing a larger and more diverse amount of scenarios using a scenario generator that is implemented with regard to the proposed metamodel. Those generated scenarios are be used, on the first hand, to gather future user's needs and preferences through multidisciplinary meeting, and on the other hand, to generate event logs as an input for process mining: scenario being modeled as process. This mining step, using the ProM[2] tool, allows us to analyze the system interactions and discover its behavior models.

2.2   Meaning and Example of Scenario

Since scenario is the first class-citizen component of our design, we now explain the meaning of a scenario as it is used in our approach and we present in details a simplified example of a scenario that describes the functionality of Sadikikoi system in its environment. We follow the scenario classification framework of [8] that represents four different views for a scenario: purpose, content, life-cycle and form. The *purpose* is to specify the system by simplifying the system functioning description and facilitating the understanding of the system behavior. The *content* of each scenario is a description of the system behavior in its ambient environment. It is represented as a process that describes the different possible interactions between the system and the stakeholders. The description of several different situations, the associated system actions for each situation, and the interactions with the user allow us to specify the needs and expectations of older people. Each scenario has a *life-cycle* i.e. any scenario is first created from discussion or automatically generated by our generator, possibly updated/refined after discussions and used for learning and test purposes. A scenario can be expressed in different *forms*: either as a text which is a standard form that is understandable by all actors or a tabular form (log files) compliant with our metamodel and that enables to get a covering, comprehensive and suitable format for designers.

Let us now present an example of scenario. It involves three actors which are the user, the family and the system. The scenario presents in Fig. 2 by a BPMN[3] model is a scenario where the user starts his day by *going out* of the house and then gets a *notification* from the system to take an umbrella because it's raining. Then, the user goes to *buy a newspaper*. He receives a *goal reminder* message that describes its current planning so that he won't forget his appointments (i.e. doctor, meeting...). The user then goes *shopping* but suddenly he falls down. The system Sadikikoi detects this abnormal situation because the user stops moving for a certain period of time. Even though it is an abnormal situation, the system cannot be sure if the user is just taking a break or if he *fell down* so it *sends a message* to the user and seeks for a response.

---

[2] ProM is a generic open-source framework for implementing process mining tools in a standard environment. All information concerning this extensible framework can be found on the URL: *http://www.processmining.org*.

[3] Business Process Model and Notation is a graphical representation for specifying business processes in a business process model

Because the user *ignores the message*, the system reinforces the fact that an abnormal situation occurred. So, the system *sends a message* to inform *the user's family* of the situation. The family *comes to rescue* meanwhile the system continues to *call the user* because this action would help the user to get his consciousness back or may alarm people around him so they can help. Finally, the user *goes back home*.

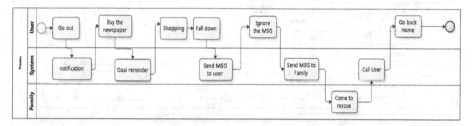

Fig. 2. A descriptive workflow of a scenario example

Obviously, this is a simplified version of a possible scenario. Real scenarios are more complex and contain more interactions between actors and the environment. In order to produce those examples, many interactions leading to several updates have been done either because something was missing or for deleting possible ambiguities. Because we work in an interdisciplinary context, including HSS, medicine and computer science thanks to scenarios a common language among them was created. After we got several validated possible scenarios, we define a scenario metamodel that describes all the required elements to create a complete and comprehensive scenario.

3   A Scenario Metamodel

By formalizing the scenario structure, we will be able to generate more different scenarios that may be presented in different forms (tabular, textual…) without missing an important detail or interaction.

The proposed scenario metamodel describes the key elements and their relationships required to describe the functionalities and behavior of an ambient system that assists seniors into their daily outdoor activities.

This model enables the instantiation of various possible daily scenarios.

Fig. 3 shows the conceptual metamodel, which represents the scenario structure, as an UML package diagram. We have a model composed of three packages: Stakeholder Description, Context Description, and System Action.

1.   *The Stakeholder Description* package contains all the information about the human being involved in the system: main user (person with cognitive impairments), preferences, calendar with these various constraints and information about other actors involved in the system (e.g. godfather, family, neighbours ...).

2.   *The Context Description* package concerns the different variable elements over time and space. It includes all the user activities, unexpected events and the space-time environment in which these actions take place. It describes the observations and various perceptions of our system.

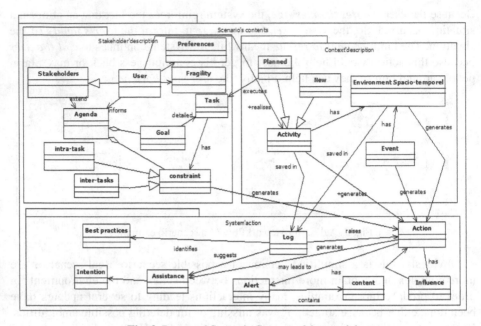

Fig. 3. Proposed Scenario Structure Metamodel

3.   *The System Action* package describes the possible actions (suggestions, alerts) of the system and the various interactions between them.

Based on the above scenario structure metamodel, several scenarios where instantiated in order to validate the completeness and accuracy of our proposed model. Some of the concepts used in our proposed model are defined in order to remove any possible ambiguity and provide a better understanding. These concepts are complex and difficult to formalize because their definitions may vary depending on the context of their use.

-TASK: it represents the tasks set by the user in his calendar. These tasks are not necessarily well detailed but can be just goals as they can be well defined tasks with a fixed duration and precise location. So, the task presents the planning of an activity

-CONSTRAINT: it describes a restriction that may have one of these two types: 1.Intra-task constraint that contains constraints within the task itself such as fixing the place and time of its realisation. 2.Inter-task constraint that describes the scheduling between tasks (example: such as TaskA should be done before TaskB).

-ACTIVITY: it represents the realisation of a planned task or an unexpected task. The succession of activities describes the user daily activities outdoors.

-EVENT: the event can be defined through a behavior (e.g. hit someone) or a signal observed in the environment (e.g. it starts to rain) or even a signal from the system itself (e.g. no battery). All these types of events are interpreted by the system perceptions (i.e. data recorded by the sensors of the system). Most events have the characteristic of being short in time, in other words, they usually occur suddenly and in an unexpected way. They can be detected by a change in the value of a variable or the appearance of a new variable in a given instant. An event can cause different impacts (positive or negative) according to its type and its spatiotemporal environment.

-ENVIRONMENT: the environment taken into account is a temporal and geographic location compliant with the definition of Salembier and al. "The environment describes a stable structure that describes the position with respect to time and space". [9]

-ACTION: it describes the actions of the system. Each action is defined by a type, level and a recipient. Moreover, each action has an impact / effect on the recipient. An action is described by a state change of system effectors (e.g. the action "Send an unobtrusive signal to the user" can be represented by changing the state of the effector from the stat {vibrator = OFF}) to the {vibrator = ON}).

4   Building the System Behavior by Means of Process Mining

The idea here is to synthesize several scenarios in a single process able to play each scenario. The process produced represents a possible model of the system behavior including the interactions with users and the environment.

In order to use process mining technique, the generated scenarios should be converted into event logs which are used as input of the process mining algorithm called alpha [10]. We create a log file including all the generated scenarios and we deduce a single process model, represented by Petri Nets, synthesizing all the possible combinations. When we apply the alpha algorithm on real scenarios, we got a complex diagram ("spaghetti" like, see Fig. 7) difficult to present in this paper for clarity reasons. Also, we choose to illustrate the results through basic scenarios. For this, let us consider the tasks shown in the Table 1.

Table 1. Description of used tasks in the described scenario

TASK	DESCRIPTION	ACTORS	INTERACTIONS	TASK	DESCRIPTION	ACTORS	INTERACTIONS
A	Go out	User	--	I	Send MSG	System	User
B	notification	System	User	J	Respond to the MSG	User	--
C	Buy the newspaper	User	--	K	Go to the pharmacy	User	--
D	Goal reminder	System	User	L	Ignore MSG	User	--
E	Go to the library	User	--	M	Call User	System	User
F	Walk out	User	--	N	Send MSG	System	Family
G	Shopping	User	--	O	Come to rescue	Family	User
H	Fall down	User	--	P	Go back home	User	--

From these tasks, six different scenarios were generated (ABCDEHIJKP, ABCDFHILMNOP, ABCDGHILNOMP, ABCDEHILNOMP, ABCDFHIJKP, ABCDGHILMNOP). Using the PROM tool, the alpha algorithm applied to these six

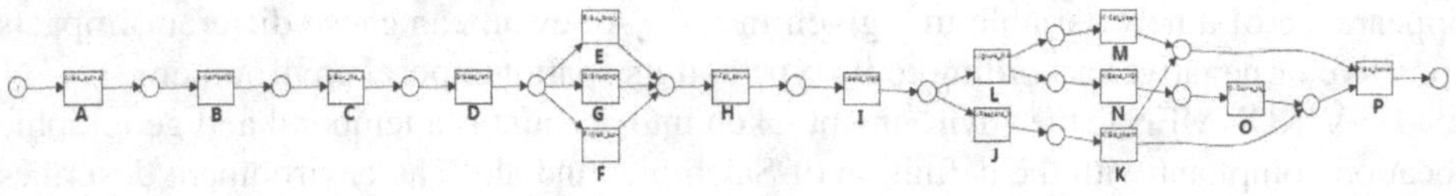

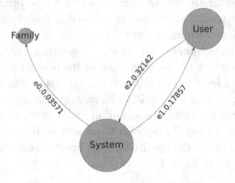

Fig. 4. Petri Net representation of the system behavior derived from scenario examples

scenarios produces the Petri Net shown in Fig. 4. Let us notice that it is a dynamic model that can execute and simulate all the six scenarios previously defined.

In this model, we could differentiate the common repeated elements of the scenarios, possible parallelism and choice actions. This diagram gives us a clear picture of our system behavior in its environment. We also apply a social network analysis on the event log, as shown in the sociogram (Fig. 5) that visualizes the different interactions between the actors of the scenarios.

Fig. 5. The social network of the scenarios examples

As you can see, the size of each actor is proportional to the number of his actions and we can clearly see that the user keeps a reasonable level of autonomy. By analyzing those resultant diagrams (Petri net, sociogram…), we build a better understanding of our system behavior and the different required interactions. Moreover, they could be used for future formal validation purposes.

5   Principles of the Scenario Generator

The fact that our aim is to develop an intelligent device that adapts to its environment by learning, the generation of test cases is crucial to forge the learning experience of Sadikikoi system and also to evaluate our system later by analysing its behaviors in response to different scenarios. As the Sadikikoi system is targeting persons with cognitive deficiencies, it would be difficult to evaluate our assistant system by targeted users because it could be inconvenient and dangerous especially in an early stage of the development. So, the idea is that a scenario generator capable of simulating user's activities along with our system actions is needed in order to investigate our system learning capabilities and its efficiency in a dynamic and evolving environment.

From the scenario metamodel used to generate several scenarios understandable from a user point of view, we extract a subset of components needed to describe scenarios from a system point of view. Giving the fact that the generator concerns the dynamic part of the model, the stakeholder' description package is not represented in the generator. We select the following relevant elements:

— **Perceptions** which represent all recorded data and collectible by the system via sensors, external applications or internal databases. The generator generates perceptions that represent the system inputs that define the triggers of its actions. The succession of perceptions describes the progress of user daily activities through time.

— **Tasks** which define the end of the realization of a planned task in compliance with the scheduling constraints in the planning.

— **Actions** which represent the outputs of the system. An action can be a notification, a suggestion or an alert that is executed by an effector (e.g. screen, high speaker…) or an external application.

The generator goal is to create numerous virtual scenarios for a given period of time (e.g. a day, a week…) in order to evaluate the learning capabilities. A scenario is a set of mini-scenarios. A mini-scenario is a group of perceptions and actions defining a meaningful situation such as a realization of an activity or a disorientation situation. The generator works in two steps. The first one generates a set of mini-scenarios. The second one aggregates mini-scenarios to form scenarios for a given period of time. Any element of a mini-scenario may appears several times in one or different scenarios.

Fig. 6 shows a succession of perceptions, ends of activities and system actions constituting a scenario. The associated semantic is given by the scenarist and not considered by the generator. We can notice chronological changes in user's actions/reactions, in the dynamic environment in which he is located and in the system interactions.

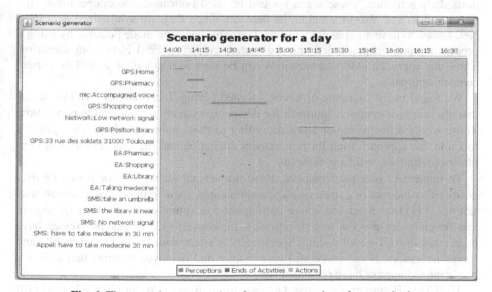

Fig. 6. The scenario generator interface: representation of a scenario day

Many parameters can be modified before generating the scenarios such as the period, which determines the total number of generated scenarios, the frailty of the user with regard of his autonomy level (determining the number of system actions in scenarios). When we applied the process mining, on the generated scenarios partially represented by fig 7, we get the Petri Net diagram (spaghetti form, shown in Fig.8).

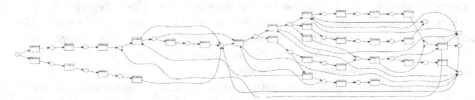

Fig. 7. The diagram discovered from the generated scenarios

For clarity reasons, we choose to represent only a three days period. We can see the complexity of the resultant discovery diagram. Despite its complexity, we can identify the emergence of repeated parts of the process and this could help us to analyse the system behavior and even facilitate the comparison between what should happen and what actually happened later on the evaluation phase of the system.

6   Discussion and Conclusion

This paper has presented a scenario metamodel enabling the generation of collaborative scenarios for defining the functionalities of an assistant ambient system that supports daily activities. These scenarios can be used both to define cooperatively the target system and also to simulate its behavior since they are synthetized as a Petri Net, known to have an operational semantics. This has been made possible by using a process mining technique deriving the system behavior (Petri Net) from scenarios. Thus, this approach based on scenarios can be seen as a guideline to follow a user-centered approach.

We found in the literature different Assisted Living Technologies (ALT) systems that also use user-centered approach for designing systems for vulnerable population. Those works differ from our work according to three points of views: the system's function, the approach used for requirement elicitation and the way scenarios are exploited (definition, structure, diversity...).

To implement user participations, there are several techniques for gathering user needs either indirect (e.g. interviews, questionnaires, focus groups, observations, user testing...) [11] or direct and active participation of future users all along the design and development process (e.g. meeting, workshops...) [12]. Many progresses have been made in ALT-based systems specifically in smart home projects and mobile and wearable sensors. Some successful case studies and deployed systems that follow a user-centered approach are presented below.

The COACH system (Cognitive Orthosis for assisting aCtivities in the Home) [13] supports users with activities of daily living at home. The process used in this work is interesting because diverse techniques are used such as group discussion, collaborative brainstorming, animated videos scenarios, paper prototypes. In comparison to our Sadikikoi system, in the COACH system neither the scenarios nor the process have been formalized. Moreover, it has been designed through limited number of scenarios (corresponding to lived experiences) indoor while we generate a covering set of scenarios for outdoors practices. The Home Care Reminder System [14] that uses a user-centered process involving formative co-design sessions with six groups of older users. They used interactions with both paper-based prototypes and prototypes running on mobile devices. The similarity with our work is that they used an iterative and inclusive process as we do, but they do not cover all the design life-cycle and no formal scenario structure is provided.

The closest work to ours is the "KITE" project (Keeping In Touch Everyday) [15] that implements a solution on notepad and armband to locate and to keep in touch and communicate with people with dementia for promoting their independence. The interesting point in this successful project is that it involves users with dementia in all the stages of the participatory design process: scoping stage, participatory design, prototype development. Scenarios are built in cooperation with users and based on the real-life experiences. However, the scenarios are not formalized and limited to the stories gathered through the interactions with users while in our case we give a formal and validated model of scenario from which we are able to derive a covering and probable set of scenarios. Moreover, two steps of our process are automated: generation of scenarios and derivation of the system behavior from the scenarios. Finally, our scenarios are a first class-citizen component of Sadikikoi design that is used in the different steps of our process life cycle from system requirement to evaluation.

In future works, we should first start the design of the interface of the Sadikikoi system in collaboration with users. Also, the learning module should be implemented before making tests in a real setting.

Acknowledgement. We would like to acknowledge both the Midi-Pyrénées Region and the IRIT laboratory that funded the Compagnon project.

References

1. United Nations: Population Division. World Population Ageing. Department of Economic and Social Affairs (2013). http://www.un.org/en/development/desa/population/publications/pdf/ageing/WorldPopulationAgeing2013.pdf (last consulted April 01, 2015)
2. Fondation of France. 2010-2014: the French increasingly only. http://www.fondationdefrance.org/Outils/Mediatheque/Etudes-de-l-Observatoire?id_theme=11344 (last consulted April 01, 2015) (in French)
3. Chan, E.C.M.: Est'eve, D.: Assessment of activity of elderly people using a home monitoring system. Int. J. Rehabil. Res. **28**(1), 69–70 (2006)
4. Aghajan, H., Augusto, J.C., Wu, C., McCullagh, P., Walkden, J.-A.: Distributed vision-based accident management for assisted living. In: Okadome, T., Yamazaki, T., Makhtari, M. (eds.) ICOST. LNCS, vol. 4541, pp. 196–205. Springer, Heidelberg (2007)

5. Rashidi, P., Mihailidis, A.: A Survey on Ambient-Assisted Living Tools for Older Adults. IEEE J. BH **17**(3), 579–590 (2013)
6. Mitzner, T., Rogers, W.: Understanding older adults' limitations and capabilities, and involving them in the design process. In: Abstracts on CHFCS, Atlanta, GA, USA (2010)
7. Alaoui, M.: Application of a Living Lab approach to the development of social TV services for the elderly. Ph.D.: networks, knowledge and organizations. University of Technology of Troyes: Institut Charles Delaunay. (2013) 142p (in French)
8. Rolland, C., Ben Achour, C., Cauvet,C.: A proposal for a scenario classification framework. Requirements Engineering, 23–47 (1998)
9. Salembier, P., Dugdale, J., Frejus, M., Haradji, Y.: A descriptive model of contextual activities for the design of domestic situations. In: Proceedings ECCE 2009 (2009)
10. Aalst,V.D.: Process mining: discovery, conformance and enhancement of business processes, p. 368. Springer Science & Business Media (2011)
11. Muller, M.J.: Participatory design: the third space. In: HCI 2010: Human-Computer Interaction: Development Process, pp. 1051–1068 (2003)
12. Newell, A.F., Gregor, P.: User sensitive inclusive design': in search of a new paradigm. In: Proceedings CUU 2000, New York, NY, USA, p. 39–44 (2000)
13. Hwang, A.S., Truong, K.N., Mihailidis, A.: Using participatory design to determine the needs of informal caregivers for smart home user interfaces. In: Proceedings of the 6th International Conference on Pervasive Computing Technologies for Healthcare, San Diego, California, USA (2012)
14. McGee-Lennon, M., Smeaton, A., Brewster, S.: Designing home care reminder systems: lessons learned through co-design with older users. In: Proceedings of the 6th ICPCTH, San Diego, California, USA (2012)
15. Robinson, L., Brittain, K., Lindsay, S., Jackson, D., Olivier, P.: Keeping In Touch Everyday (KITE) Project: Developing Assistive Technologies with People with Dementia and Their Carers to Promote Independence. International Psychogeriatrics **21**, 494–502 (2009)

Regression Methods in the Authority Identification Within Web Discussions

Kristína Machová[✉] and Jaroslav Štefaník

Department of Cybernetics and Artificial Intelligence,
Technical University, Letná 9, 042 00, Košice, Slovakia
kristina.machova@tuke.sk, jaroslav.stefanik@student.tuke.sk

Abstract. The paper describes the problem of authority identification within web discussions solving using linear and nonlinear regression methods. The goal is to find an approximation of dependency of the authority value on variables representing parameters of the structure and particularly the content of selected web discussions. The approximation function can be used at first for computation of the authority value of a given discussant, at second, for discrimination of an authoritative discussant from non-authoritative contributors to the web discussion. This information is important for web users, who search for truthful and reliable information in the process of decision making about important things. The web users would like to be influenced by some credible professionals. The various regression methods were tested. The best solution was implemented in the Application for the Machine Authority Identification.

Keywords: Authority identification · social web mining · Linear regression · Non-linear regression · Web forums

1 Introduction

We live in the information era. A volume of information, which is discovered each day, is too large and too time consuming to be processed by a human. Everybody from us needs sometimes an access to the relevant supporting information for our decision making. To know the relevance of information we have found, we need information about sources of the obtained information and their credibility. In other words it is important to know the sources, which are authoritative ones. A web forum discussion can be a repository of various kinds of useful information: facts, opinions, ideas, attitudes, and so on. However, useful information is mixed with non-useful or misleading information. Every web user can join the web discussion but many of them have not sufficient experiences or theoretical knowledge about the discussed themes. The web discussion often contains an opinion spam and an information trash. So, it is the matter of principal to search for authoritative discussants to let them influence our important decisions. And just the searching for an authority and its machine identification among all discussants of web forum is our challenge.

© Springer International Publishing Switzerland 2015
M. Núñez et al. (Eds.): ICCCI 2015, Part I, LNAI 9329, pp. 203–212, 2015.
DOI: 10.1007/978-3-319-24069-5_19

To achieve our main goal – machine authority identification, we had to do the following three steps:

1. To find such variables - parameters of the structure and content of the web discussion, which are the most related to the authoritative contributing.
2. To define a dependency of the variable "Authority" of a web discussion on the independent variables selected in the first step. We tried to find an approximation of this dependency using the Linear and Nonlinear Regression [1] based on the method of the Ordinary Least Squares (OLS) [2].
3. To use this approximation function for the discrimination of the authoritative from non-authoritative contributors to the web discussion.

Before starting the machine authority identification, we had to solve a number of technical problems. The first one was the automatic extraction of the conversation content and structure from the web page with the web discussion. The second one was to extract the values of selected independent variables from previously obtained information about the discussion. Another problem was how to obtain the values of dependent variable "Authority" for regression function training. We decided for two alternative ways – to obtain values of "Authority" from human "expert" and to extract them directly from the web discussion as so called "wisdom of a crowd".

2   Authority and Web Discussion

2.1   Web Discussion Group

Our attention was on an authority of a web discussion forum. The discussion group was developed in the society Usenet from the beginning of 80th years of 20thcentury [3]. Two computer specialists Jim Ellis and Truscott have come with a new idea to create a system of rules for the contributions creation. Nowadays, WWW society becomes the main organization, which supports and spreads various platforms for Internet discussion groups using various settings up of different web servers. The internet discussion is represented by a web page, where users insert their contributions (opinions and reactions). Within this paper, the web users joining a web discussion will be called the contributors or discussants. They add their opinions, ideas and attitudes to the web discussion and in this way they create so called "conversational content". The authority identification represents the mining of this conversational content and its internal structure. There are different types of Internet discussion forums according to their scope [4]: a discussion to web article, guestbook, discussion forum etc. The paper focuses on the web discussion dedicated to some given theme.

2.2   Authority Identification in General

The concept "authority" comes from the Latin word "augere". It denotes a person, whose opinions, attitudes or decisions are respected by other members of the group and whose decisions and advices are expected by other members of the group. The authority is derived from the relations between people (web users), positions and hierarchies [5]. There are many kinds of authorities. For example according to prestige, authority can be:

- *Formal (functional) authority*–coming from his formal position regardless of his personal properties. It is a leadership of a person who is mandated to make decisions. It is obviously the result of a position of a person within an organization.
- *Informal (natural) authority*–is based on someone's personal properties and professional assumptions. Such person has a spontaneous influence on others, because of his persuasiveness and good experiences with his advices/decisions. The people, who let an authority to lead them, reinforce the weight of the authority.

The formal authority can be at the same time the informal one. The formal authority can sometimes change his status to informal and vice versa.

2.3   Authority of a Web Discussion

The virtual web authority has different characteristics as the authority in real life. It is related to the structure of the web, which is based on hyperlinks among web pages. The Google has discovered very complicated relations among web pages and references. Well known tool for the web page authority calculation is PageRank [6]. Other known approaches to the web page authority calculating are HITS algorithm [7] and SALSA [8]. These approaches are also based on an input and output hyperlinks of the evaluated web page. There are also tools of the respected portal "Seomoz", for example MozTrust [9] and Open Site Explorer [10]. All these tools cannot be easily used for calculating of an authority of the web discussion forum.

The authority identification from web discussion forums is a similar problem as web page authority calculation, because authority identification from web discussion is concentrated on web page, the discussion runs on. On the other hand, it is also a different problem, because no input or output references between this page and other pages are taken into account. Only references inside this page between various discussants are considered. These references are represented by reactions on contributions. All mentioned methods (PageRunk, HITS, SALSA, MozTrust and Open Site Explorer) calculate authority of each web page separately. One page leads to one measure of authority. Within the authority mining from the web conversation, not only one but all contributions of the given discussant are evaluated. All information about all contributions related to one discussant has to be concentrated and used for the authority estimation. Nevertheless, we can inspire ourselves by these techniques and take into account the number of references as reactions on an actual contribution.

In our previous work [11], we have taken into account mentioned number of reactions on all contributions of evaluated discussant, but also the number of all contributions of this discussant, the number of reaction of the discussant on the bottom level of the conversation tree (Fig. 3), the polarity matching between opinion of the discussant and opinion of all discussion, the positions of contributions in the discussion tree and the length of his/her contributions. Some of these variables have appeared to be not so important for the precise estimation of the authority. Another problem of this approach was in way of the estimation function generation. For these reasons, we decided to modify the set of variables - arguments of the conversational structure and to use the regression methods for training the authority estimation function.

3   Used Methods

We tried to solve the problem of the authority estimation within the web discussion forum using a machine learning method based on regression analysis. The regression analysis can be a simple $Y = f(x)$ or a multiple regression, when we are searching the dependency of one dependent variable(Y) on more other independent variables (x_1, x_2, ...x_N) -see equation (1). These variables are called "regresses" or "predictors".

$$Y = f(x_1, x_2, ...,x_N). \tag{1}$$

Within the regression analysis, it is very important to realize, which one of variables is dependent and which are independent. The goal is to describe this relation by a suitable mathematic model, for example by linear or nonlinear function. The result will be a regression curve, which should optimally match the empirical polygon [12].

3.1   Linear and Non-linear Regression

Within the two dimensional space, the linear regression can be described by the equation (2) and is illustrated in the Fig.1 for two dimensional space.

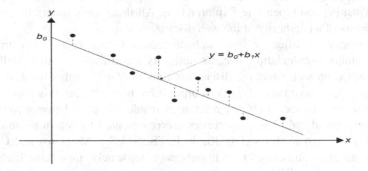

Fig. 1. Linear regression in two dimensional space

The goal is to find such values of constants b_0, b_1, ..., b_n (in two dimensional space b_0 and $b_1$of the linear line, see Fig.1) to achieve the optimal matching with the point graph consist of m points (observations). These constants can be dedicated from the point estimation using the Ordinary Least Squaresmethod (OLS) [2].

$$y_i = b_0 + b_1 x_{i1} + ... + b_n x_{in} + \varepsilon_i \tag{2}$$

Sometimes, it is not possible to find a satisfactory precise linear relation. In this case, the relation can be modeled by somenon-linear function, the most frequently exponential function ($y = ae^{bx}$) or logarithmic function ($y = a + b \ell n\, x$) [1].

3.2   Specification of the Variables of a Discussion Structure

We have selected 120 discussants from the portal "www.sme.sk". Consequently, the following variables for each discussant were extracted from all his contributions:

- AE– Average Evaluation of the contribution
- K– value of the Karma of the user, which is the contribution author
- NCH– Number of Characters within his/her contributions
- AL–Average Layer in the conversation tree (see Fig.3)
- ANR – Average Number of Reactions on his/her contributions
- NC – Number of Contributions of given discussant

These variables were used to form the training set (is illustrated in the Fig.2) for selected regression method.

Nickname	AE	K	NCH	AL	ANR	NC
Peter Z	60	108	26	0	1	1
V12	80	182	220	2	0,5	2
fer	80	171	548,5	3	2,5	2
sandokan555	80	162	57,5	4	0,5	2
Peter_S	50	99	112,5	6	0	2
Darkman	80	167	117	3	0	1
Jesse Pinkman	40	74	210,5	1,5	1,5	2
mm	60	108	22	1	1	1

Fig. 2. Each line of the training set represents one discussant and contains the values of variables *AE, K, NCH, AL, ANR and NC.*

Average evaluation of the contribution (*AE*) is represented by the ratio of the sum of all reactions (agree (+) and disagree (-)) on the contributions of given discussant to the number of all his contributions. This average evaluation is available on the web discussion page. The range of the *AE* is the number from 0 to 80.

Value of *karma* (*K*) of the discussant is also available on the discussion web page. The karma is a number from 0 to 200, which represents activity of the discussant from last 3 months (within the portal "www.sme.sk").

Number of characters (*NCH*) represents the length of discussant contributions. It penalized authors with too short and so less informative contributions.

Average Layer (*AL*) in the conversation tree (see Fig. 3.) is the average number of all layers, which the contributions of the discussant are situated in. The conversation tree is a graphical representation of the web discussion. The *AL* represents the information, when the discussant joined the discussion, from the beginning or at the end.

Average number of reactions (ANR) on the all contributions of the given discussant is the number of reactions per one his contribution.

Number of contributions (NC) is simply the whole number of contributions of the given discussant.

It may happen that a good contribution of already well-known authority finishes the discussion on the Web. It is truth that in such a case there is no reaction on this contribution. It does not disturb the measure of the authority, because of high probability that there were more previous contributions of this contributor with many reactions within the given discussion. These reactions can balance the lack of reactions on the finishing contribution.

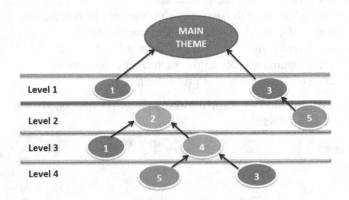

Fig. 3. The conversation tree has 4 levels. The main theme is in the root and reactions are situated on levels 1 – 4. All reactions of the same discussant have the same color.

All these variables were considered to be independent variables. The dependent variable of the regression function Y was dedicated from:

1. evaluation of each discussant by "human expert",
2. evaluation of each discussant by other discussants and it represents "wisdoms of the crowd".

4   Implementation and Testing

The authority value $A \equiv Y$ was estimated by a linear and non-linear function of selected variables (AE, K, NCH, AL, ANR and NC). The four regression functions for authority estimation were generated in the process of machine learning:

1. Linear function learned from the "human expert" (L-EXPERT) is represented by formula (3):

$$A = 0,4383AE + 0,0746K + 0,0281NCH - 2,1932AL - 3,4386ANR + 8,0102NC \qquad (3)$$

2. Linear function learned from the "wisdoms of the crowd" (L-CROWD) is represented by formula (4):

$$A = 0,4385AE + 0,325K + 0,002NCH - 0,2928AL - 0,0853ANR + 1,0728NC \qquad (4)$$

3. Non-linear function learned from the "human expert" (NL-EXPERT) is represented by formula (5):

$$A = 0,0382AE^{1,7192} - 0,3295K^{0,959} + 0,4470NCH^{0,681} + 0,1825AL^{0,0001} - 0,6269ANR^{3,2394} \qquad (5)$$
$$+ 20,2509NC^{0,2977}$$

4. Non-linear function learned from the "wisdoms of the crowd" (NL-CROWD)is represented by formula (6):

$$A = 0,0185AE^{1,8135} + 141,5704K^{-78,39} + 0,0018NCH^{1,0457} - 0,0011AL^{3,7717} - \qquad (6)$$
$$0,5562ANR^{0,0001} + 37,6642NC^{0,0038}$$

All these functions were created using standard MATLAB functions: "regress" in the case of linear and "lsqnonlin" in the case of non-linear relations. No auxiliary regularization method was used, because the input data matrix was regular. The input data can hardly be considered as noise-data obtained for example from a device. These used input data map the structure of the given web discussion using defined variables. In the case of nonlinear regression, also exponential parameters were elicited from the training data using the function "lsqnonlin". It solves nonlinear least-squares (nonlinear data-fitting) problems and uses numerical optimization method "Trust-Region-Reflective Least Squares Algorithm". The default settings were used, only the number of iterations was extended.

We had considered also polynomial functions to be used for solving the problem of authority identification, but we decided to use a more general form of the function with parameters in its exponents, where exponents need not to be integer values.

All the four functions (from (3) to (6)) were tested. The concise results of these tests are illustrated in Tab.1 and Tab.2.

Table 1. Average deviation of four versions of authority estimation function

Version	Average deviation
L-EXPERT	17,3489
L-CROWD	3,2998
NL-EXPERT	18,1131
NL-CROWD	6,5618

At first, the average deviations were calculated. According to the results in Tab.1, the better functions were obtained by learning from the "crowd" than by learning from the "expert". The deviations for some of tested discussants for the best version L-CROWD are illustrated in Fig.4.

At second, these four versions of regression function were tested using obvious measures of a machine learning efficiency: precision and recall. The regression problem, when the value of A (authority) attribute should be estimated from the interval <0, 100> using formulas (3-6), was adopted to classification problem in the following way. A threshold T has been stated experimentally (T=70) and discussants were classified into categories: "authority" and "non-authority". The discussants were classified to the class "authority" when their value of A was equal to or greater than T and they were classified to the class "non-authority" when their value of A was smaller than T. The precision π and recall ρ were computed according to formulas (7) and (8):

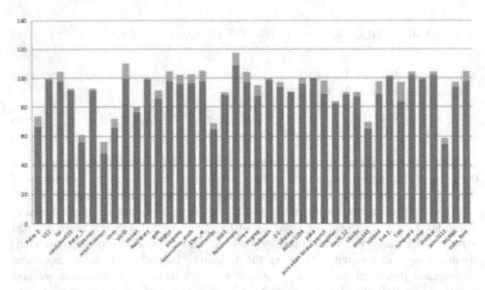

Fig. 4. The values of Authority (red colour) and deviations (green colour) for some of tested discussants for the best version L-CROWD

$$\pi_j = \frac{TP_j}{TP_j + FP_j} \tag{7}$$

$$\rho_j = \frac{TP_j}{TP_j + FN_j} \tag{8}$$

Where:

TP is the number of True Positives (the method classifies these examples as positive (authority) and they are truly positive according to the expert's (crowd's) opinion).

FP is the number of False Positives (the method classifies these examples as positive (authority) but they are not positive according to the expert's or crowd's opinion).

FN is the number of False Negatives (the method classifies the examples as negative (non-authority) but they are positive according to the expert's (crowd's) opinion).

Some key and the most important achieved results of tests are presented in Tab.2.

The linear regression learned from the "crowd", with best results of testing, was implemented in the Application for the Machine Authority Identification (AMAI). This application provides the list of all discussants with the actual value of their Authority. The AMAI also displays the value of the authority of the discussant, which was selected by a user. This value is from the interval <0, 100>. The application provides not only the binary decision whether the discussant is or is not the authority, but also it provides a precise numeric value of its authority.

Table 2. Values of precision and recall of four versions of regression functions were obtained in the three-time cross validation.

		PRECISION		RECALL	
Test	Version	EXPERT	CROWD	EXPERT	CROWD
Cross val. 12_3	Linear regression	0.78	0.99	0.69	0.99
	Non-linear regression	0.72	0.99	0.66	0.88
Cross val. 13_2	Linear regression	0.65	0.98	0.65	0.93
	Non-linear regression	0.67	0.97	0.67	0.86
Cross val. 23_1	Linear regression	0.68	0.97	0.67	0.67
	Non-linear regression	0.69	0.97	0.69	0.67
Average	Linear regression	**0.70**	**0.98**	**0.67**	**0.80**
	Non-linear regression	**0.67**	**0.97**	**0.67**	**0.80**

5   Conclusions

The design of solving the problem of the authority identification from conversational content using the linear and nonlinear regression was presented. The measure of the authority A was estimated as dependency on variables (AE, K, NCH, AL, ANR and NC) - parameters of the structure and content of given web discussions. The four generated estimation functions were tested. According to the values of average deviations (see Tab.1) the best solution is the linear function learned from crowd (L-CROWD). The second one is the nonlinear function learned for crowd (NL-CROWD). Linear and non-linear functions learned from a single human evaluator – expert – seem to be worse. The same conclusions can be deduced from the resulting average values of precision and recall in Tab.2. It can be hardly said who is the expert on the authority identification. Also an opinion of a psychologist may be subjective. On the other hand, combined opinion of many discussants can be objective.

There are other existing authority identification methods, as Klout, TwentyFeet, My Web Carrer [13] and our previous work [11]. All these methods use formulas for authority estimation, but these formulas were generated more experimentally without considering a theoretically based way. For this reason, we tried to generate the relation between the authority and the structure of web discussion using the classic mathematical approach based on the linear and nonlinear regression.

For the future we plan to elicit the constants of linear and nonlinear equations using evolutionary algorithms [14, 15] in order to calculate not only constant values but the form of a non-linear regression function as well.

Acknowledgements. The work presented in this paper was supported by the VEGA project 1/1147/12 "Methods for analysis of collaborative processes mediated by information systems".

References

1. Pazman, A., Lacko V.: Lectures from Regression Models. University of Comenius Bratislava, Bratislava, Slovakia, 132 (2012) (in Slovak). ISBN: 978-80-223-3070-1
2. Pohlman, J.T., Leitner, D.W.: A Comparison of Ordinary Least Squares and Logic Regression. The Ohio Journal of Science. **103**(5), 118–125 (2003)
3. What is Usenet? www.usenet.org
4. Internet forum. http://en.wikipedia.org/wiki/Internet_forum
5. Chavalkova, K.: Authority of a teacher. Philosophic faculty of the University of Pardubice, Pardubice, Czech Republic (2011). (in Czech)
6. Fiala, D.: Time-aware PageRank for bibliographic networks. Journal of Infometrics **6**(3), 370–388 (2012)
7. Li, L., Shang, Y., Zhang, W.: Improvement of HITS-based algorithms on web documents. In: 11th International Conference on the WWW, pp.527-535. ACM, Hawaii, USA (2002)
8. Lempel, R., Moran, S.: The stochastic approach for link structure analysis (SALSA) and the TKC effect. Computer Networks: The International Journal of Computer and Telecommunication Networking. **33**(1-6), 387–401 (2000)
9. Hallur, A.: MozRank and MozTrust: Everything you Should Know. http://www.gobloggingtips.com/mozrank-and-moztrust/
10. Fishkin, R.: Open Site Explorer News Link Building Opportunity Section. http://moz.com/blog/open-site-explorers-new-link-building-opportunities-section
11. Machová, K., Sendek, M.: Authoritative authors mining within web discussion forums. In: 9th International Conference on Systems, pp.154-159. International Academy, Research and Industry Association, Nice, France (2014)
12. Introduction to regress analysis (in Czech). http://www.statsoft.cz/file1/PDF/newsletter/2014_26_03_StatSoft_Uvod_do_regresni_analyzy.pdf
13. Štefaník, J.: Aproximation of the relation of an authority on the parameters of the structure of web discussion. Technical University of Košice, Košice, Slovakia (2015). (in Slovak)
14. Mach, M.: Evolution algorithms – problems solving. FEI Technical University, Košice, 135 ps. (2013) (in Slovak). ISBN 978-80-553-1445-7
15. Ćádrik, T., Mach, M.: Evolution classifier systems (in Slovak). Electrical Engineering and Informatics IV. In: Proc. of the FEI Technical University of Košice, Košice, pp. 168-172 (2013). ISBN 978-80-553-1440-2

Sem-SPARQL Editor: An Editor
for the Semantic Interrogation of Web Pages

Sahar Maâlej Dammak[✉], Anis Jedidi, and Rafik Bouaziz

MIRACL Laboratory, FSEGS, Sfax University, Airport Road,
BP 1088, 3018 Sfax, Tunisia
maalej_sahar@yahoo.fr, jedidianis@gmail.com, raf.bouaziz@fsegs.rnu.tn

Abstract. The semantic Web is an infrastructure that enables the inter-
change, the integration and the reasoning about information on the Web.
In our annotation approach, we have proposed metadata of the semantic and fuzzy annotation in RDF to describe pages in a semantic Web
environment. Our annotation of a page represents an enhancement of
the first result of annotation done by the "Semantic Radar" Plug-in on
the page. Now, we want to achieve a semantic and fuzzy interrogation of
the annotated Web pages in order to improve the interrogation results
for the domain experts. We propose in this paper a new editor named
"Sem-SPARQL Editor", which is based on the SPARQL language, to
query the semantic and fuzzy annotations in RDF of pages.

Keywords: Semantic web · Annotation · Fuzzy semantic interrogation ·
Domain ontologies · Web pages · SPARQL

1 Introduction

The primary goal of the semantic Web is to improve the interrogation process
by constructing and using metadata for annotating semantically the Web pages.
Achieving this goal requires to automate the annotation process by using appropriate ontologies. This constitutes a promising research orientation in which we
have positioned our works. So, we have proposed a fuzzy semantic annotation approach of the Web pages [6], [9]. This approach described the Web pages by particular
metadata, called "*Fuzzy semantic annotations*", in a semantic Web environment.
The annotations are stored in RDF documents. In fact, our annotation method is
an enhancement of the first result of the annotation done by the Semantic Radar
Plug-in [2], [3], [4] on the Web pages, using an enriched domain ontology and the
FOAF[1] (friend-of-a-friend) and SIOC[2] (semantically-interlinked online communities) ontologies [7], [9]. The concepts of the result of Semantic Radar are connected
to several concepts of the ontology, but the connections may be uncertain. We have
then proposed to associate each line of annotation with a weight indicating the possibility degree of the annotation concept [8].

---

[1] http://xmlns.com/foaf/spec/
[2] http://rdfs.org/sioc/spec/

© Springer International Publishing Switzerland 2015
M. Núñez et al. (Eds.): ICCCI 2015, Part I, LNAI 9329, pp. 213–222, 2015.
DOI: 10.1007/978-3-319-24069-5_20

Furthermore, to improve the relevance of the interrogation results after the semantic annotation of the Web pages, we have proposed a new method for filtering the indexed Web pages (annotated and/or non-annotated). This method sorts the retrieved Web pages, after the interrogation, before their display to the user. In addition, it groups and places the most relevant Web pages at the top of the extracted list. Thus, we are interested in the construction of an annotation, interrogation and filtering framework in order to contribute to improve the performance of the interrogation process. This framework is entitled *"FSAI Web-Pages"* (*Fuzzy Semantic Annotation and Interrogation of the Web Pages*). It includes two components: the annotation component, entitled *"Annotation of Web pages"* and the querying and filtering component, entitled *"Filtering Web Pages"* [9]. The first component automates the semantic and fuzzy annotation process of the Web pages. The second component automates the filtering process of the pages by generating the relevant interrogation results.

Now, we intend to respond to the requirements of the domain experts by providing them with the semantic and fuzzy results for their study domains. In this case, our component *"Filtering Web Pages"* must first find the relevant pages, including the annotated pages. Then, it must support the semantic and fuzzy interrogations to address the requirements of the experts. Finally, it must be based on an interrogation language to query the annotation RDFs. Since SPARQL [3] has been rapidly adopted as a standard for querying semantic Web data, we have chosen to adopt this language for our semantic and fuzzy environment. Thus in this paper, we are interested in the development of a new editor, entitled *"Sem-SPARQL Editor"* (*Semantic-SPARQL Editor*), in order to automate the generation of the semantic and fuzzy interrogation results that respond to queries. This editor is included in the *"Filtering Web pages"* component. It is developed under *"NetBeans IDE"*, for improving the interrogation results, employing the *"Apache Jena[4]"*, to ensure that domain experts can execute the queries with SPARQL.

The rest of the paper is structured as follows. The second section presents briefly the architecture of our framework, the stages of our approach to the semantic annotation, interrogation and filtering of the Web pages and the annotation RDF of the pages. We show in the third section the development details of the fuzzy semantic interrogation in our editor. The fourth section shows an overview of the related works on the annotation and the interrogation of the Web pages. Finally, in the fifth section we present our conclusion and some further work.

2   Backgrounds

2.1   Architecture of Our Framework "FSAI Web-Pages"

Our proposal begins with a fuzzy semantic annotation of the Web pages by RDF documents containing the annotation descriptors, which we proposed in

---

[3] http://www.w3.org/TR/rdf-sparql-query/
[4] https://jena.apache.org/

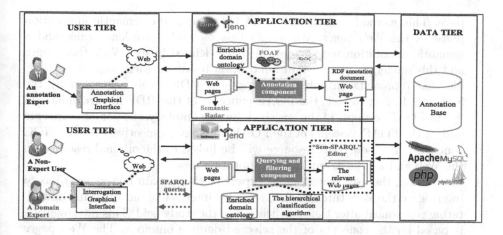

Fig. 1. Architecture of our "FSAI Web-Pages" framework

equivalence rules and in an annotation model [6]. We have created for this annotation process an annotation component, entitled "*Annotation of Web pages*". By exploiting the proposed annotations on the pages, our framework aims to improve the Web interrogation results. We have created in this sense a querying and filtering component, entitled "*Filtering Web Pages*". This component presents the relevant Web pages (annotated and/or non-annotated) to the user after the interrogation [9]. In case the user is a domain expert, the new editor returns the semantic and fuzzy information, which is found in the annotation RDFs of the annotated pages, to respond to his/her query. Fig. 1 shows the global architecture (represented in three-tier) of our assistance framework to the annotation, interrogation and filtering of the Web pages.

2.2 An Approach to Semantic Annotation, Interrogation and Filtering

We have proposed a new approach to proceed to the semantic annotation, the interrogation and the filtering of the Web pages [9]. The stages of our approach are as follows (cf. Fig. 1):

1. The annotation expert has to interrogate the Web through the graphical interface of our annotation component after the specification of the study field. This interrogation is based on the concepts of the selected domain ontology.
2. Then, the retrieved Web pages must be transferred to the Semantic Radar for an automatic and semantic analysis. This analysis retrieves an RDF file for each page containing the descriptors of FOAF and SIOC. The set of RDF files extracted by the Semantic Radar represents the input of the annotation process.
3. Subsequently, the analyzed pages have to be automatically annotated by our method, shown in [7] and [9], to produce RDF metadata for each Web

page. This method assists the expert throughout the semantic annotation phase of the Web pages. We use in this method equivalence rules and a semantic annotation model that we have defined in [6]. With these rules and this model, we produce an annotation result for each page as an RDF file that represents an enhancement of the RDF result generated by the Semantic Radar. In fact, the FOAF concepts of this RDF file are enhanced by the FOAF concepts of the enriched domain ontology, the domain concepts and/or the FOAF concepts of the FOAF ontology (respectively for the SIOC concepts). The new RDF resource will be linked to the original resource on the Web and will be saved in the annotation base.

4. In addition, the user (the domain expert or the domain non-expert) has to interrogate the Web through the graphical interface of our querying and filtering component after the specification of the study field. This interrogation is based on the concepts of the selected domain ontology. The Web pages retrieved after the user interrogation pass to the filtering system. This system sorts these Web pages, and returns them to the user. Also, it groups the most relevant pages (including the annotated ones) and appears them at the top of the extracted list. The filtering system checks the annotation of the pages through the access to the annotation base. Thus, we have proposed to use an appropriate method for filtering Web pages (based on a scores calculation of the pages and a hierarchical classification of these pages for grouping the relevant ones). This method helps to improve the relevance of the interrogation results.

5. If the user is a domain expert, he/she can interrogate the proposed RDF documents (containing semantic and fuzzy descriptors) for the relevant pages with the SPARQL language, through the graphical interface of our editor. This interrogation can be of the semantic and fuzzy type. The objective of this paper is to detail the feasibility of the semantic and fuzzy interrogation in our editor.

2.3   The RDF of the Semantic and Fuzzy Annotation

In our approach, we have defined a process model for the semantic annotation of Web pages [6]. The main activity of this model consists in the search of equivalence between the concepts of the result of Semantic Radar and the concepts of the selected enriched domain ontology. Following this model, we have obtained an annotation document in RDF, automatically generated for each Web page, by our annotation system. Each annotated concept, in an annotation RDF, is now linked with the semantic and fuzzy descriptors of the annotation that we have defined in the annotation model. The semantic descriptor (such as <HasChild>) frames the annotation concept while the fuzzy descriptor (such as <FuzzyValue>) frames the weight (*i.e.*, the possibility degree) of this concept, in relation to the concept to annotate. We have already defined two formulas to calculate the concept weights of the ontology [8]. The following is an extract of an annotation document in RDF for the page http://journal.webscience.org/532/. This page has been displayed to a domain expert [8], [9], as a relevant page, after the interrogation in the "Network of scientists" domain [13]. If we follow this RDF document, we can generate semantic and fuzzy results concerning the study field of the domain expert. We will show an interrogation example of this RDF document in the next section.

```
<?xml version='1.0' encoding='UTF-8'? > <!DOCTYPE rdf:RDF [....
<rdf:RDF>
<rdf:Description rdf:about="&epid;person/ext−binghe@indiana.edu?">
<foaf:familyName rdf:datatype="&xsd;string">He</foaf:familyName>
<foaf:givenName rdf:datatype="&xsd;string">Bing</foaf:givenName>
<rdf:type rdf:resource="&foaf;Person" />
<rdfs:HasChild rdf:datatype="&xsd;string">FacultyMember</rdfs:HasChild>
<rdfs:FuzzyValue rdf:datatype="&xsd;decimal">0.0015</rdfs:FuzzyValue>
<rdfs:HasChild rdf:datatype="&xsd;string">EmeritusProfessor
</rdfs:HasChild>
<rdfs:FuzzyValue rdf:datatype="&xsd;decimal">0.0047</rdfs:FuzzyValue>
</rdf:Description> .... </rdf:RDF>
```

3 Development of the Fuzzy Semantic Interrogation in Our Editor

We have developed a component called *"Filtering Web Pages"* for our framework, under *"NetBeans IDE"*, to improve the relevance of the interrogation results. In the graphical interface of *"Filtering Web Pages"*, the user must specify the study domain (left part *Domain ontologies*, cf. Fig. 2). After specifying his/her domain, we automatically show to the user the hierarchy of the ontology (*Hierarchy of Ontology*, cf. Fig. 2) to assist him/her in writing the semantic query. We have actually imported the *"Jena API"* in the java application of our querying and filtering component to read the ontologies and to extract their concepts. The user can now select the concepts for writing his/her query. After writing the semantic query (in the text area, cf. Fig. 2), he/she can start his/her search (after the click on the *"Search"* button, cf. Fig. 2). A list of URLs of the Web pages automatically appears (a list is a set of ten (10) retrieved pages by Google). After the display of a URLs set, we automate the scores calculation (cf. the left table in Fig. 2) and the hierarchical classification (cf. the right table in Fig. 2) for these pages by our filtering system. Indeed, in the background, we automate the calculation of the concept weights for the selected domain ontology, following the proposed formulas in [8]. The user can then take an idea on the weighting of the selected ontology concepts after the click on the *Weight of concepts* button (cf. Fig. 2). These weights are considered in the scores calculation for the retrieved Web pages. Actually, the filtering system displays, on the one hand, the sorted URLs to the user (in the column *Classification by sorting* of the table *Classification* in the right of Fig. 2), and, the other hand, the URLs of the most relevant Web pages (in the column *Hierarchy classification* of the table *Classification*, cf. Fig. 2). In the example of the Fig. 2, our filtering system gives, after a Web interrogation in the domain *"Network of Scientists"*, a single Web page, which is the

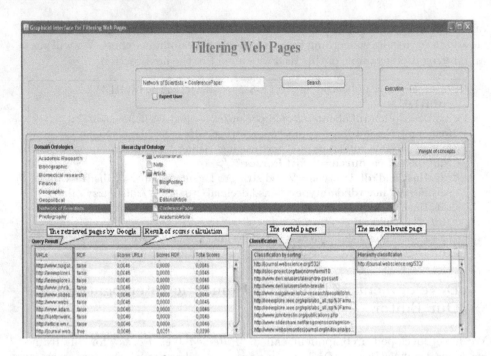

Fig. 2. Example of the interrogation by the "Network of Scientists" ontology

unique most relevant page compared to the others: http://journal.webscience. org/532/. This page is an annotated page by our annotation component (an extract of the annotation RDF of this page in Section 2.3). In other cases, we can obviously have several relevant pages.

After the improvement of the relevance of the interrogation results, a domain expert may have more information about the study field. Such information can be of semantic and fuzzy type. They are extracted from the annotation RDFs of the annotated Web pages. We have then developed a new editor, called "Sem-SPARQL Editor", which supports the semantic and fuzzy queries in SPARQL. In fact, the editor interface will appear after the click on the "Expert User" check box of the "Filtering Web Pages" interface (cf. Fig. 2). To assist the expert to write the query, we have displayed the classes and properties of the FOAF and SIOC ontology (necessary to query the metadata in the annotation RDF in question) in the interface of our editor (cf. Fig. 3). In addition, the domain expert should follow a well-defined syntax in the writing of his/her query in order to have a correct semantic and fuzzy result for the study domain (cf. The syntax in the following). We also note that our editor executes the query only after a syntactic checking: (i) the checking of the general syntax of a SPARQL query and (ii) the checking of the syntax of a triple in the condition of the query. In addition, a lexical checking is done by the editor in order to verify that the used properties and classes in the query existing in the concerned ontologies. The query must be written in the "Query Text" area of the editor interface (cf. Fig. 3). We propose to display the interrogation result in two ways. The table on the left details the semantic and fuzzy result and the display area on the right summarizes the interrogation result (cf. the *"output"* part of the Fig. 3).

The general syntax of the semantic and fuzzy query

SELECT ∗
WHERE { [?subject ?predicate ?object]*. // a triple in the RDF graph
[FILTER(?variable *operator* value)]* // Arithmetic expressions }

N.B:
[...]* : indicates that the specification can be multiple.
Subject: the FOAF (or SIOC) class or variable.
Predicate: the FOAF (or SIOC) property or the annotation descriptor (semantic or fuzzy).
Object: variable.
Operator: =, >, <, >=, <=.

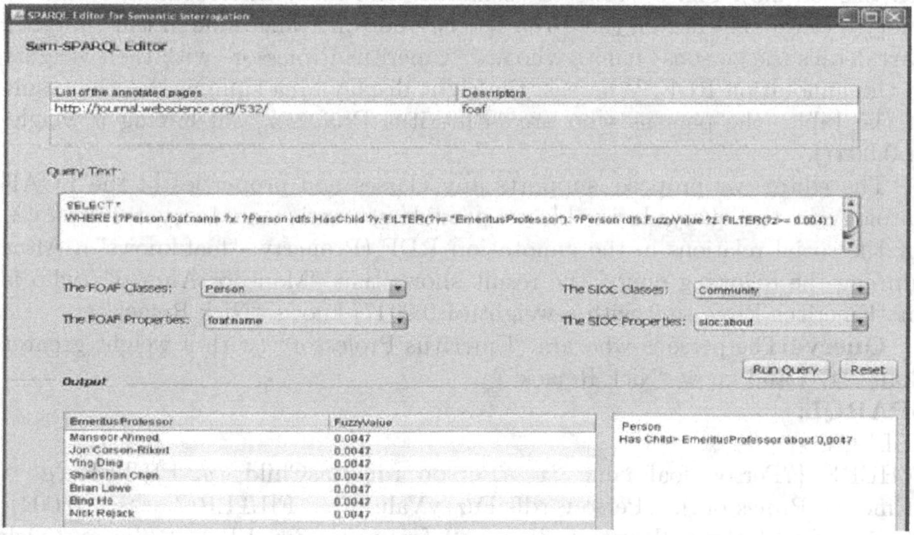

Fig. 3. Example of the fuzzy semantic interrogation in "Sem-SPARQL"

We remain in the same study field *"Network of scientists"*; the following are a few queries that a domain expert can ask with SPARQL in our editor, given that the query already asked for the interrogation in *"Filtering Web Pages"* is: *Network of scientists + ConferencPaper* (cf. Fig. 2).

Query1: The persons who have participated in the conference papers (Name and Family Name)?
SPARQL:

SELECT ∗
WHERE {?Person foaf:name ?x. ?Person foaf:familyName ?y}
   Query2: The persons who are "Faculty Member"?
SPARQL:
SELECT ∗
WHERE {?Person **rdfs:HasChild** ?x. FILTER(?x = "FacultyMember"). ?x foaf:name ?y}
   Query3: The persons who are "Emeritus Professor" with a weight (percentage) greater than 0.4%?
SPARQL:
SELECT ∗
WHERE {?Person foaf:name ?x. ?Person **rdfs:HasChild** ?y. FILTER(?y = "EmeritusProfessor"). ?Person **rdfs:FuzzyValue** ?z. FILTER(?z >= 0.004)}

We conclude that in order for a domain expert to have a semantic and fuzzy result, he/she should use the descriptors already defined in our annotation model (which are also automatically extracted in the annotation RDFs). In this case, our editor is able to return semantic and fuzzy results from the existing data in the annotation RDF of the annotated page: http://journal.webscience.org/532/ (cf. Section 2.3). Fig. 3 shows the semantic and fuzzy interrogation result by query 3 (after the click on the "*Run Query*" button). The table in the "*output*" part details the persons? names who are "Emeritus Professor" with their weights in the annotation RDF. Whereas, the right display area summarizes the result of the table (the persons who are "Emeritus Professor" all having a weight of 0.0047).

The editor, we propose, supports any classes and properties of the FOAF ontology in the query. It is able to provide semantic and fuzzy results even for the social relations in the annotation RDF (Property: "foaf:knows"). After running the following query, the result shows that "Mansoor Ahmed" (who is an "Emeritus Professor" with a weight of 0.0047) knows "Nick Rejack".

   Query4: The persons who are "Emeritus Professor" (with a weight greater than 0.4%) and know "Nick Rejack"?
SPARQL:
SELECT ∗
WHERE {?Person foaf:name ?x. ?Person rdfs:HasChild ?y. FILTER(?y = "EmeritusProfessor"). ?Person rdfs:FuzzyValue ?z. FILTER(?z >= 0.004). ?Person **foaf:knows** ?Person2. ?Person2 foaf:name ?w. FILTER(?w = ?Nick Rejack?)}

After a study of the most recent query editors based on the SPARQL language, such as "Flint sparql editor[5]" and "Virtuoso sparql query editor[6]", we note that the raw results to SPARQL queries are very hard to read. The results typically contain URIs, which require the navigation of the RDF graph in question [11]. While our editor not only supports the semantic and fuzzy queries in

---

[5] http://openuplabs.tso.co.uk/demos/sparqleditor
[6] http://dbpedia.org/sparql

SPARQL, but also it is able to meet the requirements of the experts with a clear and understandable result in the semantic and fuzzy aspects.

4 Related Works

This paper recalls our approach to the semantic annotation of the Web pages. Different solutions with a similar goal have been proposed before. The annotation approaches are limited to only a part of a Web page (text, images, etc.), or to the semi-structured documents, to extract the annotation metadata in RDF [1], [12], [14], etc. These works have adapted the interrogation tools (by SPARQL) to their needs in order to query the annotation RDFs (proposed by the annotation process). The most commonly used tools are (i) Corese [5], which is a semantic search engine to query by SPARQL the annotations in RDF; (ii) Jena [10], which is a Java API to build the applications for the Semantic Web. It provides a programming environment for RDF, RDFS, OWL and SPARQL. In our case, the semantic and fuzzy interrogation result is based on the annotation RDFs of the relevant Web pages, returned by our querying and filtering component. We then propose to adapt the "*NetBeans*" environment of our component by the integration of the *Jena API* ("*Jena ARQ SPARQL*[7]"). Therefore, we create our own interrogation editor that represents a constituent of the querying and filtering component. Our editor then supports the queries in SPARQL and interrogates the semantic and fuzzy descriptors proposed in the annotation RDFs of the pages. In addition, it proposes understandable semantic and fuzzy results by the domain experts.

5 Conclusion and Some Further Work

In this paper, we have proposed a semantic and fuzzy interrogation of the annotated Web pages through a new editor, called "*Sem-SPARQL Editor*". This editor is included in the "*Filtering Web pages*" component. It supports the semantic and fuzzy queries with SPARQL. These queries contain the FOAF classes, the FOAF properties and/or the semantic and fuzzy descriptors for querying the annotation RDFs of the Web pages. Our editor improves then the interrogation results for the domain experts, by providing them with the semantic and fuzzy results. The results of the development of our editor have shown the feasibility and benefits of our proposals in comparison with (i) the results we have found in the literature; (ii) the interrogation results by key words which do not satisfy the requirements of the experts. As future works, we want to improve the proposed semantic and fuzzy interrogation by using all the classes and the properties of the SIOC ontology. Also, we will propose an automatic translation approach of the semantic and fuzzy queries in natural language into queries in SPARQL. This task is in progress and will be published later. In fact, the translation approach that we are currently developing is based on the information extraction tools.

---

[7] ARQ is a query engine for Jena that supports the SPARQL RDF Query language. http://jena.apache.org/documentation/query/index.html

References

1. Abacha, A.B, Zweigenbaum, P.: Annotation et interrogation sémantiques de textes médicaux. In: Atelier Web Sémantique Médical, pp. 61–70. Nîmes (2010)
2. Bojars, U., Breslin, J., Peristeras, V., Tummarello, G., Decker, S.: Interlinking the social web with semantics. J. IEEE Intelligent Systems **23**, 29–40 (2008)
3. Bojars, U., Passant, A., Giasson, F., Breslin, J.: An architecture to discover and query decentralized RDF data. In: 3rd Workshop on Scripting for the Semantic Web, Innsbruck, Austri (2007)
4. Bojars, U., Fernández, S.: Semantic Radar (2009). https://addons.mozilla.org/en-US/firefox/addon/semantic-radar/
5. Corby, O., Dieng-Kuntz, R., Faron-Zucker, C., Gandon, F.: Ontology-based approximate query processing for searching the semantic web with corese. In: INRIA Report (2005)
6. Maâlej Dammak, S., Jedidi, A., Bouaziz, R.: Semantic annotation framework for web resources. In: The Fifth International Conference on Internet Technologies and Applications (ITA), Wrexham, North Wales, UK , pp. 106–113 (2013)
7. Maâlej Dammak, S., Jedidi, A., Bouaziz, R.: Automation and evaluation of the semantic annotation of web resources. In: The 8th International Conference for Internet Technology and Secured Transactions (ICITST), London, UK, pp. 448–453 (2013)
8. Maâlej Dammak, S., Jedidi, A., Bouaziz, R.: Fuzzy semantic annotation of Web resources. In: World Symposium on Computer Applications and Research (WSCAR), Sousse, Tunisie, pp. 253–258 (2014)
9. Maâlej Dammak, S., Jedidi, A., Bouaziz, R.: Automation of the semantic annotation of Web resources. The 8th International Journal of Internet Technology and Secured Transactions (IJITST) **5**, 133–148 (2014)
10. McBride, B.: Jena: implementing the rdf model and syntax specification. In: The Second International Workshop on the Semantic Web (SemWeb), Hong Kong, China (2001)
11. Rietveld, L., Hoekstra, R.: YASGUI: feeling the pulse of linked data. In: Janowicz, K., Schlobach, S., Lambrix, P., Hyvönen, E. (eds.) EKAW 2014. LNCS, vol. 8876, pp. 441–452. Springer, Heidelberg (2014)
12. Salvadores, M., Horridge, M., Alexander, P.R., Fergerson, R.W., Musen, M.A., Noy, N.F.: Using SPARQL to query BioPortal ontologies and metadata. In: Cudré-Mauroux, P., Heflin, J., Sirin, E., Tudorache, T., Euzenat, J., Hauswirth, M., Parreira, J.X., Hendler, J., Schreiber, G., Bernstein, A., Blomqvist, E. (eds.) ISWC 2012, Part II. LNCS, vol. 7650, pp. 180–195. Springer, Heidelberg (2012)
13. Stella, M., Shanshan, C., Mansoor, A., Brian, L., Paula, M., Nick, R., Jon, C., Bing, H., Ying, D.: The VIVO collaboration.: the VIVO ontology: enabling networking of scientists. In: The ACM WebSci 2011, Koblenz, Germany, pp. 1–2 (2011)
14. Thiam, M.: Annotation Sémantique de Documents Semi-structurés pour la Recherche d'Information. PhD thesis. Paris-Sud University, 91405 Orsay Cedex, France (2010)

Computational Intelligence
and Games

Enhancing *History*-Based Move Ordering in Game Playing Using Adaptive Data Structures

Spencer Polk[✉] and B. John Oommen

School of Computer Science, Carleton University, Ottawa, Canada
andrewpolk@cmail.carleton.ca, oommen@scs.carleton.ca

Abstract. This paper pioneers the avenue of enhancing a well-known paradigm in game playing, namely the use of *History*-based heuristics, with a totally-unrelated area of computer science, the field of Adaptive Data Structures (ADSs). It is a well-known fact that highly-regarded game playing strategies, such as alpha-beta search, benefit strongly from proper move ordering, and from this perspective, the *History* heuristic is, probably, one of the most acclaimed techniques used to achieve AI-based game playing. Recently, the authors of this present paper have shown that techniques derived from the field of ADSs, which are concerned with query optimization in a data structure, can be applied to move ordering in multi-player games. This was accomplished by ranking opponent threat levels. The work presented in this paper seeks to extend the utility of ADS-based techniques to two-player and multi-player games, through the development of a new move ordering strategy that incorporates the *historical* advantages of the moves. The resultant technique, the History-ADS heuristic, has been found to produce substantial (i.e, even up to 70%) savings in a variety of two-player and multi-player games, at varying ply depths, and at both initial and midgame board states. As far as we know, results of this nature have not been reported in the literature before.

1 Introduction

The problem of intelligently playing a game against a human player using AI techniques has been studied extensively, leading to many well-known techniques, such as the alpha-beta search scheme, capable of achieving excellent performance in a wide variety of games [1]. The performance of alpha-beta search can be further improved by proper move ordering, which can lead to improved tree pruning and thus permit deeper search or a farther look-ahead in the game [2,3]. Specifically, the best pruning is obtained when the best move is searched first, thus leading to a wide range of move ordering heuristics, such as the Killer Moves and the History

B.J. Oommen—*Chancellor's Professor*; *Fellow: IEEE* and *Fellow: IAPR*. The second author is also an *Adjunct Professor* with the Dept. of ICT, University of Agder, Grimstad, Norway.

M. Núñez et al. (Eds.): ICCCI 2015, Part I, LNAI 9329, pp. 225–235, 2015.
DOI: 10.1007/978-3-319-24069-5_21

heuristic, that seek to obtain this arrangement in an environment where perfect information is naturally unavailable [3,4].

Originally unrelated to game playing, the field of ADSs deals with the dynamic optimization of a data structure based on the object access frequencies [5,6]. ADSs attempt to solve the problem of the uneven accesses of objects by reorganizing them in response to queries. The field provides a wide range of methods to accomplish this, such as the Move-to-Front and Transposition rules for adaptive lists [7]. As the objective is to improve the performance of the data structure, these operations, by necessity, must be inexpensive to implement [7].

Previously, the authors of this present paper, demonstrated that techniques derived from ADSs had applications in game playing, particularly within the context of move ordering, leading to the development of the Threat-ADS heuristic [8] for multi-player games. The Threat-ADS heuristic operates within the context of the Best-Reply Search (BRS) for *multi*-player games, where opponents' moves are grouped together to form a "super-opponent", who minimizes the perspective player's gains [9]. The Threat-ADS places the opponents within an adaptive list, and groups opponent moves based on the list, achieving statistically significant improvements in terms of tree pruning in a wide variety of cases [8,10]. Based on the success of the Threat-ADS, our submission is that it is worthwhile to investigate if ADSs can achieve improvements to move ordering using a metric other than opponent threats. This is the focus of this work.

The rest of the paper is laid out as follows. Section 2 describes, in more detail, the motivation for this research endeavour. Section 3 details the History-ADS, our new technique described in this work. Section 4 describes the experiments we have performed to demonstrate the History-ADS' capabilities. Section 5 presents the results from the experiments we have conducted, and Section 6 contains the discussion and analysis of the results we have obtained. Finally, Section 7 concludes the paper.

2   Motivation

The well-known and historically successful Killer Moves and History heuristics, among others, make use of the concept of move history to achieve move ordering [3]. Specifically, they are based on the hypothesis that if a move produced a cut earlier in the search of the game tree, it is likely to be a strong move if it is encountered again, and it should thus be searched first. The proven performance of these techniques shows that this hypothesis is valid, and we know that move history is a metric by which move ordering can be achieved.

The previous success of the Threat-ADS heuristic demonstrates that ADS-based techniques can provide tangible benefits, in terms of move ordering, within game tree searches. Given that it can achieve this using opponent threats, and move history is known to be a worthwhile metric for move ordering, it is reasonable to believe that ADSs can use move history as well as opponent threats. This can intuitively be achieved by having an ADS hold moves, rather than opponents, and update its structure when a move produces a cut, thereby keeping moves that frequently produce cuts near the head of the data structure.

An ADS-based technique using move history would provide a number of benefits including:

- Provide an inexpensive move-ordering strategy for a wide variety of games;
- Demonstrate another metric, aside from opponent threats, that ADSs can employ to achieve move ordering;
- Be applicable to two-player games, along with multi-player games, unlike the Threat-ADS heuristic;
- Potentially achieve greater savings in terms of tree pruning, compared to the Threat-ADS, given that many more possible moves exist in a game than opponents, in all but the most trivial of cases.

The development of a heuristic that accomplishes this, which we have named the History-ADS heuristic, is described in the next section.

3   The History-ADS Heuristic

3.1   Developing the History-ADS Heuristic

The development of the History-ADS heuristic is achieved by explaining the analogous operations of the Threat-ADS heuristic [8]. First of all, we must understand how to update the ADS at an appropriate time, i.e., through querying it with the identity of a move that has led to a cut. The identity of such a move will vary depending on the game, however it would normally be represented in the same way as it is in the context of the Killer Moves or the History heuristics. When the ADS is queried with a move, the move is repositioned according to its update mechanism, or added to the data structure if it is not already present. This is analogous to querying the ADS with the most threatening opponent within the context of the Threat-ADS. Then, at each Max and Min node, we must somehow order the moves based on its order within the ADS.

Fortunately, within the silhouette of alpha-beta search, there is a very intuitive location to query the ADS, which is where an alpha or beta cutoff occurs, before terminating that branch of the search. To actually accomplish the move ordering, when we expand a node and gather the available moves, we explore them in the order proposed by the ADS, as alluded to before.

The last issue that must be considered is that a move that produces a cut on a Max node may not be likely to produce a cut on a Min node, and vice versa. This would occur, for example, if the perspective player and the opponent do not have analogous moves, such as in Chess or Checkers, or if they are some distance from each other. We thus employ two list-based ADSs within the History-ADS heuristic, one of which is used on Max nodes, while the other is used on Min nodes. This parallels how the History heuristic maintains separate values for each player [3].

To clarify matters, a simple example of how the History-ADS updates its data structures and applies that knowledge to achieve move ordering is provided in Figure 1. The formal execution of the enhanced alpha-beta search is provided in Algorithm 1.1. The reader must observe that the enhanced algorithm merely adds a couple of extra lines of pseudo-code to the original alpha-beta search scheme. The efficiency of this inclusion and the marginal additional time footprint encountered, is obvious!

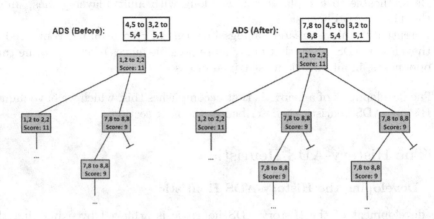

Fig. 1. A demonstration of how an ADS can be used to manage move history over time. The move (7,8) to (8,8) produces a cut, and so it is moved to the head of the list, and informs the search later.

3.2   Features of the History-ADS Heuristic

As the reader will observe, the History-ADS heuristic is very similar, in terms of its design, construction and implementation, to the Threat-ADS heuristic. Like the Threat-ADS heuristic, it does not in any way alter the final value of the tree, a trait common to many schemes that enhance tree pruning. As in the case of the Threat-ADS, the ADSs are added to the algorithm's memory footprint, and their update mechanisms must be considered with regard to its running time.

Compared to the Threat-ADS, the History-ADS can be expected to employ a much larger data structure, as there will be many more possible moves than total opponents in any non-trivial game. One must further observe that, within its ADSs, the History-ADS remembers any move that produces a cut, for the entire search, even if it never produces a cut again and lingers near the end of the list. However, as duplicate moves are never added to the list, the maximum possible length is equal to the number of viable moves, and thus the History-ADS heuristic's memory requirements are bound by those of the highly-regarded History heuristic. Furthermore, as it only maintains memory on moves that have, indeed, produced a cut, it is very likely that the size of the data structures will be far less than the number of possible moves.

Algorithm 1.1. Mini-Max with History-ADS

Function BRS(node, depth, player)

1: **if** node is terminal or depth ≤ 0 **then**
2:     **return** heuristic value of node
3: **else**
4:     **if** node is max **then**
5:         **for all** child of node in order of MaxADS **do**
6:             $\alpha = max(\alpha, minimax(child, depth - 1)$
7:             **if** $\beta \leq \alpha$ **then**
8:                 break (Beta cutoff)
9:                 query MaxADS with cutoff move
10:            **end if**
11:        **end for**
12:        **return** α
13:    **else**
14:        **for all** child of node in order of MinADS **do**
15:            $\beta = min(\alpha, minimax(child, depth - 1)$
16:            **if** $\beta \leq \alpha$ **then**
17:                break (Alpha cutoff)
18:                query MinADS with cutoff move
19:            **end if**
20:        **end for**
21:        **return** β
22:    **end if**
23: **end if**

End Function Mini-Max with History-ADS

However, unlike the History heuristic, which requires moves to be sorted based on the values in its arrays, the History-ADS shares the advantageous quality of the Threat-ADS in that it does not require sorting. One can simply explore moves, if applicable, in the order specified by the ADS, thus allowing it to share the strengths of the Killer Moves heuristic, while simultaneously maintaining information on all those moves that have produced a cut.

Lastly, unlike the Threat-ADS, which was specific to the BRS, the History-ADS works within the context of the two-player Mini-Max algorithm. However, since the BRS views a multi-player game as a two-player game by virtue of it treating the opponents collectively as a single grouped entity, the History-ADS heuristic is also applicable to it, and it thus functions in both two-player and multi-player contexts. We will thus be investigating its performance in both these avenues in the following sections.

4   Experimental Model

Given the similarities between the two techniques (i.e., the History-ADS and the Threat-ADS heuristic), we consider it useful to employ a similar set of experiments in analyzing their performances. This will permit a fair comparison of the schemes on a level playing field. We are interested in learning the improvement gained from using the History-ADS heuristic, when compared to a search that does not employ it. We accomplish this by taking an aggregate of the Node Count (NC) over several turns. The NC measure is defined as the number of nodes that are expanded during the search, i.e. excluding those generated but then pruned before being visited. Historically, this metric has been shown to be highly correlated to runtime, while also being platform-agnostic [3]. For a variety of games, we average this value over fifty trials, with different ADS update mechanisms, and varying game states.

As in our previous work, we will employ the Virus Game, Focus, and Chinese Checkers when considering the multi-player case. Focus and Chinese Checkers are both well-known multi-player games, and the Virus Game is a territory control game described in detail in [8]. However, since the History-ADS is applicable to two-player games as well, we require an expanded set of games to handle these. We have opted to use the two-player version of Focus, Othello, and the very well-known Checkers, or Draughts. However, the requirement in Checkers that forces jumps when possible, often leads to a game with a very small branching factor. This factor can vary greatly especially in different midgame states. Thus, for our experiments, we choose to relax this rule, and do not require that a player must necessarily make an available jump. We shall refer to this game as "Relaxed Checkers". While Checkers has been solved, it still serves as a useful testing environment for the general applicability of a domain-independent strategy, given how well-known and documented the game is in the literature [11].

In testing the Threat-ADS heuristic, we had varied the update mechanism used, the ply depth of the search, and most recently, the starting position of the game, to test its performance in midgame states [12]. We have elected to perform a subset of these experiments, using the History-ADS heuristic, to provide as much information about its performance as possible. Specifically, we compare the Move-to-Front and Transposition update mechanisms, perform trials at both initial and midgame board states, and vary the depth of the search between games. Trials with Othello, Relaxed Checkers, and the Virus Game are done to a 6-ply depth, and trials with Chinese Checkers, and both variants of Focus to a 4-ply depth. Midgame states were generated by playing games a set number of turns using intelligent players, then taking measurements from this position. The details of how midgame states are generated is described in detail in [12].

The number of turns we aggregate the NC over is five for Relaxed Checkers, Othello, and Chinese Checkers, three for Focus, and ten for the Virus Game. To determine statistical significance, we employ the Mann-Whitney test, as we do not make the assumption of normalcy. This also provides the Effect Sizes – an easily-readable indication of the degree of savings given by the History-ADS.

Our results are presented in the next section.

5   Results

Our results are presented in the following subsections, first for two-player games, and after that, for multi-player games. In every case, the highest percentage advantage gleaned by the History-ADS heuristic is highlighted in **bold face**.

5.1   Results for Two-Player Games

Table 1 presents our results for the game of Othello. The History-ADS produced noticeable gains in terms of NC in all cases, in both initial and midgame states, and the Move-to-Front rule outperformed the Transposition rule.

Table 1. Results of applying the History-ADS heuristic for Othello.

Midgame?	Update Mechanism	Avg. NC	Std. Dev	P-Value	Effect Size
No	None	5,061	2,385	-	-
No	Move-to-Front	**3,827 (24%)**	1,692	6.0×10^{-3}	0.51
No	Transposition	4,057	2,096	0.015	0.42
Yes	None	20,100	9,899	-	-
Yes	Move-to-Front	**14,500 (28%)**	6,303	1.2×10^{-3}	0.59
Yes	Transposition	15,200	6,751	0.014	0.45

Consider Table 2, where we showcase our results for Relaxed Checkers. Again, we observed a very large reduction in NC, and again, the Move-to-Front achieved better performance than the Transposition rule.

Table 2. Results of applying the History-ADS heuristic for Relaxed Checkers.

Midgame?	Update Mechanism	Avg. NC	Std. Dev	P-Value	Effect Size
No	None	78,600	10,600	-	-
No	Move-to-Front	**40,800 (48%)**	5,619	$< 1.0 \times 10^{-5}$	3.58
No	Transposition	48,600	6,553	$< 1.0 \times 10^{-5}$	2.84
Yes	None	64,000	25,700	-	-
Yes	Move-to-Front	**36,100 (44%)**	12,700	$< 1.0 \times 10^{-5}$	44%
Yes	Transposition	43,600	16,600	$< 1.0 \times 10^{-5}$	0.79

Table 3 shows our results for two-player Focus, where we again see the same pattern observed in the previous cases, although with an even larger reduction in NC, for reasons that will be discussed below.

Table 3. Results of applying the History-ADS heuristic for two-player Focus.

Midgame?	Update Mechanism	Avg. NC	Std. Dev	P-Value	Effect Size
No	None	5,250,000	10,600	-	-
No	Move-to-Front	**1,290,000 (75%)**	88,000	$< 1.0 \times 10^{-5}$	10.39
No	Transposition	1,800,000	158,000	$< 1.0 \times 10^{-5}$	9.07
Yes	None	10,600,000	-	-	
Yes	Move-to-Front	**2,420,000 (77%)**	637,000	$< 1.0 \times 10^{-5}$	2.37
Yes	Transposition	2,910,000	760,000	$< 1.0 \times 10^{-5}$	2.22

5.2   Results for Multi-Player Games

Table 4 holds our results for the Virus Game. As we have witnessed in the two-player case, the History-ADS obtained substantial gains in pruning, and the Move-to-Front rule outperformed the Transposition rule marginally.

Table 4. Results of applying the History-ADS heuristic for the Virus Game.

Midgame?	Update Mechanism	Avg. NC	Std. Dev	P-Value	Effect Size
No	None	10,500,000	1,260,000	-	-
No	Move-to-Front	**4,690,000 (55%)**	1,010,000	$< 1.0 \times 10^{-5}$	4.57
No	Transposition	4,850,000	739,000	$< 1.0 \times 10^{-5}$	4.45
Yes	None	12,800,000	1,950,000	-	-
Yes	Move-to-Front	**5,940,000 (54%)**	832,000	$< 1.0 \times 10^{-5}$	3.51
Yes	Transposition	6,060,000	974,000	$< 1.0 \times 10^{-5}$	3.45

Table 5. Results of applying the History-ADS heuristic for Chinese Checkers.

Midgame?	Update Mechanism	Avg. NC	Std. Dev	P-Value	Effect Size
No	None	3,370,000	1,100,000	-	-
No	Move-to-Front	**1,250,000 (63%)**	316,000	$< 1.0 \times 10^{-5}$	1.92
No	Transposition	1,320,000	338,000	$< 1.0 \times 10^{-5}$	1.87
Yes	None	8,260,000	-	-	
Yes	Move-to-Front	**3,400,000 (59%)**	899,000	$< 1.0 \times 10^{-5}$	1.84
Yes	Transposition	3,650,000	920,000	$< 1.0 \times 10^{-5}$	1.74

Table 6. Results of applying the History-ADS heuristic for multi-player Focus.

Midgame?	Update Mechanism	Avg. NC	Std. Dev	P-Value	Effect Size
No	None	6,970,000	981,000	-	-
No	Move-to-Front	**2,180,000 (69%)**	184,000	$< 1.0 \times 10^{-5}$	4.88
No	Transposition	2,740,000	271,000	$< 1.0 \times 10^{-5}$	4.32
Yes	None	14,200,000	8,400,000	-	-
Yes	Move-to-Front	**3,240,000 (77%)**	1,730,000	$< 1.0 \times 10^{-5}$	1.30
Yes	Transposition	3,570,000	1,860,000	$< 1.0 \times 10^{-5}$	1.26

Table 5 presents our results for Chinese Checkers. We saw the same pattern emerge as before, with a very large reduction in NC, and the Move-to-Front rule again achieved better performance than the Transposition rule.

Finally, Table 6 has our results for multi-player Focus. The same pattern, seen in all the other cases, is observed yet again.

6   Discussion

Our results strongly reinforce the hypothesis that an ADS managing the move history, employed by the History-ADS heuristic, can achieve improvements in tree pruning through better move ordering, in both two-player and multi-player games. This confirms that such an ADS does, indeed, correctly prioritize the most effective moves, based on their previous performance elsewhere in the tree, as initially hypothesized.

Our results confirm that the History-ADS heuristic is able to achieve a statistically significant reduction in NC in the three two-player games, Othello, Relaxed Checkers, and the two-player variant of Focus, as well as three multi-player games, the Virus Game, Chinese Checkers, and the multi-player variant of Focus. We observed a drastic variation in the reduction between cases, i.e., from 24% to a value as high as 77%, depending on the game. Despite this wide range, these savings are very high, substantially exceeding those obtained by the Threat-ADS heuristic [8]. Savings remain high in both initial and midgame board positions, demonstrating that the History-ADS heuristic can perform well over the course of a game.

Our second observation is that the Move-to-Front rule outperformed the Transposition rule in all cases. This is a reasonable outcome, as the adaptive list may contain dozens to hundreds of elements, and it would take quite a bit of time for the Transposition rule to migrate a particularly strong move to the head of the list. As opposed to this, the Move-to-Front would migrate the move to the front quickly, and would likely keep it there. This phenomenon also confirms that the order of elements within the list matters to the move ordering. In other words, merely maintaining an unsorted collection of moves that have produced a cut will not perform as well as employing an adaptive list, further supporting the use of the History-ADS heuristic.

When considering the wide range of savings between the various games, we observe that in general, the larger the overall game tree, the greater are the proportional savings. There are a number of possible reasons why this is the case. Firstly, we note that the History-ADS retains information on all moves that produce a cut, and in a larger tree, it is likely that there will be more moves of this nature. Thus, the History-ADS has, available to it, more information to work with. Secondly, in games with a very large branching factor, such as Focus or Chinese Checkers, many of the available moves tend to not be very good, and the prioritization of a small number of strong moves can thus prune huge "chunks" of the search space. Lastly, in the case of deeper trees, the History-ADS has more opportunities to apply its knowledge of strong moves gleaned from other levels of the tree, similar to the reasoning behind the History heuristic.

As expected, the History-ADS heuristic achieved noticeably greater savings compared to the Threat-ADS heuristic, which obtained between a 5% and 20% reduction in tree size, with most advantages being around 10% [8,10]. This is not particularly surprising, however, as there are many more possible moves than opponents, and so the History-ADS achieves that improved saving with a larger adaptive list. That the Threat-ADS can, in some cases achieve half of the History-ADS' savings with a list of size 3-4 is, in and of itself, impressive. Thus, it remains a viable option when the more complex History-ADS is not necessary, perhaps due to other move ordering strategies being employed.

7   Conclusions and Future Work

Our work presented in this paper introduces the History-ADS heuristic, a powerful move ordering technique, which employs an ADS to rank moves based on their historical performance within the search. The development of the History-ADS builds upon our previous work applying ADSs to move ordering, specifically the Threat-ADS heuristic. Compared to it, the History-ADS is able to achieve substantially greater savings in terms of tree pruning in a wide variety of cases, and serves to extend the applicability of ADS-based techniques to the scope of two-player games. Given the domain-independent nature of the History-ADS heuristic and its excellent performance in those games tested in this work, we hypothesize that it will perform well in a very wide variety of two-player and multi-player games beyond those explored here.

The History-ADS heuristic represents our initial attempt to extend ADS-based techniques to two-player games, and employ the powerful move history metric. As this is a first attempt, the intuitive idea of storing all moves that produce a cut within a single adaptive list was employed. However, it may be that only a subset of these stored moves are particularly important, and limiting the length of the ADS may preserve the majority of its performance, while using much less memory and time to traverse the list. Furthermore, applying the knowledge of the cuts obtained from all levels of the tree, equally suffers from the issue of disproportionately favouring those moves that are strong near the leaf nodes. Thus another avenue of research is to explore more complex ADSs that mitigate this issue. Alternatively, it is possible that maintaining separate ADSs for different levels of the tree can mitigate this as well. Both of these concepts are the subjects of current and ongoing research.

References

1. Russell, S.J., Norvig, P.: Artificial Intelligence: A Modern Approach, 3rd edn., pp. 161–201. Prentice-Hall Inc., Upper Saddle River (2009)
2. Knuth, D.E., Moore, R.W.: An analysis of alpha-beta pruning. Artificial Intelligence **6**, 293–326 (1975)
3. Schaeffer, J.: The history heuristic and alpha-beta search enhancements in practice. IEEE Transactions on Pattern Analysis and Machine Intelligence **11**, 1203–1212 (1989)
4. Reinefeld, A., Marsland, T.A.: Enhanced iterative-deepening search. IEEE Transactions on Pattern Analysis and Machine Intelligence **16**, 701–710 (1994)
5. Gonnet, G.H., Munro, J.I., Suwanda, H.: Towards self-organizing linear search. In: Proceedings of FOCS 1979, the 1979 Annual Symposium on Foundations of Computer Science, pp. 169–171 (1979)
6. Hester, J.H., Hirschberg, D.S.: Self-organizing linear search. ACM Computing Surveys **17**, 285–311 (1985)
7. Albers, S., Westbrook, J.: Self-organizing data structures. In: Fiat, A., Woeginger, G.J. (eds.) Online Algorithms 1996. LNCS, vol. 1442, pp. 13–51. Springer, Heidelberg (1998)
8. Polk, S., Oommen, B.J.: On applying adaptive data structures to multi-player game playing. In: Proceedings of AI 2013. Thirty-Third SGAI Conference on Artificial Intelligence, pp. 125–138 (2013)
9. Schadd, M.P.D., Winands, M.H.M.: Best Reply Search for multiplayer games. IEEE Transactions on Computational Intelligence and AI in Games **3**, 57–66 (2011)
10. Polk, S., Oommen, B.J.: On enhancing recent multi-player game playing strategies using a spectrum of adaptive data structures. In: Proceedings of TAAI 2013, the 2013 Conference on Technologies and Applications of Artificial Intelligence (2013)
11. Schaeffer, J., Burch, N., Bjornsson, Y., Kishimoto, A., Muller, M., Lake, R., Lu, P., Sutphen, S.: Checkers is solved. Science **14**, 1518–1522 (2007)
12. Polk, S., Oommen, B.J.: Novel AI strategies for *multi*-player games at intermediate board states. In: Ali, M., Kwon, Y.S., Lee, C.-H., Kim, J., Kim, Y. (eds.) IEA/AIE 2015. LNCS, vol. 9101, pp. 33–42. Springer, Heidelberg (2015)

Text-Based Semantic Video Annotation for Interactive Cooking Videos

Kyeong-Jin Oh, Myung-Duk Hong, Ui-Nyoung Yoon, and Geun-Sik Jo[✉]

Department of Computer and Information Engineering, Inha University,
100 Inha-ro Nam-gu, Incheon, Republic of Korea
{okjkillo,hmdgo,entymos13}@eslab.inha.ac.kr, gsjo@inha.ac.kr

Abstract. Videos represent one of the most frequently used forms of multimedia applications. In addition to watching videos, people control slider bars of video players to find specific scenes and want detailed information on certain objects in scenes. However, it is difficult to support user interactions in current video formats because of a lack of metadata for facilitating such interactions. This paper proposes a text-based semantic video annotation system for interactive cooking videos to facilitate user interactions. The proposed annotation process includes three parts: the synchronization of recipes and corresponding cooking videos based on a caption-recipe alignment algorithm; the information extraction of food recipes based on lexico-syntactic patterns; and the semantic interconnection between recognized entities and web resources. The experimental results show that the proposed system is superior to existing alignment algorithms and effective in semantic cooking video annotation.

Keywords: Interactive cooking video · Caption-recipe alignment · Lexico-syntactic pattern · Interconnection · Semantic video annotation

1 Introduction

With the rise of the Internet and the proliferation of smart devices, huge amounts of video data have been generated and consumed by users. People not only watch videos but also want to interact with them. They control slider bars of video players to find and repeat certain scenes, and if they are interested in certain objects in a video, they try to search information on the object by using search engines with keywords. It is difficult to directly support this user interaction in existing video formats because of a lack of metadata applications that can facilitate such interactions. To support user interactions, some annotation tasks for videos are required.

Cooking videos are popular. Many people refer to cooking videos to prepare their meals. In addition, they scroll cooking videos to find specific scenes and seek information on objects appearing in cooking videos, such as ingredients and cooking tools. However, just a few cooking video services provide interaction functions within cooking videos. This paper proposes a text-based semantic video annotation system for interactive cooking videos that can support user interactions. To annotate cooking videos, the proposed system automatically synchronizes each recipe instruction to the

© Springer International Publishing Switzerland 2015
M. Núñez et al. (Eds.): ICCCI 2015, Part I, LNAI 9329, pp. 236–244, 2015.
DOI: 10.1007/978-3-319-24069-5_22

corresponding part of the cooking video through a caption-recipe alignment algorithm. Once the alignment process is complete, the system recognizes ingredients and cooking tools used in each recipe instruction based on lexico-syntactic patterns (LSPs) [8] and semantically interconnects each extracted entity to corresponding semantic web entity having detailed information on the entity. Through semantic video annotation, existing cooking videos are transformed into interactive cooking videos.

The rest of this paper is organized as follows. Section 2 provides the background. Section 3 presents the proposed text-based semantic video annotation system for interactive cooking videos. Section 4 evaluates the proposed system, and Section 5 concludes with some avenues for future research.

2   Background Knowledge

2.1   Video Annotation

With rapid advances in web technologies, huge amounts of multimedia data are generated on the Web. Because of rapid increases in the number of videos, handling video data has become an important topic. Many researchers have focused on handling videos and video annotation plays an essential role this handling.

Approaches based on image processing apply machine learning techniques such as object recognition and object tracking [1][15]. These techniques can successfully achieve high-level features by recognizing objects appearing in videos. However, their annotation performance depends on the accuracy of object recognition with domain characteristics. Other approaches apply text-based information [3][13]. Textual metadata are considered an important factor in the information extraction of video resources. However, because few video resources provide textual information, there is a need to create textual metadata before a video annotation task.

2.2   Interactive Videos

Interactive videos represent a type of video supporting user interactions within the video. Interactive videos incorporate some information management and decision-making capacity as part of the video capability [2][7]. Interactive videos contain clickable areas called "hotspots," which allow users to interact with the video. User interactions through hotspots can be classified into two types. When a user clicks a hotspot for an object, the interactive video displays information on the object. If a hotspot is not for an object, then the interactive video moves to a different part of the video or opens another video. Video annotation is a core technique for generating interactive videos. To annotate videos, it is essential to acquire timestamps and instructions as textual descriptions. Using interactive video annotation tools such as Wirewax [14] allows the user to make definitive interactive videos manually, but the accountability of annotation tasks is delegated to the user.

3   Text-Based Semantic Video Annotation for Interactive Cooking Videos

3.1   System Architecture

This section describes the proposed semantic video annotation system. If perfect closed captions on cooking videos are given, then cooking video annotation without any alignment between captions and recipes is possible. However, automatically generated captions through machine learning techniques are not complete. Consequently, through caption-recipe alignment, each recipe instruction corresponding to an incomplete caption sentence should be identified, and recipe-related information belonging to the instruction is annotated in the appropriate part of the cooking video. Fig. 1 shows the system architecture for the proposed system. After the caption-recipe alignment, recipe-related information such as ingredients and cooking tools is extracted, and this information is interconnected with the semantic entity having detailed information. Based on identified information, cooking videos are semantically annotated.

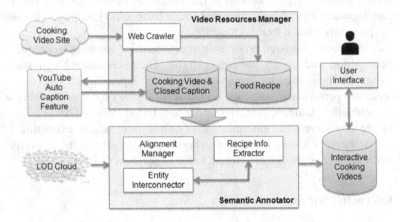

Fig. 1. The system architecture for the proposed interactive cooking video system

3.2   Caption-Recipe Alignment Algorithm

Fig. 2 provides an example of a closed caption and a recipe. A closed caption includes time information, but its text is not complete and has misrecognized words. Recipe instructions are complete sentences but have no time information in the cooking video. To annotate a cooking video, each part of the cooking video corresponding to each instruction of a food recipe should be identified.

Closed Caption	Recipe Instructions
7 00:00:44,030 --> 00:01:00,469 drain the pasta and toss it well with one generous tablespoon olive oil and set aside	In a large pot of boiling salted water, cook spaghetti pasta until al dente. Drain well. Toss with 1 tablespoon of olive oil, and set aside.
8 00:01:03,050 --> 00:01:18,700 now days slices uncooked bacon then in a large skillet cook the bacon into a slightly Press	Meanwhile in a large skillet, cook chopped bacon until slightly crisp

Fig. 2. An example of closed captions and recipes

The caption-recipe alignment task entails finding subsequences that are virtually identical. The task is mapped to the identification of Longest Common Subsequence (LCS) between two different documents. The LCS problem is solved by applying Dynamic Programming (DP) technique [3][13]. Fig. 3 shows the caption-recipe alignment algorithm [11]. The algorithm uses sentences in closed captions and recipes as its input and produces an aligned pair list. Functions on lines 15 and 16 are from the traditional DP algorithm for identifying the LCS. A similarity matrix is constructed using sentences. Each element of the matrix is filled with similarity values.

```
Input: cooking video caption sentence sequence C;
recipe sentence sequence R;
Output: aligned pair list L;
01 generateAlignedPairList(C, R)
02 let α be a threshold value
03 let S be a m x n matrix
04 let c be a m x n matrix // computing the length of an LCS
05 let b be a m x n matrix // simplifying construction of an optimal solution
06 m ← length[C]
07 n ← length[R]
08 call computeCRSimilarity(C, R, S) // caption-recipe similarity
09 α ← threshold // using (1)
11 for each c ∈ C
12    for each r ∈ R
13       if S[c, r] >= α then c[c, r] ← 1
14       else c[c, r] ← 0
15 call LCS-LENGTH(C, R, c, b)
16 call PRINT-LCS(b, C, m, n)
01 computeCRSimilarity(C, R, S)
02 set all elements to 0 in matrix S
03 for each c ∈ C
04    for each r ∈ R
05       S[c, r] ← Sim(c, r) // cosine similarity
```

Fig. 3. The caption-recipe alignment algorithm

To measure the similarity value between two sentences, a cosine similarity measure for the vector space model can be used. In contrast to the traditional LCS process, each element of the sentence-based matrix has a value between 0 and 1. A threshold value of cosine similarity should be determined to determine whether a similarity value in an element is valid. The threshold value can be determined as follows:

$$\text{Min\_thres} = \text{avg(sim)} + \alpha \times \sigma \text{ (sim)} \tag{1}$$

where, avg(sim) refers to the average value of all cosine similarity values and σ is the standard deviation. In addition, α is a parameter, and 0.75 is assigned through parameter adjustment. Optimal paths in the similarity matrix are detected by applying the DP algorithm based on the threshold value. Once the caption-recipe alignment process is complete, the time for each recipe instruction is identified.

3.3   Information Extraction Using Lexico-Syntactic Patterns

Interactions with objects appearing in videos represent an important feature for users watching the video. For this, interactive videos should provide relevant information when the user interacts with an object in an interactive video. In the case of cooking videos, information such as ingredients, ingredient portions, and cooking tools should be provided. Traditional rule-based Named Entity Recognition (NER) techniques generally rely on a small set of patterns to identify relevant entities from normal text and often require a gazetteer list for the NER technique. LSP is applied for information extraction to overcome the limitations [8]. LSPs are based on text tokens and syntactic structures of text for a string matching pattern. The information extraction task typically achieves high precision if LSPs are applied to the task [10][12].

Table 1 shows some LSPs used to extract information.

Table 1. Examples of LSPs for the information extraction

LSP Category	LSP	Examples
Chef	Recipe by {NNP}*	Recipe by Peter
Ingredients & Ingredient Portions	CD pound {VBD}{NN}* CD tablespoon {NN}*	1 pound uncooked spaghetti 1 tablespoon olive oil
Cooking Tools	IN {DT}{JJ}{NN}*	In a large pot

To extract useful information, sentences of recipes were analyzed, and 23 LSPs were defined for the proposed method. Patterns of chefs, ingredients, ingredient portions and cooking tools are defined in the LSPs.

3.4   Semantic Interconnections for Offering Detailed Information

Interactive videos should provide detailed information on objects such as ingredients and cooking tools when the user interacts with the video. To provide this information, entities which are extracted based on LSPs, should be connected external sources with detailed information on them. The proposed system interconnects the entities to Wikipedia and DBPedia resources as the linked data format of Wikipedia. DBPedia can be queried through SPARQL Endpoint with a SPARQL query made using a certain entity. However, SPARQL Endpoint sometimes returns no results because the query is sensitive at the morphology. To solve this limitation, the proposed approach uses the fact that all entities of DBPedia are created based on each page of Wikipedia and that DBPedia and Wikipedia use the same local identifier. Wikipedia provides a redirection function. This function handles capital and small letters and synonyms and shows redirected pages with a notice that the page is redirected from a given URL.

4   Experiments

The proposed text-based semantic video annotation system requires resources such as cooking videos, food recipes, and text captions. Through the developed web crawler, cooking videos and recipes are extracted from allrecipes.com[1]. Text captions for cooking videos are generated using YouTube's auto-caption feature.

To verify the proposed system, a caption-recipe alignment algorithm, a LSP-based information extraction method, and a semantic interconnection method were evaluated. The caption-recipe alignment algorithm was evaluated using benchmark algorithms associating videos with related text data. The evaluation was performed using 20 cooking videos and food recipes. Table 2 shows alignment accuracy according to each algorithm.

Table 2. Experimental results for caption-recipe alignment

Approach	Proposed Method	Naïve DP Method	Cooking Navi
Recall	0.857	0.779	0.757
Precision	0.89	0.809	0.787
F-Measure	0.873	0.793	0.772

The accuracy of each alignment approach was measured by calculating the ratio of successfully aligned instructions for recipes to sentences in the closed captions. The naïve DP method associates recipe instructions to sentences in closed captions based on words [13]. The Cooking Navi method associates each instruction for recipes with cooking videos by using the sum of the scores derived through three steps [6]. The proposed method was found superior to both of these methods. The experimental results reveal the reason why these two methods produced lower accuracy than the proposed method. The naïve DP method constructs a similarity matrix based on words and finds an optimal path. An optimal path should form a diagonal line near to the 45-degree angle to obtain better alignment performance. However, the difference between declarative sentences in captions and imperative sentences in recipes produced an optimal path with a serpentine course. In addition, alignment performance was reduced because incomplete captions had a negative impact on the construction of an optimal path. The Cooking Navi method uses combined information derived from the ordinal structure of a recipe, background information of video scenes, and co-occurrences of words in captions and recipes. However, ordinal structures not well built in some recipes had unfavorable effects. The background information and verbs in instructions was associated with four types: board, table, range, and others. This dimension reduction led to the loss of information and the deterioration of alignment performance. The proposed caption-recipe alignment algorithm showed greater accuracy than the compared algorithms by eliminating winding paths and information loss.

Table 3 shows the accuracy of information extraction based on LSPs. The LSP-based information extraction method was highly precise, but LSPs remained incom-

---

[1]   http://allrecipes.com/

plete. In the case of "dry white wine", the LSP-based method handled "dry" as an adjective, which reduced its performance. For cooking tools, two cooking tools "skillet" and "pan" were different syntactically, but they were same tool in the cooking video. This indicates a need for a method that can address the synonym problem.

Table 3. The Accuracy of Information Extraction Based on LSPs

	Chef	Ingredient	Ingredient Portion	Cooking Tool
Precision	20 / 20 (100%)	208 / 227 (91.6%)	219 / 227 (96.5%)	93 / 93 (100%)
Recall	20 / 20 (100%)	208 / 244 (85.2%)	227 / 253 (89.7%)	93 / 105 (88.6%)

As shown in Table 4, using the redirection function of Wikipedia produced high accuracy in the semantic interconnection.

Table 4. The accuracy of semantic interconnection

Type	Ingredient	Cooking Tool
Precision	73 / 73 (100%)	11 / 11 (100%)
Recall	73 / 76 (96%)	11 / 13 (84.6%)

However, cooking tools such as "non-stick skillet" and "9*13 inch baking dish" were not interconnected through the proposed approach.

Fig. 4 shows an example of relevant information annotated to a cooking video through the proposed system. A recipe instruction is annotated to a designated cooking video part through the proposed caption-recipe alignment algorithm. Ingredients and cooking tools extracted from recipe instructions are interconnected with Wikipedia resources by the semantic interconnection, and are annotated to cooking videos.

Fig. 4. An Example of Semantic Cooking Video Annotation

5   Conclusions and Future Research

This paper proposes a text-based semantic video annotation system that can generate interactive cooking videos and support user interactions. For this, the system automatically synchronizes each instruction of recipes to each part of the cooking video through the proposed caption-recipe alignment algorithm. The system extracts information on ingredients and cooking tools by using LSPs and interconnects each extracted entity to the Wikipedia entity as things of the semantic web. Based on this information, the system automatically generates interactive cooking videos by semantically annotating cooking videos. The experimental results show that the proposed system is superior to existing algorithms from an alignment accuracy perspective. The LSP-based information extraction method achieved high accuracy (over 95%) for given LSPs. The semantic interconnection method also produced high accuracy for interconnecting ingredients and cooking tools. Although the proposed method produced satisfactory results for semantic video annotation, the proposed caption-recipe alignment algorithm needs to be improved through weighting techniques [4][9].

Acknowledgement. This work was supported by Institute for Information & communications Technology Promotion (IITP) grant funded by the Korea government (MSIP) (No. R0101-15-0054, WiseKB: Big data based self-evolving knowledge base and reasoning platform)

References

1. Ballan, L., Bertini, M., Bimbo, A.D., Seidenari, L., Serra, G.: Event detection and recognition for semantic annotation of video. J. Multimedia Tools and Applications **51**(1), 279–302 (2011)
2. Bellman, S., Schweda, A., Varan, D.: A Comparison of Three Interactive Television AD Formats. Journal of Interactive Advertising **10**(1), 14–34 (2009)
3. Cour, T., Jordan, C., Miltsakaki, E., Taskar, B.: Movie/script: alignment and parsing of video and text transcription. In: Proceeding of the 10th European Conference on Computer Vision: Part IV, pp. 158–171 (2008)
4. Guo, W., Diab, M.: A simple unsupervised latent semantics based approach for sentence similarity. In: Proceeding of First Joint Conference on Lexical and Computational Semantics, pp. 586–590 (2012)
5. Hamada, R., Okabe, J., Ide, I., Satoh, S., Sakai, S., Tanaka, H.: Cooking navi: assistant for daily cooking in kitchen. In: Proceeding 13th annual ACM International Conference on Multimedia, pp. 371–374 (2005)
6. Hamada, R., Miura, K., Ide, I., Satoh, S., Sakai, S., Tanaka, H.: Multimedia Integration for Cooking Video Indexing. In: Aizawa, K., Nakamura, Y., Satoh, S. (eds.) PCM 2004. LNCS, vol. 3332, pp. 657–664. Springer, Heidelberg (2004)
7. Homer, B.D., Plass, J.L.: Level of Interactivity and executive functions as predictors of learning in computer-based chemistry simulations. Journal of Computers in Human Behavior **26**, 365–375 (2014)
8. Jacobs, P.S., Krupka, G.R., Rau, L.F.: Lexico-semantic pattern matching as a companion to parsing in text understanding. In: Proceeding of the Workshop on Speech and Natural Language, Collocated with the 6th Human Language Technology Conference, pp. 337–341 (1991)
9. Liu, Y., Liang, Y.: A Sentence Semantic Similarity Calculating Method based on Segmented Semantic Comparison. Journal of Theoretical and Applied Information Technology **48**(1), 231–235 (2013)

10. Maynard, D., Funk, A., Peters, W.: Using lexico-syntactic ontology design patterns for ontology creation and population. In: Proceeding of the Workshop on Ontology Patterns, Collocated with the 8th International Semantic Web Conference, pp. 39–52 (2009)
11. Oh, K.J., Hong, M.D., Sim, S.Y., Jo, G.S.: Automatic indexing of cooking video by using caption-recipe alignment. In: Proceeding of IEEE International Conference on Behavior, Economic and Social Computing (BESC), pp. 1–6 (2014)
12. Panchenko, A., Morozova, O., Naets, H.: A semantic similarity measure based on lexico-syntactic patterns. In: Proceeding of the 11th Conference on Natural Language Processing (KONVENS), pp. 174–178 (2012)
13. Turetsky, R., Dimitrova, N.: Screenplay alignment for closed-system speaker identification and analysis of feature films. In: Proceeding of IEEE International Conference on Multimedia and Expo, pp. 1659–1662 (2004)
14. WireWax: Interactive Video Annotation Tools. https://www.wirewax.com
15. Wu, J., Worring, M.: Efficient Genre-Specific Semantic Video Indexing. IEEE Transactions on Multimedia 14(2), 291–302 (2012)

Objects Detection and Tracking on the Level Crossing

Zdeněk Silar[✉] and Martin Dobrovolny

Faculty of Electrical Engineering and Informatics,
University of Pardubice, Studentska 95, 53002 Pardubice, Czech Republic
{zdenek.silar,martin.dobrovolny}@upce.cz

Abstract. In this article is presented algorithm for obstacle detection and objects tracking in a railway crossing area. The object tracking is based on template matching and sum of absolute differences. The object tracking was implemented for better reliability of presented system. For optical flow estimation is used a modified Lucas-Kanade method. The results of proposed algorithm were verified in a real traffic scenarios consisted of two railway crossings in Czech Republic during 2013–14 under different environmental conditions.

Keywords: Level crossing monitoring · Objects detection · Optical flow velocity vectors · Template matching · Objects tracking · K-means clustering

1   Introduction

This article describes a method for object detection in the railway crossing area. The presented methods were evaluated in project SGS FEI_2014002, "The System for Image Analysis of Space Occupancy and Unknown Space Exploration". For better reliability objects tracking was implemented in the system during 2015 (SGSFEI 2015002). The solution of level crossings security problem is a very important aspect to reduce accidents at level crossings by finding new methods such as vision based method.

Proposed occupancy status detection is based on image processing methods only. For this purpose a new algorithm was developed for reliable objects and background separation. This algorithm uses a modified optical flow estimation based on the Lucas-Kanade method. Velocity vectors obtained from the optical flow estimation are processed by the clustering algorithm.

The original algorithm for the moving objects mask estimation is based on the cumulative method and its detailed description is given in Section 3. In second part of the article is presented the proposed method for effective clustering of the optical flow vectors using K means method [1], [2]. Finally, results of the experiments performed on real traffic data are described using suitable input data.

2   Current State

The problematic of clearance detection and object detection in 3D railway crossing area is a primary aim of the railway crossing security. Lot of partial researches has

© Springer International Publishing Switzerland 2015
M. Núñez et al. (Eds.): ICCCI 2015, Part I, LNAI 9329, pp. 245–255, 2015.
DOI: 10.1007/978-3-319-24069-5_23

been conducted in this area, for example people detection in the train track [3], [4] or general obstacle detection [5], [6] etc. with common aim of the object detection. Unfortunately a system for complex solving of clearance detection in railway crossing area is currently missing. The objective of this research is to detect incoming, staying and outgoing objects reliably in the area of level crossing.

Usually for the object detection are used methods based on a radar sensors [3], [7], [8], analysis of infra-red images [5], [9], multi sensor fusion or a vision based systems in fusion with other sensors [10], [11], [12], [13]. The common problem for vision based detection methods is the background estimation. Application of standard edge detectors on the complicated background usually leads to the over-segmentation [14]. For the background estimation is usually used the cumulative method [15], [16]. This method is suitable for the situations with stationary background - our case. For the classification is necessary to determine the direction and speed of moving objects. Our approach uses the optical flow estimation [17] and clustering for robust obstacle detection as well as for the object trajectory determination.

Currently, the authors do not know about a similar system that comprehensively and automatically evaluates the railway crossing occupancy.

3   The System for Occupancy Detection at Level Crossing

The proposed system was realised as a modular system (Fig. 1). After basic pre-processing we first process input data by optical flow. For reliable occupancy detection we also estimate the objects masks in the same time. This solution significantly improves the processing time. The objects masks are detected by the background estimation method after first step. The calculations are based on acquired and adequately pre-processed data.

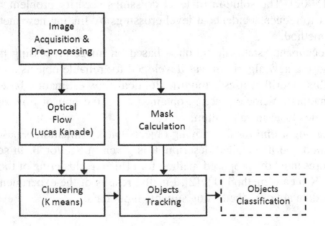

Fig. 1. Architecture of Clearance Detection system.

3.1   Image Acquisition and Pre-Processing

The image data in the proposed system are acquired by CCTV camera. The camera continually takes images of railway crossing. Images were transferred over rail network technology [20]. Images were resized and normalized in range <0;1> and processed in the IEEE754 format. Next approach was completely processed in brightness channel.

3.2   Optical Flow Estimation for Dynamic Objects Detection

Optical flow estimation is computationally demanding. In our case we chose the Lucas-Kanade (L-K) method for its advantages described in [17]. The L-K method is among the fastest and therefore most widely used methods for calculation of optical flow. The L-K method introduce the error term ρ_{LK} for each pixel (2).

$$\rho_{LK} = \sum_{x,y \in \Omega} [\nabla I(x,y,t) \cdot \vec{v} + I_t(x,y,t)]^2 .$$

(1)

where Ω is neighbourhood of the pixel.

To find a minimal error it is necessary to compute derivation of the error term ρ_{LK} by individual components of velocity and put the result equal to zero. After finding a minimal error, several adjustments and after transfer to a matrix form the expression for the optical flow calculation is as follows:

$$\vec{v} = [A^T A]^{-1} A^T \vec{b} .$$

(2)

where for N pixels ($N = n^2$, for n x n of Ω neighbourhood) and $(xi,yi) \in \Omega$ in time t holds

$$A = [\nabla I(x_1, y_1), ..., \nabla I(x_N, y_N)],$$
$$\vec{b} = -(I_t(x_1, y_1), ..., I_t(x_N, y_N)).$$

(3)

So we will obtain the resultant velocity for one pixel by solution of the system (3). Instead of the calculation of the sums, the convolution was used to reduce the algorithm complicacy. The Fig. 2 shows an example of optical flow vectors computed by L-K for different size of picture as well as the computation time.

However, the problem is reliable distinction between passing objects in a traffic scene. This problem was solved by modification based on clustering of optical flow vectors.

Fig. 2. The optical flow vectors estimated by algorithm L-K (2).

3.3   Mask Calculation for Clustering Centroids Determination

We used the cumulative method for background estimation (Fig. 3). In our case the background image is an image without the cars, pedestrians and other obstacles.

N = 100

Fig. 3. Estimated background (left picture) and moving objects (N represents the number of signal realizations).

The principle of the mask obtaining is based on the subtraction of the background model from the current image I_{i+t}. Using the above-obtained background matrix B_N and the current frame I_{i+t} is thus possible to obtain the matrix M at any time. The matrix M (mask) represents the dynamic objects in a scene.

$$M = B_N - I_{i+t}.$$ (4)

The main disadvantage of this approach is the significant over-segmentation [18]. Therefore, next process is applied.

Fig. 4. Determination of the centroid by a mask (the original image, the mask according to equation (1), the binary image with a marked median-centroid and frame).

The transformation of the mask M into a binary image was performed using thresholding (removes noise from the image). Subsequently there was performed morphological. The whole centroids determination procedure (for next K-means clustering application) is shown in Fig. 4.

The disadvantage of this procedure is the lack of information about the movement of objects. For this purpose we used the method of optical flow estimation.

3.4 Separation of Detected Objects Using Clustering

The clustering was performed on the optical flow vectors obtained by *L-K* method. For these purposes the transpose matrix of complex optical flow vectors V^{MxN} was reshaped (stacked) on vector of dimension $(M*N) \times 1$.

The result is a reshaped column vectors with the parameters (amplitude and angle). Subsequently there was created the four-column matrix A (5), where the first two columns contains the pixels coordinates, the second and third the amplitude mag_{ij} and angle φ_{ij} of optical flow vectors, which was computed for the corresponding pixel of the original reshaped matrix V.

$$A = \begin{bmatrix} 1 & 1 & mag_{11} & \varphi_{11} \\ 1 & 2 & mag_{12} & \varphi_{12} \\ & & \vdots & \\ 1 & N & mag_{1N} & \varphi_{1N} \\ 2 & 1 & mag_{12} & \varphi_{21} \\ & & \vdots & \\ M & N & mag_{MN} & \varphi_{MN} \end{bmatrix} \tag{5}$$

Each row in the matrix A represents a one four-dimensional object that contains the suitable data for K-means clustering algorithm. The A matrix and the center coordinates thus represent the input data for the K-means algorithm. The outputs are then the specified coordinates for n centroids and the assignment of cluster vectors.

3.5 Criterion for Formation of the New Clusters (Euclidean Distance)

The used criterion for the new clusters formation was the degree of similarity between objects (cars, pedestrians, etc.). It was represented by the Euclidean distances between parameters in matrix A, for which the calculation was:

$$d_E(o_k, o_l) = \sqrt{\sum_{i=1}^{n} (o_{ki} - o_{li})^2} . \tag{6}$$

We used the parametric non-hierarchical clustering. For the set of n vectors x_j ($j=1,\ldots,n$) is performed division in to the c clusters $G_i.(i=1,\ldots,c)$. The assessment function, the minimum we are looking for, is defined as:

$$J = \sum_{i=1}^{c} J_i = \sum_{i=1}^{c} \left(\sqrt{\sum_{k, x_k \in G_i} (x_k - c_i)^2} \right). \tag{7}$$

where J_i are distances within one group i.

To minimize the number of iteration steps in the K-means were the initial centroids determined precisely as two-dimensional medians of detected objects mask (using the above-described Background Estimation method).

3.6   Passing Objects Separation

The algorithm worked correctly only in the case of completely separated objects. When more objects overlaps, there origin problem with detection. For these objects only one median – the centroid was computed. For this reason the described method based on directional properties of the optical flow vectors has been proposed.

The calculated direction φ determines optical vector memberships to one of selected groups. The groups are defined by auxiliary vector h:

$$h_i = \begin{cases} 1 & for \quad \varphi(p_i) \in \langle \alpha_1; \alpha_2 \rangle \\ 2 & for \quad \varphi(p_k) \in \langle \beta_1; \beta_2 \rangle \end{cases} \quad i \neq k .$$

(8)

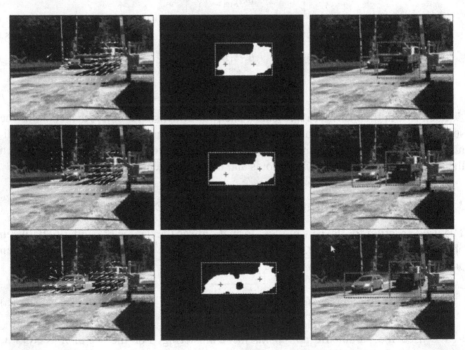

Fig. 5. K-means clustering and separation of passing objects (mask, optical flow clustering, objects separation).

In this 1 or 2 means selected group, α_1 and α_2 define the angle range for searching in one direction, β_1 and β_2 conversely. The angle ranges are disjunctive to each other.

Functionality of the proposed algorithm is presented in Fig. 5. Clustering works correctly even in the case of overlapping objects for which the algorithm based on the directional optical flow is able to detect objects moving in the opposite directions.

3.7   Objects Tracking

For the reliability improvement was during 2014 implemented objects tracking in our system. By objects tracking we are able precisely monitoring cars or pedestrians in the railway crossing area. The optical flow estimation usually fails in the situations, when objects suddenly stop in the area. The only information source in these cases was presented background estimation algorithm. By objects tracking we achieved complex information about objects trajectory from the point of entry into the scene to the moment of leaving.

For objects tracking we chose the Template Matching based on Motion Estimator approach [19]. The template – the tracked object - was estimated from background estimation. (Chapter 3.3) The criterion was in first approach based on two-dimensional correlation:

$$R_{corr^{2D}} = \frac{\sum_x \sum_y (S_{xy} - u_T)(T_{xy} - u_T)}{\left(\sum_x \sum_y (S_{xy} - u_S)^2\right)\left(\sum_x \sum_y (T_{xy} - u_T)^2\right)}. \tag{9}$$

where: S – Represents ROI in processed image, T – The template with located object, u_S and u_T – represent mean values of S and T matrix.

For the speed improvement was the searching realized only in neighbouring region. The object tracking works well even in poor light conditions. Later was in the system implemented Template Matching based on estimation of Sum of Differences:

$$R_{SAD^{2D}} = \sum_x \sum_y abs(I^i(x + u, y + v) - T^i(x, y)). \tag{10}$$

This solution brings noticeable speed improvement without impact on localization accuracy.

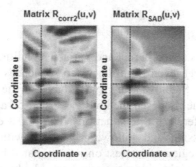

Fig. 6. Examples of $R_i(u,v)$ matrix calculated by 2D correlation and SAD (Blue colour represents minimum, red colour maximum).

In both cases we are locating for a global extreme of R^{2D} function. On the Fig. 6 is presented resulting accumulator for selected ROI. The localized global extreme represent object corner coordinates. The localization is performed repeatedly for every object and in every new picture. By this approach is possible estimated new trajectory of all tracked objects. The calculation is well parallelized.

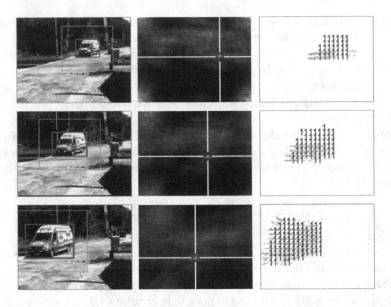

Fig. 7. Example of $R_i(u,v)$ matrix with global extreme and corresponding tracked object.

On the Fig. 7 is presented result of tracking algorithm for selected object. The ROI is depicted by green color. Size of ROI is selected according to expected maximal velocity of objects in scene. New mask is calculated and resized.

4   Experiment

The extended method was tested in real and variable conditions. For this purpose, the individual images were retrieved from the video sequences that have been captured from the railroad crossing Steblova near city Pardubice in the Czech Republic.

The functionality of the proposed method was tested on a wide range of video sequences representing different traffic situations. The Fig. 8 demonstrates example of correct detection of passing and following vehicles with tracking implemented. On the left is presented region of interest on the estimated background calculated by cumulative method. The right side presents detected objects with highlighted and tracked car in the railway crossing area.

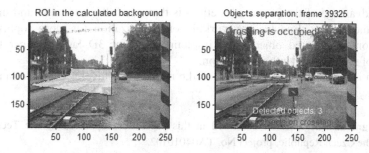

Fig. 8. The objects separation example.

We had also the opportunity to test the robustness of the developed method, on the data acquired from industrial camera watching level crossing in the city district Pardubice-Slovany. The camera watches this dangerous level-crossing in continuous operation and the low quality images are stored on backup media for future use. The obtained results were acceptable. The Fig. 9 shows that the proposed method can be used for lower quality data.

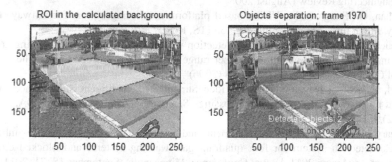

Fig. 9. Estimation of the optical flow and results of clustering operations over lower quality data.

5   Conclusion

The proposed procedure allows overview and documentation detection of the railway crossing clearance. An adaptive background learning method is proposed and applied to a real life traffic video sequence. We are able to obtain more accurate information about the intrusion objects. The system reliability was increased by object tracking implementation. The system maintains information about all objects. The benefit of this method is a significant improvement of objects outlines detection accuracy. The method also provides information about the speed and the direction of objects by application of clustering on optical flow vectors, thus it is possible reliably distinguish between objects moving in different directions.

A key advantage of the proposed method is that it is fully automatic and unsupervised. At the present time we are working on optimization for more sophisticated classification of detected objects and implementation of 3D laser scanner for fast overview of scene with depth information.

Our aim is to increase the reliability of the method to a level acceptable in railway signalling systems.

Acknowledgment. The research described in this paper was supported by the Technology Agency of the Czech Republic, project No. TA04010102.

References

1. Barron, L., Fleet, D.J.: Performance of Optical Flow Techniques. International Journal of Computer Vision **12**(1), 43–77 (1994)
2. Lucas, B.D., Kanade, T.: An iterative image registration technique with an application to stereo vision. In: DARPA Image Understanding Workshop, pp. 121–130 (1981)
3. Hisamitsu, Y., Sekimoto, K.: 3-D Laser Radar Level Crossing Obstacle Detection System. IHI Engineering Review (August 2008)
4. Sehchan, O., Sunghuk, P.: Vision based platform monitoring system for railway station safety. In 7th International Conference on ITS, Korea, June 6–8, 2007
5. Takeuchi, H., Shozawa, T.: Obstacle detection with automatic pedestrian tracking at level crossings using multiple single-row laser range scanners. In: Proceedings of The International Conference ASPECT2006, No. 8 (2006)
6. Hazel, E.: Recent Developments in Video Surveillance: Intelligent Surveillance System Based on Stereo Vision for Level Crossings Safety Applications. InTech, USA, chap. 5 (2012). ISBN 978-953-51-0468-1
7. Hilleary, T.N., John, R.S.: Development and testing of a radar-based non-embedded vehicle detection system for four quadrant gate warning systems and blocked crossing detection. In: Arema 2011 Annual Conference, Minneapolis, September 18–21, 2011
8. Xue, J., Cheng, J.: Visual monitoring-based railway grade crossing surveillance system. In: CISP 2008 Proceedings of the 2008 Congress on Image and Signal Processing, vol. 2, pp. 427–431. Washington (2008)
9. Mockel, S., Schefer, F.: Multi-sensor obstacle detection on railway tracks. In: Intelligent Vehicles Symposium, pp. 42–46, June 9–11, 2003
10. Horimatsu, T.: ITS Sensor for Railroad Crossing Safety. Fujitsu Journal, Japan **57**(5), 545–550 (2006)
11. Iwata, K., Nakamura, K.: Objects recognition in traffic scenes by using multiple laser range scanners. In: The 25th Asian Conference on Remote Sensing (2004)
12. Fakhfakh, N., Khoudour, L.: A Video-Based Object Detection System for Improving Safety at Level Crossings. Open Transportation journal, Safety at Level Crossings 15 (2010)
13. Zhang, Ch., Chen, S.: Adaptive background learning for vehicle detection and spatio-temporal tracking. In: ICICS-PCM 2003, Singapore, Decenber 15–18, 2003
14. Reddy, V., Sanderson, C.: A Low-Complexity Algorithm for Static Background Estimation from Cluttered Image Sequences in Surveillance Contexts. EURASIP Journal on Image and Video Processing (2011)

15. Dang, T., Hoffmann, Ch.: Fusing optical flow and stereo disparity for object tracking. In: The IEEE 5th International Conference on Intelligent Transportation Systems, Singapore, September 3–6, 2002
16. Moon, H., Chellappa, R.: Optimal Edge-Based Shape Detection. IEEE Transactions on Image Processing **11**(11), 1209–1226 (2002)
17. Fakhfakh, N., Khoudour, L.: A Video-Based Object Detection System for Improving Safety at Level Crossings. Open Transportation Journal, Supplement on "Safety at Level Crossings" 15 (2010)
18. Horalek, J., Matyska, J., Sobeslav, V.: Communication protocols in substation automation and IEC 61850 based proposal. In: Proceedings of the CINTI 2013–14th IEEE International Symposium on Computational Intelligence and Informatics, art. no. 6705214, pp. 321–326 (2013)

Spatial-Based Joint Component Analysis Using Hybrid Boosting Machine for Detecting Human Carrying Baggage

Wahyono and Kang-Hyun Jo[✉]

Graduate School of Electrical Engineering, University of Ulsan, Ulsan 680-749, Korea
wahyono@islab.ulsan.ac.kr, acejo@ulsan.ac.kr

Abstract. This paper introduces a new approach for detecting and classifying baggage carried by human in images. The human region is modeled into several components such as head, body, foot and bag. This model uses the location information of baggage relative to human body. Features of each component is extracted. The features are then used to train boosting support vector machine (SVM) and mixture model over component. In experiment, our method achieves promising results in order to build automatic video surveillance system.

Keywords: Carried baggage detection and classification · Joint component · Video surveillance · HOG · Boosting machine · Mixture model · Support vector machine

1 Introduction

In automatic video surveillance system, detecting human carrying baggage is potentially important objective for security and monitoring in public space. In practice, the human-baggage detector can be used for doing such a task. However, the task is inherently difficult due to the wide range of baggage that can be carried by a person and the different ways in which they can be carried. In the literature, there have been several approaches proposed for detecting baggage that abandoned by the owners [1,2] or still being carried [3–5] by them. Tian [1] proposed a method to detect abandoned and removed object using background subtraction and foreground analysis. In their approach, the background is modeled by three Gaussian mixture that combining with texture information in order to handle lighting change conditions. The static region obtained by background subtraction is then analyze using region growing. This region is then classified as abandoned or removed object by some rules. However, in some cases this method produces many false alarm due to imperfect background subtraction. To overcome this problem, Fan [2] proposed relative attributes schema to prioritize alerts by ranking candidate region. However, in real implementation to know who is the owner of abandoned baggage is very important. Therefore, as prior process, the system should capable to detect the person who carried baggage. The authors from [3,4] proposed same concept to detect carried object by

© Springer International Publishing Switzerland 2015
M. Núñez et al. (Eds.): ICCCI 2015, Part I, LNAI 9329, pp. 256–264, 2015.
DOI: 10.1007/978-3-319-24069-5_24

people. They utilized the sequence of human moving to make spatial temporal template. It was then aligned against view-specific exemplar generated offline to obtain the best match. Carried object was detected from the temporal protrusion. The author in [4] extend the framework such that the system can also classify the baggage type based on the position in relevance to the human body carrying it. However, the method assumes that parts of the carried objects are protruding from the body silhouettes. Due to its dependence on protrusion, the method cannot detect non protruding carried object. The protruding problem can be solved by method from [5]. This method utilized ratio color histogram. Using assumption of the color of carried object is different with clothes, it will achieve good result in accuracy. However, this method is dependence on event where the bag being transferred or left. The assumption of observing the person before and after the change in carrying status is application specific and cannot be used as a general carried object detector.

This paper proposes a novel approach for detecting human carrying baggage. Our approach utilizes the strong connection between baggage and body component. Instead of constructing model for entire of object, our approach builds a model for each component [8] of body including the baggage component according to possible placement. Adopted from [4] with some improvements, spatial model of baggage is used for detecting human carrying baggage. Overall, the major contributions of this paper are following; 1) Part-based model schema and the relationship among them specific for detecting person carrying baggage. 2) Bag part mixture model for solving strong variation problem of baggage (e.g., location, size, shape, and color).

2   The Proposed Approach

This section addresses the detail of our approach in order to detect human carrying baggage. Our approach utilizes spatial information of bag on the human body. First, body proportions is used for estimating spatial parameters. The human body is then modeled into several component of body including bag component that placed in possible placement. Feature extraction is applied on full body, as root, and all components. The placement of bag component is utilized for classifying bag categories. As bag may have variation in size, color, and appearance, the bag component is modeled as a mixture component. Detection involves searching over arrangement of component based on geometric relationship among components relative to full body. This is done separately by scoring each joint component model.

2.1   Body Proportions

Regarding human body proportions, as shown Fig. 1(a), the average of the height and weight of a person are defined by $H = 8h$, $W = 2h$, respectively, where h is the height of head, such that $h = H/8$. The center of the body in vertical axis denotes bend line B position. Vertical line C is denoted as the center of body in

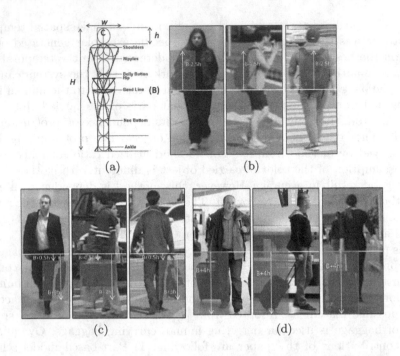

Fig. 1. Spatial model of baggage. Human body proportions (a). The category of bag is divided into three major categories relative to bend line location (white line); (b) backpack or hand bag, (c)tote or duffle bag, and (d)rolling luggage.

horizontal axis that traverse the centroid of body. Let define T be the position of the top of the head in the image, and L be the most left location of body in the image, then $B = T + 4h$ and $C = L + h$. These spatial parameters are employed for making our human carrying baggage model described in detail in the next subsection.

2.2   Spatial-Based Human Carrying Baggage Model

The main idea of our spatial model is by determining the possible placement of baggage correspond to the body proportion and the direction of view. Training images are collected, and the location of baggage is determined manually. Regarding to our training images, backpack and handbag are mostly located around or the top of bend line. Tote bag, duffle bag and rolling bag are located in the bottom of bend line with different average of height. Thus, our spatial model is divided in into three major categories, 1) backpack or hand bag, 2) tote bag or duffle bag and 3)rolling luggage. As shown in Fig. 1(b)-(d), spatial models of baggage defines the set of conditions for checking whether the bag exists or not. In practical, if our model detect that location depicted in Fig. 1(b) as bag with high probability value, then the bag is classified as a backpack; if not, then

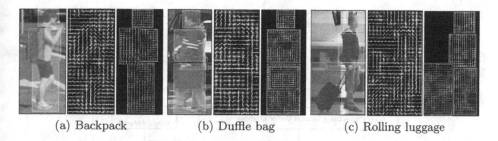

(a) Backpack	(b) Duffle bag	(c) Rolling luggage

Fig. 2. Visualizations of HOG features of sample models. For each category, first image is input image, the second image is features of full body defined as root filter, and the last image is feature of components, where white and green line correspond to body and baggage component, respectively.

it is placed the bag in location of other categories. If no baggage is identified in all spatial model, it is deducted that the human is not carrying any baggage.

2.3 Joint Component Analysis

In our approach, components is defined as small part of object that we interest to determine whether or not the object is present on the image. By the fact that, if several parts are detected , the object of interest may exist by analyzing their location relative to object. Many object detection and recognition problems have been successfully implemented using joint component analysis, such as face [6], human [7], and general object detection [8] with incredible result.

In our work, the human carrying baggage is modeled based on observations of the small component and their relative position among them. These model intuitive the relationship between component. Human object is divided into four components c_i; head, torso, leg and baggage component, as shown in Fig. 2. Height of head, torso, and leg component are $1.5h$, $2.5h$ and $4h$, respectively, while the height of bag component is varying according to our model.

The feature vector is extracted from full body and all components, independently. A low dimension feature vector approximately covers an entire object, while higher dimension feature vector cover smaller regions of the object. The approach described here is independent of the specific choice of feature. In our implementation, the histogram of oriented gradient (HOG) [9] feature is used for description the model. HOG features introduce invariances to small image deformations and photometric transformation.

Formally, the component is represented by $c_i = \{x, y, s, v\}$, where specifying an anchor position (x, y) relative to full body in the sth scale of image and feature vector v. The combining set of component detection using interpolation is

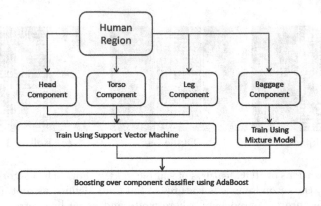

Fig. 3. Training strategy of our approach.

proposed to analyze the model on the image. The score of component interpolation for model m based on hybrid of boosting [10] is defined as follows

$$J(m) = \sum_{i=1}^{N} \omega_i \varphi_i(x, y, s, v) \tag{1}$$

where ω_i is the weighting, $\omega_i > 0$ and φ_i is the output probability of classifier of the component c_i, N is the number of components, which are used for current classifier. The results of each of component are then clustered into each group for combining component process. Our goal is to find the weighting value using boosting machine.

2.4 Final Decision

For detecting human carrying baggage in an image, the score for each model is computed according to the best possible placement of bag part and other parts relative to full body. Let M be the number of possible model trained in our framework. The final score of object hypothesis being human-baggage object is the maximum value among the score of each model that formulated as Eq.(2). In addition, if the value of J_{final} is less than a fixed threshold, the object hypothesis is classified as other objects.

$$J_{final} = \max_{m_1, \ldots, m_M} \left(J(m_1), J(m_2), \ldots, J(m_M) \right) \tag{2}$$

3 Training

Our training strategy is shown in Fig. 3. The body components are trained using support vector machine. Baggage component is trained using mixture model. The boosting machine is the applied for combining these two classifier.

3.1   Body Component Learning Using Support Vector Machine

The support vector machine (SVM) is applied to learn body component as a weak classifier. The standard SVM technique [12] is used to learning of partial part model. The output probability of SVM classification is computed by

$$\Phi(x) = P(y = 1|x) = \frac{1}{1 + exp(-h(x))} \tag{3}$$

where $h(x)$ is the signed distance of feature vector x to the margin of the SVM model.

3.2   A Mixture Model of Baggage Component

For handling the variation of color, shape and size, mixture model is used for training baggage component. More formally, let define a distribution f is a mixture of K component distribution $f_1, f_2, ..., f_K$ if

$$f(x) = \sum_{k=1}^{K} \lambda_k f_k(x; \theta_k) \tag{4}$$

with the λ_k being the mixing weights, $\lambda_k > 0$, $\sum_k \lambda_k = 1$, x is feature vector of observations, and θ_k specify as parameter vector of the kth component. The overall parameter vector of the mixture model is thus $\theta = (\lambda_1, ..., \lambda_K, \theta_1, ..., \theta_K)$. Our goal now is to estimate these all parameters using Expectation-Maximization (EM) algorithm. The EM algorithm is starting by guessing mixture component $\theta_1, ..., \theta_K$ and the mixing weights $\lambda_1, ..., \lambda_K$. Using current parameter guesses, the weight w_{ij} is calculated by Eq.(5), called (E-step). Then using current weights, maximize the weighted likelihood to get new parameter estimates, called M-step. The result of EM algorithm is the final parameter estimates with certain stopping criteria.

$$w_{ij} = \frac{\lambda_j f(x_i; \theta_j)}{\sum_{k=1}^{K} \lambda_k f(x_i; \theta_k)} \tag{5}$$

3.3   Hybrid Boosting Machine

Boosting is an approach to machine learning based on the idea of constructing a highly accurate classifier by combining many relatively weak and inaccurate classifier. Here, the output probability of SVM over body component and mixture model over baggage component are considered as weak classifier. As defined in Eq.(1), boosting machine is used for determining the weighting value of component interpolation. In practical, the AdaBoost algorithm, found by Freund and Schapire [14] is implemented.

(a) Category 1   (b) Category 2   (c) Category 3

Fig. 4. Some samples used for training. Category 1 includes backpack and handbag, category 2 consists of tote bag and duffle bag, and category 3 contains rolling luggage. First , second and third row for each category are representing viewing direction from front, side and back viewing direction. Thus, in total, nine models of human-baggage are built.

4   Experiments

4.1   Datasets

Our model was tested on human-baggage dataset. It was collected from small subset of INRIA [9], Caltech [13] and our own images. Since our work was focused on human-baggage detection, only human carrying baggage images were selected. Our dataset is divided into two groups, training and testing groups. Each group is classified into 3 categories manually for creating ground truth. The training group contains 338 human-baggage object consisted of 132, 111, and 95 data for category 1, 2, and 3, respectively. The testing group contains 202 human-baggage object distributed as 78, 67, and 57 data for category 1, 2, and 3, respectively. Data from category 1 and 2 is resized to be 128x64 pixels resolution, while for category 3 is more wider becomes 128x72 pixels resolution. Figure 4 shows several samples of our dataset used in our implementation.

4.2   Results

First, our model was evaluated for classified object into human with or without baggage. The human carrying baggage is set to be positive samples. Human

Fig. 5. Some typical detection results. The human-baggage object is detected as red bounding box.

without baggage is set to be negative samples. All samples are collected manually. It cropped to fit object region and annotated its part components based on our model. For first evaluation, 338 positives samples and 1,352 negative sample were used for training. Our method achieves detection rate of 86.13% and 77.02% for training and testing data, respectively. Since, as our knowledge, there is no method researching this specific task, we do not compare our method with others yet. However, we have tried to just use original HOG and SVM on full body [9], but the result is not promising around 45.62%.

Next, our method was evaluated on full image databases. Sliding window mechanism in any position and scale are used to detect human-baggage region. This technique was used for each body component and baggage component according to our models. In practical, the same human-baggage region is usually detected several times with overlapped bounding boxes. Therefore, it is necessary to combine the overlapped regions for unifying detection and rejecting miss detection. The detected regions are grouped based on their position, and prediction probability. Some typical results of our method are shown in Fig. 5.

5 Conclusion

This paper proposed joint component analysis for detecting and classifying baggage carried by human. First, the human region was modeled into four component, head, body, leg and baggage component. The model utilized the spatial information of baggage relative to human body. Histogram of oriented gradient (HOG) features were extracted on each component. The features of body component were trained using support vector machine (SVM). In contrary, A mixture model was applied for modeling the baggage component for handling variation of baggage size, shape and color. Boosting machine based on linear combination of these two classifier over component was performed. After conducting extensive experiment, our method achieves 77.02% for detection rate. Nevertheless, our method has some limitations for detecting and classifying baggage carried by human. First, it may fail to detect multiple baggage carried by the same person. The additional model should be considered in our future work for handling this problem. Second, our method fail to detect overlapping human-baggage region. Increasing the number of part body can be one of the solutions for solving this issue. Third, our method may fail to detect baggage that has same color with clothes. Combining several features may solve this problem.

Acknowledgments. This research was supported by Basic Science Research Program through the National Research Foundation of Korea (NRF) funded by the Ministry of Education (NRF-2013R1A1A2009984).

References

1. Tian, Y.L., Feris, R., Liu, H., Humpapur, A., Sun, M.-T.: Robust detection of abandoned and removed objects in complex surveillance video. IEEE Trans. SMC Part C **41**(5) (2011)
2. Fan, Q., Gabbur, P., Pankanti, S.: Relative attributes for large-scale abandoned object detection. In: International Conference on Computer Vision (2013)
3. Damen, D., Hogg, D.: Detecting carried ibject from sequences of walking pedestrians. IEEE Trans. PAMI 34(6), June 2012
4. Tzanidou, G., Edirishinghe, E.A.: Automatic baggage detection and classification. In: IEEE 11th International Conference on Intelligent Systems Design and Applications (2011)
5. Chuang, C.-H., Hsieh, J.-W., Chen, S.-Y., Fan, K.-C.: Carried object detection using ratio histogram and its applications to suspicious event analysis. IEEE Trans. on Circuit and System for Video Technology **19**(6), June 2009
6. Tzimiropoulos, G., Pantic, M.: Gauss-Newton deformable part models for face alignment in-the-wild. In: ICIC 2014, Taiyuan, China, August 3, 2014
7. Hoang, V.-D., Hernandez, D.C., Jo, K.-H.: Partially obscured human detection based on component detectors using multiple feature descriptors. In: Huang, D.-S., Bevilacqua, V., Premaratne, P. (eds.) ICIC 2014. LNCS, vol. 8588, pp. 338–344. Springer, Heidelberg (2014)
8. Felzenszwalb, P.F., Girshick, R.B., McAllester, D., Ramanan, D.: Object Detection with Discriminatively Trained Part Based Models. IEEE Trans. PAMI **32**(9), September 2010
9. Dalal, N., Triggs, B.: Histogram of oriented gradients for human detection. In: Conference on Computer Vision and Pattern Recognition, pp. 886–893 (2005)
10. Hoang, V.-D., Ha, L.M., Jo, K.-H.: Hybrid Cascade Boosting Machine using Variant Scale Blocks based HOG Features for Pedestrian Detection. Neurocomputing **135**, 357–366 (2014)
11. Home office scientific development branch: imagery library for intelligent detection systems (i-LIDS). In: The Institution of Engineering and Technology Conference on Crime and Security, pp. 445–448, (2006)
12. Chih-Chung, C., Chih-Jen, L.: LIBSVM: a Library for Support Vector Machine. ACM Transaction on Intelligent Systems and Technology **2**, 1–27 (2011)
13. Dollar, P., Wojek, C., Schiele, B., Perona, P.: Pedestrian detection: a benchmark. In: International Conference on Computer Vision and Pattern Recognition (2009)
14. Freund, Y., Schapire, R.E.: A decision-theoretic generalization of on-line learning and an application to boosting. Journal of Computer and System Sciences **55**(1), 119–139 (1997)

Multi-population Cooperative Bat Algorithm for Association Rule Mining

Kamel Eddine Heraguemi[1]([✉]), Nadjet Kamel[1], and Habiba Drias[2]

[1] Department of Computer Science, Faculty of Sciences, University-Setif,
Setif, Algeria
k.heragmi@yahoo.fr, nkamel@univ-setif.dz
[2] USTHB, LRIA, Algiers, Algeria
hdrias@usthb.dz

Abstract. Association rule mining (ARM) is well-known issue in data mining. It is a combinatorial optimization problem purpose to extract the correlations between items in sizable data-sets. According to the literature study, bio-inspired prove their efficiency in term of time, memory and quality of generated rules. This paper investigates multi-population cooperative version of bat algorithm for association rule mining (BAT-ARM) named MPB-ARM which is based on bat inspired algorithm. The advantage of bat algorithm is the power combination between population-based algorithm and the local search, however, it more powerful in local search. The main factor to judge optimization algorithms is ensuring the interaction between global diverse exploration and local intensive exploitation. To maintain the diversity of bats, in our proposed approach, we introduce a cooperative master-slave strategy between the subpopulations. The experimental results shows that our proposal outperforms other bio-inspired algorithms already exist and cited in the literature including our previous work BAT-ARM.

Keywords: Association rules mining · Bat algorithm · BAT-ARM · Bat algorithm for association rule mining · Support · Confidence

1 Introduction

Data mining is a set of techniques for knowledge discovery that aims to extract the most accurate, potentially useful and understandable information from a huge database. An important research domain in data mining is association rule mining (ARM)[1], which aims to extract the correlations between the attributes among databases. Since the traditional algorithms give a large number of rules, ARM is still an open problem. Formally, association rule problem is defined as follow: Let $I = \{i_1, i_2, ..., i_n\}$ be a set of literals called items, let D be a transactional database where each transaction T contains a set of items. An association rule is implication like $X \implies Y$ where $X, Y \in I$ and $X \cap Y = \emptyset$. The item-sets X, Y are named antecedent and consequent, respectively.

© Springer International Publishing Switzerland 2015
M. Núñez et al. (Eds.): ICCCI 2015, Part I, LNAI 9329, pp. 265–274, 2015.
DOI: 10.1007/978-3-319-24069-5_25

With the huge quantity and the rapid increase of the saved data in our world, users no longer take care to the large number of association rule, but care about the quality of generated rules and its usefulness. Two main measures are used in the field of ARM to evaluate the quality of rules, namely Support and Confidence. The support of an item X is the proportion of transactions of D that contains X. The support of rule $X \rightarrow Y$ is $Support(X \cup Y)$. The confidence of a rule $X \rightarrow Y$ is the proportion of the transaction covering X and Y and equal to $\frac{Support(X \cup Y)}{Support(X)}$. In other word, support implies frequency of occurring patterns, and the confidence means the strength of the implication of the rule. So, The ARM process concerning the extraction of the interesting rule satisfied a minimum support (MinSup) and minimum confidence (MinConf), which are specified by the user[1].

Several algorithms are proposed in the literature to address this issue. Three major methods have dominated associated rules mining: Apriori [2], FP-growth[10] and Charm [21]. The main drawbacks of these conventional algorithms are: firstly, they are too resources consuming especially with a huge number of transactions in the datasets when there is not enough of physical memory. Secondly, they need too much execution time when it comes to sizable databases. On the other hand, optimization algorithms provide robust and efficient results in ARM problem. The major aim of optimization process is to find the global solutions that maximize/minimize the objective function. The main mechanism to achieve the success of optimization algorithm is the interaction between global diverse exploration and the local intensive exploitation. The most useful strategy used to maintain the diversity of the solution in the population of the algorithm is cooperation strategy between sub-populations in the main population. In this paper, we propose a new modified multi-population bat algorithm with cooperative strategy that deals with association rule mining. The main advantage of bat algorithm[20] concerns the link between the population based algorithms and the local search which is a step to achieve the exploration and exploitation. To maintain the diversity of the bats in the population, we propose a master-slave cooperative strategy between sub-populations. An experimental study is carried out on several well-known datasets in the domain of association rule mining. To prove the efficiency of our proposed approach, the results are compared to those of similar approaches in the literature.

The rest of this paper is organized as follows. The next section leads with summary of the existing ARM algorithms. Section 3, we present the modified bat algorithm for association rule mining (BAT-ARM). In section 4, we present formally our approach in term of rule encoding, fitness function and the cooperative strategy between the subpopulations. Section 5, reports on the experimental results for our approach and the comparison with other ARM existing algorithms. Finally, we conclude with our prospective for a future work.

2   Related Work

In the literature, several works propose genetic algorithms with different fitness and genetic operations. In[9], the authors propose to optimize association rules and generate the best rules. In this work, first general Apriori algorithm is used to generate the rules, then the genetic algorithm is used to optimize the rules set and generate the strong rules with a new fitness function. In [19], the authors propose a genetic algorithm to identify association rules without specifying minimum support called ARMGA. The main inconvenience with this algorithm is the generation of invalid chromosomes and the production of many rules. In addition to genetic algorithm, there are many other bio-inspired approaches which are proposed to extract association rules. In [15] the authors developed G3APRM algorithm, which is based on genetic programming. The authors use the grammar guided genetic programming (G3P) to avoid invalid individuals found by GP process. Also G3PARM permits multiple variants of data by using a context free grammar.

In [17] binary particle swarm optimization based association rule miner called (BPSO) is provided. This algorithm generates the best M association rules without specifying the minimum support and minimum confidence, where M is a given number. In [3] we find a review on application of Particle Swarm Optimization in Association Rule Mining. ACO_R is an algorithm based on ant colony optimization (ACO) for ARM. It is developed in[16] for the continuous domains. The drawback of this algorithm is its high consuming time to search only one solution.

The authors of [6] developed a new algorithm called BSO-ARM. This algorithm is inspired from bees behavior and based on BSO algorithm. The results show that BSO-ARM performs better than all genetic algorithms. As extension to their work, the authors present amelioration to BSO-ARM in [5], where three strategies to determinate the search area of each bee are proposed (modulo, next, syntactic). These improvements give its fact on the quality of rules extracted by BSO-ARM, but in the same time the algorithm takes more CPU time. The same authors present another Hybrid approach called (HBSO-TS)in[4] for mining association rules based on Bees swarm algorithms and Tabu-search. The results show that HBSO-TS extract useful rules in reasonable time. Recently, in [12], the authors present a hybrid method combining both genetic algorithm and particle swarm optimisation called hybrid GA/PSO (GPSO) which is used to bring out the balance between exploration and exploitation, which will result in accurate prediction of the mined association rules and consistency in performance. GA reduces the exploitation tasks and PSO taken care on the exploration.

In [13], a multi-objective genetic algorithm approach to mine association rules for numerical data was proposed. In this work confidence, interestingness and comprehensibility are used to define the fitness function. The results show that the generated rules by this method are more appropriate than similar approaches.

3   ARM Based on Bat Algorithm (BAT-ARM)

In[11], we proposed a new algorithm for ARM inspired from bat behavior, which aims to generate the best rules in defined dataset starting from minimum support and confidence with reasonable execution time. First, we present the encoding of each solution X which represents a rule and contains k item. Therefore, the solution X has $k+1$ positions, where the first position separates between the antecedent and the consequent of the rule. For the rest of the rule, if i^{th} item exists in the rule then the position k contains i else it take-in 0.

In association rule mining, the rule is accepted if its support and confidence satisfy user minsup and minconf. Based on this definition we describe a simple objective function based on the support and the confidence to evaluate the solution and never generate invalid rules. Based on the definition of bat algorithm in[20] new formal description for bat motion is described related to association rule mining bases, where the frequency, velocity and position are defined as follow:

- **Frequency** f_i: presents how many items can be changed in the actual rule, where the maximum frequency f_{max} is the number of attributes in the dataset and the minimum frequency f_{min} is 0.
- **Velocity** v_i: indicates where the changes will be started.
- **Position** x_i: it is the new generated rule based on new frequency, velocity and the loudness.

The generation of new positions (rules) is extracted based on the frequency, velocity of each virtual bat which are updated at each iteration by Eq. 1,2

$$f_i^t = 1 + (f_{max})\beta,$$ (1)

$$v_i^t = f_{max} - f_i^t - v_i^{t-1},$$ (2)

After that, the rule is generated, according to the loudness of each bat, if it was less than a random value *rand*, the value of the actual bit in the rule will increase else it will decrease by 1. This method can lead to generate values out of range or duplicated bits. The solution of these two main problems was proposed, where all the values out of range and one of the duplicated are deleted. BAT-ARM provides a great performance in term of CPU-time and memory usage in the face of FP-growth algorithm, thanks to the echolocation concept of the bat algorithm that can determine which part of the best rule have changed to get a better position (rule) for the actual bat.

4   Multi-Population Bat Algorithm

4.1   Rule Encoding

In this paper, unlike BAT-ARM which scans the whole dataset to calculate the support of each item-set, we use vertical layout database. In this way, the support of an item-set can be easily computed by simply intersecting any two subsets[18]. For rules, each solution X represents a rule and contains k items. Therefore, the solution X is represented with a vector S which contains $k+1$ positions where:

1. S [0] separates between the antecedent and the consequent of the rule,

Algorithm 1. BAT-ARM Algorithm

objective function(fitness function)
Initialize the bat population x_i and v_i
Define pulse frequency f_i at x_i
calculate the fitness of each initial position f_i at x_i
Initialize pulse rates r_i and the loudness A_i
while ($t <$ *Max number of iterations*) **do**
    for all Bats in the population **do**
        Generate new solutions by adjusting frequency f_i,
        and updating velocities and locations/solutions **Eq.1,2**
        Generate a new solution x_i
        if ($rand > r_i$) **then**
            Generate a local solution around the selected
            best solution by changing just one item in the rule A^*.
        end if
        if ($f(x_i) > f(x_i*)$) **then**
            Accept the new solutions
            $x_i* = x_i$
            Increase r_i and reduce A_i
        end if
        Rank the bats to the best solution
    end for
end while
Post-process results and visualization the best detected rules

2. S[i] = j where i>0 If the jth item in the database is in the rule, else the position contains 0.

For example, let $I = \{i_1, i_2, ..., i_{10}\}$ be a set of items :

- $X1 = \{3, 1, 5, 0, 6, 2, 0, 0, 7, 0, 0\}$ represents the rule $i_1, i_5 \Rightarrow i_6, i_2, i_7$,

4.2 Fitness Function

As mentioned above, in association rule mining, the rule is accepted if its support and confidence satisfy user minsup and minconf. If α and β are two empirical parameters, the fitness function of a rule R is presented as follow:

$$f(R) = \begin{cases} \alpha conf(R) + \beta supp(R) & \text{if accepted rule R} \\ -1 & \text{otherwise} \end{cases}$$

4.3 Master-Slave Cooperative Strategy for MPB-ARM

In optimization algorithms there is competition between the solutions of the population. A solution with better fitness has more chance to improve the actual best solution. Usually, much better improvement can be gained through cooperative algorithms. A multi-population cooperative (Master-Slave) strategy

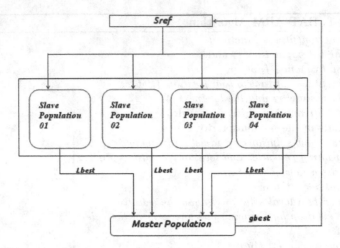

Fig. 1. Master-Slave cooperative strategy for MPB-ARM

for PSO was proposed in[14]. An alternative strategy is presented in this study for bat algorithm. This approach works based on one master population and several slave ones such that all populations are in the same bat algorithm. Slaves are evolved in parallel to generate the best local *lbest* solution of each one and send it to the Master population. This last finds the best general among local best *gbest* of all the slaves with aid of bat algorithm. Bats use the same generation method created in BAT-ARM. After, all the slaves are notified by the master population by the new global solution *gbest*, which will be a new reference (Sref) for next iteration. The process of strategy is illustrated schematically in Fig. 1. This relationship between the master population and slave populations maintains the right balance of exploitation and exploration, which is required for successful optimization process. Algorithm 2 illustrates the pseudo code of our approach.

5   Experimentation and Results

In order to perform experimentations, several well-known and frequently used real world datasets in data mining, such as Frequent and mining dataset Repository[7], Bilkent University Function Approximation Repository [8], are used in this section for several tests. The results are compared to different approaches in term of fitness, and to the BAT-ARM in term of execution time.

5.1   Benchmarks and Setup Description

The above described cooperative strategy are scripted using Java and all tests were performed on computer Intel core I5 machine with 4Go of memory running on Linux Ubuntu. We examined our approach on seven well known datasets

---

Algorithm 2. MPB-ARM Algorithm

Require: *Number of Slave populations, Number of bats in each slave population*
 Initialize the bat algorithm paramaters
 calculate the fitness of each initial position
 while ($t < Max$ *number of iterations*) **do**
  for all Slave-population **do**
   for all Bat in population **do**
    Apply A^* in algorithm 1 related to the global best
   end for
   Send the local best solution to master population
  end for
  for all Bat in Master-population **do**
   Apply A^* in algorithm 1 to generate the global best solution gbest
  end for
  Notify All slave population with the new gbest
  Zirorize the mater population
 end while
 Post-process results and visualization the best detected rules

---

with different sizes of transactions, items and average size per transaction. For instance, Chess dataset has 3196 transactions with 75 items when the average per transaction is 37, unlike mushroom dataset which has much more transactions and items when it has just 23 items per transaction. The used datasets *Bolts, Sleep, Basketball, IBM-Quest-standard, Quak, Chess, Mushroom* are taken from [7,8].

Table 1. Data sets description

Data set Name	Transactions size	Item size	Average size
Bolts	40	8	8
Sleep	56	8	8
Basketball	96	5	5
IBM-Quest-standard	1000	40	20
Quak	2178	4	4
Chess	3196	75	37
Mushroom	8124	119	23

Table 1 illustrates the different datasets used in our tests. To get the best performance in our cooperative strategy, we change the principal parameters: Number of population, bats per population and iterations.

5.2   Results and Comparisons

To prove the effectiveness of the cooperative version proposed for BAT-ARM, we start the comparison between these two approaches (MPB-ARM, BAT-ARM)

Table 2. Comparing our approach to BAT-ARM w.r.t CPU time (sec)

Dataset Name	#Iteration	MPB-ARM			BAT-ARM	
		#Pupulation	#bats	Time(Sec)	#Bats	Time(Sec)
IBM-Quest-standard	100	2		1	10	2.03
		5		3	25	5
		10		7	50	10
	200	2		2	10	3
		5		6	25	9
		10		13	50	19
Chess	100	2		12	10	13
		5		37	25	72
		10	5	70	50	67
	200	2		21	10	28
		5		68	25	73
		10		130	50	141
Mushroom	100	2		14	10	26
		5		30	25	86
		10		68	50	151
	200	2		27	10	68
		5		66	25	199
		10		144	50	341

in term of execution time. The implementation results proof that the new version never generate non admissible or false rules. Table 2 summarizes the results obtained from the average of 100 executions of each algorithm with the same number of bats. To make the comparison totally fair, the number of bats in each population in MPB-ARM is fixed to 5, and the numbers of populations vary which aims to get the same number of bats. For instance, when we have 50 bats in BAT-ARM, we use 10 populations in MPB-ARM. From Table 2 we notice that the cooperative strategy prove its effectiveness face to BAT-ARM, where the CPU time is reduced for all tests on the datasets and specially minimized till the half with mushroom dataset. This can be explained by the converting of dataset from horizontal to vertical layout which guaranty a fast way to compute the support of each rule. In addition to this, the parallel execution of the slave populations can give more gain in time. These interesting results obtained in term of CPU time are inadequacy to judge the approach. In addition to time the algorithm need to generate rules with high quality. In bio-inspired algorithms the fitness function evaluates the quality of the solutions so it needs to be maximized to extract the best rules. Table 3 illustrates all the results obtained by the execution of *MPB-ARM, BAT-ARM, $ACO_R, G3APARM$ and *BSO-ARM* with different strategies(modulo, next, syntactic). We note that MPB-ARM outperforms all other algorithms including BAT-ARM. This can be explained by the limit of BAT-ARM which is the absence of communication between bats when this drawback was canceled in the cooperative version. In addition to this, the search strategy of the slave populations guided by the master one can gives more improvement to the solutions.

Table 3. Comparing our approach to existing approaches w.r.t fitness

Dataset Name	MPB-ARM	BSO-ARM			ACO$_R$	G3APRM	BAT-ARM
		Modulo	Next	Syntatic			
Bolts	1.0	1.0	1.0	1.0	0.69	0.92	0.86
Sleep	1.0	1.0	1.0	1.0	0.67	0.90	0.73
Basketball	1.0	0.97	1.0	0.92	0.61	0.93	0.54
IBM-Quest standard	0.98	0.94	0.9	0.93	0.45	0.88	0.41
Quak	1.0	1.0	1.0	1.0	0.73	0.90	0.52
Chess	0.97	0.88	0.88	0.60	0.3	0.86	0.92
Mushroom	0.73	0.52	0.75	0.72	0.1	0.85	0.73

6   Conclusion

Our earlier proposed algorithm named BAT-ARM has two main problems making the algorithm losing time: horizontal layout of database, and absence of communication between bats. The horizontal layout of database takes much time to compute any measure, and the absence of communication between bats allows different bats to generate the same rules. In this paper, we proposed a multi-population cooperative bat algorithm for association rule mining, where the drawbacks of BAT-ARM are solved. First, vertical layout is used for the database where a simple intersection can compute the support of any item-set. Then, we investigated the master-slave strategy for the cooperation between the sub-populations. This strategy makes the search in slave populations guided by the master one. The presented experimentations prove the efficiency of our approach MPB-ARM which outperforms BAT-ARM. We can see a decrease in execution time to almost half. It outperforms the similar bio-inspired methods in terms of quality of extracted rules. As future work, we plan to generalize our approach to the numerical association rules. Our approach needs to be improved and tested on large database specially *WebDocs* which exceeds a million and a half of transactions. We think also about parallelizing the algorithm and implement it on a GPU to improve both the solution quality and its running time.

References

1. Agrawal, R., Imieliński, T., Swami, A.: Mining association rules between sets of items in large databases. In: ACM SIGMOD Record, vol. 22, pp. 207–216. ACM (1993)
2. Agrawal, R., Srikant, R., et al.: Fast algorithms for mining association rules. In: Proc. 20th Int. Conf. Very Large Data Bases, VLDB, vol. 1215, pp. 487–499 (1994)
3. Ankita, S., Shikha, A., Jitendra, A., Sanjeev, S.: A review on application of particle swarm optimization in association rule mining. In: Satapathy, S.C., Udgata, S.K., Biswal, B.N. (eds.) Proceedings of Int. Conf. on Front. of Intell. Comput. AISC, vol. 199, pp. 405–414. Springer, Heidelberg (2013)
4. Djenouri, Y., Drias, H., Chemchem, A.: A hybrid bees swarm optimization and tabu search algorithm for association rule mining. In 2013 World Congress on Nature and Biologically Inspired Computing (NaBIC), pp. 120–125. IEEE (2013)

5. Djenouri, Y., Drias, H., Habbas, Z.: Bees swarm optimisation using multiple strategies for association rule mining. International Journal of Bio-Inspired Computation **6**(4), 239–249 (2014)
6. Djenouri, Y., Drias, H., Habbas, Z., Mosteghanemi, H.: Bees swarm optimization for web association rule mining. In: 2012 IEEE/WIC/ACM International Conferences on Web Intelligence and Intelligent Agent Technology (WI-IAT), vol. 3, pp. 142–146. IEEE (2012)
7. Goethls, B., Zaki, M.J.: Frequent itemset mining dataset repository(2003). http://fimi.ua.ac.be/data/
8. Guvenir, H.A., Uysal, I.: Bilkent university function approximation repository (2000). http://funapp.cs.bilkent.edu.tr/DataSets/
9. Haldulakar, R., Agrawal, J.: Optimization of association rule mining through genetic algorithm. International Journal on Computer Science & Engineering **3**(3) (2011)
10. Han, J., Pei, J., Yin, Y.: Mining frequent patterns without candidate generation. In: ACM SIGMOD Record, vol. 29, pp. 1–12. ACM (2000)
11. Heraguemi, K.E., Kamel, N., Drias, H.: Association rule mining based on bat algorithm. In: Pan, L., Păun, G., Pérez-Jiménez, M.J., Song, T. (eds.) BIC-TA 2014. CCIS, vol. 472, pp. 182–186. Springer, Heidelberg (2014)
12. Indira, K., Kanmani, S.: Mining association rules using hybrid genetic algorithm and particle swarm optimisation algorithm. International Journal of Data Analysis Techniques and Strategies **7**(1), 59–76 (2015)
13. Minaei-Bidgoli, B., Barmaki, R., Nasiri, M.: Mining numerical association rules via multi-objective genetic algorithms. Information Sciences **233**, 15–24 (2013)
14. Niu, B., Zhu, Y., He, X.-X.: Multi-population cooperative particle swarm optimization. In: Capcarrère, M.S., Freitas, A.A., Bentley, P.J., Johnson, C.G., Timmis, J. (eds.) ECAL 2005. LNCS (LNAI), vol. 3630, pp. 874–883. Springer, Heidelberg (2005)
15. Olmo, J., Luna, J., Romero, J., Ventura, S.: Association rule mining using a multi-objective grammar-based ant programming algorithm. In: 2011 11th International Conference on Intelligent Systems Design and Applications (ISDA), pp. 971–977 (November 2011)
16. Parisa, M., Behrouz, M., Mahdi, N., Afshin, S.: Multi-objective numeric association rules mining via ant colony optimization for continuous domains without specifying minimum support and minimum confidence. International Journal of Computer Science Issues **8**(1) (2011)
17. Sarath, K., Ravi, V.: Association rule mining using binary particle swarm optimization. Engineering Applications of Artificial Intelligence **26**(8), 1832–1840 (2013)
18. Savasere, A., Omiecinski, E.R., Navathe, S.B.: An efficient algorithm for mining association rules in large databases (1995)
19. Yan, X., Zhang, C., Zhang, S.: Genetic algorithm-based strategy for identifying association rules without specifying actual minimum support. Expert Systems with Applications **36**(2), 3066–3076 (2009)
20. Yang, X.-S.: A new metaheuristic bat-inspired algorithm. In: González, J.R., Pelta, D.A., Cruz, C., Terrazas, G., Krasnogor, N. (eds.) NICSO 2010. SCI, vol. 284, pp. 65–74. Springer, Heidelberg (2010)
21. Zaki, M.J., Hsiao, C.-J.: Charm: An efficient algorithm for closed itemset mining. In: SDM, vol. 2, pp. 457–473. SIAM (2002)

Supervised Greedy Layer-Wise Training for Deep Convolutional Networks with Small Datasets

Diego Rueda-Plata[1], Raúl Ramos-Pollán[1]([✉]), and Fabio A. González[2]

[1] Universidad Industrial de Santander, Bucaramanga, Colombia
sandiego206@gmail.com, rramosp@uis.edu.co
[2] Universidad Nacional de Colombia, Bogotá, Colombia
fagonzalezo@unal.edu.co

Abstract. Deep convolutional neural networks (DCNs) are increasingly being used with considerable success in image classification tasks trained over large datasets. However, such large datasets are not always available or affordable in many applications areas where we would like to apply DCNs, having only datasets of the order of a few thousands labelled images, acquired and annotated through lenghty and costly processes (such as in plant recognition, medical imaging, etc.). In such cases DCNs do not generally show competitive performance and one must resort to fine-tune networks that were costly pretrained with large generic datasets where there is no a-priori guarantee that they would work well in specialized domains. In this work we propose to train DCNs with a greedy layer-wise method, analogous to that used in unsupervised deep networks. We show how, for small datasets, this method outperforms DCNs which do not use pretrained models and results reported in the literature with other methods. Additionally, our method learns more interpretable and cleaner visual features. Our results are also competitive as compared with convolutional methods based on pretrained models when applied to general purpose datasets, and we obtain them with much smaller datasets (1.2 million vs. 10K images) at a fraction of the computational cost. We therefore consider this work a first milestone in our quest to successfully use DCNs for small specialized datasets.

Keywords: Convolutional networks · Deep learning · Greedy layer-wise training

1 Introduction

Deep representation learning, an approach to learn hierarchical data representations using multilayer neural networks, has shown to be very successful when addressing different computer vision tasks such as object detection, object identification and scene understanding, improving, to a large extent, the state-of-the-art. Among these methods, one of the most successful is deep convolutional networks (DCNs). A DCN is a neural network that combines different types

© Springer International Publishing Switzerland 2015
M. Núñez et al. (Eds.): ICCCI 2015, Part I, LNAI 9329, pp. 275–284, 2015.
DOI: 10.1007/978-3-319-24069-5_26

of layers (convolutional, pooling and classification) in a multilayer architecture, which is trained in a supervised way using gradient descent optimization.

An important milestone in image classification was reached by Krizhevsky et al. [8] with a DCN which was able to improve the state-of-the-art in the ImageNet challenge[1], reducing the error from 26.2% to 15.3%. The network consists of 8 layers, the first 5 layers correspond to combinations of convolution and pooling layers and the last three correspond to fully connected layers. The network receives as input a 224 × 224 pixels image and has 1,000 output corresponding to the target concepts. This architecture has 650,000 neurons and their connection weights account for more than 60 million parameters, which must be adjusted from the training data.

One key factor in the success of this network is the fact that is was trained with more than a million images. This helped to capture the high visual variability of the 1,000 concepts, but also prevented the overfitting of a model with that huge amount of parameters. In contrast, when such a network is trained with a small datset it does not show a competitive performance. An alternative is to use an approach called fine-tuning, which starts from a network trained with a large dataset (such as ImageNet) and then retrains the network with the target small dataset. The problem of this approach is that large datasets might largely differ from the task in hand and the resulting trained DCNs might not necessarily be found adequate. Table 1 shows the performance we obtained by using the pretrained ImageNet model and by training it from scratch with the datasets used in this experiment. The importance of the weight initialization achieved with large datasets becomes evident. When this weight initialization it is not possible due to unavailability of large amounts of relevant data, using DCNs might become unpractical.

Nevertheless, we face the fact that in certain application areas there might not exist an appropriate pretrained model and we are left with the available data. Thus, in this work, we propose to use a supervised layer-wise strategy to mitigate the problems caused by the combination of a large network and a small training dataset. Our goal is to show how this strategy improves the results obtained by a conventional full network trained from scratch, approaching a performance close to the one reached with a fine-tuned network. We choose to use for this purpose still generic datasets of small size (1K to 15K images) so that we can compare to fine-tuning the pretrained ImageNet model (with roughly 1.2 million images and 1000 categories) before using specialized datasets in future work.

We show how this layer-wise strategy outperforms both full DCN training without initialized weights and other methods in the literature. When we use the fine-tuned ImageNet model for the same tasks our performance remains competitive although lower. However, we achieve our results at a fraction of the computational cost using still generic datasets which are not so distinct from ImageNet. We expect this work to constitute a first proof-of-concept in our quest to successfully use DCNs for small specialized datasets.

---

[1] The ImageNet challenge consisted in classifying 1.2 million images from the ImageNet image collection into 1,000 different classes.

The rest of this paper is structured as follows. Section 2 describes DCNs and the layer-wise approach we follow. Section 3 describes the datasets and DCNs we used in our experiments. Section 4 shows the results we obtained and, finally, Sections 5 draws some conclusions.

2 Methods

2.1 Deep Neural Networks

Convolutional neural networks were first proposed by Fukushima [4] and later improved with great succes by Yan LeCun [10] in the famous LeNet-5 network that could classify digits and used in several contexts (the US postal service among one of them). Since then, DCNs have been sucessfully used to classify a variety of signals lying at the core of many applications ranging from speech recognition to object detection [13]. DCNs are composed of stacked neuron layers of different types building altogether distributed, hierarchical representations, from low level concepts such as colors, borders, gradients, etc. (in images) to higher level ones such as components and objects. These layers include: (1) **Convolutional layers**, where neurons are tiled together so that insted of responding to the whole signal or image they only respond to overlapping regions, this way similar features or concepts can be detected across the signal spectrum; (2) **Pooling layers**: detected features are averaged or summarized over a certain signal region to reduce variance and provide robustness to small translations; and (3) **Fully connected layers**: the last stages of the network contain a traditional neural network structure yielding to the prediction and classification tasks. Fully connected layers typically use the *dropout method* to randomly discard neurons in the optimization process to avoid overfitting.

Others kinds of layers might also be found for normalizing data, etc. Typically a DCN is composed of several stages of combined convolutional and pooling layers, followed by a few fully connected layers. See [13] or [1] among others.

DCNs are trained in a supervised manner, giving to the network pairs of input and expected output (such as labelled images) and allowing an opimization process to take place over the whole network to minimize some error measure in classification. Other deep architectures are used and trained in an unsupervised manner to learn automatically new data representations with a high degree of success, namely stacked autoencoders [14] and restricted boltzman machines [5].

2.2 Greedy Layer-Wise Training

As deep architectures grow larger the number of parameters to tune in the optimization process also increases reaching easily the order of the millions. Weight initialization then becomes an issue [3], and without the appropriate strategy deep architectures are bound to failure. When using deep networks for supervised learning this is usually addressed by starting off with a network model that has been pretrained with a large standard set of data in a costly process.

This pretrained model is then fine tuned by training it with our data of interest in a process which is much more computationally affordable.

A successful approach to train unsupervised deep architectures is to do it layer-wise [2] where instead of training the whole set of stacked layers in one single optimization process through the available data, this is done in several stages, one per layer. First, all layers but the first one are removed from the network architecture and a full training cycle is performed initializing connection weights randomly and yielding a set of new weights for this first layer after processing the input data. Then, the second layer is added, keeping the weights obtained for the first layer in the previous step and initializing randomly the weights of the second layer. A full training cycle with all data is again performed on this two-layer network. The resulting network is then appended the third layer, preserving the weights of the previous two layers and so on. Through this process what happens is that subsequent training processes for new layers refine the weights obtained earlier for previous layers.

In this work we use the greedy layer-wise method for supervised training of a DCN as explained in Section 3.3. For this we used the CAFFE deep learning framework [6] over a GPU computing infrastructure.

The application of a layer-wise strategy to supervised learning has ben previously explored, but these early works, in contrast to ours, found it as too greedy for supervised training as it might hinder generalization performance [9].

3   Experimental Setup

3.1   Datasets

We used four different datasets to test the performance of layer-wise training of a DCN and compare it with full training its complete structure. The properties of these datasets are described in Table 1. See also Figure 1 for a few sample images of each.

Table 1. Image datasets used and ImageNet accuracy by finetuning the pretrained model and by training it from scratch with the validation procedure used in this work.

	number of images	number of classes	images per class	avg image resolution	finetune ImageNet	ImageNet scractch
Flowers17[2]	1369	17	80	750x500	92%	66%
Flowers102[3]	8187	102	40-258	750x500	86%	66%
Caltech101[4]	9144	101	40-80	300x200	85%	45%
Scene 67[5]	15620	67	100+	500x400	55%	26%

Table 1 also includes the performance we obtained on validation data (see Section 3.2) by finetuning the pretrained ImageNet with each dataset and the result of training the same ImageNet network structure from scratch. The importance of the weigth initialization achieved with large datasets becomes evident.

Fig. 1. Sample images from the four datasets selected for experimentation

3.2   Validation Procedure

For training and validation splits we tried for each dataset to stick as much as possible to the standards set by their authors. In Flowers17 three splits are provided, each one using a 50/25/25 rate for train, validation and test respectively, evenly split for all classes. We therefore used 50% of the data for train (680 images) and 25% for validation (340), discarding the remaining 25%. As three splits were provided we performed all experiments three times. In Flowers102 the authors provide a split in which each class has 10 images in validation, ten images in test and the rest in train. Again, we discarded the images in test and used 6149 images for train and 1020 for validation. Most classes in this dataset have in total between 50 and 100 images. Nine classes contain more than 150 images and no class has less than 40 images.

In Caltech101 there is no fixed split provided by the authors but, due to the imbalance of classes in the dataset, they recommend the following. If a class has less than 100 images, then 60 are randomly picked for training, 20 for validation and 20 for testing. If the class has more than 100 images, then 100 are taken for training and the rest is evenly split for train and validation. Again, after doing the random split following this schema, we discarded the images chosen for test and repeated the process and experimentation twice. In total, we used 4042 images for training and 1111 for validation.

In Scene67 there is a good balance of classes and, therefore, 80% was selected for training and the rest for validation. This process and the corresponding experimentation was performed twice.

3.3   Convolutional Networks

Then, we devised a base DCN with five layers and defined different setups gradually removing layers so that each setup could be trained separately for each dataset. The base network is shown as L5 in Figure 2 containing five layers of convolution and pooling followed by 3 fully connected layers which remain fixed. Then, each setup was defined as follows:

Setup L1: From the base network we removed layers 2 to 5, initialized it with random weights and trained resulting network

Setup L2: We added layer 2 to Setup L1, keeping the weights obtained from the first layer, initialized randomly the rest of the weights and trained the resulting network.

Setup L3 to L5 were likewise configured adding a layer at each step, keeping the weights learnt the previous setup and initializing randomly the rest.

Setup FULL: We used the same structure as in Setup L3 but initialized the full network with random weights training, therefore, the network from scratch without using any learnt weights from other setups.

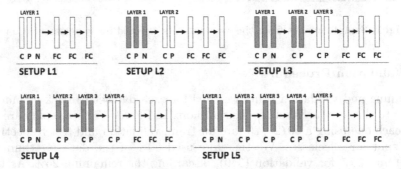

Fig. 2. Setups for layer-wise training of convolutional networks. In gray the layers initialized with weights obtained by training previous setups. 'C' stands for a convolution layer, 'P' for pooling, 'N' for normalization and 'FC' for fully connected.

In all setups the last fully connected layer had one output neuron per class in the dataset (thus, it was different for each dataset). Additionally, the first fully connected layer was configured with a dropout ratio of 0.5 and we used a SOFTMAX loss function at the last fully connected layer.

Thus, we have five convolutional networks which are gradually trained layerwise and L3 is also trained from scratch (setup FULL). The structure of each network can be seen as a subset of the following one (L2 is L1 plus a new layer, likewise for L3 and L2, etc.) and Table 2 shows the full structure of L5 and the total neuron and connection count for the Flowers17 dataset. As mentioned, the last fully connected layer differs for each dataset, resulting is a slightly different number of neurons and connections for different datasets. Just as reference, using the Flowers17 dataset, L1 has 2,161 neurons and 1,199,136 connections. L2 has 2417 neurons and 1,977,376 connections. L3 has 2,673 neurons and 2,567,200 connections and L4 has 2,929 neurons and 3,157,024 connections.

All datasets and setups were run with a minibath size of 16 images through 100K iterations and we obtained the loss curve (with training data) and the accuracy curve (with validation data) in each case. Additionally, for visual inspection, we extracted the 96 features learned at the first layer in all cases. We ran our experiments at the GPU infrastructure at the Center for High Performance and Scientific Computing at Universidad Industrial de Santander (Colombia), servicing a cluster with 128 Tesla M2050 and M2075 NVIDIA GPUs.

Table 2. Convolutional network configuration for Setup L5. Setups L1 to L4 remove gradually convolution and pooling layers.

Layer	Input	Filter	Stride	Pad	Output	Connections per Neuron	Neurons	Total Connections
conv1	227x227x3	11x11	4	0	55x55	363x96	96	34,848
pool1	55x55x96	3x3	2	0	27x27			
conv2	27x27x96	5x5	1	2	27x27	2400x256	256	614,400
pool2	27x27x256	3x3	2	0	13x13			
conv3	13x13x253	3x3	1	1	13x13	2304x256	256	589,824
pool3	13x13x253	3x3	2	0	6x6			
conv4	6x6x256	3x3	1	1	6x6	2304x256	256	589,824
pool4	6x6x256	3x3	2	0	3x3			
conv5	3x3x256	3x3	1	1	3x3	2304x256	256	589,824
pool5	3x3x256	3x3	2	0	1x1			
FC1	256				1024	1024x256	1024	262,144
FC2	1024				1024	1024x1024	1024	1,048,576
FC3	1024				17	1024x17	17	17,408
TOTALS							**3,185**	**3,746,848**

4 Results

Table 3 shows the average accuracy obtained with validation data for each dataset and setup. All runs of the same setup and dataset produced accuracies within 1% of the average. In bold we can see that our best results are achieved with the L4 setup for all datasets whereas the FULL setup achieves poor results overall and, specially, as compared with L3 which is exactly the same DCN structure but trained layer-wise. Furthermore, we also see that, for this DCN, adding more layers is not likely to improve performance and with L5 it starts to drop systematically on all datasets.

Table 3. Average accuracy in validation data for each dataset and setup. The column 'Literature' includes the best results reported in the literature without using DCNs.

	FULL	L1	L2	L3	L4	L5	Literature
Flowers 17	71%	69%	76%	82%	**86%**	84%	81.3% [12]
Flowers 102	45%	57%	74%	82%	**83%**	79%	82.5% [11]
Caltech 102	56%	43%	56%	62%	**64%**	61%	66.2% [15]
Scene 67	34%	23%	32%	37%	**41%**	36%	60.8% [7]

Figure 3 shows the outcome of the training process in terms of the loss function as measured in train data (driving the optimization function during training) together with the accuracy in the validation data which is measured every 10K iterations during training. Training time for the each experiment varied between 3 and 4.5 hours for the 100K iterations with which experiments

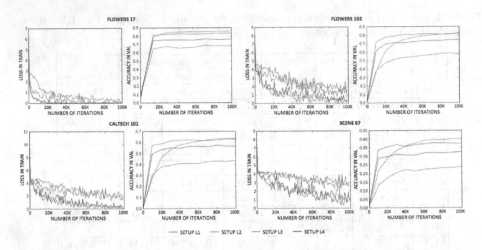

Fig. 3. Loss in train and accuracy in validation for layer-wise trained setups per dataset.

were originally configured. As it can be observed in L1, L2 and L3 stable results are obtained after roughly 30K iterations whereas in L4 this happens at about 60K iterations. With this, we assume as around 2 hours the effective computing time required for each experiment which allows us finer planning for further experiments with other datasets or network structures.

In order to further understand the layer-wise process we visually inspect the 96 features learnt at the first convolutional layer at each setup. This is shown in Figure 4 for setups L1, L4 and FULL. It can be observed that L1 learns a set of features which remain very stable throughout the layer addition and re-training process. At L4 we see basically the same features as in L1 but with less noise and better defined. In all, although not shown here, layer-wise training from L1 to L4 gradually provides cleaner features and Figure 4 simply shows the end points of this process. In all cases, the network managed to learn distinctive features

However, the features extracted at the first convolutional layer in the setup FULL are much less informative and more noisy, as it can be seen also in Figure 4. This further supports the convenience of the supervised layer-wise approach when using datasets of these sizes. Note that in the Flowers17 dataset (which is the smallest and with less classes) the features at the setup FULL are extremely uninformative whereas in the other datasets some structure seems to be emerging but could not evolve enough, probably due to the lack of more data. Observe as well the fact that in general setup L1 yields worse results than setup FULL but the features learnt at the first layer have higher quality.

5   Conclusions and Future Work

In this paper we showed how a greedy layer-wise training strategy can be advantagous for training DCNs with small size datasets, yielding cleaner and more interpretable visual features, as well as improved accuracy. We used generic datasets to

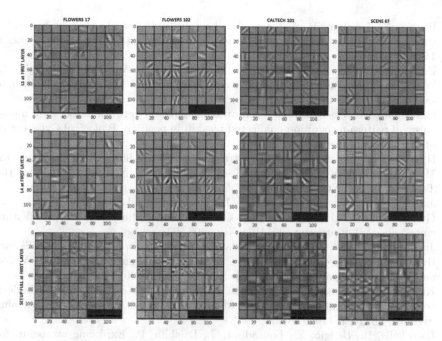

Fig. 4. Features at the first convolutional layer for setups L1, L4 and FULL per dataset.

understand the performance of this method and compare as well it with fine-tuning the ImageNet pretrained model. We show how this method outperforms traditional training of DCNs and, in certain cases, also other methods reported in the literature. Even with generic cases, as the ones used in our experiments, accuracy can be within 3% to 10% of that of the ImageNet finetuned network at a fraction of the computing cost and time. Future work is based on gradually using more specialized datasets (such as in medical imaging) to further understand when and how the greedy layer-wise method for supervised training of DCN can be advantageous.

Acknowledgments. This work was partially funded by project 1367 from VIE at Universidad Industrial de Santander and project "Diseño e implementación de un sistema de cómputo sobre recursos heterogéneos para la identificación de estructuras atmosféricas en predicción climatológica" number 1225-569-34920 through Colciencias contract number 0213-2013. Experiments presented in this paper were run on the GridUIS-2 experimental testbed at Universidad Industrial de Santander High Performance and Scientific Computing Centre (http://www.sc3.uis.edu.co).

References

1. Bengio, Y., Courville, A., Vincent, P.: Representation learning: A review and new perspectives. IEEE Transactions on Pattern Analysis and Machine Intelligence **35**(8), 1798–1828 (2013)

2. Bengio, Y., Lamblin, P., Popovici, D., Larochelle, H., et al.: Greedy layer-wise training of deep networks. Advances in Neural Information Processing Systems **19**, 153 (2007)
3. Erhan, D., Bengio, Y., Courville, A., Manzagol, P.-A., Vincent, P., Bengio, S.: Why does unsupervised pre-training help deep learning? The Journal of Machine Learning Research **11**, 625–660 (2010)
4. Fukushima, K.: Neocognitron: A self-organizing neural network model for a mechanism of pattern recognition unaffected by shift in position. Biological Cybernetics **36**(4), 193–202 (1980)
5. Hinton, G.E.: A practical guide to training restricted boltzmann machines. In: Montavon, G., Orr, G.B., Müller, K.-R. (eds.) Neural Networks: Tricks of the Trade, 2nd edn. LNCS, vol. 7700, pp. 599–619. Springer, Heidelberg (2012)
6. Jia, Y., Shelhamer, E., Donahue, J., Karayev, S., Long, J., Girshick, R., Guadarrama, S., Darrell, T.: Caffe: Convolutional architecture for fast feature embedding. arXiv preprint (2014). arXiv:1408.5093
7. Juneja, M., Vedaldi, A., Jawahar, C.V., Zisserman, A.: Blocks that shout: Distinctive parts for scene classification. In: 2013 IEEE Conference on Computer Vision and Pattern Recognition (CVPR), pp. 923–930. IEEE (2013)
8. Krizhevsky, A., Sutskever, I., Hinton, G.E.: Imagenet classification with deep convolutional neural networks. In: Advances in Neural Information Processing Systems, pp. 1097–1105 (2012)
9. Larochelle, H., Bengio, Y., Louradour, J., Lamblin, P.: Exploring strategies for training deep neural networks. The Journal of Machine Learning Research **10**, 1–40 (2009)
10. LeCun, Y., Bottou, L., Bengio, Y., Haffner, P.: Gradient-based learning applied to document recognition. Proceedings of the IEEE **86**(11), 2278–2324 (1998)
11. Nilsback, M-E., Zisserman, A.: Automated flower classification over a large number of classes. In: Proceedings of the Indian Conference on Computer Vision, Graphics and Image Processing (December 2008)
12. Nilsback, M.-E., Zisserman, A.: A visual vocabulary for flower classification. In: 2006 IEEE Computer Society Conference on Computer Vision and Pattern Recognition, vol. 2, pp. 1447–1454. IEEE (2006)
13. Schmidhuber, J.: Deep learning in neural networks: An overview. Neural Networks **61**, 85–117 (2015)
14. Vincent, P., Larochelle, H., Lajoie, I., Bengio, Y., Manzagol, P.-A.: Stacked denoising autoencoders: Learning useful representations in a deep network with a local denoising criterion. The Journal of Machine Learning Research **11**, 3371–3408 (2010)
15. Zhang, H., Berg, A.C., Maire, M., Malik, J.: Svm-knn: Discriminative nearest neighbor classification for visual category recognition. In: 2006 IEEE Computer Society Conference on Computer Vision and Pattern Recognition, vol. 2, pp. 2126–2136. IEEE (2006)

An Equilibrium in a Sequence of Decisions
with Veto of First Degree

David Ramsey[1] and Jacek Mercik[2(✉)]

[1] Wroclaw University of Technology, Wrocław, Poland
david.ramsey@pwr.edu.pl
[2] Wroclaw School of Banking & Gdansk School of Banking, Wrocław, Poland
jacek.mercik@wsb.wroclaw.pl

Abstract. In this paper we model a sequence of decisions via a simple majority voting game with some players possessing an unconditional or conditional veto. The players vote (yes, no or abstain) on each motion in an infinite sequence, where two rounds of voting take place on each motion. The form of an equilibrium with retaliation is introduced, together with necessary and sufficient conditions for an equilibrium in such a game. A theorem about the form of the equilibrium is proved: given that one veto player abstained in the first round, in the second round of voting on a motion: (1) a veto player should vote for the motion if the value of the j-th motion to the i-th player (measured relative to the status quo) is greater than t_2, veto if it is less than t_1 and otherwise abstain; (2) a non-veto player should vote for the motion if the value of the j-th motion to the i-th player (measured relative to the status quo) is greater than t_3 and otherwise abstain; (3) the thresholds t_1, t_2 and t_3 satisfy given conditions.

Keywords: Majority voting game with retaliation · Veto · Equilibrium

1   Introduction

Decision making under majority voting has been analysed using various game theoretic approaches for many years. In the classical approach to majority decisions (see for example [2, 6]), all the decisions made by a committee are assumed to be made in isolation from other votes. This is only true to some extent. The results of a particular vote may well affect the future behaviour of decision makers. Therefore, it is reasonable to consider decision making as a dynamic game with a sequence of votes on various motions, rather than a set of independent voting games. Moreover, introducing a veto (unconditional or conditional) into such a process seriously complicates the analysis required to find an equilibrium (see [5]). In this paper we construct the form of an equilibrium when players can vote yes, no or abstain on each motion in an infinite sequence and veto is allowed.

This paper is arranged as follows. Section 2 presents preliminaries connected with the game theoretical language used when modelling a simple voting game and how to introduce vetoes into these models. Section 3 presents the model for a game in which voters have equal voting weights, but some have an unconditional or conditional veto

© Springer International Publishing Switzerland 2015
M. Núñez et al. (Eds.): ICCCI 2015, Part I, LNAI 9329, pp. 285–294, 2015.
DOI: 10.1007/978-3-319-24069-5_27

(voting yes/no/abstain). This section describes the possible consequences of abstaining in a decision making process. Section 4 presents the form of an equilibrium with retaliation. We assume that, in general, a majority of voters prefer a motion to the *status quo* and a heavy tailed probability distribution is used to reflect the value of a motion to a voter. Section 5 is devoted to the derivation of an equilibrium. We present conditions for the thresholds t_1, t_2 and t_3 based on the consequences of a veto or non-veto player changing a "yes" vote to a "no" vote or abstention. Finally, we make some conclusions and suggestions for future research.

2   Preliminaries

Let $N = \{1, 2, ..., n\}$ be a finite set of committee members, q be a quota, where $q \leq n - 1$ and w_j be the voting weight of member j, where $j \in N$. It will be assumed that the quota is low enough to ensure that unanimity is not required to pass a

In this paper, we consider a special class of cooperative games called weighted majority games. A weighted majority game G is defined by a quota q and a sequence of nonnegative numbers w_i, $i \in N$, where we may think of w_i as the number of votes, or weight, of player i and q as the threshold, or quota of votes, needed for a coalition to win. We assume that q and w_j are nonnegative integers. Any subset of the players is called a coalition.

A cooperative game with the set of players N is given by a map $v : 2N \rightarrow R$ with $v(\emptyset) = 0$. The space of all such games on N is denoted by G. The domain $SG \subset G$ of simple games on N consists of all maps $v \in G$ such that

(i) $v(S) \in \{0,1\}$ for all $S \in 2^N$;
(ii) $(N) = 1$;
(iii) v is monotonic, i.e. if $S \subset T$ then $v(S) \leq v(T)$.

A coalition S is said to be winning for $v \in SG$ if $v(S) = 1$. Otherwise, a coalition is said to be losing. Therefore, passing a bill, for example, is equivalent to forming a winning coalition consisting of voters. A simple game (N,v) is said to be proper, if and only if the following is satisfied: for all $T \subset N$, if $v(T) = 1$ then $v(N \backslash T) = 0$.

We only analyse simple and proper games where players may vote either yes-no or yes-no-abstain, respectively.

If a given committee member can transform any winning coalition into a non-winning one by using a veto, then that veto is said to be of first degree.

If the veto of a given committee member turns some, but not all, winning coalitions not including that member into non-winning coalitions, then that veto is defined to be of second degree [4].

We shall denote a committee (weighted voting body[1]) with set of members N, quota q and weights w_j, $j \in N$ by $(N, q, \mathbf{w}) [= (N, q, w_1, w_2, ..., w_n)]$. We shall assume that the w_j are nonnegative integers. Let $t = \sum_{j=1}^{n} w_j$ be the total weight of the committee.

---

[1] This comes directly from the definition of a weighted game.

Let V denote the set of all committee members equipped with a veto. We assume that the cardinality of the set V, $V \subset N$, is equal to or greater than 1, i.e. $card(V) = c_v \geq 1$.

3   The Model of the Game

Let us consider a simple game with veto of first or second degree. There are two types of players: those with veto $\{N_v\}$ and those without veto $\{N_{\sim v}\}$, $\{N_v\} \cup \{N_{\sim v}\} = \{N\}$. One of these sets of players could be empty. If $\{N_v\} = \emptyset$, then the game is a standard simply majority game.

Within the framework of such games, abstention by a non-veto player is equivalent to voting against an issue. This is not the case for veto-players. When a veto player abstains, then there exists a coalition not including this player which is winning when no veto is applied, but becomes a losing coalition when a veto of either type is applied.

In proper veto games at least one player should be a veto player. Let $card\{N_v\} = n_v \geq 1$. In addition, it is assumed that each player has the same voting weight, which can be set to be equal to 1 without loss of generality. Let K_v and k_v denote the set and number of veto-players choosing to abstain, respectively $(K_v \subseteq N_v, 0, k_v = card\{K_v\})$. This means that the game (N, q, w) with abstention requires consideration of the set of games $(N \backslash K_v, q, w)$ with just yes-no voting for $K_v \subseteq N_v$. Note that the game $(N \backslash N_v, q, w)$ is a simple voting game without power of veto.

Example. The United Nations Security Council (UNSC) has 15 voters $(N = 15)$. Five of the voters $(n_v = 5$; permanent members of the UNSC) have veto power (veto of first degree which cannot be overruled). There are two rounds of voting. No power of veto is available in the first round and nine "yes" votes are required for the vote to pass through to the second round, where players have three possible votes: "yes", "no" and "abstain". A "no" vote from a permanent member is interpreted as a veto. Nine "yes" votes are required for a motion to be passed $(q = 9)$. A veto from any of the 5 permanent members is sufficient to block a motion, regardless of the number of "yes" votes. Since a "no" vote from a non-veto player is equivalent to an abstention, we will only refer to "no" votes specifically in the case of permanent members applying their power of veto. Any other vote which is not a "yes" vote will be referred to as an abstention.

4   The Form of an Equilibrium with Retaliation

The players vote on a infinite sequence of motions which are ordered in time. Assume that they maximize their average payoff per round. The value of the j-th motion to the i-th player (measured relative to the status quo) is denoted $X_{i,j}$ and is not a centrally distributed normal random variable [3], but has a non-central Laplace distribution with density function $f(x) = \frac{1}{2} e^{-|x-1|}$. Note that $E(X_{i,j}) = 1$ and $P(X_{i,j} < 0) = \frac{e^{-1}}{2} \approx 0.1839$. Hence, normally a majority of voters prefer a motion to the status quo. Also, the tails of this distribution are heavy. It is assumed that the $X_{i,j}$ are independent and identically distributed. We denote a random variable with this distribution by X.

Example. The adoption of such a non-central Laplace distribution reflects, in our opinion, the nature of the UN security council, i.e. one that considers motions regarding regions of conflict, where action is required, but sometimes a motion may be clearly against the interests of a small number of council members. It is assumed that a motion is introduced by the player who values the motion most highly and that the players can accurately assess who a motion most favours (so a player cannot cheat by finding an ally to introduce a bill).

For convenience, it is assumed that in the first round of voting the players send honest signals expressing whether they prefer a motion to the status quo. Thus member i initially votes "yes" to motion j if $X_{i,j} > 0$ and otherwise abstains. Intuitively, these signals will be interpreted as a signal that a member wishes to vote for (when $X_{i,j} > 0$), a signal that a non-veto member wishes to abstain (when $X_{i,j} < 0$), or a signal that a veto member is willing to veto a motion if he believes there is no chance of repercussion or feels sufficiently strongly against the motion. It is assumed that the beliefs of the members about how the other members will vote is based on these signals[2].

It is also assumed that players who support a motion that is vetoed by a single player, but would have been passed if the vetoer had abstained rather than veto, react against the vetoer by abstaining on the next motion introduced by the vetoer. This means that the next motion introduced by the vetoer will not be passed. Note that in practice other means of negative reciprocation may be used (e.g. a trade block). However, this approach is used to internalize the costs of repercussion within the framework of a dynamic voting game [1]. It should be noted that the frequency with which a member introduces motions may be greater or lower than the frequency with which he is the lone vetoer of a motion that would otherwise be successful. In the first case, it is assumed that each such veto by that member is met by a reaction from the remaining members only when the lone vetoer introduces its next motion. In the second case, it is assumed that each motion introduced by such a member is vetoed. In order for this convention to be stable, the average reward obtained by the reciprocators reacting to such lone vetoes is greater than the average reward obtained when veto player i vetoes any motion that is unfavourable to them, i.e. when $X_{i,j} < 0$. Regarding the constraints on the players, it is assumed that negative reciprocation aimed at multiple members who together block a motion is either impractical or impossible. Given the above assumptions, when motion j is being voted upon, the game can be in one of the two following states: either the motion was introduced by a veto member who is presently a target for repercussion (State 1) or the motion was introduced by a member who is not presently a target for repercussion (State 0).

It will be assumed that repercussions against a lone vetoer are on average costly to the remaining members, any veto member introduces motions more frequently than it blocks motions due to a lone veto and that no veto member who prefers a motion to

---

[2] Thus we ignore the possibility of tactical voting in the first round. This may be considered in a future paper.

the status quo will veto a motion (note that if a threat to veto is always carried out, it would pay such a member to also veto the motion, thus ensuring that there are no repercussions). In this case, when the game is in State 0, the voting only affects the outcome of the present vote and the outcome of the next vote introduced by a lone vetoer. Hence, players should maximize the sum of their payoffs from these two votes. In order to keep this paper short, these assumptions will be tested in a future paper.

The following results follow from these assumptions:

Result 1: If $X_{i,j} < 0$, then member i prefers abstaining to voting for.

Proof: Suppose that member i can only choose between abstaining and voting for. Given the actions of the other members, member i may be either pivotal (i.e. when member i abstains the motion will be blocked and when member i votes for the motion will be passed) or non-pivotal. In the first case, there will be no repercussions and thus member i strictly prefers the motion being blocked to the motion being passed. In the second case, member i is indifferent between voting for and abstaining. It follows that when $X_{i,j} < 0$, member i's action "abstain" dominates his action of voting for.

Result 2: Given that no veto member abstains in the first round, member i should vote for a motion when $X_{i,j} > 0$ and abstain when $X_{i,j} < 0$. It should be noted that in this case all the veto members will vote for a motion.

Proof: Given that no veto member abstains in the first round, no member believes that there will be a veto by another member. Given this belief, each veto member strictly prefers the motion being passed to the motion being blocked and thus voting for dominates both abstaining and vetoing (particularly since it is believed that vetoing would lead to repercussions). Similarly, in the case of non-veto member i, the action of voting for dominates abstaining when $X_{i,j} > 0$. Result 2 then follows from Result 1. It should be noted that given the beliefs of the players and the fact that voting proceeded to the second round, a non-veto member i is indifferent between voting for and abstaining, because he believes that the motion will be passed regardless of how he votes. However, if we assume that voters make errors with some small probability, a non-veto member always prefers abstaining to voting for when $X_{i,j} < 0$.

Result 3: Given that at least two veto members initially vote against a motion, such veto members should veto in the second round.

Proof: The strictly best outcome for such veto members is for the motion to be passed without any repercussions. Given that $X_{i,j} < 0$ and veto member i believes that at least one other veto member will veto the motion, then he maximizes his expected payoff by vetoing.

Note: In practice, if negotiations indicate that two veto members are against a motion, then it is likely that such a motion will be either modified or withdrawn before a vote occurs (see for example [7] where the frequency of the vetoes occurring in the Security Council is presented).

Hence, given the assumptions made, it suffices to consider votes in which initially one veto player votes against the motion. In this case, it is assumed that the beliefs of players regarding the voting patterns of other players in this case come from historical data on voting patterns. We look for an equilibrium at which all the veto members use the same strategy and all the non-veto members use the same strategy. Explicitly, assume that initially one veto player votes against a motion and let k be the number of non-veto players that initially vote against a motion. We may assume that $k \in \{0,1,...,N-q\}$, since otherwise the motion would not continue to the second round. We look for an equilibrium where:

Each veto member votes for motion j when $X_{i,j} > t_2(k)$, abstains when $t_1(k) < X_{i,j} < t_2(k)$ and vetoes when $X_{i,j} < t_1(k)$.

Each non-veto member votes for motion j when $X_{i,j} > t_3(k)$.

The intuition behind such a strategy profile is as follows: suppose that there is a threat of a motion being vetoed. Any player will only support such a motion when he is reasonably strongly in favour of such a motion (in this case the possible gain from passing the motion is enough to overcome the threat of a loss resulting from possible repercussion). A veto member will only veto when he is strongly against a motion (in this case the gain from being able to veto such a motion outweighs the threat of retaliation).

Definitions: Let V_P and V_N be the expected value of a veto player and a non-veto player, respectively, from voting on a motion. Let $V_{P,1}$, $V_{P,2}$, $V_{N,1}$ and $V_{N,2}$ be the expected value from voting of: a) a veto player introducing a motion, b) a veto player who is not introducing a motion, c) a non-veto player introducing a motion, d) a non-veto player who is not introducing a motion, respectively.

Result 4: Given that only one veto member, denoted i, initially votes against motion j, he should veto it if and only $X_{i,j} < -V_{P,1}$.

Proof: Player i should maximise the expected sum of the payoffs obtained in the vote on the present motion and the vote on the next motion that he introduces. Let p_k be member i's estimate of the probability that a motion gains at least q votes given that k non-veto members initially voted against the bill. By vetoing motion j, player i obtains a payoff of 0 in the present vote. With probability p_k there will be repercussions when member i introduces his next bill and then his payoff in that vote will be 0, otherwise his expected payoff will be $V_{P,1}$. Assume that player i abstains in the vote on motion j. In the present vote, by conditioning on whether the motion is passed or not and using the law of total probability, member i obtains an expected payoff of $p_k X_{i,j} + (1 - p_k) \times 0$. Since there will be no repercussions, his expected payoff from the next motion he introduces will be $V_{P,1}$. Comparing the sum of the expected payoffs in these two votes, player i should veto motion j when $(1 - p_k)V_{P,1} > p_k X_{i,j} + V_{P,1} \Rightarrow X_{i,j} < -V_{P,1}$.

Note: It follows that at equilibrium, the probability of such a veto player vetoing a motion does not depend on the number of non-veto players initially voting against the motion. One might think that a veto member is less likely to veto a motion when no-one else initially votes against a bill, since the probability of retaliation is very large. However, this is compensated for by the fact the expected present gain from being able to veto is also larger, since it is very likely that an unfavourable bill can only be stopped by using the power of veto.

Result 5: Non-veto player i should vote for motion j if and only if $X_{i,j} > \frac{sV_{N,2}}{1-s}$, where s is player i's prior probability that the veto player who initially voted against a motion will veto it in the second round.

Proof: Here we assume that the cost of punishing a lone vetoer is positive, i.e. $V_{N,2} > 0$. It follows that any non-veto player who voted against a motion in the initial round will also vote against it in round 2. Hence, we consider a non-veto player who initially voted for the motion. Suppose his prior probability that the motion obtains at least q votes given that k non-veto players initially voted against the motion and: a) he votes for in round 2 and b) he abstains in round 2, are equal to r_k and $\tilde{r}_k$, respectively, where $r_k > \tilde{r}_k$.

For the motion to be passed, it is necessary to obtain q votes in round 2 without the veto of the veto player who initially voted no. If player i votes for in the present round, the probability that the motion will be passed is $(1 - s)r_k$. Hence, player i's expected reward from the present vote is $X_{i,j}(1 - s)r_k$. For repercussions to occur when the veto player threatening to veto next introduces a bill, it is necessary for both the motion to obtain q votes and the threat of veto to be realized. It follows that player i's expected reward from the next motion introduced by the potential vetoer is $(1 - sr_k)V_{N,2}$.

Now consider the expected sum of payoffs when this non-veto player abstains in the second round of voting. Arguing as above, player i's expected reward from the present vote is $X_{i,j}(1 - s)\tilde{r}_k$. Player i's expected reward from the next motion introduced by the potential vetoer is $(1 - s\tilde{r}_k)V_{N,2}$. It follows that player i prefers voting for motion j to abstaining if and only if

$$X_{i,j}(1 - s)r_k + (1 - sr_k)V_{N,2} > X_{i,j}(1 - s)\tilde{r}_k + (1 - s\tilde{r}_k)V_{N,2}.$$

It follows that voting for is preferred if and only if

$$X_{i,j}(r_k - \tilde{r}_k)(1 - s) > s(r_k - \tilde{r}_k)V_{N,2} \Rightarrow X_{i,j} > \frac{sV_{N,2}}{1-s}.$$

Result 6: Veto player i should vote for motion j if and only if $X_{i,j} > \frac{sV_{P,2}}{1-s}$, where s is player i's prior probability that the veto player who initially voted against a motion will veto it in the second round.

Proof: The proof of this result is analogous to the proof of Result 5 and is thus omitted.

Theorem (Form of the Equilibrium): At equilibrium, given that one veto player abstained in the first round, in the second round of voting on motion j,:

1) A veto player should vote for the motion if $X_{i,j} > t_2$, veto if $X_{i,j} < t_1$ and otherwise abstain.
2) A non-veto player should vote for the motion if $X_{i,j} > t_3$ and otherwise abstain.
3) The thresholds t_1, t_2 and t_3 satisfy the following condtions:

$$t_1 = -V_{P,1}; \qquad t_2 = \frac{P(X<t_1)V_{P,2}}{P(X>t_1)}; \qquad t_3 = \frac{P(X<t_1)V_{N,2}}{P(X>t_1)}.$$

Note that this theorem is a corollary of Results 4-6.

5   Derivation of the Equilibrium

We now attempt to derive such an equilibrium as described above using policy iteration. Using such an approach, we define initial thresholds $t_{1,1}$, $t_{2,1}$ and $t_{3,1}$. Given the
present set of thresholds: $t_{1,k}$, $t_{2,k}$ and $t_{3,k}$, we calculate the corresponding $V_{P,1}$, $V_{P,2}$ and
$V_{N,2}$ (which are each dependent on the strategy profile). The set of thresholds used is
then updated using

$$t_{1,k+1} = -V_{P,1}; \quad t_{2,k+1} = \frac{P(X < t_{1,k+1})V_{P,2}}{P(X > t_{1,k+1})}; \quad t_{3,k+1} = \frac{P(X < t_{1,k+1})V_{N,2}}{P(X > t_{1,k+1})}.$$

If this iterative process converges, then the limit of this process gives the equilibrium thresholds that voting decisions are based on. We now derive a procedure for
estimating $V_{P,1}$, $V_{P,2}$ and $V_{N,2}$ given the strategies used by the members. Note that it
follows from the form of the strategy profiles considered that if a motion obtains at
least q votes in the second round, then it receives at least q votes in the initial vote.
Hence, obtaining at least q votes with no veto in the second round is a necessary and
sufficient condition for a motion to be passed. A motion will be passed if:

1) All veto players initially vote for the motion and at least $q - n_v$ non-veto
   players vote for the motion. In this case, the pattern of voting in the second
   round is exactly the same as in the first round.
2) $n_v - 1$ veto players initially vote for the motion. In the second round of voting, no veto player (specifically, the one who initially abstained) vetoes the
   motion and at least q players vote for the motion.

The process proposed for deriving the equilibrium consists of the following
elements:

1) Defining the probability that a player votes for the motion in the second round
   given that no veto player abstained in the first round (which in this case is
   equivalent to a player voting for the motion in the first round) and the probability that the value of a motion to a player exceeds the appropriate threshold
   given the player voted for the motion in the first round.
2) Deriving the expected values of a motion to
   a) a veto player introducing a bill,
   b) a veto player not introducing a bill, and
   c) a non-veto player not introducing the bill.

The players maximize their average payoff per round. The exact form of the equilibrium depends on the probability distribution describing the value of a motion to a
player (here, assumed to be a non-central Laplace distribution), but the general form
of the equilibrium will be unaffected by this distribution (as long as it satisfies the
general conditions assumed for such a distribution, as outlined above, e.g. in general,
a clear majority of the members are in favor of a motion).

6   Conclusions

We have described the equilibrium conditions for a sequential voting game with ve-toes in which players may retaliate against a player who vetoes a motion which clearly has popular support. When no veto player abstains in the first round (i.e. no-one threatens to veto in the second round) or more than two veto players abstain, then in the second round each player votes for a motion if and only if he prefers the motion to the status quo. Given that one veto player abstained in the first round, then the other players only vote for the motion when they see that it is significantly better than the status quo (i.e. the value of the motion relative to the status quo is above a given positive threshold). This threshold is chosen to ensure that the expected gains from supporting such a motion are large enough to outweigh the expected loss from any form of retaliation. Analogously, a veto player who was the only such player to ab-stain in the first round will only veto a motion in the second round when the motion is clearly worse to him than the status quo. An iterative procedure to derive such an equilibrium was described.

It should be noted that it has been assumed that this equilibrium satisfies several conditions, namely,

- Veto players introduce motions more frequently than they are lone vetoers.
- Retaliation is expected to be costly to each of the players.
- No member favouring a motion to the status quo will veto that motion.
- Players do not vote tactically in the first round.

In order to test these assumptions, it is necessary to derive a candidate for an equi-librium strategy using the iterative procedure described above and then check whether the conditions described above are satisfied. This will be described in a future paper.

The model presented above is clearly a simplified version of the reality seen in the UNSC. For example, since the United Kingdom and the United States often have common interests, the values of a motion to these members will be thus correlated and not independent as assumed in the model presented here. However, the authors feel that the qualitative form of the equilibrium presented here gives an intuitive picture of the behaviour of such decision making bodies.

References

1. Başar T., Olsder, G.J.: Dynamic Noncooperative Game Theory, 2nd edn., SIAM Classics in Applied Mathematics, 23 (1999)
2. Brams, S.J.: Negotiation games. Routledge, New York (1990)
3. Clauset, A., Shalizi, C.R., Newman, M.E.J.: Power-Law Distributions in Empirical Data. SIAM Rev. **51**(4), 661–703 (2009)
4. Mercik, J.: Classification of committees with vetoes and conditions for the stability of power indices. Neurocomputing **149**(Part C), 1143–1148 (2015)
5. Mercik, J., Ramsey, D.: On a simple game theoretical equivalence of voting majority games with vetoes of first and second degrees. In: Nguyen, N.T., Trawiński, B., Kosala, R. (eds.) ACIIDS 2015. LNCS, vol. 9011, pp. 284–294. Springer, Heidelberg (2015)

6. von Neumann, J., Morgenstern, O.: Theory of Games and Economic Behaviour. Princeton Univ. Press (1944)
7. Ramsey, D., Mercik, J.: A formal a priori power analysis of the security council of the united nations. In: Kersten, G., Kamiński, B., Szufel, P., Jakubczyk, M. (eds.) Proceedings of the 15th International Conference on Group Decision & Negotiation. Warsaw School of Economics Press, Warsaw (2015)

DC Programming and DCA for Dictionary Learning

Xuan Thanh Vo[1]($\boxtimes$), Hoai An Le Thi[1], Tao Pham Dinh[2],
and Thi Bich Thuy Nguyen[1]

[1] Laboratory of Theoretical and Applied Computer Science EA 3097,
University of Lorraine, Île du Saulcy, 57045 Metz, France
{xuan-thanh.vo,hoai-an.le-thi,thi-bich-thuy.nguyen}@univ-lorraine.fr
[2] Laboratory of Mathematics, National Institute for Applied Sciences-Rouen,
Avenue de l'Université, 76801 Saint-Etienne-du-Rouvray, Cedex, France
pham@insa-rouen.fr

Abstract. Sparse representations of signals based on learned dictionaries have drawn considerable interest in recent years. However, the design of dictionaries adapting well to a set of training signals is still a challenging problem. For this task, we propose a novel algorithm based on DC (Difference of Convex functions) programming and DCA (DC Algorithm). The efficiency of proposed algorithm will be demonstrated in image denoising application.

Keywords: Dictionary learning · Sparse representation · DC programming · DCA

1 Introduction

The research of sparse representation of signals has drawn more and more interest in recent years. Modeling a signal using a linear combination of a few atoms or bases from an overcomplete dictionary has shown to be very effective in many signal processing applications such as image compression, image denosing, texture synthesis, etc. Up to now, there are two methods for building dictionaries: the first uses predefined dictionaries based on various types of wavelets ([14]), and the second learns from training sets of data ([16],[3], [15]). It has been proved that the use of learned dictionaries instead of predefined dictionaries gives better results for many image processing tasks ([3]).

Most algorithms for dictionary learning ([1],[15],[22]) iteratively alternate between two phases: sparse coding and dictionary updating. In the sparse coding phase, a sparse representation of signals is performed while the currently learned dictionary is fixed. In the dictionary updating phase, the learned dictionary is recomputed using the new sparse representation of signals.

The problem of sparse representation (sparse coding) is often modeled using the ℓ_0-norm. Unfortunately, this leads to an NP-hard problem. The ℓ_0-norm can be dealt directly by using greedy algorithms such as Matching Pursuit (MO) [13] and Orthogonal Matching Pursuit (OMP) [18]. Another

© Springer International Publishing Switzerland 2015
M. Núñez et al. (Eds.): ICCCI 2015, Part I, LNAI 9329, pp. 295–304, 2015.
DOI: 10.1007/978-3-319-24069-5_28

alternative for treating the ℓ_0-norm is to replace it with some tractable convex/nonconvex relaxations. A widely used convex relaxation of the ℓ_0-norm is the ℓ_1-norm, also known as Lasso ([23]) or basis pursuit ([2]). However, the ℓ_1-norm is, in certain cases, inconsistent for variable selection and biased ([4]). A number of nonconvex relaxations such as the smoothly clipped absolute deviation (SCAD) penalty [4], the capped-ℓ_1 penalty ([19]), etc. have been proposed to overcome this drawback and shown to be very efficient in sparse optimization. All of these nonconvex relaxations belong to a family call DC approximation that was thoroughly studied in ([6]). Following the results in [6], in this paper, we propose to use the capped-ℓ_1 function ([19]) for modeling sparsity.

Among the methods for tackling sparse optimization problems, methods based on DC (Difference of Convex functions) programming ad DCA (DC Algorithm) have appeared to be very efficient ([7],[8],[9],[17],[6],[10],[11],[12] among others). Motivated by these success, we will investigate DC programming and DCA for solving the dictionary learning problem.

Our contributions in this paper are three folds. First, we use the capped-ℓ_0 function to relax ℓ_0-norm for modeling sparsity in the dictionary learning problem. Second, we develop an algorithm based on DC programming and DCA to solve the new formulation of dictionary learning problem. Third, we apply the new algorithm for the problem of image denoising.

The rest of the paper is organized as follows. The next section provides the formulation of the dictionary learning problem. Section 3 will present a brief overview of DC programming and DCA, then present the algorithm based on DC programming and DCA for solving the dictionary learning problem. Some experiments of image denosing is conducted in section 4 to evaluate the performance of our proposed method. Finally, section 5 concludes the paper.

2 Problem Formulation and General Schema Solution

Given a training set $X = [x_1, \ldots, x_L] \in \mathbb{R}^{n \times L}$ of L signals of n dimension. Dictionary learning aims to find a dictionary matrix $D = [d_1, \ldots, d_k] \in \mathbb{R}^{n \times k}$ of k atoms (columns) of n dimension with $n < k$, and corresponding coefficient matrix $W = [w_1, \ldots, w_L] \in \mathbb{R}^{k \times L}$ such that $x_i \approx Dw_i$ and w_i is as sparse as possible, $\forall i = 1, \ldots, L$. This task can be modeled as the following optimization problem

$$\min_{D \in \mathcal{C}, W \in \mathbb{R}^{k \times L}} \sum_{i=1}^{L} \left\{ \frac{1}{2} \|x_i - Dw_i\|_2^2 + \lambda \Phi(w_i) \right\}, \tag{1}$$

where $\mathcal{C} = \{D \in \mathbb{R}^{n \times k} : \|d_j\|_2 \leq 1 \ \forall j = 1, \ldots, k\}$, $\Phi(\cdot)$ is a penalty function encouraging sparsity, and $\lambda > 0$ is a trade-off parameter between the reconstruction/fitness terms $\frac{1}{2}\|x_i - Dw_i\|_2^2$ and the sparsity terms $\Phi(w_i)$. In our case, the function $\Phi(\cdot)$ is the capped-ℓ_1 penalty defined by

$$\Phi(w) = \sum_{j=1}^{k} \min(1, \alpha|w_j|), \quad w = (w_1, \ldots, w_k) \in \mathbb{R}^k, \tag{2}$$

where $\alpha > 0$ is a parameter.

The schema solution of the problem (1) is as follows

Schema Solution:

Initialized from an initial dictionary D^0, we iteratively alternate these two phases until convergence of D.

- Sparse coding phase: fix D, update W. This phase will leads to solve the problem:

$$w_i \in \arg\min \left\{ \frac{1}{2}\|x_i - Dw\|^2 + \lambda\Phi(w) : w \in \mathbb{R}^k \right\} \quad \forall i = 1, \dots, L. \quad (3)$$

- Dictionary updating phase: fix W, update D by solving the problem

$$\min_{D \in \mathcal{C}} \frac{1}{2}\langle A, D^T D \rangle - \langle B, D \rangle, \quad (4)$$

where $A = WW^T$, $B = XW^T$, and $\langle A, B \rangle$ denotes the inner product of two matrices A, B: $\langle A, B \rangle = \text{trace}(A^T B)$.

In the next section, we will reformulate the subproblems (3) and (4) as DC programs and present DCA based algorithms for solving them.

3   DC Programming and DCA for Dictionary Learning

In this section, we will study the use of DC programming and DCA in solving the dictionary learning problem. Throughout the section, the absolute matrix $|X|$ is defined as $[|X|]_{ij} = |X_{ij}|$ for all i, j. X_{IJ} indicates sub–matrix with row (resp. column) indices taken from I (resp. J). $X_{i:}$ (resp. $X_{:j}$) denotes the ith row (resp. j^{th} column) of the matrix X. The notation $\circ$ (resp. $\frac{[\cdot]}{[\cdot]}$) denotes the component–wise product (resp. division) of matrices. The projection of a vector $x \in \mathbb{R}^n$ on a subset $\Omega \subset \mathbb{R}^n$, denoted by $P_\Omega(x)$, is the nearest element of Ω to x w.r.t. Euclidean distance.

3.1   Outline of DC Programming and DCA

A general DC program is that of the form:

$$\alpha = \inf\{F(x) := G(x) - H(x) \mid x \in \mathbb{R}^n\} \quad (P_{dc}),$$

where G, H are lower semi-continuous proper convex functions on $\mathbb{R}^n$. Such a function F is called a DC function, and $G - H$ a DC decomposition of F while G and H are the DC components of F.

A point x^* is called a *critical point* of $G - H$, or a generalized Karush-Kuhn-Tucker point (KKT) of (P_{dc})) if

$$\partial H(x^*) \cap \partial G(x^*) \neq \emptyset, \quad (5)$$

where $\partial\theta(x)$ denotes the subdifferential of θ at x.

Based on local optimality conditions and duality in DC programming, the DCA consists in constructing two sequences $\{x^k\}$ and $\{y^k\}$ (candidates to be solutions of (P_{dc}) and its dual problem respectively). Each iteration k of DCA approximates the concave part $-H$ by its affine majorization (that corresponds to taking $y^k \in \partial H(x^k)$) and minimizes the resulting convex function.

Generic DCA scheme

Initialization: Let $x^0 \in \mathbb{R}^n$ be an initial guess, $k \leftarrow 0$.

Repeat

- Calculate $y^k \in \partial H(x^k)$
- Calculate $x^{k+1} \in \arg\min\{G(x) - \langle x, y^k \rangle : x \in \mathbb{R}^n\}$ (P_k)
- $k \leftarrow k+1$

Until convergence of $\{x^k\}$.

Convergences properties of DCA and its theoretical basic can be found in [5,20]. It is worth mentioning that

- DCA is a descent method *without linesearch*.
- If the optimal value α of problem (P_{dc}) is finite and the infinite sequences $\{x^k\}$ and $\{y^k\}$ are bounded then every limit point x^* of the sequences $\{x^k\}$ (resp. $\{x^k\}$) is a critical point of $G - H$.
- DCA has a *linear convergence* for general DC programs, and has a *finite convergence* for polyhedral DC programs.

A deeper insight into DCA has been described in [5]. For instant it is crucial to note the main feature of DCA: DCA is constructed from DC components and their conjugates but not the DC function f itself which has infinitely many DC decompositions, and there are as many DCA as there are DC decompositions. Such decompositions play a critical role in determining the speed of convergence, stability, robustness, and globality of sought solutions. It is important to study various equivalent DC forms of a DC problem. This flexibility of DC programming and DCA is of particular interest from both a theoretical and an algorithmic point of view. For a complete study of DC programming and DCA the reader is referred to [5,20,21] and the references therein.

3.2 Sparse Coding Phase: Update W

For convenience, we omit the subscript of x, the problem (3) becomes one optimization problem as follows:

$$\min_{w \in \mathbb{R}^k} \left\{ f_D(w) = \frac{1}{2}\|x - Dw\|^2 + \lambda\Phi(w) \right\} \tag{6}$$

where $D \in \mathbb{R}^{n \times k}$ and fixed, $x \in \mathbb{R}^n$.

Note that $\Phi(w)$ is a DC function with a DC decomposition given by

$$\Phi(w) = k + \alpha\|w\|_1 - \sum_{i=1}^{k} \max(1, \alpha|w_i|).$$

So f_D is also a DC function and has a DC decomposition $f_D = g_D - h_D$, where

$$g_D(w) = \frac{1}{2}\|x - Dw\|^2 + \lambda\alpha\|w\|_1 + k\lambda, \quad h_D(w) = \lambda\sum_{j=1}^{k}\max(1, \alpha|w_i|).$$

Then DCA for solving problem (6) is simply as follows (for $l = 0, 1, 2, \dots$):
- Calculate $y^l \in \partial h_D(w^l)$.
- Calculate $w^{l+1} \in \arg\min\{\frac{1}{2}\|x - Dw\|^2 + \lambda\alpha\|w\|_1 - \langle w, y^l\rangle : w \in \mathbb{R}^k\}$. (P_l).
Computation of $y^l \subset \partial h_D(w^l)$ is explicitly given by:

$$y \in \partial h_D(w) \Leftrightarrow \begin{cases} y_j = 0 & \text{if } |w_j| < \frac{1}{\alpha} \\ y_j \in \text{sign}(w_j)[0, \lambda\alpha] & \text{if } |w_j| = \frac{1}{\alpha} \quad \forall j = 1, \dots, k. \\ y_j = \text{sign}(w_j)\lambda\alpha & \text{otherwise,} \end{cases} \quad (7)$$

We will discuss on how to solve problem (P_l) below.

DCA for Solving Problem (P_l).

If we omit the subscript of y, then the problem (P_l) takes the form:

$$\min_{w \in \mathbb{R}^k} \overline{f}_D(w) := \frac{1}{2}\|x - Dw\|^2 + \lambda\alpha\|w\|_1 - \langle w, y\rangle. \qquad (P_l)$$

Let $\rho \in \mathbb{R}^k_{++}$ such that $D^T D \preceq \text{diag}(\rho)$, then $\overline{f}_D$ has a DC decomposition $\overline{f}_D = \overline{g}_D - \overline{h}_D$ given by

$$\overline{g}_D(w) = \sum_{j=1}^{k}\left(\frac{1}{2}\rho_j w_j^2 + \lambda\alpha|w_j| - y_j w_j\right), \quad \overline{h}_D(w) = \sum_{j=1}^{k}(\frac{1}{2}\rho_j w_j^2) - \frac{1}{2}\|x - Dw\|^2.$$

DCA for solving problem (P_l) consists of (for $t = 0, 1, 2, \dots$):

- Compute $z^t = \nabla\overline{h}_D(w^t) = \text{diag}(\rho)w^t - D^T(Dw^t - x)$
- Compute:

$$w^{t+1} \in \arg\min\left\{\sum_{j=1}^{k}\left(\frac{1}{2}\rho_j(w_j)^2 + \lambda\alpha|w_j| - y_j w_j - z_j^t w_j\right) : w \in \mathbb{R}^k\right\}$$

$$\Leftrightarrow w_j^{t+1} = \arg\min_{w_j}\frac{1}{2}\left(w_j - \frac{y_j + z_j^t}{\rho_j}\right) + \frac{\lambda\alpha}{\rho_j}|w_j| = \frac{S(y_j + z_j^t, \lambda\alpha)}{\rho_j} \quad \forall j = 1, \dots, k,$$

where $S(u, \beta) = \text{sign}(u)(|u| - \beta)_+$ is the soft thresholding formula. For simplicity, we rewrite the above updating rule in vector form as follows:

$$w^{t+1} = \frac{[S(y + z^t, \lambda\alpha)]}{[\rho]}, \qquad (8)$$

where the operation S is component-wise, i.e. $S(a, b) = (S(a_i, b_i))_i$.

A simple choice of ρ can be $\rho = |D^T D| 1_{k \times 1}$. However, we can do in a more effective way. Observe that if $w_j^t = 0$ and $|D_{:j}^T(Dw - x) - y_j| \leq \lambda\alpha$, we have $S(y_j + z_j^t, \lambda\alpha) = 0$ for any choice of ρ. Thus, the updating rule (8) makes no change on the component j^{th} of w^{t+1}. If we define

$$I(w, y) = \{j = 1, \ldots, k : w_j \neq 0 \text{ or } |D_{:j}^T(Dw - x) - y_j| > \lambda\alpha\}, \quad w, y \in \mathbb{R}^k,$$

then at the iteration t^{th}, we only need to consider variables $\{w_j : j \in I\}$ (here $I = I(w^t, y)$ for short). Repeat the above procedure with w_I (resp. $D_{:I}$ and y_I) replacing w (resp. D and y), we compute

$$z_I^t = \text{diag}(\rho_I)w_I^t - D_{:I}^T(D_{:I}w_I^t - x_I)$$

and

$$w_I^{t+1} = \frac{[S(y_I + z_I^t, \lambda\alpha)]}{[\rho_I]}, \quad w_j^{t+1} = 0, \quad \forall j \notin I,$$

where $\rho_I = |D_{:I}^T D_{:I}| 1_{|I| \times 1}$ and $\rho_j = 0 \; \forall j \notin I$.

These are equivalent to compute

$$\omega = \frac{[p]}{[|D^T D|p]},$$

where $p \in \mathbb{R}^k$, that is defined by $p_j = 1$ if $j \in I$ and $p_j = 0$ otherwise.

Then we compute

$$z^t = w^t - (D^T(Dw^t - x) - y) \circ \omega, \tag{9}$$

$$w^{t+1} = S(z^t, \lambda\alpha\omega). \tag{10}$$

Following the general convergence properties of DCA, we can prove that under the updating rules (9)–(10), the function $\overline{f}_D$ is decreasing.

Note that, in the context of dictionary learning, w is expected to be very sparse. This implies that very few components of w need to be updated (corresponding to $\omega_j \neq 0$). To exploit this fact for solving problem (P_l), we will not calculate ω after each iteration. Instead, we only compute ω from the beginning and keep using it later on. This means that we do not actually solve problem (P_l). We are now in a position to describe the DCA for solving problem (6).

DC Algorithm for the Sparse Coding Phase:

Initialization: Initialize $w^0 \in \mathbb{R}^k$, $T > 0$ (maximum number of inner-iterations), $\epsilon > 0$ (stopping tolerance), $l \leftarrow 0$

Repeat

1. Compute $y^l \in \partial h(w^l)$ by: $y_i^l = 0$ if $|w_i^l| \leq \frac{1}{\alpha}$ and $y_i^l = \text{sign}(w_i^l)\lambda\alpha$ otherwise, for all $i = 1, \ldots, k$

2. Compute p^l by: $p_i^l = 1$ if $i \in I(w^l, y^l)$ and $p_i^l = 0$ otherwise, for all $i = 1, \ldots, k$

3. Compute $\omega^l = \frac{[p^l]}{[|D^T D|p^l]}$, $w^{(l,0)} = w^l$, and set $t = 0$

   Repeat

   - $t = t + 1$

- Compute $z^t = w^{(l,t-1)} - (D^T(Dw^{(l,t-1)} - x) - y^l) \circ \omega^l$
- Compute $w^{(l,t)} = S(z^t, \lambda \alpha \omega^l)$
    Until $t = T$ or $\|w^{(l,t-1)} - w^{(l,t)}\| / \max(1, \|w^{l,t}\|) < \epsilon$
4. Set $w^{l+1} = w^{(l,t)}$
5. Set $l \leftarrow l+1$
Until $\|w^l - w^{l-1}\| / \max(1, \|w^l\|) < \epsilon$

3.3   Dictionary Updating Phase: Update D

For updating D we solve the optimization of the form

$$\min_{D \in \mathcal{C}} f_W(D) := \frac{1}{2}\langle A, D^T D\rangle - \langle B, D\rangle, \tag{11}$$

where $A = WW^T, B = XW^T$.

Let $\gamma = |A| \mathbf{1}_{k \times 1}$, we can decompose f_W as $f_W = g_W - h_W$, where g_W and h_W are given by

$$g_W(D) = \frac{1}{2}\sum_{j=1}^{k}\gamma_j \|D_{:j}\|^2, \quad h_W(D) = \frac{1}{2}\sum_{j=1}^{k}\gamma_j\|D_{:j}\|^2 - \left(\frac{1}{2}\langle A, D^T D\rangle - \langle B, D\rangle\right).$$

DCA applied for the problem (11) then consists of two steps
- Compute $\overline{D}^{(t)} = \nabla h_W(D^{(t)}) = \Gamma \circ D^{(t)} - (D^{(t)}A - B)$, where $\Gamma \in \mathbb{R}^{n \times k}$ is the matrix defined by $\Gamma_{i:} = \gamma, \forall i = 1, \dots, n$.
- Compute

$$D^{(t+1)} = \arg\min\left\{g_W(D) - \langle\overline{D}^{(t)}, D\rangle : D = [d_1, \dots, d_k] \in \mathcal{C}\right\}$$

$$= \arg\min_{D \in \mathcal{C}}\sum_{j=1}^{k}\left(\frac{1}{2}\gamma_j\|d_j\|^2 - \langle\bar{d}_j^{(t)}, d_j\rangle\right) = \arg\min_{D \in \mathcal{C}}\sum_{j=1}^{k}\left\|d_j - \frac{1}{\gamma_j}\bar{d}_j^{(t)}\right\|^2$$

$$\Leftrightarrow \quad d_j^{(t+1)} = \operatorname*{Proj}_{\|d_j\|\le 1}\frac{\bar{d}_j^{(t)}}{\gamma_j} = \frac{\bar{d}_j^{(t)}}{\max\{\gamma_j, \|\bar{d}_j^{(t)}\|\}}, \quad \forall j = 1, \dots, k. \tag{12}$$

We summarize this procedure in the following algorithm.
DC algorithm for the dictionary updating stage
Initialization: Initial matrix $D^{(0)} \in \mathcal{C}$, $t \leftarrow 0$
Repeat
    - Compute $\overline{D}^{(t)} = \Gamma \circ D^{(t)} - (D^{(t)}A - B)$.
    - Compute $D^{(t+1)}$ by (12).
    - Set $t \leftarrow t + 1$.
Until $\|D^{(t-1)} - D^{(t)}\| < \epsilon$.

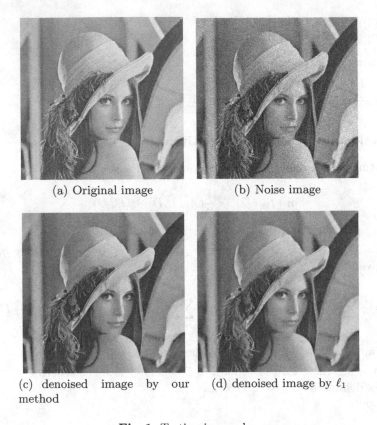

(a) Original image   (b) Noise image

(c) denoised image by our   (d) denoised image by ℓ_1
method

Fig. 1. Testing image: lena

Table 1. PSNR comparison with KSVD and ℓ_1 penalty

Method	barbara	boat	house	lena	peppers
our method	**31.00**	**30.48**	**33.42**	**32.54**	**32.39**
KSVD	30.86	30.37	33.18	32.42	32.22
ℓ_1	29.70	29.75	32.04	31.55	31.59

4   Numerical Experiment

In this section, we carry out some experiments of image denosing to demonstrate
the efficiency of our dictionary learning method. For the sake of comparison,
we also implemented the closely related dictionary learning method based on
ℓ_1-norm. Two these algorithms were implemented in the Matlab R2007a, and
executed on a PC Intel i5 CPU650, 3.2 GHz of 4GB RAM. We also compare with
the well-known standard algorithm K-SVD [1]. The experiments were conducted
on five gray scale images: lena, barbara, boat, house, and peppers. The code of

K-SVD and the testing images are taken from KSVD-Box (v3) package (http://www.cs.technion.ac.il/~ronrubin/software.html).

For image denosing, we followed the procedure described in [3]. We generated noisy images by adding zero-mean white and homogeneous Gaussian noise with deviation $\sigma = 20$. For each testing image, the training set includes $L = 40000$ patches of size 8×8, which are regularly sampled from the original noisy image in an overlapping manner. Each patch is converted to a vector of size $n = 64$ and then normalized to have zero mean. The size of the dictionary was set to $k = 256$. After the dictionary learning process is finished, the denoising process is as follows. We first sampled all the patches from the noisy image. Each patch is normalized to have zero mean, then we computed its sparse representation coefficients w.r.t the learned dictionary. We reconstructed each patch using its sparse representation coefficients. Further the denoised image was formed by packing the denoised patches together.

For methods based on ℓ_0 and ℓ_1 norms, the value of the trade-off parameter λ was chosen in the set $\{500, 600, \ldots, 1500\}$, and the parameter α of the function Φ was set to 1.

Figure 1 shows visual results for one selected image. We can observe that our dictionary learning method can provide better denoising result compared with the ℓ_1-norm method. Specifically, our results tend to be smoother and clearer.

Table 1 shows the peak signal-to-noise ratio (PSNR) of our method and the methods using ℓ_1 and K-SVD. It can be noticed that, our method achieves a significantly better performance of denoising in terms of PSNR than ℓ_1-based method. While it is a little better than K-SVD.

5   Conclusion

In this paper, we have studied the DC programming and DCA for dictionary learning problem and applied it on the denoising image problem. The dictionary learning process have been done with the training set including the overlapping patches taken from the noisy images. The efficiency of DCA has been approved experimentally on the set of gray images and the results show that our method is promising. In future works, we will extend our method to online learning manner and apply in other applications of image processing.

References

1. Aharon, M., Elad, M., Bruckstein, A.: K-SVD: An algorithm for designing of over-complete dictionaries for sparse representation. IEEE Transactions on Signal Processing **54**(11), 4311–4322 (2006)
2. Chen, S., Donoho, D., Saunders, M.: Atomic decomposition by basis pursuit. SIAM Journal on Scientific Computing **20**, 33–61 (1999)
3. Elad, M., Aharon, M.: Image denoising via sparse and redundant representations over learned dictionaries. IEEE Trans. Image Process. **54**(12), 3736–3745 (2006)
4. Fan, J., Li, R.: Variable selection via nonconcave penalized likelihood and its oracle properties. J. Am. Stat. Assoc. **96**(456), 1348–1360 (2001)

5. Thi, L.: H.A. and Pham Dinh, T.: The DC (difference of convex functions) Programming and DCA revisited with DC models of real world nonconvex optimization problems. Annals of Operations Research **133**, 23–46 (2005)
6. Le Thi, H.A., Pham Dinh, T., Le, H.M., Vo, X.T.: DC approximation approaches for sparse optimization. Eur. J. Oper. Res. **244**(1), 26–46 (2015)
7. Le Thi, H.A., Nguyen, V.V., Ouchani, S.: Gene selection for cancer classification using DCA. In: Tang, C., Ling, C.X., Zhou, X., Cercone, N.J., Li, X. (eds.) ADMA 2008. LNCS (LNAI), vol. 5139, pp. 62–72. Springer, Heidelberg (2008)
8. Le Thi, H.A., Le, H.M., Nguyen, V.V., Pham Dinh, T.: A DC Programming approach for feature selection in support vector machines learning. Adv. Data Analysis and Classification **2**(3), 259–278 (2008)
9. Le Thi, H.A., Nguyen Thi, B.T., Le, H.M.: Sparse signal recovery by difference of convex functions algorithms. In: Selamat, A., Nguyen, N.T., Haron, H. (eds.) ACIIDS 2013, Part II. LNCS, vol. 7803, pp. 387–397. Springer, Heidelberg (2013)
10. Le, H.M., Le Thi, H.A., Nguyen, M.C.: Sparse Semi-Supervised Support Vector Machines by DC Programming and DCA. Neurocomputing **153**(4), 62–76 (2015)
11. Le Thi, H.A., Nguyen, M.C., Pham Dinh, T.: A DC programming approach for finding Communities in networks. Neural Computation **26**(12), 2827–2854 (2014)
12. Le Thi, H.A., Vo, X.T., Pham Dinh, T.: Feature Selection for linear SVMs under Uncertain Data: Robust optimization based on Difference of Convex functions Algorithms. Neural Networks **59**, 36–50 (2014)
13. Mallat, S., Zhang, Z.: Matching pursuits with time-frequency dictionaries. IEEE Trans. Signal Process. **41**(12), 3397–3415 (1993)
14. Mallat, S.: A wavelet tour of signal processing, 2nd edn. Academic Press, New York (1999)
15. Mairal, J., Bach, F., Ponce, J., Sapiro, G.: Online learning for matrix factorization and sparse coding. Journal of Machine Learning Research **11**, 19–60 (2010)
16. Olshausen, B.A., Field, D.J.: Sparse coding with an overcomplete basis set: A strategy employed by V1? Vision Research **37**, 3311–3325 (1997)
17. Ong, C.S., Le Thi, H.A.: Learning sparse classifiers with difference of convex functions algorithms. Optimization Methods and Software **28**(4), 830–854 (2013)
18. Pati, Y.C., Rezaiifar, R., Krishnaprasad, P.S.: Orthogonal Matching Pursuit: recursive function approximation with application to wavelet decomposition. In: Asilomar Conf. on Signals, Systems and Comput., pp. 40–41 (1993)
19. Peleg, D., Meir, R.: A bilinear formulation for vector sparsity optimization. Signal Processing **88**(2), 375–389 (2008)
20. Pham Dinh, T., Le Thi, H.A.: Convex analysis approach to DC programming: Theory, algorithms and applications. Acta Math. Vietnamica **22**(1), 289–357 (1997)
21. Pham Dinh, T., Le Thi, H.A.: Dc optimization algorithms for solving the trust region subproblem. SIAM. J Optimization **8**, 476–505 (1998)
22. Skretting, K., Engan, K.: Recursive least squares dictionary learning algorithm. IEEE Transactions on Signal Processing **58**(4), 2121–2130 (2010)
23. Tibshirani, R.: Regression shrinkage and selection via the lasso. Journal of the Royal Statistical Society, Series B **58**(1), 267–288 (1996)

Evolutionary Algorithm for Large Margin Nearest Neighbour Regression

Florin Leon[1(✉)] and Silvia Curteanu[2]

[1] Department of Computer Science and Engineering, "Gheorghe Asachi" Technical
University of Iaşi, Bd. Mangeron, 700050 Iaşi, Romania
fleon@cs.tuiasi.ro
[2] Department of Chemical Engineering, "Gheorghe Asachi" Technical University of Iaşi,
Bd. Mangeron, 700050 Iaşi, Romania
scurtean@ch.tuiasi.ro

Abstract. The concept of a large margin is central to support vector machines
and it has recently been adapted and applied for nearest neighbour classification. In this paper, we suggest a modification of this method in order to be used
for regression problems. The learning of a distance metric is performed by
means of an evolutionary algorithm. Our technique allows the use of a set of
prototypes with different distance metrics, which can increase the flexibility of
the method especially for problems with a large number of instances. The proposed method is tested on a real world problem – the prediction of the corrosion
resistance of some alloys containing titanium and molybdenum – and provides
very good results.

Keywords: Large margin · Nearest neighbour regression · Evolutionary algorithm · Prototypes

1 Introduction

Regression analysis includes any technique for modelling different kinds of processes
with the goal of finding a relationship between a dependent variable and one or more
independent variables, given a set of training instances or vectors in the form of
$(\mathbf{x}_i, y_i)$ pairs, where $\mathbf{x}_i$ are the inputs and y_i is the output of a sample. The general
regression model is [8, 10, 11]:

$$y_i = f(\mathbf{x}_i) + e_i, \tag{1}$$

where f is the regression model (the approximating function), y_i is the desired output
value that corresponds to the $\mathbf{x}_i$ input of the training set and e_i is a residual whose
expected error given the sample point $\mathbf{x}_i$ is $E(e_i \mid \mathbf{x}_i) = 0$.

Presently, there are many methods that can be used for regression. Beside analytical models, where the task is to find adequate values for the coefficients usually by
means of differential optimization, we can mention several machine learning techniques such as neural networks, support vector machines (ε-SVR, ν-SVR), decision

© Springer International Publishing Switzerland 2015
M. Núñez et al. (Eds.): ICCCI 2015, Part I, LNAI 9329, pp. 305–315, 2015.
DOI: 10.1007/978-3-319-24069-5_29

trees (M5P, Random Forest, REPTree) or rules (M5, Decision Table), etc. k-Nearest Neighbour (kNN) [5] is a simple, efficient way to estimate the value of the unknown function in a given point using its values in other points. Let $S = \{\mathbf{x}_1, ..., \mathbf{x}_m\}$ be the set of training points. The kNN estimator can be simply defined as the mean function value of the nearest neighbours [13]:

$$\tilde{f}(\mathbf{x}) = \frac{1}{k} \sum_{\mathbf{x}' \in N(\mathbf{x})} f(\mathbf{x}'), \tag{2}$$

where $N(\mathbf{x}) \subseteq S$ is the set of k nearest points to $\mathbf{x}$ in dataset S.

Another, more elaborate version of the method considers a weighted average, where the weight of each neighbour depends on its proximity to the query point:

$$\tilde{f}(\mathbf{x}) = \frac{1}{z} \sum_{\mathbf{x}' \in N(\mathbf{x})} w_d(\mathbf{x}, \mathbf{x}') \cdot f(\mathbf{x}'), \tag{3}$$

where z is a normalization factor, and the inverse of the squared Euclidian distance is usually employed to assess the weights:

$$w_d(\mathbf{x}, \mathbf{x}') = \frac{1}{d(\mathbf{x}, \mathbf{x}')^2} = \frac{1}{\sum_{i=1}^{n} (x_i - x'_i)^2}. \tag{4}$$

A comprehensive review of locally weighted regression and classification methods is given in [3]. Even if kNN is a very simple method, it usually performs very well for a wide range of problems, especially when the number of instances is large enough and there is little noise in the data. One of the most important aspects of the method is the choice of the distance function. While Euclidian distance is the most commonly encountered, other particularisations of the general Minkowski distance can be used, e.g. the Manhattan distance. Some experiments in cognitive psychology suggest that humans use an exponential negative distance function for certain types of classification tasks [1]. However, the classical approach does not take into account any information about a particular problem. Beside the common practice of normalizing the instance values independently on each dimension, there is little domain knowledge incorporated into the method. The present work investigates the use of the concept of a large margin (best known in the context of support vector machines) for regression problems, by adapting its existing formulation for classification problems. We organize our paper as follows. Section 2 presents the large-margin nearest neighbour method and some of its extensions. Section 3 describes the proposed method, including the use of prototypes and evolutionary algorithms for optimization. Section 4 focuses on a case study, and section 5 contains the conclusions.

2   Related Work

Because of the importance of the distance metric, researchers sought to find ways to adapt it to the problem at hand in order to yield better performance. This is the issue of distance metric learning [15, 14, 4, 6]. The idea of a large margin, one of the fundamental ideas of support vector machines, was transferred to the kNN method for classification tasks [16-18], resulting in the large margin nearest neighbour method (LMNN). In this case, learning involves the optimization of a convex problem using semidefinite programming. The LMNN technique was also extended to incorporate invariance to multivariate polynomial transformations [9]. Since our regression method builds on the LMNN method for classification [18], we will present it with more details as follows.

In general, distance metric learning aims at finding a linear transformation $\mathbf{x}' = \mathbf{L}\mathbf{x}$, such that the distance between two vectors $\mathbf{x}_i$ and $\mathbf{x}_j$ becomes:

$$d_L(\mathbf{x}_i, \mathbf{x}_j) = \left\| \mathbf{L}(\mathbf{x}_i - \mathbf{x}_j) \right\|_2 . \tag{5}$$

Since all the operations in k-nearest neighbour classification or regression can be expressed in terms of square distances, an alternative way of stating the transformation is by means of the square matrix: $\mathbf{M} = \mathbf{L}^T \mathbf{L}$, and thus the square distance is:

$$d_M(\mathbf{x}_i, \mathbf{x}_j) = (\mathbf{x}_i - \mathbf{x}_j)^T \mathbf{M}(\mathbf{x}_i - \mathbf{x}_j). \tag{6}$$

For a classification problem, the choice of $\mathbf{L}$, or equivalently $\mathbf{M}$, aims at minimizing the distance between a vector $\mathbf{x}_i$ and its k target neighbours $\mathbf{x}_j$, where a target signifies a neighbour that belongs to the same class. At the same time, the distance between a vector and the impostors $\mathbf{x}_l$, i.e. neighbours that belong to a different class, should be maximized. In order to establish a large margin between the vectors that belong to different classes, the following relation is imposed:

$$d_M(\mathbf{x}_i, \mathbf{x}_l) \geq 1 + d_M(\mathbf{x}_i, \mathbf{x}_j). \tag{7}$$

Here, the value of 1 is arbitrary; the idea is to have some minimum value for the margin that separates the classes. However, it was proven that other minimum values for the margin would not change the nature of the optimization problem, but will only result in the scaling of the matrix $\mathbf{M}$.

Overall, the optimization problem is defined as follows:

$$\begin{aligned}
\min \quad & \sum_{ij} \eta_{ij} d_{ij} + \lambda_h \sum_{ijl} v_{ijl} \xi_{ijl} \\
\text{such that} \quad & (\mathbf{x}_i - \mathbf{x}_j)^T \mathbf{M}(\mathbf{x}_i - \mathbf{x}_j) = d_{ij} \\
& (\mathbf{x}_i - \mathbf{x}_l)^T \mathbf{M}(\mathbf{x}_i - \mathbf{x}_l) - d_{ij} \geq 1 - \xi_{ijl} \\
& \mathbf{M} \succeq 0, \ \xi_{ijl} \geq 0, \ \forall i, j, l
\end{aligned} \tag{8}$$

where $\eta_{ij} \in \{0, 1\}$ is 1 only when $\mathbf{x}_j$ is a target neighbour of $\mathbf{x}_i$, $v_{ijl} = \eta_{ij}$ if and only if $y_i \neq y_l$ and 0 otherwise, d_{ij} is the distance between $\mathbf{x}_i$ and $\mathbf{x}_j$, ξ_{ijl} is the hinge loss [12] and $\lambda_h \geq 0$ is a constant.

The two objectives, of minimizing the distance to targets and maximizing the distance to impostors, are conflicting. Following an analogy to attraction and repulsion forces in physics, Weinberger and Saul [18] introduce two forces, ε_{pull} and ε_{push}, whose balance is reached by setting the value of λ_h. In their work, the authors consider the importance of these forces to be equal, i.e. $\lambda_h = 1$.

Other researchers (e.g. [9]) suggest the use of regularization to avoid the overfitting caused by large values of the elements of $\mathbf{L}$ or $\mathbf{M}$. Regularization can be performed for example by minimizing the Frobenius norm of the parameters, and thus adding a third term to the optimization problem:

$$\min \sum_{ij} \eta_{ij} d_{ij} + \lambda_h \sum_{ijl} v_{ijl} \xi_{ijl} + \lambda_r \sqrt{\sum_p \sum_q m_{pq}^2} , \tag{9}$$

where $\lambda_r \geq 0$ is another constant.

Although there are several studies concerned with LMNN classification, so far not many researchers have addressed the issue of regression. One recent paper that presents some modifications of the LMNN metric learning for regression is [2].

3   Description of the Proposed Method

A binary classification problem can be considered as a special case of a regression problem where the desired function only takes two values: 0 and 1. In this section, we will generalize the concepts of the LMNN method and adapt it for regression. As an optimization tool we will consider an evolutionary algorithm, which can provide more flexibility in situations when we want to learn not a single distance metric, but several.

3.1   Distance Metric Learning Using Prototypes

In our work, we considered $\mathbf{L}$ (and thus $\mathbf{M}$) to be diagonal matrices. This increases the clarity of the results, because in this case the elements m_{ii} can be interpreted as the weights of the problem input dimensions, and is also a form of regularization. Given the fact that an evolutionary algorithm is used for finding the elements of the matrix, it can be easily applied to find a full matrix $\mathbf{L}$. In the unrestricted case, the elements of $\mathbf{M}$ can be directly found only while satisfying some constraints, because they should comply with the relation $\mathbf{M} = \mathbf{L}^T \mathbf{L}$, i.e. $\mathbf{M}$ should be symmetric and positive semidefinite.

By using the $\mathbf{M}$ matrix, the relation between the neighbour weights and the distance in equation (4) still holds, but now we have:

$$w_{d_M}(\mathbf{x},\mathbf{x}') = \frac{1}{d_M(\mathbf{x},\mathbf{x}')} = \frac{1}{\sum_{i=1}^{n} m_{ii} \cdot (x_i - x'_i)^2}. \tag{10}$$

In this formulation, there is a single, global matrix $\mathbf{M}$ for all the instances. However, it is possible to have different distance metrics for the different instances or groups of instances. We introduce the use of *prototypes*, which are special locations in the input space of the problem, such that each prototype P has its own matrix $\mathbf{M}(P)$. When computing the distance weight to a new point, an instance will use the weights of its nearest prototype, i.e. $m_{ii}(P)$ instead of m_{ii} in equation (10).

3.2   The Evolutionary Algorithm

The following parameters of the evolutionary algorithm were used in the experiments: 40 chromosomes in the population, tournament selection with 2 candidates, arithmetic crossover with a probability of 95% and mutation with a probability of 5% in which a gene value is reinitialized to a random value in its corresponding domain of definition. Elitism was also used such that the best solution in a generation is never lost. As a stopping criterion, a maximum number of generations, 500 in our case, was used. The number of genes in a chromosome depends on the number of prototypes and the dimensionality of the problem, namely: $n_g = n_p \cdot n_i$, where n_g is the number of genes, n_p is the number of prototypes and n_i is the number of inputs.

The domain of the genes, which represent the elements of $\mathbf{L}$, is $[10^{-3}, 10]$. All the operations in the software implementation described later deal with the elements of $\mathbf{M}$, i.e. their squares. Thus it can be considered that the corresponding values of the $\mathbf{M}$ elements lie in the $[10^{-6}, 10^2]$ domain.

The fitness function F, which is to be minimized, takes into account 3 criteria:

$$F = w_1^F \cdot F_1 + w_2^F \cdot F_2 + w_3^F \cdot F_3, \tag{11}$$

where the weights of the criteria are normalized: $w_1^F + w_2^F + w_3^F = 1$.

In order to simplify the expressions of the F_i functions, let us make the following notations, where d_M means the weighted square distance function using the weights we search for: $d_{ij} = d_M(\mathbf{x}_i, \mathbf{x}_j)$, $d_{ik} = d_M(\mathbf{x}_i, \mathbf{x}_k)$, $g_{ij} = |f(\mathbf{x}_i) - f(\mathbf{x}_j)|$ and $g_{ik} = |f(\mathbf{x}_i) - f(\mathbf{x}_k)|$.

Thus, the first criterion is:

$$F_1 = \sum_{i=1}^{n} \sum_{j \in N(i)} d_{ij} \cdot (1 - g_{ij}), \tag{12}$$

where $N(i)$ is the set of the nearest k neighbours of instance i, in our case $k = 3$. Basically, this criterion says that the nearest neighbours of i should have similar values to the one of i, and more distant ones should have different values. This criterion tries to minimize the distance between an instance i and its neighbours with similar values. If a neighbour j has a dissimilar value, the second factor, $1 - g_{ij}$, becomes small and the distance is no longer necessary to be minimized.

The second criterion is expressed as follows:

$$F_2 = \sum_{i=1}^{n} \sum_{j \in N(i)} \sum_{l \in N(i)} \max\left(1 + d_{ij} \cdot (1 - g_{ij}) - d_{il} \cdot (1 - g_{il}), 0\right). \qquad (13)$$

It takes into account a pair of neighbours, j and l, by analogy to a target and an impostor. However, for our regression problem we do not have these notions because we do not have a class which could be the same or different. We can only take into account the real values of the instance outputs. The reasoning is the same as for the first criterion, but now we try to minimize the distance to the neighbours with close values (the positive term), while simultaneously trying to maximize the distance to the neighbours with distant values (the negative term). By analogy to equation (8), we consider that a margin of at least 1 should be present between an instance with a close value and another with a distant value. The *max* function is used by analogy to the expression of the hinge loss. If we consider that j has a value close to the value of i and l has a distant one, the corresponding hinge loss will be 0 only when $d_{il} \geq 1 + d_{ij}$. The condition $j \neq l$ is implicit because when the terms are equal they cancel each other out.

The third criterion is used for regularization, to avoid large values of the weights:

$$F_3 = \sum_{j=1}^{n_p} \sum_{i=1}^{n_i} m_{ii}(j). \qquad (14)$$

However, for our case study presented in section 4, the weights are very small and this term is not needed. Overall, we use the following values for the weights of the criteria: $w_1^F = w_2^F = 0.5$ and $w_3^F = 0$.

The advantage of using an evolutionary algorithm for optimization is that prototypes can be used, with different weight values, instead of a single, global set of weights. As mentioned above, when computing the distance from a certain training instance, the distance is weighted by the values corresponding to the nearest prototype of the training instance.

We compute the positions of the prototypes using k-means, a simple clustering algorithm which tends to favour (hyper)-spherical clusters. This is why it is very appropriate for our problem, where distances are computed with variations of the Euclidian metric. However, the scope of the evolutionary algorithm can be expanded to also find the optimal number of clusters as well as their positions in the input space.

4   Case Study

In this section we will present the results of applying the regression method to a practical problem, namely the prediction of the corrosion of some alloys containing titanium and molybdenum (TiMo).

Titanium (Ti) and its alloys are widely used in dental applications due to the excellent corrosion resistance and mechanical properties. However, it has been reported that Ti is sensitive to fluoride (F^-) and lactic acid. Consequently, the samples were examined using electrochemical impedance spectroscopy (EIS) in acidic artificial saliva with NaF and/or caffeine. The material corrosion was quantified by the polarization resistance (R_p) of the TiMo alloys (output of the model) which was modelled depending on the immersion time, caffeine concentration, NaF concentration, type of alloy (Ti content), and solution pH (inputs of the model). The experimental data covered a large domain, corresponding to the following conditions: 12, 20 and 40 wt.% of Mo, 0.1, 0.3 and 0.5 wt.% NaF, 0, 0.5, 1 and 1.5 mg/mL caffeine and 3-8 for solution pH.

In order to compare different algorithms and models, we consider 2 separate problems. The first is to assess the performance on the testing set, after randomizing the dataset and selecting 2/3 of the instances as the training set and 1/3 as the testing set. The second one is to perform cross-validation with 10 folds.

We use the coefficient of determination (r^2), the squared coefficient of correlation, as a metric to compare the performance of different models. Table 1 presents the results obtained with various algorithms implemented in the popular collection of machine learning algorithms Weka [7], for the training-testing split and for cross-validation. The results are sorted in decreasing order of the cross-validation results. On each column, the maximum value is emphasised with bold characters.

Table 1. Performance of different algorithms implemented in Weka

Algorithm	Training Set	Testing Set	Cross-validation
REPTree	0.9789	0.8789	**0.8983**
n-SVR RBF	0.9109	0.8140	0.8962
M5 rules	0.8962	0.8127	0.8953
Random Forest	0.9833	**0.9181**	0.8915
kNN, k = 10	0.9994	0.8974	0.8363
e-SVR, RBF	0.7583	0.6726	0.7910
n-SVR P2	0.8187	0.7268	0.7739
e-SVR, P2	0.7903	0.7022	0.7533
NN	**1.0000**	0.8516	0.6655
Additive Regression	0.6336	0.5869	0.5895

When using our software implementation for the regression problem, 10 neighbours are considered when computing the output, because this value gave the best results for the kNN method in Weka. Fewer neighbours do not provide enough information to compute a precise value. If more neighbours are used, the influence of the more distant ones becomes negligible, since the neighbour weights are proportional to the inverse of the square distance.

In the computation of the fitness function, 3 reference neighbours are used. This number should be at least 2, in order to provide some contrast between an instance with a close value and another with a distant value (by analogy with the target and impostor instances for classification). With 2 reference neighbours, the results are far worse than with 3, with the best coefficient of determination of 0.8984. With more reference neighbours, the computation time greatly increases, while the results are no better than in case of 3 reference neighbours.

Table 2 shows the average and best results for the case of the training set – testing set split and also for cross-validation with 10 folds, while considering 4 different configurations: with 1, 2, 5 and 10 prototypes. Theoretically, each instance could have its own metric, but in our case, with a dataset of 1152 instances, the computation time would be too high for any practical purpose. The positions of the clusters are obtained with the k-means algorithm taking into account only the inputs of the instances. The results presented are obtained after 50 runs. The results with $r^2 < 0.5$ were eliminated because they were considered to be outliers. The results on the training set are omitted, because they are 1 in all cases.

Table 2. Performance of the proposed method with different numbers of prototypes

No. Prototypes	Testing Set		Cross-validation	
	Average	Maximum	Average	Maximum
1	0.8436	**0.9703**	0.8056	0.9160
2	0.8349	0.9394	0.8123	**0.9207**
5	0.7914	0.8517	0.7432	0.8273
10	0.6997	0.8270	0.6439	0.7917

Table 3. The best weights obtained for the training-testing problem with: a) 1 prototype; b) 2 prototypes

Input	m_{ii} x 1000
x_1	0.0719
x_2	0.0589
x_3	0.8551
x_4	0.0463
x_5	0.1479

Input	Prototype 1 position	Prototype 2 position	$m_{ii}(1)$ x 1000	$m_{ii}(2)$ x 1000
x_1	0.4507	0.4612	0.3889	0.0902
x_2	0.3450	0.3639	1.8994	0.0421
x_3	0.4987	0.5142	0.0193	0.0027
x_4	0.5109	0.4833	0.0716	2.5454
x_5	0.0987	0.7964	0.2239	0.0205

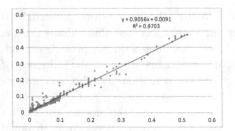 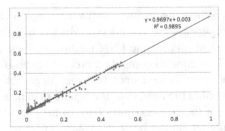

Fig. 1. Comparison between the predictions of the best model and the expected data: a) for the test set; b) for the whole dataset

While cross-validation is a standard way to compare several algorithms, it cannot provide a unique set of "good" values for the internal parameters, because it generates 10 different models. Therefore, the single split into training and testing sets is useful in this respect.

The best performance for the testing set is given by the configuration with 1 prototype, i.e. a global set of weights for all the instances of the problem. The actual values of the weights are presented in table 3a.

Figures 1a and 1b present the correlation between the desired values and the values provided by the model, for the testing set alone and for the whole dataset. Beside the high value of the coefficient of determination, one can also graphically observe the good fit of the data.

Table 3b presents the weights for the best fit on the training-testing problem, corresponding to 2 prototypes.

Beside the variation in performance between different runs, which is normal especially since the number of generations (500) and individuals in the population (40) may be a little too small for our problem, we hypothesise that there may be another reason why 1 prototype provides the best results for the training-testing case and 2 prototypes provide the best results for cross-validation. In one fold of cross-validation 90% of the data are used, compared to 67% for training in the first case. Therefore, the data diversity is larger and 2 prototypes provide more flexibility. Also the positions of the prototypes are currently fixed, pre-computed, taking into account the whole dataset, and thus the 90% situation better matches the distribution of the full data. However, when the number of prototypes increases, the problem space becomes too finely partitioned, because the evolutionary algorithm evolves the weights independently and it is almost certain that the sets of weights corresponding to different prototypes will be different, although the actual topology of the problem may not require such distinctions in the distance metric of the instances.

5   Conclusions

The results obtained for a rather difficult problem using a large margin nearest neighbour regression method are quite promising. The weights are obtained using an evolutionary algorithm, which provides simplicity and flexibility and allows the use of several distance metrics in different regions of the problem space, corresponding to different prototypes. The evolutionary paradigm can be easily applied to also find the optimal number of prototypes and their positions, at the expense of a large increase in the search space and thus computation time. The presented method allows the user to change the number of neighbours that are considered for distance calculation and the weights of the criteria of the composite fitness function, in order to adapt these parameters to the particular characteristics of the problem. The modelling procedure applied for the evaluation of the polarization resistance of the TiMo alloys as a function of process conditions contributes to a better understanding of the process and can partially replace a series of experiments that are time, material and energy consuming.

Acknowledgment. This work was supported by the "Partnership in priority areas – PN-II" program, financed by ANCS, CNDI - UEFISCDI, project PN-II-PT-PCCA-2011-3.2-0732, No. 23/2012.

References

1. Aha, D.W., Goldstone, R.L.: Concept learning and flexible weighting. In: Proceedings of the 14th Annual Conference of the Cognitive Science Society, pp. 534–539 (1992)
2. Assi, K.C., Labelle, H., Cheriet, F.: Modified Large Margin Nearest Neighbor Metric Learning for Regression. IEEE Signal Processing Letters **21**(3), 292–296 (2014). doi:10.1109/LSP.2014.2301037
3. Atkeson, C.G., Moore, A.W., Schaal, S.: Locally weighted learning. Artificial Intelligence Review **11**(1-5), 11–73 (1997)
4. Chopra, S., Hadsell, R., LeCun, Y.: Learning a similarity metric discriminatively, with application to face verification. In: Proceedings of the IEEE Conference on Computer Vision and Pattern Recognition, CVPR 2005, pp. 349–356, San Diego, CA, USA (2005)
5. Cover, T.M., Hart, P.E.: Nearest neighbor pattern classification. IEEE Transactions in Information Theory, IT-13, 21–27 (1967)
6. Goldberger, J., Roweis, S., Hinton, G., Salakhutdinov, R.: Neighbourhood components analysis. In: Advances in Neural Information Processing Systems, vol. 17, pp. 513–520. MIT Press, Cambridge (2005)
7. Hall, M., Frank, E., Holmes, G., Pfahringer, B., Reutemann, P., Witten, I.H.: The WEKA Data Mining Software: An Update. ACM SIGKDD Explorations **11**(1), 10–18 (2009)
8. Hansen, B.E.: Nearest Neighbor Methods. NonParametric Econometrics. University of Wisconsin-Madison, pp. 84-88 (2009). http://www.ssc.wisc.edu/~bhansen/718/NonParametrics10.pdf (accessed March 15, 2015)
9. Kumar, M.P., Torr, P.H.S., Zisserman, A.: An invariant large margin nearest neighbour classifier. In: Proceedings of the IEEE 11th International Conference on Computer Vision, ICCV 2007, Rio de Janeiro, Brazil (2007). doi: 10.1109/iccv.2007.4409041
10. Leon, F., Piuleac, C.G., Curteanu, S., Poulios, I.: Instance-Based Regression with Missing Data Applied to a Photocatalytic Oxidation Process. Central European Journal of Chemistry **10**(4), 1149–1156 (2012). doi:10.2478/s11532-012-0038-x
11. Mareci, D., Sutiman, D., Chelariu, R., Leon, F., Curteanu, S.: Evaluation of the corrosion resistance of new TiZr binary alloys by experiment and simulation based on regression model with incomplete data. Corrosion Science, Elsevier **73**, 106–122 (2013). doi:10.1016/j.corsci.2013.03.030
12. Moore, R.C., DeNero, J.: L1 and L2 regularization for multiclass hinge loss models. In: Proceedings of the Symposium on Machine Learning in Speech and Language Processing (2011)
13. Navot, A., Shpigelman, L., Tishby, N., Vaadia, E.: Nearest neighbor based feature selection for regression and its application to neural activity. In: Advances in Neural Information Processing Systems, vol. 19 (2005)
14. Shalev-Shwartz, S., Singer, Y., Ng, A.Y.: Online and batch learning of pseudo-metrics. In: Proceedings of the 21st International Conference on Machine Learning, ICML 2004, pp. 94–101, Banff, Canada (2004)
15. Shental, N., Hertz, T., Weinshall, D., Pavel, M.: Adjustment learning and relevant component analysis. In: Heyden, A., Sparr, G., Nielsen, M., Johansen, P. (eds.) ECCV 2002, Part IV. LNCS, vol. 2353, pp. 776–790. Springer, Heidelberg (2002)

16. Weinberger, K.Q., Blitzer, J., Saul, L.K.: Distance metric learning for large margin nearest neighbor classification. In: Advances in Neural Information Processing Systems, vol. 18, pp. 1473–1480. MIT Press, Cambridge (2006)
17. Weinberger, K.Q., Saul, L.K.: Fast solvers and efficient implementations for distance metric learning. In: Proceedings of the 25th International Conference on Machine Learning, pp. 1160–1167, Helsinki, Finland (2008)
18. Weinberger, K.Q., Saul, L.K.: Distance Metric Learning for Large Margin Nearest Neighbor Classification. Journal of Machine Learning Research **10**, 207–244 (2009)

Artificial Immune System: An Effective Way to Reduce Model Overfitting

Waseem Ahmad[1(✉)] and Ajit Narayanan[2]

[1] International College of Auckland, Auckland, New Zealand
wahmad@ica.ac.nz
[2] AUT University, Auckland, New Zealand
ajit.narayanan@aut.ac.nz

Abstract. Artificial immune system (AIS) algorithms have been successfully applied in the domain of supervised learning. The main objective of supervised learning algorithms is to generate a robust and generalized model that can work well not only on seen data (training data) but also predict well on unseen data (test data). One of the main issues with supervised learning approaches is model overfitting. Model overfitting occurs when there is insufficient training data, or training data is too simple to cover the structural complexity of the domain being modelled. In overfitting, the final model works well on training data because the model is specialized on training data but provides significantly inaccurate predictions on test data due to the model's lack of generalization capabilities. In this paper, we propose a novel approach to address this model overfitting that is inspired by the processes of natural immune systems. Here, we propose that the issue of overfitting can be addressed by generating more data samples by analyzing existing scarce data. The proposed approach is tested on benchmarked datasets using two different classifiers, namely, artificial neural networks and C4.5 (decision tree algorithm).

Keywords: Artificial immune system · Antibodies · Model overfitting · Artificial neural networks · Vaccination · Ensemble learning

1   Introduction

Data Mining is a subfield of machine learning where hidden patterns and rules are discovered from raw data. Data mining techniques can mainly be grouped into two categories, namely, supervised and unsupervised learning. In supervised learning, a model is built based on mapping input patterns to a desired output. Data consists of a set of features or variables as well as output (desired) variables. The task of any supervised learner is typically to build a model that both best fits a number of training examples (training data, which can include secondary training data for recalibrating a neural network once initially trained) and predicts the class into which new or unseen data falls, where the class information of the predicted instances may not be known (test data). On the other hand, in unsupervised learning, data without any class information is grouped in such a way that instances present in each cluster has a maximum similarity and retain a maximum dissimilarity among different clusters.

© Springer International Publishing Switzerland 2015
M. Núñez et al. (Eds.): ICCCI 2015, Part I, LNAI 9329, pp. 316–327, 2015.
DOI: 10.1007/978-3-319-24069-5_30

The effectiveness of supervised learning algorithms has been well demonstrated in he literature [1-3]. These algorithms perform well given reasonably large sets of raining data. However, the predictive capabilities of supervised learning algorithms leteriorate substantially when there is small pool of training data. In these situations, generated models suffer from data overfitting. Models generated by considering only fewer training samples hold poor generalization capabilities (high degree of specialization on seen data), hence they demonstrate higher accuracy on training data and poor predictive accuracy on unseen or test data. There are a number of approaches to deal with this data ovefitting such as regularization, early-stopping and different forms of cross-validation [4, 5]. In this paper, we are proposing a novel approach to deal with data overfitting that is inspired by the Natural Immune System (NIS) and more specifically from the process of vaccination.

Artificial Immune System (AIS) is a computing paradigm inspired by the processes and metaphors of the Natural Immune System (NIS). In recent years researchers have developed many computational algorithms by borrowing concepts of Natural Immune Systems [6-11]. NIS protects us from various harmful diseases throughout our life span. The increasing interest of computational community to explore NIS is due to its adaptation and generalization capabilities. NIS learns from captured harmful instances in such a way that it provides lifelong immunity as well as provides protection against any mutated and stronger versions of the pathogens that can cause deadly diseases. Vaccination is an example of such processes of NIS that has adaptation and generalization capabilities. The underpinning principle of vaccination is to learn from weaker strands of viruses and generate a pool of antibodies and memory cells so that the human body is well equipped to handle a stronger version of the virus.

In NIS, memory cells are solutions for keeping a history of past pathogens and viral attacks. On the other hand, antibodies, which are cloned and mutated copies of stimulated B-cells, provide protection against variants of invading pathogens. Memory cells and antibodies are regarded as specialized and generalized cells, respectively. This process of generation of generalized antibodies provides a way for our NIS to avoid overfitting. In this paper, we are proposing that one way of reducing overfitting is through the generation of antibody equivalents as part of modelling.

All AIS approaches presented in literature produce memory cells that are a summarized form of training data. Memory cells can be defined as given a data consisting of n instances, find g number of instances that provide minimum information loss ($g < n$). These memory cells are proven to have efficient specialization capabilities and results in better model accuracy. This data specialization leads to better generalization capabilities on unseen data, given an adequate amount of training data. The contribution of memory cells is well demonstrated in AIS literature, but less explored is the role of antibodies in AIS approaches. Antibodies are cloned and mutated copies of existing data samples. The population of antibodies are known to have great generalization capabilities, but little research work has been undertaken to prove this claim. The work presented in this paper is based on the generation of antibodies to achieve robust and generalized classification models.

In [12] it was demonstrated that concepts of vaccination can be applied to computing algorithms to obtain robustness and an increasing degree of generalization.

A hybrid architecture was proposed where AIS algorithm was used to generate vaccine (in the form of memory cells or antibodies) and Artificial Neural Network (ANN) was used as a learning system. ANN was primed (trained) on vaccine before injecting and retraining on original data. In experimental results, it was successfully demonstrated that primed ANN was well equipped to mount a faster and more effective response when original data was introduced to the learning system. This means ANN required fewer learning cycles to converge to a local solution and often results in better models (fewer learning errors on training data). That work concluded that memory cells and antibodies have excellent data summarization capabilities and can be used as vaccination agents for other learning systems.

The proposed architecture in [12] was tested and showed superior classification results only against training data. No efforts were made to explore the generalization capabilities of such models on unseen or test data. In this paper, we are extending earlier work done in [12] to demonstrate that models generated using antibodies have good generalization capabilities and provide superior predictive results on unseen data. In the original published work [12], random mutation was used to create antibodies. Our initial experiments demonstrated that random mutation introduced a high degree of noise and resulted in poor predictive results. Therefore, this paper proposes a directed mutation approach based on the local and global structural information present in the training data.

2   Literature Review

In the last decade, AIS has been a focus of many computational researchers and many novel machine learning algorithms have been proposed. The main idea is to represent data instances as antigens and antibodies or B-cells as clusters. One of the first AIS algorithms was proposed by De Castro and Zuben [13] called Artificial Immune Network (aiNet). It utilizes the concepts of memory cells, antibodies, clonal selection and hypermutation. The proposed algorithm was used to perform data summarization by producing memory cells and to find natural clusters in the data. The AIS algorithms in the area of unsupervised learning can be found [9, 14-16]. The most significant work in the AIS supervised learning domain is proposed by Watkins *et al* [6] that used the concept of artificial recognition balls (ARBs), resource limitation, memory cell and hypermutation. The population of memory cells compete for survival and principles of natural selection are used to evolve the population of memory cells. Finally, a population of evolved memory cells is obtained, which are used to predict class membership on unseen data using a KNN algorithm. HAIS [17] is another AIS based classifier for supervised learning and is inspired by the humoral mediated response triggered by an adaptive immune system. HAIS uses core immune system concepts such as memory cells, plasma cells, Igs, antibodies and B-cells as well as parameters such as negative clonal selection and affinity thresholds.

In the field of optimization, Woldemariam [18] introduced the notion of vaccination in an AIS algorithm to solve function optimization problems. Vaccine is extracted by dividing the search space into equal subspaces. These vaccines are then introluced into the algorithm to enhance the exploration of global and local solutions. The work done in the field of optimization using AIS algorithms can be found in [19-21]. In [12], concepts of vaccination were used to improve the accuracy and learning capabilities of artificial neural networks. A hybrid architecture was proposed based on AIS and ANN. The ANN was introduced to mimic the behavior of a learning system. AIS algorithm was used to extract memory cells or antibodies (called vaccine) from the data set. This vaccine was later introduced into the learning system (ANN) to demonstrate that the vaccine primed the learning system to handle a stronger version of the viral attack and as a consequence learning on data set not only became much faster but also provided better classification accuracy.

3   Proposed Approach

The components of NIS such as B-cells, memory cells and antibodies play an important role in keeping us alive and eliminating dangerous invaders. This process of antibody generation forms the basis of the computing algorithm proposed in this paper. In earlier published work [12], the robustness of the model built using memory cells and antibodies was measured using a three step methodology. In the first step, memory cells or antibodies were obtained from full labelled data using AIS algorithm. In step 2, ANN classifier was used to train on memory cells/antibodies and stopped once convergence was achieved on error sum of square (ESS). Finally, ANN was further trained on original full data until convergence was achieved. This three step methodology demonstrated that the final model had superior classification results on full (seen) data than the model built directly on full data.

According to this methodology, classification results are derived only from training data (model fitting) and no unseen data was used to evaluate the model's predictive capabilities. Generally, algorithms and data models are compared based on their predictive accuracy on unseen data. The common practice is to divide the data into training data (including recalibration data) and genuine, unseen test data. Models are built by single-shot or repeated exposure on training data while its accuracy and robustness are observed and measured on test data. In order to validate the contribution of vaccination concepts in the field of machine learning, the model's robustness or generalization capabilities must be evaluated on test data and not just on training data. In this paper, full data is split into training data and test data, and the same three step approach is followed. In step 1, antibodies are generated using only training data. In step 2 and 3, a classifier is used to build a model with these antibodies and the model is evaluated on full data. The rational of using full data and not just test data for the final evaluation is to incorporate any training errors obtained during the process of antibody generation. This process is shown in the form of following three steps.

$$Full\ data = training\ data + test\ data$$
$$\textbf{Step 1}: \quad antibodies = AIS\ (training\ data)$$
$$\textbf{Step 2}: \quad Model = Classifier\ (antibodies)$$
$$\textbf{Step 3}: \quad Evaluate = Model\ (Full\ data)$$

The results obtained from Step 3 are compared against step B, where the model is built on training data (Step A) and then evaluated on full data (Step B). The following section will explain the algorithm for generating antibodies.

Step A: *Model* = *Classifier* (*training data*)
Step B: *Evaluation* = *Model* (*Full data*)

3.1   Generation of Antibodies

The generation of antibodies is an important component of the proposed approach. The complete process is divided into three main stages to be described in more detail below:

Stage 1: *Generate antibodies based on localised knowledge* (*Local expansion*)
Stage 2: *Generate antbodies based on global knowledge* (*Global Expansion*)
Stage 3: *Perform negative clonal selection*

In stage 1, a clustering algorithm is used to group each class of data samples into predefined number of clusters. This step is followed by generating antibodies based on each obtained cluster (performing local expansion on each cluster). In stage 2, new data instances are generated based on the structure of data present in each class. Finally, in stage 3, the data obtained in earlier two stages is aggregated and negative clonal selection to prune similar data or clusters [17] is performed to obtain a final set of antibodies (new data set). The antibodies generated by this approach are used to build a classification model. Table 1 states the parameters used in the proposed algorithm. Stages 1 and 2 are important to capture the structural composition of each class, separately.

Table 1. Description of parameters used in the proposed algorithm

Parameters	Description
n	Number of antibodies produced against each data sample
k	Number of clusters for each class
w	Localized cluster size threshold
globalExp	Number of antibodies generated by Global Expansion Step
glVariation	Controls each class's covariance matrix values
singlePointVariation	Controls variance values on non-cluster data instances
T	Negative Clonal Threshold

Stage 1: Local Expansion

The aim of local expansion is to generate new data samples (Fig. 1) against the training data of each class (generating localized antibodies). K-means clustering algorithm is used to find natural groupings within each class. Here, each cluster can be seen as an activated B-cell that produces antibodies based on its localized state. A Matlab function of multivariate normal random numbers generator (*mvnrnd*) is used to generate new data samples (antibodies) against each cluster (Fig. 1). The function *mvnrnd* takes the mean and covariance of data and returns a matrix of random numbers chosen from a normal distribution. However, if the size of cluster is less than 3, then *normrnd* function is used to generate data samples. The function *normrnd* takes into account the mean and variance of data and returns a matrix of random numbers chosen from normal distribution.

Input: *data: k, w, n, singlePointVariation, glVariation*
Output: *newData1*
Kmeans Clustering (data, k)
For *each data cluster x*
   If size(x) > w
      newData1 = mvnrnd(mu(x), cov(x)*lmVariation, size(x)*n)
   Else
      newData1 = normrnd(mu(x), singlePointVariation, n)
   End If
End For

Fig. 1. Pseudo-code for generating data samples based on local expansion

Stage 2: Global Expansion

The aim of global expansion is to generate data for each labelled class separately. The Matlab function of multivariate normal random number generator (*mvnrnd*) is used for global expansion (Fig. 2). The data from stage 1 and 2 (local and global expansion) are aggregated to obtain a pool of antibodies (Eq. 1).

$$newData = newData1 \cup newData2 \tag{1}$$

Input: *data, gmVariation, globalExp*
Output: *newData2*
For *each class*
   newData2 = mvnrnd(mu(data), cov(data)*gmVariation, size(data)* globalExp)

Fig. 2. Pseudo-code for generating data samples based on global expansion

Stage 3: Negative Clonal Selection

Negative clonal selection is performed on all newly generated data instances of each class, separately. This stage starts by assigning full labelled data to *self-data* and all instances of *newData* (Eq. 1) are compared one-by-one. If any instance of *newData* is too similar to existing *self-data*, it is removed, otherwise added into pool of *self-data*. After the end of this process, the final pool of *self-data* is the final set of antibodies used to build classification model. The similarity measure is calculated using pairwise Euclidean distance measure.

3.2　Example

A simulated data is used to demonstrate the proposed approach. The simulated data contains a total of 30 instances, where 12 instances belong to class 1 and 18 to class 2 (Fig. 3-A). Local and Global expansions are performed in the proposed approach using parameter values of $k = [5, 5]$, $n = [5, 5]$, $w = 3$, $globalExp = [2, 2]$, $glVariation = [1, 1]$ and $T = [0.25, 0.25]$ and can be seen in Fig. 4-B and 4-C, respectively. The final pool of antibodies (*newData*) is shown in Fig. 3-D that is aggregation of local and global expansion followed by performing negative clonal selection. Local expansion is responsible for the generation of data instances based on internal clusters found in each class, whereas, global expansion is driven by the structure (mean and covariance) of the respective classes. In this paper, our emphasis is to demonstrate with empirical findings that models obtained using data instances of Fig. 3-D are more robust and generalized than model obtained using data of Fig. 3-A.

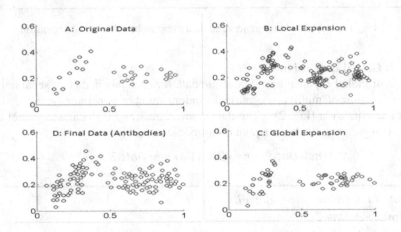

Fig. 3. Simulated 2-dimesional data highlighting the steps of proposed approach

4　Experimental Results

Four well-known and researched datasets have been selected to demonstrate the effectiveness of the proposed approach. The details of these datasets are provided in Table 2. The datasets are divided into training and test data. For experimental purposes, only those training and test data partitioning are selected which have the highest number of test errors while generating a classification model on training data. The antibodies are generated only from training data and then these antibodies are used to generate classification models. The model evaluation is conducted by obtaining classification accuracy on full data that is the aggregation of training data and test data. As stated previously, the rational of this arrangement is to avoid better classification accuracy on the test data at the cost of higher training error. Two classifiers, namely, artificial neural networks and C4.5 (J48) are used in these experiments. These experiments are conducted in Weka-3.6 [22]. The experimental methodology can be highlighted in following equations.

$$full\ data = training\ data\ + test\ data$$
$$antibodies = AIS\ (training\ data)$$
$$Model = ANN\ (antibodies)\quad OR\quad J48\ (antibodies)$$
$$Evaluation = Model\ (Full\ data)$$

The contribution of ANN in the field of machine learning and pattern recognition is well established [23, 24]. Artificial neural networks (ANNs) are computational models inspired by the functionality of biological neuron [25]. In the paper, Backpropagation algorithm is used in ANN. In Backpropagation, neurons are arranged in layers and send their signals forward and then error is calculated at the output layer that is propagated backward. Another classifier used here is J48 or C4.5 that is a decision tree based classification algorithm [26]. J48 builds the decision tree on the training data using the concept of information entropy.

All four data sets, namely, Heart, Breast cancer, Gene expression and Iris were used to demonstrate the applicability of the proposed approach. These datasets contain varying degree of complexity in terms of the number of features, instances and classes. These datasets (except gene expression dataset) were taken from UCI Machine Learning repository and further detail regarding these datasets can be found at [27]. Gene expression dataset originally had 254 features, excluding class labels. A feature selection algorithm was used to select 14 features. The datasets were divided into training and test data. It can be seen from Table 2 that after the split the training data has two to three times fewer instances than test data. Approximately 30% of the total data was used to build a classification model and the remaining data was used to evaluate the classification accuracy. The rationale of using this arrangement is to demonstrate that given fewer training samples, the proposed approach can provide good classification models. The number of features stated in the Table 2 are the number of attributes without class label. The values of parameters used are described in Table 3. Various runs are performed to finally select these parameter values. Multiple values for each parameter describe values based on each class.

Table 2. Datasets information explaining training and test data split used in this paper

Datasets	Training data	Test data	Full data	Features	Classes
Heart	80	187	267	44	2
Breast Cancer	58	136	194	34	2
Gene Expression	31	71	102	14	2
Iris	45	105	150	4	3

The empirical results of all four datasets are presented in Table 4 using ANN classifier. The second row in Table 4 represents ANN classifier results when trained on training data alone and evaluated on full data. Row 3 onwards represents ANN classifier results when antibodies are used to train ANN and full data is used for evaluation. On Iris dataset, 14 classification errors were found on full data when ANN was trained only on training data. The selected training data was used to obtain 50 separate pools of antibodies. The results obtained from each pool of antibodies are presented in Table 4. The results are summarized and compared using mean, standard deviation, maximum and minimum classification errors obtained. The maximum and minimum classification errors found for Iris data were 15 and 6, respectively, whereas mean and standard deviation were 10.2 and 2.6, respectively. The minimum classification errors obtained on Iris data is lower when antibodies were used to build the model than when the model was built only using the training data. The higher fluctuation in the classification errors of the Iris data is due to the stochastic nature of the generation of antibodies. Similar empirical results can be seen on other datasets. The best results were obtained on the Heart dataset, where the classification accuracy was improved by more than 50%. The classification errors decreased from 119 to 64. Another noticeable result was the maximum errors obtained on all datasets that were close to the errors obtained when only training data was used to build the ANN classifier.

Table 3. Parameter settings used for each dataset

Parameters	Heart	Breast Cancer	Gene Expression	Iris
K	9, 9	10, 10	5, 5	5, 5, 5
n	10, 7	15, 10	12, 12	5, 5, 5
w	3	3	3	3
globalExp	10, 5	5, 0	2, 2	2, 2, 2
glVariation	1, 0.5	0.5, 0.25	0.65, 0.65	0.5, 0.5, 0.5
singlePointVariation	5%, 2.5%	5%,2.5%	5%, 5%	5%, 5%, 5%
T	0.35, 0.25	0.35, 0.35	0.75, 0.75	0.05, 0.05, 0.05

Table 4. Classification results (errors) obtained using ANN on all datasets

	Iris	Heart	Breast cancer	Gene expression
ANN on Training data	14	119	49	12
Mean	10.2	85.6	39.4	10.1
STD	2.6	10.1	3.2	1.7
Maximum	15	109	47	14
Minimum	6	64	31	6

The classification errors on Iris data using J48 classifier are shown in Table 5. The classifier produced 16 classification errors on full data when the decision tree was built on training data (Row 2, Table 5). Row 3 onwards represents J48 classifier results when antibodies were used to build classification tree and full data was used for evaluation. Mean, standard deviation, maximum and minimum classification errors over 50 iterations were 9.8, 2.76, 18 and 4, respectively. These results are consistent with earlier results of the ANN classifier. The number of leaves and tree sizes increased when the model was built using antibodies. Antibodies add complexity to the training data and therefore the resulting tree will have more leaves and a larger tree size than trees generated by training data only. Similar empirical results can be observed on Heart, Breast cancer and Gene expression datasets in Table 6.

There has been a significant variance in the classification results obtained by both classifiers (J48 and ANN). This variance can be reduced using the ensemble method. The ensemble experiments were conducted on the Iris dataset. Fifty models using J48 algorithm are generated and a simple majority vote method is used to assign class membership to full data instances. Eight classification errors were recorded when the ensemble based J48 classifier was used. When the same ensemble principles were used on ANN, the classifier achieved nine classification errors. However, when an ensemble classifier was built on aggregating both ANN and J48 (100 populations), seven (7) classification errors were obtained (Fig. 4). The classification error value of seven is still higher than the minimum errors obtained in above experiments (4 and 6 errors for J48 and ANN, respectively). However, this ensemble method has removed a higher degree of variance in classification results.

Table 5. Classification errors obtained using J48 classifier on Iris data

	Errors	Leaves	Tree Size
Training data	16	3	5
Mean	9.18	6.04	11.08
STD	2.76	1.55	3.10
Maximum	18	9	17
Minimum	4	4	7

Table 6. Classification errors using J48 on Heart, Breast cancer & Gene expression data

	Heart Errors	Cancer Errors	Gene Errors
Training data	95	43	26
Mean	68.30	40.76	20.40
STD	11.55	6.03	5.50
Maximum	89	32	40
Minimum	25	56	10

5   Conclusion

In this paper, a novel computing algorithm inspired by the generation of generalized antibodies is proposed to introduce robustness and generalization in data. These antibodies have been able to reduce overfitting on the final model on four well-established datasets. Two classifiers, namely, ANN and C4.5, are successfully used to validate the effectiveness of the proposed approach. We have demonstrated with experimental results that both classifiers produced poor prediction accuracy on test data when a small number of training data was used. However, the pool of antibodies using the same classifiers leads to improved prediction accuracy and to more robust and generalized models.

In the future, a three way split (training, validation and test data split scheme) must be used to further evaluate the effectiveness of the proposed algorithm. The validation set can also be used to optimize parameter values. This proposed approach can form the basis of novel semi-supervised learning algorithm. Semi supervised learning can be an ideal platform to develop this approach further to evaluate its strengths.

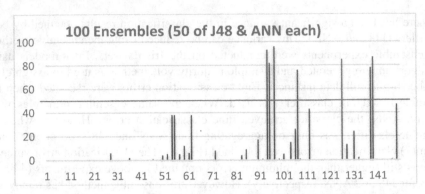

Fig. 4. Ensemble classifier obtained by using 100 models of J48 and ANN. Each horizontal bar represents counts of classification error obtain against each data instance across all models.

References

1. Kotsiantis, S.B.: Supervised machine learning: A review of classification techniques. Informatica (Ljubljana) **31**, 249–268 (2007)
2. Burges, C.: A Tutorial on Support Vector Machines for Pattern Recognition. Data Mining and Knowledge Discovery **2**(2), 121–167 (1998)
3. Murthy, S.K.: Automatic construction of decision trees from data: A multi-disciplinary survey. Data Mining and Knowledge Discovery **2**(4), 345–389 (1998)
4. Moore, A.W.: Cross-validation for detecting and preventing overfitting. School of Computer Science Carneigie Mellon University (2001)
5. Larose, D.T.: k-Nearest Neighbor Algorithm. Discovering Knowledge in Data: An Introduction to Data Mining, 90–106 (2005)
6. Watkins, A., Timmis, J., Boggess, L.: Artificial Immune Recognition System (AIRS): An Immune-Inspired Supervised Learning Algorithm. Genetic Programming and Evolvable Machines **5**(3), 291–317 (2004)
7. Ahmad, W., Narayanan, A.: Outlier detection using humoral-mediated clustering (HAIS). In: Proceedings of IEEE World Congress on Nature and Biologically Inspired Computing, NaBIC2010, pp. 45–52 (2010)
8. Castro, L.N.D., Zuben, J.: The clonal selection algorithm with engineering applications. In: Workshop Proceedings of GECCO, Workshop on Artificial Immune Systems and Their Applications, Las Vegas, pp. 36–37 (2000)
9. Li, X., et al.: ICAIS: a novel incremental clustering algorithm based on artificial immune system. In: International Conference on Internet Computing in Science and Engineering, pp. 85–90 (2008)
10. Liu, Z., et al.: FAISC: a fuzzy artificial immune system clustering algorithm. In: Third International Conference on Natural Computation (ICNC 2007) (2007)
11. Yap, F.W., Koh, S.P., Tiong, S.K.: Mathematical Function Optimization using AIS Antibody Remainder method. International Journal of Machine Learning and Computing **1**(1), 13–19 (2011)
12. Ahmad, W., Narayanan, A.: Principles and methods of artificial immune system vaccination of learning systems. In: Liò, P., Nicosia, G., Stibor, T. (eds.) ICARIS 2011. LNCS, vol. 6825, pp. 268–281. Springer, Heidelberg (2011)

13. Castro, L.N.D., Zuben, F.J.V.: aiNet: an artificial immune network for data analysis. In: Abbass, H., Sarker, R., Newton, C. (eds.) Data Mining: A Heuristic Approach, ch012, pp. 231–260. Idea Group Publishing (2002). doi:10.4018/978-1-930708-25-9

14. Younsi, R., Wang, W.: A new artificial immune system algorithm for clustering. In: Yang, Z.R., Yin, H., Everson, R.M. (eds.) IDEAL 2004. LNCS, vol. 3177, pp. 58–64. Springer, Heidelberg (2004)

15. Brownlee, J.: Clonal Selection Theory and CLONALG: The Clonal Selection Classification Algorithm (CSCA). Technical Report 2-02, CISCP, Swinburne University of Technology (2005)

16. Khaled, A., Abdul-Kader, H.M., Ismail, N.A.: Artificial Immune Clonal Selection Algorithm: A Comparative Study of CLONALG, opt-IA and BCA with Numerical Optimization Problems. International Journal of Computer Science and Network Security 10(4), 24–30 (2010)

17. Ahmad, W., Narayanan, A.: Humoral artificial immune system (HAIS) for supervised learning. In: Proceedings of IEEE World Congress on Nature and Biologically Inspired Computing, NaBIC2010, pp. 37–44 (2010)

18. Woldemariam, K.M., Yen, G.G.: Vaccine-Enhanced Artificial Immune System for Multimodal Function Optimization. IEEE Transactions on Systems, Man and Cybernetics 40(1), 218–228 (2010)

19. Castro, P.A.D., Von Zuben, F.J.: MOBAIS: a bayesian artificial immune system for multiobjective optimization. In: Bentley, P.J., Lee, D., Jung, S. (eds.) ICARIS 2008. LNCS, vol. 5132, pp. 48–59. Springer, Heidelberg (2008)

20. Castro, L.N.D., Timmis, J.: An artificial immune network for multimodal function optimisation. In: Proceedings of IEEE World Congress on Evolutionary Computation, pp. 669–674 (2002)

21. Kelsey, J., Timmis, J.: Immune inspired somatic contiguous hypermutation for function optimisation. In: Cantú-Paz, E., et al. (eds.) GECCO 2003. LNCS, vol. 2723, pp. 207–218. Springer, Heidelberg (2003)

22. WEKA, Machine Learning Group at the University of Waikato. Weka 3: Data Mining Software in Java

23. Zhang, G.: Neural networks for classification: a survey. IEEE Transactions on Systems, Man, and Cybernetics, Part C 30(4), 451–462 (2000)

24. Jain, A.K., Mao, J., Mohiuddin, K.M.: Artificial Neural Networks: A Tutorial. IEEE Computer, 31–44 (1996)

25. Ripley, B.D.: Pattern Recognition and Neural Networks. Cambridge University Press, Cambridge (1996)

26. Quinlin, J.R.: C4. 5: programs for machine learning. Morgan Kaufmann (1993)

27. UCI Machine Learning Repository. http://archive.ics.uci.edu/ml/

Malfunction Immune Wi–Fi Localisation Method

Rafał Górak and Marcin Luckner[✉]

Faculty of Mathematics and Information Science, Warsaw University of Technology,
ul. Koszykowa 75, 00–662 Warszawa, Poland
{R.Gorak,mluckner}@mini.pw.edu.pl

Abstract. Indoor localisation systems based on a Wi–Fi local area wireless technology bring constantly improving results. However, the whole
localisation system may fail when one or more Access Point (AP) malfunctions. In this paper we present how to limit the number of observed
APs and how to create a malfunction immune localisation method. The
presented solutions are an ensemble of random forests with an additional
malfunction detection system. The proposed solution reduces a growth
of the localisation error to 4 percent for the floor detection inside a
six floor building and 2 metres for the horizontal detection in case of a
gross malfunction of an AP infrastructure. The system without proposed
improvements may give the errors greater than 30 percent and 7 metres
respectively in case of not detected changes in the AP's infrastructure.

1 Introduction

Whereas outdoor localisation systems are already a part of our daily routine, an
indoor localisation systems are still being developed. Global Positioning System
(GPS) fails inside buildings and Received Signal Strength (RSS) mapping is used
instead. Measuring Wi–Fi signal strengths from multiple Access Points (AP)
in various localisations we can create a map of fingerprints.In the localisation
process one can find their position by comparison current signal strengths to the
created map. Various approaches to localisation can be found in [6,8,5,7].

However, the created map may be sensitive to some modifications of AP's
infrastructure. As a result, the localisation system may return gross errors. The
situation when the operator of the localisation system is different that the admin-
istrator of the infrastructure is especially critical, leading to major errors.

In this paper we present a localisation method that reduces errors created
by malfunction of an AP infrastructure. The system is an ensemble of random
forests with an additional malfunction detection system that works on Wi–Fi
signal from the AP infrastructure.

The remaining part of the paper is organised as follows: Section 2 presents
basic facts about the analysed data and localisation methods. Section 3 describes
how to reduce the number of observed APs, Section 4 examines a stability of the
created detection system. The approach that detects malfunctions is presented
in Section 5. Finally, Section 6 presents the self correcting system. The work is
concluded in Section 7.

© Springer International Publishing Switzerland 2015
M. Núñez et al. (Eds.): ICCCI 2015, Part I, LNAI 9329, pp. 328–337, 2015.
DOI: 10.1007/978-3-319-24069-5_31

2   Preliminaries and Notation

This section will briefly describe the fingerprinting method, which is most common in indoor localisation based on Received Signal Strength (RSS). The analysis will be conducted on data collected in the publicly accessible areas of a six floor academic building (including the ground floor). The building has irregular shape and its outer dimensions are around 50 by 70 metres and its height is 24 metres.

The data was collected in two independent series. One of them was used as the learning set, the second one as the testing set. The data was collected using the Android application created for the localisation system [4].

Data sets consist of vectors of the signal strengths from $\mathbb{R}^{570}$ labelled by the position of the point where each vector of signals was measured i.e. (x, y, f), where x, y are horizontal coordinates and f is a floor. The elements of the data sets are commonly called fingerprints.

The data set consists of a 107040 training fingerprints and 111760 test fingerprints. The training fingerprints were taken in 2676 points gathered in a 1.5 x 1.5m grid. The test fingerprints were taken in 2794 test points gathered on another day in the grid shifted by 0.75m in each horizontal direction. There were 40 fingerprints taken in every point.

The fingerprinting approach for indoor localisation is to build a model based on data set of training fingerprints with its labels, that given a vector of signal strengths (fingerprint) will give us its position.

The model can be built by several machine learning methods. In work [3] multilayer perceptron was used and in [2] k Nearest Neighbours was tested.

In this paper we will consider another technique, which is random forests method [1]. While preparing this paper we tested all three already mentioned methods. It appeared that random forrest approach gave us as good results as multilayer perceptron and far better than kNN methods. There is also another reason to choose random forrest, which is crucial for the topic we want to cover in this paper. We want to build the models as fast as possible what we explain later in Section 6. It should be mentioned that we build the models using random forests independently for all three coordinates i.e. x, y, and f. By $\hat{h}_x$, $\hat{h}_y$ and $\hat{h}_f$ we denote the estimators given by the built random forrest models.

We test our localisation model on the training set described above. In this paper we analyse the localisation model in two situations. One when the whole Wi–Fi infrastructure works without fault and another, when some of the APs are missing. The second situation is analysed in Section 4 and in Section 6 we propose the method of the model's update to take into account the changes in the infrastructure. To compare both situations, we introduce the following measures of localisation accuracy.

Horizontal error for the fingerprint $m \in \mathbb{R}^{570}$ is defined as

$$he(m) = \sqrt{(x_m - \hat{h}_x(m))^2 + (y_m - \hat{h}_y(m))^2}$$

where x_m and y_m are the horizontal coordinates of the point where the fingerprint m was taken.

Analogically, floor error is defined as

$$fe(m) = |f_m - \hat{h}_f(m)|$$

where f_m is the floor where the fingerprint m was taken.

For the test set $\mathcal{T}$ we define horizontal mean error (**HME**):

$$\mathbf{HME} = \frac{\sum_{m \in \mathcal{T}} he(m)}{|\mathcal{T}|}$$

The floor detection is evaluated by accuracy (**ACC**):

$$\mathbf{ACC} = \frac{\sum_{m \in \mathcal{T}} (1 - \mathrm{sgn}(fe(m)))}{|\mathcal{T}|}$$

In this formula sgn denotes the signum function, i.e. $\mathrm{sgn} : \mathbb{R} \to \{-1, 0, 1\}$, $\mathrm{sgn}(x) = 0 \Leftrightarrow x = 0$.

Finally, let us mention that we select a number of grown trees to be 30 as we checked that growing more trees does not improve the accuracy of the localisation algorithm.

3   Selection of Key Access Points

During the measurement we observed 570 access points. It is extremely hard to analyse signals from all sources. Therefore, we selected the most important signals at the beginning.

3.1   Analysis of Access Points

The proposed method is based on an estimation of predictor importance for decision trees. Feature importance is calculated for a split defined by the given feature. Importance is computed as the difference between Mean Squared Error (MSE) for the parent node and the total MSE for the two children in the regression task. In the classification task the Gini coefficient is used instead to estimate how the data space in the node is divided among classes. The Gini coefficient equals $2(AUC) - 1$. Where AUC is the area underneath the Receiver Operating Characteristic Curve (ROC Curve).

For a random forest, the used function computes estimates of predictor importance for all weak learners. For every tree the sum of changes in the MSE is calculated due to splits on every feature used in the recognition process. Next, the sum is divided by the number of branch nodes.

Importance is normalised to the range $[0, 1]$ with 0 representing the smallest possible importance.

For each observed feature a separate random forest was created using all sources of signals as an input data. For the created random forest feature importance is calculated.

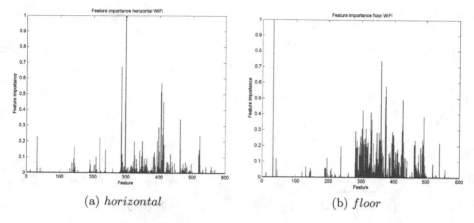

(a) *horizontal*   (b) *floor*

Fig. 1. Feature importance for Wi–Fi signals

Figure 1 presents feature importance for the sources of signals. It is very clear that the classification task (floor detection) needs bigger number of signals that the regression tasks (horizontal detection). Figure 1b shows more important features than Figure 1a.

3.2   Selection of Access Points

Sources of signals are eliminated on the base of the calculated importance.

We start with a selection of all observed features. This set will be reduced by a threshold selected from the range $[0, 1]$ with a step 0.1.

The algorithm – for each value of the threshold – selects the features with greater importance. If the number of features is less than selected before, a new set of features is created.

Selected features are used to create a new random forest. To evaluate quality of a created localiser we use statistics such as mean, median or accuracy.

For the created random forest statistics are calculated in a way that allows us to evaluate quality of a created localiser. As the statistic mean, median, or accuracy can be used.

Evaluations of localisers based on reduced number of features are compared with the evaluation of the base random forest. If the evaluation of any of the created random forests is greater or equal than the base one, we select the set of the features used to construct the random forest.

When none of the new created random forest can be selected, we repeat the whole algorithm once again but we change the range to $[0, 0.1]$ and change the threshold with the step 0.01.

In practice it is hard to obtain a better quality, but we can get a random forest that works with the same quality but on a smaller number of features.

Figure 2 shows how the reduction of features influence errors. Unfortunately, for the horizontal detection (Figure 2a) the minimal error was obtained for the

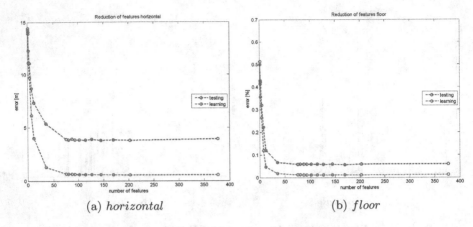

(a) *horizontal*   (b) *floor*

Fig. 2. Threshold selection for Wi–Fi signals

full set of features. In such cases the reduction was limited to the features with importance on the zero level. In such case, the feature set was reduced to less than 400 features.

For the the floor classification task (Figure 2b) the thresholds were fixed on the level that reduces the number of features to nearly 100.

We can still reduce the number of features with permission for slightly worse results. The error for both tasks – the horizontal detection and the floor detection – starts to grow when the number of features is less than 76. Therefore, we can also test this set on the testing data.

Figure 2 shows that the error curve for the testing set is very similar to the curve for the learning set. The error is bigger but the characteristic is nearly the same. That should bring good result of the regression detection models builded on the reduced sets of features.

The selected thresholds were used once again to create the solution that will be verified on the testing data.

Figure 3 presents the results for the testing set. Three detector were tested. The first one used all features. The second one used the number of features that reduced the error calculated on the learning test. The third one were created using 76 features, which is the smallest set without a rapidly growing error.

In Figure 3b we can see that for the floor detection task the accuracy stays the same for the whole range of threshold and the number of signals could be reduced to 76.

In cases of the horizontal detection tasks (Figure 3a), the differences between errors for the following thresholds are measured in centimetres. In this case 76 features it is enough to detect the two dimensional point with a precision of a few metres.

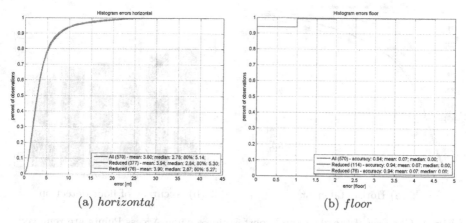

(a) *horizontal*　　　　　　　　　　　　　　(b) *floor*

Fig. 3. Results for reduced number of Wi-Fi signals

The automatic selection of the main features brings good result, but in practice it can be exchanged with an opinion made by an expert (for example on the base of the knowledge about Wi–Fi infrastructure).

4　Stability of Localisation System

Since the Wi–Fi network maintenance may actually be separated from the localisation system, it is natural to ask how the localisation algorithm performs when some Access Points are turned off and this change is not detected. It is clearly very much related to the problem of features selection and their importance, what was discussed in previous sections. We propose the following stress tests of the localisation system:

In the first step we select only 46 Access Points that constitute academic Wi–Fi network. The reason for such a choice is that it is reasonable to build a model using the network that is as stable within longer periods of time as possible. This way we exclude all the mobile or temporary Access Points used by the third parties (private people, networks build by the companies renting office space inside the building etc.).

In the second step, after removing (turning off) Access Points we select them starting from the most important ones as it was described in the previous sections and then apply the model. This way we obtain as big decrease in accuracy of the algorithm as possible.

The results of our stress tests are presented in Figure 4. We can notice a major decrease in accuracy even if only one AP is removed. However we can see in Section 6 that once such a missing AP is detected we can update our model to perform almost as well as before the removal.

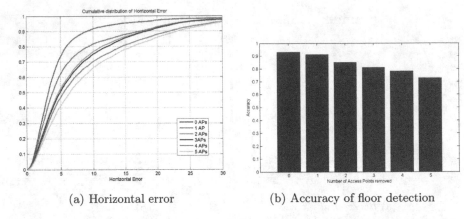

(a) Horizontal error   (b) Accuracy of floor detection

Fig. 4. Changes of the algorithm's performance when Access Points are removed

5   Detection of Infrastructure's Malfunction

As we already mentioned at the end of the previous section the problem of detecting automatically missing Access Points is very important for the localisation system to perform well regardless of the infrastructure's malfunction. Such a successful detection may give a chance to build a system that will update the localisation algorithm. It appears that this problem can be solved efficiently in the following way:

(i) For every Access Point s that is used for localisation, we create a predictor $\widehat{f}_s$ which predicts whether there is a signal from Access Points s based on the readings of the signal strength from all the remaining Access Points. Hence we have here a classification problem with two classes and there are 46 predictors (each for every Access Point). For creating $\widehat{f}_s$ we use a random forrest algorithm. We take a number of grown trees as 10.

(ii) For every Access Point s we gather N readings for which the predictor $\widehat{f}_s$ predicts that there should be a signal from Access Point s. If there was no signal for all N readings then, we detect s as missing one. Otherwise we assume that s is still present.

Obviously one of the alternative approaches can be quite simple - we could wait for a while and if there is no signal from s, then we detect it as missing. However, it may easily happen that most if not all the readings were gathered in the areas of the building where s is not detected. This is quite probable especially inside of big buildings which is the case were the localisation solutions are most needed. This explains why we should look for the readings for which we predict to have a signal from s. It is also worth to choose N to be big. It is to avoid the situation when most of the readings were gathered in the area of the building where the accuracy of the $\widehat{f}_s$ is low. Taking N big makes it more probable to

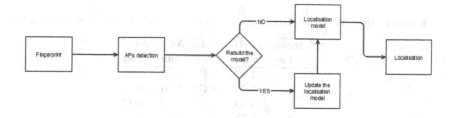

Fig. 5. Self correcting system

get a sample from most of the areas of the building where the signal from s exists. It should be mentioned at this point that parameter N depends on the occupancy of the building, that is from which parts of the building we gather the readings. Since the building of such a system is still in progress we can not experiment with parameter N. Instead we select a sample of $N = 10$ readings from all the fingerprints from the testing set for which $\widehat{f_s} = 1$ and check if we observed a signal from s. However, it should be mentioned that this choice of N corresponds to relatively greater N in the real life situation to avoid gathering the measurements from the same part of the building. Obviously when we turn off a given Access Point it will be detected with probability 1. On the other hand we checked that if the Access Point is not turned off this will also be detected with probability 1 (up 3 s.f.) for each of the 46 Access Points. We observed similar results when we turned off 5 Access Points at once.

6 Self Correcting System

In the sections above we described how the infrastructure's failure can be detected. For the sake of completeness let us recall that by the failure of the infrastructure we mean the situation when some APs are missing. Now we are ready to propose the localisation system that will remotely detect missing APs and appropriately correct the localisation model. The whole system is presented in Figure 5.

As the input of the model we take a single RSS reading. Then, for every AP we apply the predictors described in Section 5 (i). The decision if the localisation model should be rebuilt is based on the recently gathered readings as it is described and discussed in Section 5. Once we detect that some APs are missing, we rebuild the model using random forest method by excluding from training data features corresponding to the missing APs. The gain is substantial and you can analyse the results in Table 1 and in Figure 6.

Table 1 shows that before a self correction action, the localisation algorithm increases **HME** to over 7 metres and **ACC** drops by 30 percent points in case of a gross malfunction. At the same time, **HME** increases by less than 2 metres and **ACC** drops only by 4 percent points after the self correction action.

Table 1. Results of the localisation for removed APs: before and after update

Removed APs	HME before	HME after	ACC before	ACC after
0	4.47	4.47	0.93	0.93
1	4.74	4.47	0.91	0.93
2	5.93	4.72	0.85	0.92
3	6.77	4.94	0.81	0.92
4	7.36	5.07	0.78	0.92
5	7.80	5.17	0.73	0.92
6	8.09	5.34	0.71	0.91
7	9.36	5.50	0.69	0.90
8	10.66	5.98	0.68	0.90
9	11.13	6.12	0.65	0.89
10	11.95	6.41	0.63	0.89

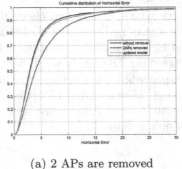

(a) 2 APs are removed

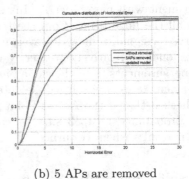

(b) 5 APs are removed

Fig. 6. Horizontal mean error when APs are removed before and after update.

Figure 6 shows two cases with removed 2 and 5 Access Points. Once again we can clearly see that after the system's update gains are substantial .

7   Conclusion

We presented the self correcting system for indoor localisation based on Wi–Fi Access Points. The system was created on the basis of the collected fingerprints. The huge number of observed signals forced the reduction of the Access Points that create the localisation system. It appeared that the number of signal's sources that should be observed can be reduced from 570 to 76 without loss of the localisation accuracy giving us a faster localisation algorithm and a system easier to control.

We also propose another natural way of features selection. We selected for building a localisation model only APs that belong to the building's Wi–Fi network. We showed that such a model with a reduced number of features is very sensitive to changes in the infrastructure such as AP removal. However, we propose a solution to this situation by firstly detecting such missing APs and secondly rebuilding the localisation model by removing APs that are missing. Finally for creating both localisation model and missing APs detection model we used the random forrest method that are very fast to create which is important when we update the model and also very fast when we want to find the actual localisation of the terminal.

The proposed solution reduces a growth of the localisation error to 4 percent for the floor localisation and 2 metres for the horizontal localisation in case of a gross malfunction of an AP infrastructure. Without proposed improvements – and the same malfunction – the floor localisation error is greater than 30 percent and the horizontal localisation error exceeds 7 metres.

Acknowledgments. The research is supported by the the National Centre for Research and Development, grant No PBS2/B3/24/2014, application No 208921.

References

1. Breiman, L.: Random forests. Machine Learning **45**(1), 5–32 (2001)
2. Grzenda, M.: On the prediction of floor identification credibility in RSS-based positioning techniques. In: Ali, M., Bosse, T., Hindriks, K.V., Hoogendoorn, M., Jonker, C.M., Treur, J. (eds.) IEA/AIE 2013. LNCS, vol. 7906, pp. 610–619. Springer, Heidelberg (2013). http://dx.doi.org/10.1007/978-3-642-38577-3_63
3. Karwowski, J., Okulewicz, M., Legierski, J.: Application of particle swarm optimization algorithm to neural network training process in the localization of the mobile terminal. In: Iliadis, L., Papadopoulos, H., Jayne, C. (eds.) EANN 2013, Part I. CCIS, vol. 383, pp. 122–131. Springer, Heidelberg (2013). http://dx.doi.org/10.1007/978-3-642-41013-0_13
4. Korbel, P., Wawrzyniak, P., Grabowski, S., Krasinska, D.: Locfusion api - programming interface for accurate multi-source mobile terminal positioning. In: 2013 Federated Conference on Computer Science and Information Systems (FedCSIS), pp. 819–823, September 2013
5. Papapostolou, A., Chaouchi, H.: Scene analysis indoor positioning enhancements. Annales des Télécommunications **66**, 519–533 (2011)
6. Roos, T., Myllymaki, P., Tirri, H., Misikangas, P., Sievanen, J.: A probabilistic approach to wlan user location estimation. International Journal of Wireless Information Networks **9**(3), 155–164 (2002)
7. Wang, J., Hu, A., Liu, C., Li, X.: A floor-map-aided wifi/pseudo-odometry integration algorithm for an indoor positioning system. Sensors **15**(4), 7096 (2015). http://www.mdpi.com/1424-8220/15/4/7096
8. Xiang, Z., Song, S., Chen, J., Wang, H., Huang, J., Gao, X.G.: A wireless lan-based indoor positioning technology. IBM Journal of Research and Development **48**(5–6), 617–626 (2004)

User Profile Analysis for UAV Operators in a Simulation Environment

Víctor Rodríguez-Fernández[1]([✉]), Héctor D. Menéndez[2], and David Camacho[1]

[1] Universidad Autónoma de Madrid (UAM), 28049 Madrid, Spain
victor.rodriguez@inv.uam.es, david.camacho@uam.es
http://aida.ii.uam.es
[2] University College London (UCL), London, United Kingdom
h.menendez@ucl.ac.uk

Abstract. Unmanned Aerial Vehicles have been a growing field of study over the last few years. The use of unmanned systems require a strong human supervision of one or many human operators, responsible for monitoring the mission status and avoiding possible incidents that might alter the execution and success of the operation. The accelerated evolution of these systems is generating a high demand of qualified operators, which requires to redesign the training process to deal with it. This work aims to present an evaluation methodology for inexperienced users. A multi-UAV simulation environment is used to carry out an experiment focused on the extraction of performance profiles, which can be used to evaluate the behavior and learning process of the users. A set of performance metrics is designed to define the profile of a user, and those profiles are discriminated using clustering algorithms. The results are analyzed to extract behavioral patterns that distinguish the users in the experiment, allowing the identification and selection of potential expert operators.

Keywords: UAVs · Human-Robot Interaction · Computer-based Simulation · Videogames · Performance metrics · Clustering · Behavioral patterns

1 Introduction

The study of Unmanned Air Vehicles (UAVs) is currently a growing area. These new technologies offer many potential applications in multiple fields such as infrastructure inspection, monitoring coastal zones, traffic and disaster management, agriculture and forestry among others [11].

The work of UAV operators is extremely critical due to the high costs involving any UAV mission, both financial and human. Thus, lot of research in the field of human factors, and more specifically, in Human Supervisory Control (HSC) and Human-Robot Interaction (HRI) systems, have been carried out, in order to understand and improve the performance of these operators [9].

© Springer International Publishing Switzerland 2015
M. Núñez et al. (Eds.): ICCCI 2015, Part I, LNAI 9329, pp. 338–347, 2015.
DOI: 10.1007/978-3-319-24069-5_32

In recent years, two topics are emerging in relation to the study of Unmanned Aircraft System (UAS). One is the effort to design systems such that the current many-to-one ratio of operators to vehicles can be inverted, so that a single operator can control multiple UAVs. The other is related to the fact that accelerated UAS evolution has now outpaced current operator training regimens, leading to a shortage of qualified UAS pilots. Due to this, it is necessary to re-design the current intensive training process to meet that demand, making the UAV operations more accessible and available for a less limited pool of individuals, which may include, for example, high-skilled video-game players [10].

This work is focused on measuring and analyzing the performance of inexperienced UAV operators using the data extracted from a multi-UAV simulator. This performance data will be used to extract behavioral patterns among users, which could be used to select potential UAS operators.

The most common metrics to assess human performance on HRI systems focus on the operator workload and its *Situational Awareness* [5]. However, it is also interesting to define some metrics that collect the performance of an operator in a direct way, as a global *score* indicating the performance quality. These metrics, also known as *Direct measures of performance quality*, create a *user profile*, and are widely used, for example, in the world of videogames [1].

The information given by the different metrics and operator interactions can help to recognize and extract some hidden information about the general use of the system. Here, *data mining and machine learning* techniques take much importance. Since multi-UAV systems are still futurist developments, it is impossible to trust any expert trying to label the operator interactions in order to make an objective supervised analysis, hence we can only work in this field by using unsupervised learning techniques [2].

For this reason, the analysis made in this work is focused on *Clustering*, a popular unsupervised technique used to group together, in a blindly way, objects which are similar to one another. In order to assess the quality of a clusterization, and to compare and decide which clustering algorithm is better for a specific dataset, the data-mining literature provides a range of different *internal validation measures*, which take a clusterization and use information intrinsic to the data to evaluate the quality of the clustering [3].

The rest of the paper is structured as follows: Section 2 gives a brief review of the simulation environment used to extract the data for this work. Section 3 details the set of metrics developed to assess the operator performance in the simulation environment used. Then, in Section 4 we make use of those metrics with the data extracted, and analyze the results using and validating some clustering techniques. Finally, Section 5 presents the conclusions and future work.

2   DWR - A Multi-UAV Simulation Environment

Retrieving data from the interactions and performance of UAV operators during a multi-UAV mission simulation is a novel task, due to the premature state of the works in this field. This is causing an impediment to expand the analysis in

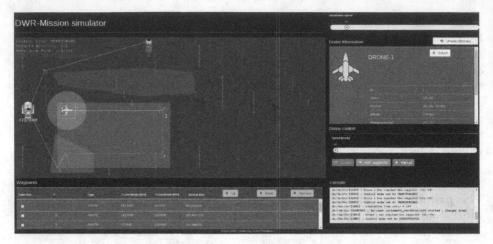

Fig. 1. Screenshot of Drone Watch And Rescue (DWR).

this field towards an accessible place, where an inexpert user could be trained to become a potential expert in UAV operations [4,10].

For this reason, the simulation environment used as the basis for this work has been designed following the criteria of accessibility and usability. It is known as Drone Watch And Rescue (DWR), and its complete description can be found in [12]. DWR gamifies the concept of a multi-UAV mission (See Figure 1), challenging the operator to capture all mission targets consuming the minimum amount of resources, while avoiding at the same time the possible incidents that may occur during a mission (e.g., Danger Areas, Sensor breakdowns). To avoid these incidents, an operator in DWR can perform multiple interactions to alter both the UAVs in the mission and the waypoints composing their mission plan.

Besides, it is remarkable how DWR saves data from a simulation. Whenever an event occurs during a simulation, DWR stores the simulation status in that moment, as a *Simulation Snapshot*. This snapshot contains all relevant information of the current status of every element taking part in the simulation. The information stored by DWR allows to reproduce the entire simulation, which is helpful for the analysis process.

3 Direct Measures of Performance Quality

The main goal of this work is to analyze the user performance quality during simulations running in the environment introduced above. That lead us to the need for defining a way to measure the performance of a user in a specific simulation.

To achieve this, five performance metrics have been defined: Agility (A), Consumption (C), Score (S), Attention (At) and Precision (P). All of them are numeric values in the range $[0, 1]$, where 0 represents the worst performance for that metric, and 1 represents the best.

Each of these metrics are computed for a specific simulation. Based on that, we can define, for a given user, his performance *profile* as the tuple (A, C, S, At, P), where each value represents an average on the set of simulations executed by the user. Below are detailed the implementation of each of the metrics.

3.1 Agility

Agility (A) measures the average speed with which the user has interacted with the simulator. In the simulation environment, a user can manipulate the simulation speed, giving values from 1 to 1000 times. A user is considered agile if he can interact when things are happening fast. Let $I(s)$ be the set of all interactions performed during a given simulation s, the Agility is computed as:

$$A(s) = \frac{\sum_{i \in I(s)} \frac{simulationSpeed(i)}{MAX\_SPEED}}{|I(s)|} \tag{1}$$

where $MAX\_SPEED = 1000$ and $simulationSpeed(i)$ gives the speed in which the simulation was running at the moment when the interaction i was made.

3.2 Consumption

Consumption (C) metric measures the fuel consumed throughout the simulation time. Given a specific instant (also called *snapshot*) sh of a simulation s, we can compute the global remaining fuel (rf) at that instant as

$$rf(s, sh) = \sum_{u \in U(s)} rf(u, sh) + \sum_{r \in R(s)} rf(r, sh) \quad,$$

where $U(s)$ is the set of UAVs participating in the simulation s and $R(s)$ is the set of refueling stations taking part in the Mission Scenario of simulation s (See DWR description in [12]). When a UAV u is destroyed during the simulation, it is considered that $rf(u, sh) = 0$ for every instant sh after the UAV destruction.

To calculate the consumption over a simulation s, we compare the remaining fuel at the end of the simulation (last snapshot, or *lSh*) with that at the beginning (first snapshot, or *fSh*):

$$C(s) = \frac{rf(s, lSh(s))}{rf(s, fSh(s))} \tag{2}$$

High values of this metric indicate that the remaining fuel at the end of the mission is high, so the consumption is considered low. On the other hand, low values mean high consumption rate. This metric also gives information about the duration of a simulation: since a user in DWR can abort a mission whenever he wants, high values of consumption will likely be associated to short missions, while low values will indicate long ones.

3.3 Score

The *Score* (*S*) metric gives a global success/failure rate of a simulation. The main goal for an operator monitoring a simulation in DWR is to capture the maximum number of targets, minimizing the resources consumed and returning all UAVs to an airport at the end of the mission. This goal can be divided into 3 sub-goals: detecting targets, minimizing resource loss and returning the UAVs to an airport to finish the mission.

Based on this description, we define the score of a simulation *s* as:

$$S(s) = \frac{1}{3} \left[\frac{|targetsDetected(s)|}{|T(s)|} + \left(1 - \frac{|destroyedUAVs(s)|}{|U(s)|} \right) + \frac{UAVsInBase(s, lSh(s))}{|U(s)|} \right] , \tag{3}$$

where $U(s)$ is the set of UAVs participating in the mission and $T(s)$ the set of mission targets. Note that $UAVsInBase(s, lSh(s))$ queries how many UAVs were positioned on an airport at the last snapshot of the simulation *(lSh(s))*.

3.4 Attention

The *Attention* (*At*) metric rates the user intensity in terms of the interactions he has performed in a simulation. Given a simulation *s*, Attention is defined as:

$$At(s) = 1 - \frac{1}{1 + \sqrt{|I(s)|}}, \tag{4}$$

where $I(s)$ is the set of all interactions performed during simulation *s*.

3.5 Precision

The *Precision* (*P*) metric measures the replanning skills of a user on a simulation, rating how he has reacted to the mission incidents. The design of this metric is based in the following assumption: A precise operator should only perform replanning interactions (add/edit/remove waypoints) when an incident occurs. Therefore, the waypoints added when no incident has happened should penalize the precision rate. Based on this, we can divide the precision computation into two parts: The precision in times of incidents (*Incident Precision*, P_I) and the precision when nothing is altering the simulation, i.e, the operator must only monitor the simulation status (*Monitoring Precision*, P_M).

$$P(s) = \frac{P_I + P_M}{2} \tag{5}$$

The *Incident Precision* (*P_I*) supposes that every waypoint added/edited/removed during a specific interval time (10 seconds for this experiment) since the beginning of an incident is placed in order to avoid that incident, so it is considered as a precise interaction. Let $In(s)$ be the set of incidents happened during the simulation s, we can compute P_I as follows:

$$P_I(s) = \frac{\sum_{i \in In(s)} p_I(i,s)}{|In(s)|} \qquad p_I(i,s) = 1 - \frac{1}{1 + |W_i(s)|},$$

where $p_I(i,s)$ gives the precision for an specific incident i. In this last equation, $W_i(s)$ is the set of all *waypoint interactions* (add/edit/remove) performed since the incident i started until 10 seconds after (i.e, interactions within the interval $[startTime(i), startTime(i)+10]$). The more waypoints are changed during that interval, the more the precision increases for that incident.

The *Monitoring Precision* (P_M) is conceptually contrary to P_I, in the sense that it penalizes the waypoint interactions performed during *monitoring time*, so the less interactions here, the more precision obtained. It is computed as

$$P_M(s) = \frac{1}{1 + |W_M(s)|}, \qquad W_M(s) = \overline{\bigcup_{i \in In(s)} W_i(s)},$$

where $W_M(s)$ is the set of all waypoint interactions performed during monitoring time, i.e, the complementary of all waypoint interactions made to avoid incidents.

4   Experimentation

The purpose of the experiment carried out using the simulation environment DWR is to analyze the behavior and performance of inexperienced UAV operators during a training session. Once we have a robust dataset, we will assess the performance of every user following the metrics defined above, and based on this evaluation, we will create and group user profiles in order to create **clusters** that indicate similar user behaviors. Finally, those clusters will be analyzed and interpreted in the context of this experiment.

4.1   Experimental Setup

In this experiment, the simulation environment (DWR) was tested with Computer Engineering students of the Autonomous University of Madrid (AUM), all of them inexperienced in HSC systems.

When a user entered the simulation environment, he did not start a simulation immediately but was prompted with a *mission selection screen*. Table 1 summarizes the main features for each of the **3 test missions** designed for this experiment. As it can be appreciated, there is an increasing order in terms of the challenge that suppose a mission.

The dataset resulted of extracting data from this experiments is composed of **127 distinct simulations**, played by a total of **25 users**. In order to achieve a robust analysis of the data extracted, we must clean the dataset by removing those simulations which can be considered as useless. Since the simulator is running in a *web-environment*, a user can "restart" (or abort) a mission simulation

Table 1. Specification summary for the test missions (T.M) of the experiment.

	T.M.01	T.M.02	T.M.03
UAVs	1	1	3
Tasks	1	1	4
Targets	1	1	4
Incidents	2	2	4
No Flight Zones	1	2	4
Refueling Stations	1	3	4

by doing a *page refresh* in his web browser. Due to that, we must remove from the dataset those simulation which have been aborted prematurely. In this work, we consider that a simulation is useless if it has been aborted before 20 seconds. From the 127 simulations composing our students dataset, only 102 of them are considered useful simulations, and will be used in the data analysis process.

Computing the metrics for all users in our dataset results in a 5-dimensional metric space, on which we can apply **clustering** methods to group together users which have similar performance profiles. We make use of four clustering algorithms: *Hierarchical, K-means, DIANA and PAM* [6–8].

For internal validation of the clusters, we selected three validation measures from the state of the art that reflect the compactness, connectedness, and separation of the cluster partitions: *Connectivity, Dunn Index and Silhouette width* [3]. All these clustering algorithms are tested using *Cluster k* values from 2 to 8.

4.2   Experimental Results

The numerical results for the clustering algorithms validation are given in Table 2. If two or more algorithms achieve the same best score for a specific measure, we check which of them is better on average and choose it as optimal. Thus, these results prove that **Hierarchical Clustering stands as the best algorithm for all validation measures**, with 2 and 7 as chosen number of clusters.

According to these results, two behavioral patterns can be distinguished, and those can be broken down into seven more specific patterns. Due to the small number of users in this experiment, and given that we have designed the missions of this experiment and the simulation environment itself, we are able to do an expert cluster analysis manually by analyzing the profiles of each cluster. Below are detailed the explanation and behavioral patterns *(B.Tags)* extracted from each of the clusters obtained by the *Hierarchical Clustering* algorithm with 7 clusters. Figure 2 shows these results graphically.

1. *Users={11}; B.Tags=[Unfocused, Impatient, Replanning Potential]*: This cluster comprises only one user. It stands out reaching maximum levels of precision and consumption, but low score levels. He represent an impatient user performing fast and precise interactions. (See Figure 2, purple profiles).

Table 2. Clustering results for the User Performance Profiles. Bolded cells represent the best results obtained for each cluster validation metric.

Clustering Method	Validation Metric	k=2	k=3	k=4	k=5	k=6	k=7	k=8
	Connectivity	**8.725**	10.037	19.326	21.593	25.561	32.650	34.883
Hierarchical	*Dunn*	0.275	0.275	0.355	0.355	0.392	**0.505**	0.505
	Silhouette	**0.382**	0.331	0.299	0.277	0.304	0.287	0.285
	Connectivity	8.725	18.015	23.638	25.630	25.760	32.650	34.883
K-Means	*Dunn*	0.275	0.355	0.255	0.255	0.492	0.505	0.505
	Silhouette	0.382	0.310	0.294	0.271	0.338	0.287	0.285
	Connectivity	9.353	17.955	21.459	21.876	24.719	27.395	32.178
DIANA	*Dunn*	0.278	0.313	0.355	0.355	0.360	0.456	0.485
	Silhouette	0.356	0.314	0.295	0.292	0.268	0.235	0.197
	Connectivity	8.725	17.170	21.683	27.508	29.119	31.352	35.012
PAM	*Dunn*	0.275	0.310	0.287	0.221	0.321	0.321	0.321
	Silhouette	0.382	0.303	0.329	0.272	0.289	0.278	0.253

2. *Users={15}; B.Tags=[Unfocused, Slow, Replanning Potential]*: This cluster comprises only one user. It stands out due to the low agility levels it features and the shortness of his missions (low consumption). Probably he played missions until the first incidence appeared, which caused him to abort the mission (See Figure 2, gray profiles).

3. *Users={1,2,19}; B.Tags=[Unfocused, Replanning Potential]*: This cluster represents a soft version of the previous one (Element 15). Users here are cautious due to its slow level of agility, and its precision/score/attention balance says that they tried to avoid the incidents making a complex *replanning* (high precision), but lost the focus on the mission targets (low scores, see Equation 3). (See Figure 2, red profiles).

4. *Users={5,18,20,21,22}; B.Tags=[Restless, Aggressive, Target-focused]*: Looking at the distribution of each metric, we can conclude that this cluster represents an average performance profile. This means that users here are a representative sample of the general behavior of the students in our dataset. Precision is low, as well as agility, and the Consumption is not high, which indicates that the missions have not been prematurely aborted. This makes sense given that all users are inexperienced (See Figure 2, yellow profiles).

5. *Users={3,4,9,10,12,13,14}; B.Tags=[Restless, Reckless, Target-focused]*: This cluster stands out for its precision-attention balance. In all cases, the Precision is very low and the Attention exceeds the average. This is clearly associated to a behavior focused on detecting targets as soon as possible, ignoring the effects of the incidents (See Figure 2, blue profiles).

6. *Users={6,8,23,24}; B.Tags=[Unconscious, Destructive, Reckless]*: This cluster clearly **represents a bad profile**, undesirable for anyone looking for good operators. The extremely low score values mean that not only none of the targets have been detected, but also the UAVs have been destroyed. (See Figure 2, green profiles).

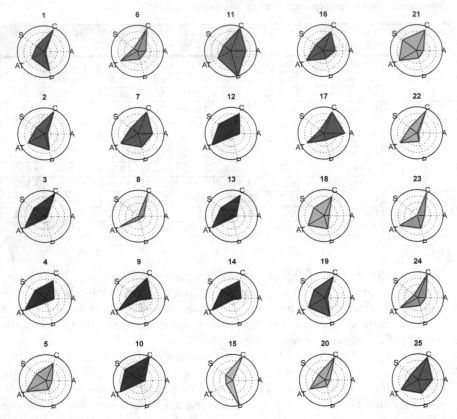

Fig. 2. User Performance Profile Clusters, from Hierarchical Clustering with k=7.

7. *Users*={ *7,16,17,25*} *B. Tags=[Fast, Responsive]*: Users in this cluster share
   a high level of agility, above the average. The rest of the metrics orbit around
   the average values, which lead us to conclude that these users can be trained
   to respond quickly to unknown situations. (See Figure 2, brown profiles).

5   Conclusions and Future Work

In this work, a multi-UAV simulation environment has been used to carry out an
experiment with inexperienced operators, in order to extract and discover behavioral patterns from the performance of each participating users . To achieve this,
three main steps have been followed: First, five performance metrics have been
designed in order to define the user performance profile Later, the user profiles
are introduced into several *clustering algorithms*, to discover some groups or patterns in the user performance. The clusterizations obtained by the algorithms

are validated, optimized and analyzed to extract the behavioral patterns hidden among the user profiles. The results show that the metrics and profiles created for this experiment characterize well the low expertise and novice behavior of the users in the experiment.

As future work, it is intended to improve the performance metrics in order to ensure that all of them offer valuable information. Besides, it would be interesting to develop the final cluster analysis and "behavioral tagging" automatically.

Acknowledgments. This work is supported by the Spanish Ministry of Science and Education under Project Code TIN2014-56494-C4-4-P, Comunidad Autonoma de Madrid under project CIBERDINE S2013/ICE-3095, and Savier an Airbus Defense & Space project (FUAM-076914 and FUAM-076915). The authors would like to acknowledge the support obtained from Airbus Defence & Space, specially from Savier Open Innovation project members: José Insenser, Gemma Blasco and Juan Antonio Henríquez.

References

1. Begis, G.: Adaptive gaming behavior based on player profiling (Aug 22 2000), uS Patent 6,106,395
2. Boussemart, Y., Cummings, M.L., Fargeas, J.L., Roy, N.: Supervised vs. unsupervised learning for operator state modeling in unmanned vehicle settings. Journal of Aerospace Computing, Information, and Communication 8(3), 71–85 (2011)
3. Brock, G., Pihur, V., Datta, S., Datta, S.: clvalid, an r package for cluster validation. Journal of Statistical Software (Brock et al., March 2008) (2011)
4. Cooke, N.J., Pedersen, H.K., Connor, O., Gorman, J.C., Andrews, D.: 20. acquiring team-level command and control skill for uav operation. Human factors of remotely operated vehicles **7**, 285–297 (2006)
5. Drury, J.L., Scholtz, J., Yanco, H.A.: Awareness in human-robot interactions. In: IEEE International Conference on Systems, Man and Cybernetics, vol. 1, pp. 912–918. IEEE (2003)
6. Hartigan, J.A., Wong, M.A.: Algorithm as 136: A k-means clustering algorithm. In: Applied statistics, pp. 100–108 (1979)
7. Kaufman, L., Rousseeuw, P.: Clustering by means of medoids (1987)
8. Kaufman, L., Rousseeuw, P.J.: Finding groups in data: an introduction to cluster analysis, vol. 344. John Wiley & Sons (2009)
9. McCarley, J.S., Wickens, C.D.: Human factors concerns in uav flight. Journal of Aviation Human Factors (2004)
10. McKinley, R.A., McIntire, L.K., Funke, M.A.: Operator selection for unmanned aerial systems: comparing video game players and pilots. Aviation, Space, and Environmental Medicine **82**(6), 635–642 (2011)
11. Pereira, E., Bencatel, R., Correia, J., Félix, L., Gonçalves, G., Morgado, J., Sousa, J.: Unmanned air vehicles for coastal and environmental research. Journal of Coastal Research pp. 1557–1561 (2009)
12. Rodríguez-Fernández, V., Menéndez, H.D., Camacho, D.: Diseño de un simulador de bajo coste vehículos aéreos no tripulados. In: Actas del X Congreso español sobre metaheursticas. algoritmos evolutivos y bioinspirados: MAEB, pp. 447–454. Mérida, Cáceres (2015)

Ontologies and Information
Extraction

Ontology-Based Information Extraction from Spanish Forum

Willy Peña[1,3] and Andrés Melgar[1,2](✉)

[1] Grupo de Reconocimiento de Patrones e Inteligencia Artificial Aplicada,
Pontificia Universidad Católica del Perú, Lima, Perú
amelgar@pucp.edu.pe
http://inform.pucp.edu.pe/grpiaa/
[2] Sección de Ingeniería Informática, Departamento de Ingeniería,
Pontificia Universidad Católica del Perú, Lima, Perú
[3] Especialidad de Ingeniería Informática, Facultad de Ciencias e Ingeniería,
Pontificia Universidad Católica del Perú, Lima, Perú

Abstract. Nowadays, institutions of higher education need to be informed about the opinion of their students about the services offered. They depend on this information to make strategic decisions. To accomplish this, many choose to hire consulting services companies that have traditional survey or focus group to obtain the required information; however, the bias that there is on this type of study, means that in some cases strategic decisions are not entirely reliable. Discussions of people in the Web 2.0 are a relevant source of information for this kind of organization due to the large amount of data that is posted every day. A case in point are the discussion forums, which made it possible for people to share their views freely on a specific topic. This is achieved thanks to comments made in publications; from which useful information can be extracted for use in other purposes. This paper will provide a prototype to exploit the information contained within the comments made in discussion forums. These sources of knowledge are often not processed for any purpose, however the proposed prototype allow us to extract relevant data from this reliable source to use as a knowledge base for a institution of higher education.

Keywords: Information extraction · Ontology · Forum · Coreference resolution

1 Introduction

Companies need to know the opinion of their customers, so that this information can help to make strategic decisions. Many organizations choose to use traditional surveys to gather information. Usually, this type of study seeks customer feedback through questionnaires or focus groups, which are limited to the number of people who choose to participate in the study; as well, they may have some bias, which depends on how the questions were phrased or interpretation

© Springer International Publishing Switzerland 2015
M. Núñez et al. (Eds.): ICCCI 2015, Part I, LNAI 9329, pp. 351–360, 2015.
DOI: 10.1007/978-3-319-24069-5_33

thereof [1]. Thanks to the Internet there are some non-traditional ways to obtain this information (*e.g.*, discussion forums) [2].

There is a need for tools that facilitate the extraction and processing of information provided by these non-traditional sources. Unfortunately, proposed solutions are not focusing on unstructured information [3]. Ontology-Based Information Extraction (OBIE) systems can be support in the processing of unstructured information [4]. This kind of system is based on the Information Extraction (IE) process, which automatically extracts certain information based on predefined templates from a text written in Natural Language (NL). To understand the knowledge in a specific domain, it uses ontologies, which provide a formal and explicit knowledge representation [4].

Although such systems present a reliable solution there are not many tools developed. The lack of such solutions, we can see reflected in four main points. First, there is a waste of the information contained in the comments made in a discussion forum [5]; however, once it has filled the space available, the administrators usually choose to create a back up of data without being processed [6]. Second, many organizations choose to meet the views of their customers through traditional surveys or focus groups; however, a system of this type can get this information at any time and present a structured way [1]. Third, traditional surveys depend on many factors to be completely reliable; however, the information contained in the comments made in discussion forums are direct point of view, which allows it to be a direct source [1]. Finally, due to the large size and number of comments, extracting information from these sources manually is expensive and time consuming [7].

This paper seeks to provide an alternative solution to the IE problem in the comments in the discussion forums. In this particular case, we will focus on extract information that come under the domain of courses offered by an institution of higher education. Furthermore, the extracted data will be represented using a database, which will facilitate the understanding of the extracted information so it can be used as a knowledge base for strategic decisions [8].

This paper is structured as follows: after this introduction, we present the literature review about IE, ontologies and related works. Subsequently, the proposed architecture is described. In the following sections, we present the the developed prototype, the results, the discussion and future works. Finally, in the last section, we present the conclusions.

2   Literature Review

2.1   Information Extraction

IE is a process that extracts unambiguous data from texts written in NL regarding a specific domain, which includes entities, relationships and events [9][10]. Typical tasks of an IE system are: Name Entity Recognition, Coreference Resolution, Template Element Production, Template Relation Production and Scenario Template Extraction. Below is a brief definition of each.

Name Entity Recognition (NER). The goal of this task is to identify all the names of people, places, organizations, dates and amounts of money. For example, as the following comment regarding who is the best teacher:

"For me the best teacher is Ruben Agapito, because it is right, teaches everything to be learned, we projected into the future and tells you everything that mathematics will serve in other courses, gives his good rest in half time, send proposed problems, send practices and examinations of past cycles with answer key provided after each practice or exam gives the answer key. I have taken Calculus 1 with that teacher and in both cases I learned a lot."

We can find the following names of entities: Ruben Agapito and Calculus 1. From this result, it can be deduced that this process depends on the domain, especially when the texts to be processed contain lots of informal language. In other words, if we try to analyse comments were regarding opinions about a beauty product would involve some adjustments to the system about the domain for which it was built [9].

Coreference Resolution (CO). This task involves identifying relationships between entities found in the previous task. Thus continuing the example; a cor-relation found in the text would be between Ruben Agapito and that teacher. Although users face this task does not add much as the other in a IE process; It is very useful when trying to develop a system of this type. The most important is that it is a key for the following tasks, allowing the association of descriptive information found through referrals to the principal entities found [9].

Template Element Production. This task is based on NER and CO obtained from the previous tasks. From which seeks descriptive information for enti-ties found. For example, following the above case, the entity found for Ruben Agapito and its coreferential that teacher is obtained that Ruben Agapito is a teacher as a descriptive information of the entity [9].

Template Relation Production. This task requires the identification of a small number of possible relationships between the identified template elements. For example, in the present case relationships between entities Ruben Agapito and Calculus 1, where each have particular properties that are related to each other are obtained; i.e. you get that Ruben Agapito teaches Calculus 1. Therefore, building relationships template provides a major feature for any IE [9].

Scenario Template Extraction. This taks shows item templates as entities with descriptions of events and relationships. For example, you can identify as Pedro Perez is in his second cycle of computer engineering career and is taking the course Calculus 2 Professor Ruben Agapito. This task depends on the domain and is related to the stage of user interest. This avoids any errors in the development of this task, leaving aside some of the occurrences on important stages [9].

2.2   Ontologies

An ontology is a conceptualization of a domain of knowledge where concepts and relationships are described, which are comprehensible to both humans and machines. This conceptualization is represented by the following components: classes, properties, data types, objects, values and relationships[11]. OBIE can process unstructured and semi-structured from text found in NL through mechanisms driven by ontologies data. Additionally, it allows present results using ontologies which facilitates the understanding of the extracted information [4]. In this kind of system there are two approaches for using ontologies. First, one is when used as input data system, which generates the ontology is built by hand so it can be used. On the other hand, there is another approach that focuses on building an ontology as part of the tasks performed by the system [4].

2.3   Related Works

Many studies on processing and IE from unstructured data were analyzed in order to establish a framework that allows for the understanding of unstructured data and the using for different purposes. [12] analyses different types of forums incorporating to IE system a fingerprint system to detect the different structures of each one. On the other hand, [13] simplifies scanning discussion forums extracting entities comments made in NL through the use of ontologies. Moreover, [14] proposes a method for automatic IE from layers. Also, [15] provides a mechanism for semantic search to find relevant answers made in on-line discussion forums regarding software development. Finally, [16], solves the problem of IE based on the assumption that the web page contains structured notes for objects; *i.e.*, contains HTML structure.

After analyzing the primary studies obtained as a result of the systematic review we observed that most investigations are devoted to IE, using methods such as layering, learning automatic rules, learning statistical models, among others; while a lesser extent, there are researches that perform the extraction of such information based on ontologies by focusing on knowledge engineering. This work will focus on OBIE, which allow the understanding and interpretation of the content of comments posted on a discussion forum. Thus, we take maximum advantage of unstructured data; whereby the educational organization may use them to generate statistics student opinion regarding their courses and professors.

3   The Architecture Proposed

The solution proposed (see figure 1) is based on the application of heuristics rules from morph-syntactic and semantic information to resolve relationships between the sources found in the comments, also it is supported by an ontology which helps to manage the knowledge about the domain that is being developed. These rules and the ontology were defined according to a specific domain using comments about courses and teachers from an university .

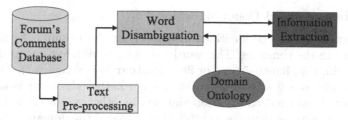

Fig. 1. Architecture Diagram

Text Preprocessing. This mechanism makes a preliminary analysis of comments, making an initial standardization that allows the NL processing needed for the following mechanisms. To achieve this, it used class libraries from Freeling[1] and Lucene[2] to perform the language analysis needed.

Domain Ontology. This artefact allow us to represent the knowledge of the chosen domain. To achieve this goal, we used Protégé[3] which is an open-source ontology editor and framework for building intelligent systems.

Word Disambiguation. This mechanism have the goal to resolve the words which have a ambiguous meaning. We used each word content in the preprocessed comments and the domain ontology to find which of them are ambiguous. The Jena API[4] allow us to manipulate the ontology.

Information Extraction. This mechanism is in charge of find the principal sources in the comments and perform the IE task, called coreference resolution which consist on finding the mentions of the sources found.

4   Prototype Implementation

4.1   Module of Text Preprocessing

In this first stage, we have grouped the comments in a publication. For each one, we labelled words containing offensive language from a dictionary compiled from the data that was used. Also, special characters and texts that do not have relevant information within the publication were ignored, then some spelling errors were corrected and words were converted to their most basic form (lemma). Finally, each word passed by a morphosyntactic analysis allowing us to identify the type, gender and number of words in the post.

---

[1] http://nlp.lsi.upc.edu/freeling/
[2] http://lucene.apache.org/core/
[3] http://protege.stanford.edu/
[4] https://jena.apache.org/

4.2 Module of Word Disambiguation

A case of ambiguity can be identified in homonymy; for example, by having the word `Bances` in the comment. This word can refer to different terms depending on the domain; *e.g.*, `BancesDiana` or `BancesRicardo` (see figure 2). First, every word will undergo a first evaluation to check whether the term is ambiguous and if is not considered as an abbreviation. In other words, it is considered ambiguous if it is more than once in different nodes of the domain ontology and is considered as an abbreviation if the number of characters from your lemma version is not less than or equal to 5. If that term is found only once, it is not necessary that the word go through the process of disambiguation, otherwise we proceed to calculate the similarity of each of these returnees matches.

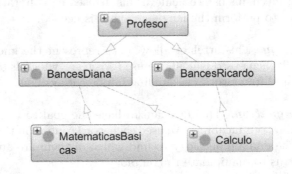

Fig. 2. Relationship diagram according to domain ontology

The nodes which has been considered the ambiguous word are obtained and for each additional terms (other words in the publication), does the following: similarity calculation is performed with the support of supplementary terms, which are represented by all the words surrounding the ambiguous term, within the comment. This calculation involves making an evaluation within each node and parent node of each of them; with their properties. Furthermore, in order to have a better precision of this mechanism, it is considered that the farther this ambiguous term complementary node is less the rating is; *i.e.*, if the additional term was found in the same ambiguous node will have a weight of 1; if it is found in the parent node will weigh 0.5.

After obtaining the similarity score for supplementary level ontology term, we proceed to assign a weight due to proximity to the complementary term is ambiguous term. For example, as can be seen in the equation 1, a weight is assigned based on the position of the complementary term; considering a t_i as the term ambiguous and t_k as the complementary term. Finally, if you could detect a higher score in candidates to resolve ambiguity, lemma form of such term shall be replaced with the comment.

$$f(x) = \begin{cases} \frac{1}{i-k} & k < i \\ \frac{1}{k-i} & k > i \end{cases} \tag{1}$$

4.3   Module of Information Extraction

This stage is based on [17], where a heuristic application from the morphosyntactic and semantic information to determine the relationship between sources of text is presented. However, unlike this solution, this module will solve the relationships between entities based on ontology. Correference analysis consists of two phases:

Identification and Recovery of Sources. At the first level, identifying sources that seek to extract from each of the comments made. For this, the following types of sources were defined: i) **primary source** (are the courses and teachers within the ontology of the sources from which he will seek to extract information); ii) **secondary source** (are the materials and comments within the ontology which allow eligible parts of the comments); iii) **missed source** (these are sources that in principle are not in the text, but are recovered through the sense of prayer); iv) **reference source** (they are nouns, verbs, adjectives, numerals, adverbs and omitted sources that are candidates for corefer with primary and secondary source).

If primary or secondary source is found by a numeral or an adverb, this will be classified as special, because in the domain, can be a source directly. As mentioned above, the operation of this phase is as follows: i) from the list of words that results in the pre-processing module, for each name (own or common), verb, adjective, adverb and numeral in the list; it is checked whether any of the words in its lemma form are within the ontology in some of the properties of the classes of professors and courses; ii) if is is qualified as a primary source. If not found, it is checked again if it comes to extracting information sought; *i.e.*, if it is in the comment properties classes or material. If this happens it will be classified as a secondary source; iii) finally, if it was not in a none of the above cases the word will be qualified as a reference source. One time on this classification is passed to identify possible sources omitted in opinions; the candidates for this type of source is the reference sources. To accomplish this task, we consider come criterias, assigning a positive or negative rating according to each case. In this case the candidate with the highest score will be the one to be replaced with references to the primary source and will be labelled as Missed source. To retrieve a source a minimum score that should be considered as a candidate for a missing font is established; which for purposes of this project was established with a value of two, because this type of source to be omitted must have a reference to gender and number with any existing font, not to be the case at least must comply with all the other rules.

Coreference Resolution. At this stage will seek to analyse the comments looking coreference from their sources; that is, when the previously identified sources refer to the same real-world entity. An approach to assigning scores so to penalize or reward the candidates to be a history of a source depending on their particular characteristics is used. For this coreference between the following sources sought: i) reference sources or sources be omitted candidates for corefer with a primary or secondary source source mentioned before that source;

ii) secondary sources are candidates for corefer with a primary source mentioned before that source.

Then, for each of the sources candidates the following steps are followed as discussed in [17]: i) scores are assigned to each of the sources that can co-refer. For this analysis, syntactic and semantic criteria way to give a score for each candidate source is applied. As shown in table 1, certain criteria are established from the raising in [17], where the categories of sources and the criteria used for each sample, also an impact is established by occurrence: true / not true; ii) based on the scores is determined by what the candidate source correfiere source. For this, it is considered as coreferential the source that has the highest score, only if, this score is greater than zero; Otherwise the candidate source has no coreference with any of the sources that were analyzed.

Table 1. Criteria for solving correference

	Special Source	Missed Source	Pronoun	Adjective	Verb	Name
Proximity	+4	+3	+2			
Concordance			+1/-3	+1/-2	+1/-2	+1/-2
Concordance number			+1/-3	+1/-2	+1/-2	+1/-2
Concordance person			+1/-3	+1/-2	+1/-1	+1/-1
Concordance semantic class				+1/-2	+1/-1	+1/-1
Concordance lemma					+2	+2

5   Discussion and Future Works

For the evaluation of OBIE system developed, we considered the following metrics: precision, recall and F-score (see table 2). At first, we observe that coreference analysis needs that the ambiguities of the domain are resolved, as this will help to differentiate concepts within the domain, which will improve performance reducing errors caused by to misidentified main sources. Furthermore, it was found that spelling errors negatively influence system performance because these words are not always the same and does not apply to the lexical domain. For this reason, it is recommended that future work can be considered as input comments that do not have such errors; this means it could be considered as input data academic documents or news items that are in text format to test and evaluate mechanisms to evaluate the performance metrics again.

On the other hand, the database obtained as the result of the execution of the system developed could be used for further tasks like sentiment analysis which will help to understand what the students need inside the institution and make decisions about that information. Also, it is known that the extracted

Table 2. Metrics evaluation

Component	Precision	Recall	F-score
Text pre-processing	79%	71%	75%
Word disambiguation	76%	76%	76%
Information extraction	74%	77%	75%
OBIE system	76%	75%	75%

data could help to improve the performance of text mining or opinion mining, because it will focus on the information that is the aim to be processed. Finally, it is important to notice that some sources may not be recovered because some of the rules defined on the domain ontology do not consider those terms; however, for future works, the domain could be extended adding more synonyms for words or lemmas analysed or other members of the major classes thus ensuring that new terms are covering.

6 Conclusion

In first place, it is concluded that the proposed prototype of an OBIE system allows to obtain and store the extracted data in a structured way with an acceptable performance. Also, that an institution of higher education could use this data for analysis and decisions making. Besides, it can be concluded that the domain analysis is of great importance in the development of such systems. This is because, the lexical and context can determine the meaning given to a word within the specified domain; that is, without a clear understanding of the domain exist great difficulty to perform a task of IE. Similarly, in the development it will required data to review and evaluate the rules generated from initial domain analysis; so that these can be modified and improved according to the results obtained.

References

1. Kongthon, A., Angkawattanawit, N., Sangkeettrakarn, C., Palingoon, P., Haruechaiyasak, C.: Using an opinion mining approach to exploit web content in order to improve customer relationship management. In: Proceedings of PICMET 2010: Technology Management for Global Economic Growth (PICMET), pp. 1–6 (2010)
2. Castellanos, M., Hsu, M., Dayal, U., Ghosh, R., Dekhil, M., Ceja, C., Puchi, M., Ruiz, P.: Intention insider: discovering people's intentions in the social channel. In: Proceedings of the 15th International Conference on Extending Database Technology. EDBT 2012, pp. 614–617. ACM, New York (2012)
3. Shu, Z., Wen-Jie, J., Ying-Ju, X., Yao, M., Hao, Y.: Opinion analysis of product reviews. In: Sixth International Conference on Fuzzy Systems and Knowledge Discovery. FSKD 2009, vol. 2, pp. 591–595 (2009)

4. Wimalasuriya, D.C., Dou, D.: Ontology-based information extraction: an introduction and a survey of current approaches. Journal of Information Science **36**, 306–323 (2010)
5. Manuel, K., Indukuri, K.V., Krishna, P.R.: Analyzing internet slang for sentiment mining. In: Second Vaagdevi International Conference on Information Technology for Real World Problems (VCON), pp. 9–11 (2010)
6. Yang, C.C., Ng, T.D., Wang, J.-H., Wei, C.-P., Chen, H.: Analyzing and visualizing gray web forum structure. In: Yang, C.C., Zeng, D., Chau, M., Chang, K., Yang, Q., Cheng, X., Wang, J., Wang, F.-Y., Chen, H. (eds.) PAISI 2007. LNCS, vol. 4430, pp. 21–33. Springer, Heidelberg (2007)
7. Galvis Carreno, L.V., Winbladh, K.: Analysis of user comments: an approach for software requirements evolution. In: 35th International Conference on Software Engineering (ICSE), pp. 582–591 (2013)
8. Soutar, G.N., Turner, J.P.: Students preferences for university: a conjoint analysis. International Journal of Educational Management **16**, 40–45 (2002)
9. Cunningham, H.: Information extraction-a user guide. arXiv preprint cmp-lg/9702006 (1997)
10. Abolhassani, M., Fuhr, N., Növert, N.: Information extraction and automatic markup for XML documents. In: Blanken, H.M., Grabs, T., Schek, H.-J., Schenkel, R., Weikum, G. (eds.) Intelligent Search on XML Data. LNCS, vol. 2818, pp. 159–174. Springer, Heidelberg (2003)
11. Gruber, T.R.: Toward principles for the design of ontologies used for knowledge sharing? International Journal of Human-Computer Studies **43**, 907–928 (1995)
12. Gang, H., Yingwei, Z., Xiaochun, W.: Information extraction of forum based on regular expression. In: 5th International Conference on Intelligent Human-Machine Systems and Cybernetics (IHMSC), vol. 2, pp. 118–122 (2013)
13. Mohajeri, S., Esteki, A., Zaiane, O.R., Rafiei, D.: Innovative navigation of health discussion forums based on relationship extraction and medical ontologies. In: IEEE International Conference on Bioinformatics and Biomedicine (BIBM), pp. 13–14 (2013)
14. YingJu, X., YuHang, Y., Fujiang, G., Shu, Z., Hao, Y.: An integrated approach for information extraction. In: 5th International Conference on New Trends in Information Science and Service Science (NISS), vol 1, pp. 122–127 (2011)
15. Gottipati, S., Lo, D., Jing, J.: Finding relevant answers in software forums. In: 26th IEEE/ACM International Conference on Automated Software Engineering (ASE), pp. 323–332 (2011)
16. Machova, K., Penzes, T.: Extraction of web discussion texts for opinion analysis. In: IEEE 10th International Symposium on Applied Machine Intelligence and Informatics (SAMI), pp. 31–35 (2012)
17. Acerenza, F., Rabosto, M., Zubizarreta, M., Rosa, A., Wonsever, D.: Coreference resolution between sources of opinions in spanish texts. In: XXXVIII Conferencia Latinoamericana En Informatica (CLEI), pp. 1–8 (2012)

Knowledge-Based Approach to Question Answering System Selection

Agnieszka Konys[✉]

Faculty of Computer Science and Information Technology,
West Pomeranian University of Technology, Szczecin, Poland
akonys@wi.zut.edu.pl

Abstract. A growth of data published on the Web is still observed. Keyword-based search, used by most search engines, is a common way of information retrieval on the Web. Subsequently, keyword-based search may provide a huge amount of retrieved valueless information. This problem can be solved by Question Answering System (QAS, QA system). One of the challenging tasks for available QA systems is to understand the natural language questions correctly and deduce the precise meaning to retrieve accurate responses. A significant role of QA and an increasing number of them may cause a problem with selection the most suitable QA system. The general aim of this paper is to provide knowledge-based approach to QA system selection. It should ensure knowledge systematization and help users to find a proper solution that meets their needs.

Keywords: Question answering system · Ontology · Information retrieval · Natural language processing

1 Introduction

The Web is enormously larger in size and, a growth of published data is still observed. Incredible amount of information and data redundancy causes it compliant to traditional techniques for answer extraction. There are many disadvantages of Web search. Keyword-based search, used by most search engines, is a common way of information retrieval on the Web. It gives lot of web pages for searching a single word. Moreover, it is time consuming and takes more time for extracting relevant document [1]. In other words, it provides the large amount of retrieved irrelevant information [2]. Question answering (QA) systems address this problem.

Question answering is the task of automatically answering a question posed in natural language. One of the challenging tasks for available QA systems is to understand the natural language questions correctly and deduce the precise meaning to retrieve accurate responses. It is possible to determine QA as a specific type of information retrieval. QA task combines techniques from artificial intelligence, natural language processing, statistical analysis, pattern matching, information retrieval, and information extraction. QA system requires understanding of natural language text, linguistics and common knowledge. Given a set of documents, a QA system attempts

© Springer International Publishing Switzerland 2015
M. Núñez et al. (Eds.): ICCCI 2015, Part I, LNAI 9329, pp. 361–370, 2015.
DOI: 10.1007/978-3-319-24069-5_34

to find out the correct answer to the question pose in natural language [3]. Natural Language Processing (NLP) technique is mostly implemented in QA system for asking user's question. Next, several steps are also followed for conversion of questions to query form for getting an exact answer.

A term of QA system becomes very popular. The role and importance of QA is emphasized by a currently existing number of QA systems. It is seemed that this research area is still developing, and in the nearest future it may bring more innovative solutions. This paper provides a comparative analysis of available QA systems, and due to a number of differentiated between each other QA systems, it presents a proposal of knowledge-based approach to QA system selection. The general aim of this is to ensure knowledge systematization and help users to find a proper QA system. The adaptation of knowledge-engineering mechanism should improve the process of knowledge acquisition of QA systems. Another problem is the assessment of existing solutions (in terms of their practical application). In literature, it is possible to find the experiments of QA system evaluations (e.g. based on recall and precision measures). Undoubtedly these researches have a significant value, but a user does not have any support in a QA system selection process.

2   The Role of Question Answering Systems

2.1   The Challenges of Question Answering Systems

One of the challenging tasks for available QA systems is to scale large amounts of data. Moreover, it should be robust in the face of heterogeneous and possibly conflicting data collected from a large number of sources [4]. Semantic Web and ontology are the key technologies of QA system. Ontology plays a key role in the Semantic Web by enabling knowledge sharing and exchanging, and is becoming the crucial methodology to represent domain-specific conceptual knowledge in order to support the semantic capability of a QA system [5]. QA system uses NLP techniques and a given ontology as input to process a question, thereafter it searches for the required information to identify the answer, and finally present the answer to the user in both non-structured and structured collection of data [2, 9]. Therefore, it does not require the user to learn the vocabulary or structure of the ontology to be queried.

Many challenges exist in adopting Semantic Web technologies for Web data. Furthermore, it is possible to point at unique challenges of the Web, in terms of scale, unreliability, inconsistency, and disruptions are largely overlooked by the current Semantic Web standards [4]. The crucial key for available QA systems is to understand the natural language questions in a correct way. It is seemed that QA system is taking an important role in current search engine optimization concept. From the technological perspective, QA uses natural or statistical language processing, information retrieval, and knowledge representation and reasoning as potential building blocks. It involves text classification, information extraction and summarization technologies. In general, QA system has three components such as question classification, information retrieval, and answer extraction. These components play a essential role in QA system [3].

The remaining challenges of QA systems encompass answering a question. It means that a search engine should analyze the question, perhaps in the context of some ongoing interaction. The important issue is that QA systems must find one or more answers by consulting on links. then it must present the perfect answer to the user in appropriate form clear or supporting materials [2]. Another inconvenience is the assurance of efficiency of the processes of searching and querying content in massive in scale and vastly heterogeneous. It is seemed that existing approaches to querying semantic data have difficulties to measure their models effectively to cope with the increasing amount of distributed semantic data available online [6].

2.2   The General Processes Existing in QA Systems

On base of an literature review it is possible to indicate the general steps existing almost in each of analyzed QA systems. It is worth to emphasize that in many cases these steps have different names with a similar meaning. For the most part, an input query is posed by a user in natural language (due to a selected QA system different forms are permissible). As a next step, the question is converted into query form. Then, the query is processed (for ex. as a bag of words or in other form) using different techniques. It attempts to match the parsed question words to the triples (e.g. Query Triple form into Onto-Triple form) stored in the lexicon. These triple patterns are matched by applying pattern matching algorithms. The process of mapping encompasses the application of the various algorithms (e.g. WordNet, String metric). Many techniques and tools are integrated to interpret parse trees of natural language queries into SPARQL. These steps occur most frequently in analyzed QA systems. The figure 1 presents the basic, general steps that exist in QA systems in a significant simplification.

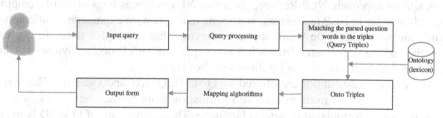

Fig. 1. The general steps existing in QAS

An essential task in machine-based understanding of questions is to enable relevant question classification, formulation of right queries, ambiguity resolution, semantic symmetry detection, identification of temporal relationship in complex questions [6]. In the similar way identification of a perfect answer requires proper validation mechanism [7]. In the simply way, the processing of a QA system generally may be composed of the following phases, encompasses question analysis (1st phase), document analysis (2nd phase) and answer analysis (3rd phase) [7].

3   Related Works - An Analysis of Selected QA Systems

Different types of question answering system based on ontology and Semantic Web model with different query format: the types of input, query processing method, input and output format of each system and its limitations. An analysis of a literature allows to identify and describe some QA systems. The characteristics of the following approaches are included below. The analysis encompasses selected QA: PANTO (Portable nAtural laNguage inTerface to Ontologies approach) [8], FREyA (Feedback Refinement and Extended Vocabulary Aggregation) [11], Querix [10], ORAKEL [12], QACID (Question Answering system applied to the CInema Domain) [7], AquaLog [14], PowerAqua [15], SWSE (Semantic Web Search engine) [4], SMART (Semantic web information Management with Automated Reasoning Tool) [13], QuestIO (Question-based Interface to Ontologies) [16], GINSENG (Guided Input Natural language Search ENGine) [17], NLP-Reduce [18], and SemantiQA [19]. They are differentiate significantly between each other, but it is possible to identify many similarities as well.

As an example, QA systems such as: PANTO, AquaLog, and PowerAqua have a similar form. They are based on triple-based model. In PANTO, multiple existing techniques and tools are integrated to interpret parse trees of natural language queries into SPARQL. Then, AquaLog allows to classify questions into 23 categories [14]. AquaLog, unilike PANTO, is based on a shallow parser and depends on handcrafted grammars to identify terms, relations for composing query-triples, while the parse tree by the deep parser provides PANTO more modification information between nominal phrases [8]. PowerAqua evolved from the Aqua-Log. It reduces the knowledge acquisition bottleneck problem typical of KB systems, and allows to answer queries that can only be solved by composing information from multiple sources [15].

A similar approach, NLP-Reduce, processes NL queries as bags of words, employing only two basic NLP techniques: stemming and synonym expansion. It attempts to match the parsed question words to the synonym-enhanced triples stored in the lexicon (the lexicon is generated from a KB and expanded with WordNet synonyms), and generates SPARQL statements for those matches [18].

Approximate capabilities are offered by QACID, Querix and QuestIO. The first of them, QACID, allows users to retrieve information from formal ontologies by using as input queries formulated in natural language. The general aim of QACID is to collect queries from a given domain. Each query is considered as a bag of words, mapping between words in NL queries into KB by using string distance metrics. QACID is applied to the cinema domain [9]. Then Querix allows to pose queries in natural language, thereby asking the user for clarification in case of ambiguities [10]. It offers a domain-independent natural language interface for the Semantic Web [6]. In contrast, QuestIO works by recognizing concepts inside the query through the gazetteers, without relying on other words in the query. Although this approach uses very shallow NLP, it is quite efficient for very small and domain-specific ontologies [16].

Different structure has QA system called FREyA. It combines methods such as feedback, query refinement and extended vocabulary. The general aim of this approach is to enable system-user interaction, in order to assist the user formulate the query and express including user's need more precisely, within the boundaries of system capabilities [11]. Another of presented solutions, ORAKEL, allows to compute intentional answers of user's queries. It computes wh-based questions as logical query form, thereby, knowledge representation exploits F-Logic and Onto broker. Moreover, a customisation is performed through the user interaction, using FrameMapper software [12]. The next of analyzed solutions, SemanticQA, makes it possible to complete partial answers from a given ontology with Web documents. It supports the users in constructing an input question as they type, by presenting valid suggestions. The answers are ranked using a semantic answer score [19].

The main feature of SMART is a semantic query with validation. It is an open-source system with integrated query form. SMART uses DL reasoners and a graphical representation of query and mapping of DL queries to SPARQL [13]. Another QA system, SWSE, supports SPARQL with RDF representation [4]. It consists of crawling, data enhancing, indexing and a user interface for search, browsing and retrieval of information. Next, GINSENG [17] guides the user through menus to specify NL queries, while systems such as PANTO, NLP-Reduce, Querix and QuestIO, generate lexicons, or ontology annotations (FREya), on demand when a KB is loaded.

3.1 A Comparison of Selected QA Systems

The general aim of a comparative analysis of selected QA systems is to provide a specified description of these systems, considering the key factors that may have essential meaning for a user. Due to limited space of this paper, it was impossible to present all the characteristics in details. The comparison encompasses the most popular and frequent cited QA systems. In many cases the names of criteria should be generalized. It helps in limitation of the total number of criteria in the ontology project. Then the reasoning mechanism computes the results more efficient and faster. It is possible to change and modify the set of criteria. This ontology project is centered on the most expressive features of the analyzed QA systems.

Table 1. A comparison analysis of QA systems

name	authors	criterion subcriterion	query entry NL Query	Keywords	F-Logic	SPARQL	output SPARQL query	additional options domain lexicons	grammar collection	full shallow grammar	bag of words	semantic search	triple-based relation	pattern-matching pattern-matching	including gazetteers	entity lexicon
PANTO	Wang et al., 2007		+				+			+			+	+		
FREyA	Damljanovic et al., 2010		+				+			+			+	+	+	
QUERIX	Kaufmann et al., 2006		+							+			+	+		
ORAKEL	Cimiano et al., 2007				+	+	+	+	+							
QACID	Ferrandez et al., 2009		+				+	+			+			+		
AquaLog	Lopez et al., 2009		+							+			+	+		+
PowerAqua	Lopez, Sabou et al., 2007		+							+			+	+		+
SMART	Battista et al., 2007				+	+										
SWSE	Hogan et al., 2011			+								+				
QuestIO	Tablan et al., 2008		+				+						+	+	+	
GINSENG	Bernstein, Kuffman et al., 2006		+											+		
NLP-Reduce	Kaufmann et al., 2007		+				+				+			+		
SemanticQA	Tartir et al., 2010		+				+							+		

On base of analysis of selected QA systems, the set of criteria and sub-criteria was constructed. The following criteria and sub-criteria were considered: query entry (NL Query, Keywords, F-Logic, SPARQL), output (SPARQL query), additional options (domain lexicons, domain grammar collection, full shallow grammar, semantic search, bag of words, triple-based relation), pattern-matching - structural lexicon (pattern-matching, including gazetteers, entity lexicon. The analysis encompasses 13 QA systems: PANTO, FREyA, Querix, ORAKEL, QACID, AquaLog, PowerAqua, SWSE, SMART, QuestIO, GINSENG, NLP-Reduce, and SemantiQA. The final result is presented in table 1. A symbol "+" informs of existence of a given sub-criterion. Any mistakes or lacks of descriptions are caused by a limited access to full functionalities of analyzed QA systems.

It is possible to enrich the set of QA systems and to extend the number of criteria and sub-criteria. As a result, the additional criteria and limitations (presented also in section 3.2) are added in the provided ontology.

3.2   Limitations of QAS

Current QA Systems are capable of evaluating answers from complex system of data. The analysis of selected QA systems allows to identify essential existing limitations. It is worth to emphasize that many of analyzed QA systems were created several years ago, and since that, many issues have changed. The most frequent constraints consider the domain of interest, user participation and interaction, reasoning processes, or using shallow NLP. Any problem with the present QA system is that they suffer from low recall. The answer to question is also limited to pre-defined categories[2].

As an example, PANTO works only with a small ontology. In FREyA, suggestions are selected by user each time. Similar situation takes place in QUERIX, where a user is asked for clarification. Furthermore, user interaction is required in SMART. Another exemplary inconvenience that occurs in AquaLog, is a lack of appropriate reasoning services defined by ontology. Moreover data heterogeneity is the problem that is appeared in SWSE. As it was mentioned formerly, more of these problems can be solved in the nearest future. It is assumed that this field of interest will be still developed and existing QA systems will be improved (e.g. PowerAqua as AquaLog improvement) or replaced by other ones [4, 6, 11, 10, 14, 15].

4   Knowledge-Based Approach to QA System Selection

The general aim of presented knowledge-based approach is to provide a possibility to select a proper QA system to a given problem. A domain of interest is limited to QA systems. It is assumed that the ontology that supports QA system selection should ensure a freedom in requirements definition processes. Moreover, it should provide the knowledge systematization of QA systems domain, and enables a time reduction for QA system selection.

The users may look for different QA systems and the information of them (and tools). The proposed knowledge-based approach should provide necessary knowledge of a selected QA system. The user do not have to have a broad knowledge of available QA systems. The proposed ontology help them to find a solution that suits the best to their preferences. It is worth to emphasize that provided results are only the recommendation for the user. Moreover, this process may be improved by using MCDA (Multiple-criteria decision analysis) methods to refine the final results or consider additional factors in selection and evaluation processes.

The users define a set of preferences the solution should have. A reasoning mechanism computes the definitions and provides a set of results (a set of QA systems) which fulfills pre-defined requirements. On the output the users obtain a set of results. It is seemed that the best option is an automatic ontology construction for QA systems, but this domain of interest is relatively small and a set of analyzed data does not require a high level of automation. The validation process of automatic ontology construction may take much more time, than manually building it from scratch.

4.1   Systemic Procedure of Ontology Construction

The basis for the ontology construction was a thorough analysis of considered solutions and then the experiment of identification of the set of criteria and sub-criteria (Table 1). On base of this, the taxonomy was built. The aim of the taxonomy is to ensure systematization and classification for particular solutions [20]. The set of criteria was created on the basis of available characteristics of QA systems.

A general procedure of an ontology construction consists of the following phases: (1) defining a set of criteria, (2) taxonomy construction, (3) ontology construction, (4) formal description, (5) defined classes creation, (6) reasoning process, (7) consistency verification, (8) a set of results. A domain of modeling encompasses the set of 13 QA systems. The ontology was built using the Protégé application. The applied technology standard is OWL (Ontology Web Language). The general aim of proposed systemic procedure is to ensure the knowledge systematization and provide a specified guideline supporting QA system selection.

5   Case Studies: Ontology Supporting QA System Selection

The case study presents practical examples of ontology for QA systems. It is divided into two parts: at first, a practical usage of ontology for QA systems is provided. Next, an exemplary case study of knowledge systematization to QA systems domain is presented. Due to the limited space of the paper only small part of ontology for QA systems is offered in this case study.

5.1   Case Study: Practical Application of Ontology for QA Systems

It is supposed that a decision-maker is looking for the QA system that fulfills a set of pre-defined requirements. The preferable QA system should satisfy at least one of the following criteria: (1) query entry: NL query, and (2) output: SPARQL query, and (3) additional options: triple-based relation. The application of the reasoning mechanism provides a set of QA systems with regard to a user preferences: QuestIO, FREyA, and PANTO. The figure 2 illustrates the obtained results. A class called Case Study 1 includes the definitions. It is also known as a defined class. It means that each entity that belongs to this class must fulfill necessary and sufficient conditions. Defined classes are signified in orange color.

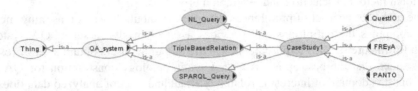

Fig. 2. Practical application of ontology for QA systems - results of 1st case study

It is worth to notice that any change in the set of criteria provides a modified set of results at the end. For example, if a user adds extra criteria as follows: (4) pattern matching: entity lexicon, or (5) additional options: query processing as a bag of words, or (6) additional options: domain lexicons) to the previous set of criteria, the final set of results presents as follows (fig. 3).

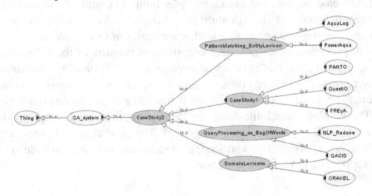

Fig. 3. Practical application of ontology for QA systems - results of 2nd case study

In Case Study 2 the set of results is bigger than in Case Study 1. It depends directly of a constructed definition. The 2nd case study was extended to 3 extra criteria, none-theless, QA system may, but not must, fulfill these criteria. It is possible to specify a non-limited set of queries. Moreover, the user does not have to have a broad knowledge of QA systems, but can still make a reasonable choice. It is worth to emphasize that smaller, domain ontology for QA systems allows to compute the results in short period of time.

5.2   Case Study: Knowledge Systematization to QA Systems Domain

The general aim of the proposed ontology is to ensure knowledge systematization in QA systems domain. A significant advantage of the ontology for QA systems is a possibility to place a description of each of analyzed QA system inside the ontology.

For each of provided QA systems ontology includes a description of a given QA system and information where a user may find more additional details. Figure 4 illustrates an example of knowledge systematization to QA systems domain, especially for AquaLog QA system.

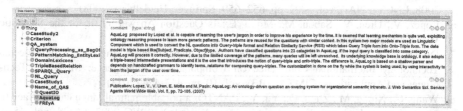

Fig. 4. An example of knowledge systematization to QA systems domain

6   Conclusion

This paper presented the comparative analysis of selected QA systems. Moreover, the problem of selection of QA system from existing solutions was undertaken. In this paper, a proposal of knowledge-based approach to Question Answering System selection was described. The main aim was to ensure knowledge systematization and to help users to find a proper QA system for a given decision problem. The adaptation of knowledge-engineering mechanism should improve the process of knowledge acquisition of QA systems.

In this paper the ontology for QA systems was presented. The analysis of literature allows to identify 13 QA systems. On base of the comparative analysis, OWL standard was used to create the ontology. Due to the limited space of publication, only the small portion of the practical application of ontology for QA systems was presented. It is worth to notice that it is possible to specify a non-limited set of queries for the ontology. Moreover, the user does not have to have a broad knowledge of QA systems, but he can still make a reasonable choice. Furthermore, the proposed ontology ensures knowledge systematization and provides a description of each of analyzed QA systems. It reduces a time necessary to gather information of a given QA system.

Future researches encompass elaboration of own complex procedure for construction QA system or enhance one of existing approaches.

References

1. Barskar, R., Gulfishan, F.A., Barskar, N.: An approach for extracting exact answers to question answering (QA) system for english sentences. In: International Conference on Communication Technology and System Design, vol. 30, pp. 1187-1194. Procedia Engineering (2012)
2. Nalawade, S., Kumar, S., Tiwari, D.: Question answering system. International Journal of Science and Research (IJSR) 3(5) (2014)
3. Gupta, P., Gupta, V.: A survey of text question answering techniques. International Journal of Computer Applications 53(4) (2012)

4. Hogan, A., Harth, A., Umbrich, J., Kinsella, S., Polleres, A., et al.: Searching and browsing linked data with SWSE: the semantic web search engine. J. Web Semantics **9**, 365–401 (2011)

5. Guo, Q., Zhang, M.: Question answering system based on ontology and semantic web. In: Wang, G., Li, T., Grzymala-Busse, J.W., Miao, D., Skowron, A., Yao, Y. (eds.) RSKT 2008. LNCS (LNAI), vol. 5009, pp. 652–659. Springer, Heidelberg (2008)

6. Lopez, V., Uren, V., Sabou, M., Motta, E.: Is Question Answering fit for the Semantic Web?: a Survey, Semantic Web, pp. 125-155 (2011)

7. Dwivedi, S.K, Singh, V.: Research and reviews in question answering system. In: Procedia Technology International Conference on Computational Intelligence: Modeling Techniques and Applications (CIMTA 2013), vol. 10, pp. 417-424 (2013)

8. Wang, C., Xiong, M., Zhou, Q., Yu, Y.: PANTO: a portable natural language interface to ontologies. In: Franconi, E., Kifer, M., May, W. (eds.) ESWC 2007. LNCS, vol. 4519, pp. 473–487. Springer, Heidelberg (2007)

9. Fernandez, O., Izquierdo, R., Ferrandez, S., Vicedo, J.L.: Addressing ontology-based question answering with collections of user queries. Information Processing & Management **45**, 175–188 (2009)

10. Kaufmann, E., Bernstein, A., Zumstein, R.: Querix: a natural language interface to query ontologies based on clarification dialogs. In: Proceedings of the 5th International Semantic Web Conference, (ISWC 2006), pp: 980-981 (2006)

11. Damljanovic, D., Agatonovic, M., Cunningham, H.: Natural language interfaces to ontologies: Combining syntactic analysis and ontology-based lookup through the user interaction. Semantic Web: Res. Appli. **6088**, 106–120 (2010)

12. Cimiano, P., Haase, P., Heizmann, J., Mantel, M., Studer, R.: Towards portable natural language interfaces to knowledge bases-the case of the ORAKEL system. Data Know. Eng. **65**, 325–354 (2007)

13. Battista, A.D.L., Villanueva-Rosales, N., Palenychka, M., Dumontier, M.: SMART: a web-based, ontology-driven, semantic web query answering application (2007)

14. Lopez, V., Uren, V., Motta, E., Pasin, M.: AquaLog: an ontology-driven question answering system for organizational semantic intranets. J. Web Semantics Sci. Service Agents World Wide Web **5**, 72–105 (2007)

15. Lopez, V., Fernandez, M., Motta, E., Stieler, N.: PowerAqua: supporting users in querying and exploring the semantic web, antic web. J. of Semantic Web **3**(3), 249–265 (2011)

16. Tablan, V., Damljanovic, D., Bontcheva, K.: A natural language query interface to structured information. In: Bechhofer, S., Hauswirth, M., Hoffmann, J., Koubarakis, M. (eds.) ESWC 2008. LNCS, vol. 5021, pp. 361–375. Springer, Heidelberg (2008)

17. Bernstein, A., Kauffmann, E., Kaiser, C., Kiefer, C.: Ginseng: a guided input natural language search engine. In: Proceedings of the 15th Workshop on Information Technologies and Systems, Bristol, UK (2006)

18. Kaufmann, E., Bernstein, A., Fischer, L.: NLP-Reduce: a"naive" but domain-independent natural language interface for querying ontologies. In: Proceedings of the 4th European Semantic Web Conference (ESWC 2007) Innsbruck, Austria (2007)

19. Tartir, S., Arpinar, I.B., McKnight, B.: SemanticQA: exploiting semantic associations for cross-document question answering. In: 2011 4th International Symposium on Innovation in Information & Communication Technology (ISIICT), pp. 1-6. IEEE (2011)

20. Konys, A.: Knowledge-based approach to COTS software selection processes. In: Wiliński, A., El Fray, I., Pejaś, J. (eds.) Soft Computing in Computer and Information Science, vol. 342, pp. 191–205. Springer, Heidelberg (2015)

Text Relevance Analysis Method over Large-Scale High-Dimensional Text Data Processing

Ling Wang[1], Wei Ding[1], Tie Hua Zhou[2(✉)], and Keun Ho Ryu[2]

[1] Department of Computer Science, School of Electrical & Computer Engineering,
Northeast Dianli University, Jilin, China
smile28671ling@gmail.com, wxj3429@126.com
[2] Database/Bioinformatics Laboratory, School of Electrical & Computer Engineering,
Chungbuk National University, Chungbuk, Korea
{thzhou,khryu}@dblab.chungbuk.ac.kr

Abstract. As the amount of digital information is exploding in social, industry and scientific areas, MapReduce is a distributed computation framework, which has become widely adopted for analytics on large-scale data. Also, the idea which is used to solve the large-scale data problem by the use of approximation algorithms has become a very important solution in recent years. Especially for solving high-dimensional text data processing, semantic Web and search engine are required to pay attention to proximity searches and text relevance analysis. The difficulties of large-scale text processing mainly include its quick comparison and relevance judgment. In this paper, we propose an approximate bit string for approximation search method on MapReduce platform. Experiments exhibits excellent performance on efficiency effectiveness and scalability of the proposed algorithms.

Keywords: High-dimensional text · Approximate nearest neighbors search · Approximation algorithm · Hamming distance

1 Introduction

Approximate Nearest Neighbors (ANN) search is a crucial part to many applications in several areas of science, industry, society and many more, which play an important role for text data preprocessing, exploration and hypotheses generation. The problem of text ANN can be formally stated as follows.

DEFINITION 1 (APPROXIMATE NEAREST NEIGHBORS SEARCH). Let D be the collection of all text records in d-dimensional space R^d, we are given a query text record q, a metric distance function d, and a distance threshold d'. The problem is to find all of text records group $G=\{g_1, g_2, ..., g_m\}$ that satisfy $d(q, g_i) \leq d'$.

The task of text ANN search is the grouping of objects by their similarity to each other, such that similar texts are located in the same group (cluster) and dissimilar texts are located in different clusters. Most of the traditional approaches consider all attributes of the texts for determining an appropriate grouping of objects. However, as recent developments show, clustering becomes less and less meaningful with a growing number of dimensions due to the so called "curse of dimensionality" [1]. The curse of di-

M. Núñez et al. (Eds.): ICCCI 2015, Part I, LNAI 9329, pp. 371–379, 2015.
DOI: 10.1007/978-3-319-24069-5_35

mensionality states that with a growing number of dimensions the distances between data objects become more and more alike, such that no meaningful grouping is possible anymore. Additionally, in datasets with a large number of dimensions the problem of irrelevant or noise dimensions occur.

MapReduce [2] is a distributed computation framework, which has become widely embraced by commercial organizations (such as Google, Facebook, Yahoo, etc.) to analyze tremendous amounts of data. For example, dozens to hundreds of TB datasets are processed daily by MapReduce at Google [3], and 75TB of compressed data is processed every day by Hadoop at Facebook [4]. MapReduce has become a promising and popular choice used for large-scale data processing due to its two main reasons. First, the framework can scale to thousands of commodity machines in a fault-tolerant manner and thus is able to use more machines to support parallel computing. Second, the framework has a simple and yet expressive programming model through which users can parallelize their programs without concerning about issues like fault-tolerance and execution strategy. Therefore, how to execute ANN search efficiently on large-scale datasets that are stored in a MapReduce cluster is an intriguing problem that meets many practical needs.

In this paper, we propose Approximate Bit String (ABS) algorithm for text proximity searches and relevance analysis on MapReduce. In the first phase, we rewrite new features of all text records, which also used for load balancing of computations in the MapReduce framework. In the second phase, we partition data based on the regions divided by our proposed, and compute candidate text for each region independently using MapReduce to perform efficient parallel ANN search on large-scale text data. In this work, we focus on Hamming metric similarity/distance function, which can be efficiently computed by using bit operations. Our contributions are summarized as follows: (1) We propose an approximation approach to implementing the data shuffling stage in MapReduce; (2) We evaluate proposed algorithm on real dataset to demonstrate it is easily and nicely approach for ANN search.

The rest of this paper is organized as follows. Section 2 addresses related work. Section 3 describes how ABS is used for ANN search on MapReduce. Section 4 reports the results of experiments and Section 5 concludes this paper.

2   Related Work

Retrieving relevant content from large-scale databases containing high-dimensional data is becoming common in many applications, such as text documents, images, videos, etc. In the last decades a large number of clustering approaches were developed. Expectation Maximization (EM) algorithm, and k-means algorithms are only a very small subset of existing approaches that are used in different applications. Recently, mapping the original high-dimensional data to similarity-preserving binary codes provides an attractive solution for search problems [5, 6, 7, 8]. Several powerful techniques have been proposed recently to learn binary codes for large-scale nearest neighbor search and retrieval, which can be applied to any type of vector data. New techniques using the approximation vector appeared to provide a solution to the dimensionality curse. VA-file (Vector Approximation File) [9], [10] is the first meth-od based on this approach. When searching vectors, the entire approximations file is scanned to select candidate vectors. Using the Hamming distance as a similarity met-ric has been studied in the theory community. The Hamming-distance computation is still performed in a linear fashion over feature vectors.

MapReduce is an established, error-tolerant framework for parallel computing and a programming paradigm developed by Google Corp. in 2004. Its open-source im-plementation Hadoop [11] is a powerful tool spreads in the industry and is successful-ly used at many internet based world player companies due to its high scalability and availability, including Facebook, Twitter and Yahoo. There has been much work on MapReduce and related distributed programming frameworks over the past several years. The active application fields of the MapReduce framework include machine learning algorithms [12, 13]. While similarity search in MapReduce has been studied [14, 15], MapReduce as a reliable distributed computing model has been adopted for handling a variety of similarity queries, e.g., [16, 17, 18]. Recently, there has been considerable interest on supporting similarity join queries over MapReduce frame-work. [19] studies how to perform set-similarity join in parallel using MapReduce. [20] proposes a general framework for processing join queries with arbitrary join conditions using MapReduce. In [21], they study how to extract k closest object pairs from two input datasets in MapReduce. Some similarities with k-NN processing can be found in work by Rares et al. [22] which describes a couple of approaches for computing set similarities on textual documents, but it does not address the issue of k-NN query processing. The pairwise similarity is calculated only for documents with the same prefixes (prefix filtering), which can be considered as the LSH min-hashing technique. Lin [23] describes a MapReduce based implementation of pairwise similar-ity comparisons of text documents based on an inverted index. However, reducing the searched dataset might not be sufficient. For data in large dimension space, compu-ting the distance might be very costly.

3   Approximate Bit String for ANN Search

3.1   MapReduce Preliminaries

In the MapReduce programming model [2], master architecture manages and moni-tors map and reduce tasks, and uses a Distributed File System (DFS) to manage the I/O files. Data is expressed through <key, value> pairs, and computations are repre-sented by map and reduce tasks.

- ✓ Map tasks: Reading input records, and execute the user defined map func-tion, which takes a <key, value> pair as input and produces zero or a list of <key, value> pairs, i.e., $<k_i, v_i> \rightarrow$ list$<k_j, v_j>$.
- ✓ Reduce tasks: Applying reduce functions to the values associated with unique key and produce a single result for that key, <key, list-of-values>. When a reduce task starts, the numbers of concurrent threads are used to copy map outputs, i.e., $<k_j, $ list$(v_j)> \rightarrow$ list $<k_l, v_l>$.

In detail, its execution consists of a map phase and a reduce phase. In the map phase, a set of mappers execute the map function that tags each tuple based on the join key. The map function is responsible for reading the data in form of <key, value> pairs from a storage system and their preprocessing. The input pairs are in general arbitrarily distributed on different computing nodes such that each record is processed separately, independent from the other data records. The outputs of the map phase are intermediate <key, value> pairs that are sorted according to their keys and pairs with equal keys are sent to reducer. Map output tuples are partitioned on the join key and shuffled across the network to reducers. Note that both the sorting and partitioning functions are customizable. In the reduce phase, each reduce task first gets its corresponding map output partitions from the map tasks and merges them. Then for each key, the reducer applies the reduce function on the values associated with that key and outputs a set of final key-value pairs. Several MapReduce jobs can be chained together, later phases being able to refine and/or use the results from earlier phases.

3.2   The Proposed Framework

Assume approximate bit string for rewriting features of text data, relevant texts can be matched each other in the Hamming space. If each ABS of text is extracted in same order, then the probability of these text data being identical will be high. The proposed ABS algorithm for ANN search can be described as follows (As illustrated in Fig. 1). 1) Using ABS to rewrite all text records into vectors space; 2) Partitioning input data based on Hamming distance and data size. 3) Intermediate results need to be shuffled and sorted from the mappers where they are created to the reducers where they are consumed. 4) Associating with each intermediate result for searching top k nearest candidates and expand them with their ANN.

ABS algorithm rewrite all text records into Hamming space and represent the approximate bit string $\{0, 1\}^d$ for each text data. In this paper, we focus on Hamming metric distance function, which is widely used in ANN search for high-dimensional data. For example, high-dimensional data is mapped into l-dimensional binary codes that are, then linearly scanned to perform Hamming distance comparisons. The matching works can be considered as a generalized k-nearest neighbors problem for which we need to ensure a properly distance metric that can be used to compare vectors based on their features. It is defined as follows:

DEFINITION 2 (HAMMING METRIC DISTANCE). We adopt the Hamming distance $d(\bullet, \bullet)$ as the distance measure between two approximate bit string: $d = \|ABS_i, ABS_j\|$, which is number of positions at which the corresponding feature bits are different. Also, the metric distance satisfying the following properties, $\forall x, y, z \in D$:

- ✧   Non-negativity: $d(x, y) \geqslant 0$
- ✧   Coincidence Axiom: $d(x, y) = 0$, iff $x = y$
- ✧   Symmetry: $d(x; y) = d(y, x)$
- ✧   Triangle Inequality: $d(x, z) \leqslant d(x, y) + d(y, z)$

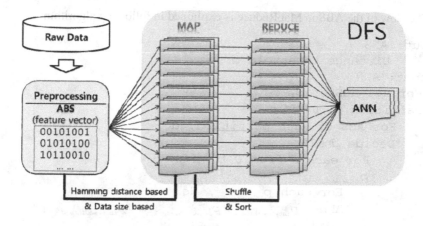

Fig. 1. The proposed ANN search on MapReduce

3.3   Approximate Bit String Algorithm

ABS based on the sparse and dense features of text vector in the high-dimensional space. And each text feature is independently rearranged followed by feature weight accordingly. The rewriting features of text $T=(f^*_1, f^*_2, ..., f^*_m)$ in D should satisfy the following conditions:

$$T = \left\{ f' \mid f'_i = \frac{f_i + f_{n-i+1}}{2}, i \leq \frac{n+1}{2} \right\} \tag{1}$$

where f is original feature of text vector. f' is an intermediate variable from f to f^*. Particularly, if the size of database $n\%2==0$, the $m= n/2$. And then

$$ABS_{f'}[i] = \begin{cases} 1, & if \quad f' \geq f_{mean} \\ 0, & otherwise \end{cases} \tag{2}$$

where f_{mean} is computed on weighted sum of text features. We summarize the ABS algorithm in the following.

```
Input: Raw text records D={T₁, T₂, …, Tₙ}
Output: Approximate Bit String
    Initialize D with all features set to 0
    For each feature f∈Tᵢ, do
        Decide f_weight ≥ f_mean
        f•1
    End for
Return ABS
```

The flow of the ABS in MapReduce is explained in following algorithm.

```
Input: ABS
       Distance threshold d'
Output: ANN
   For all ABS of D, do
   Partition p_j, 0≤j≤n
      For ABS of each partition, do
      Decide d≤d'
         For each d from 0 to d', do
         p_intermediate-d ← T_i
            For each p_intermediate, do
            ANN← p_intermediate-0 ∪ p_intermediate-1 ... ∪ p_intermediate-d'
            End for
         End for
      End for
   End for
Return ANN
```

4   Experiments

We empirically evaluated the performance of our proposed ABS algorithm on large-scale real world datasets Reuters-21578 [24] using MapReduce. In the first set of experiments, we report the retained similarity compare with Jaccard index, which is used for matching two feature vectors due to it is a simple measure that incurs low computation cost. The Jaccard index defined as follows.

$$Jacc(v_i, v_j) = \frac{|v_i \cap v_j|}{|v_i \cup v_j|} \tag{3}$$

where v_i and v_j are the feature vectors. As shown in Fig. 2, the ABS is an approximation algorithm for nearest neighbors search, but the result clearly shows the ABS keep the high quality similarity like in original text data space.

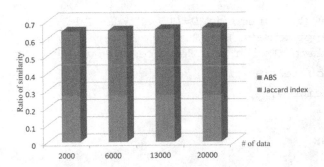

Fig. 2. The retained similarity of ABS

In the second set of experiments, the recall curves are studied in Fig. 3. Ideally, an ANN search system is measured by recall, which should be able to achieve high-quality search with high speed, while using a small amount of space. In the ideal case, the recall score is 1.0, which means all the nearest neighbors are returned. Since the entire candidate texts will be ranked based on their Hamming distances to the query text and only the top k candidates will be returned. From the reported, experiments demonstrate the superiority of the proposed ABS for ANN search.

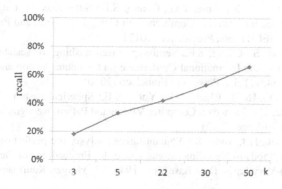

Fig. 3. Recall of ABS for ANN search

5   Conclusion

Due to the wide adoption of the MapReduce framework, we study the ANN search problem on MapReduce. In this paper, we have presented a novel ANN search approach based on ABS in large scale text records. Our approach is a classic approximation method based on the Hamming metric distance. We rewrite all text records into $\{0, 1\}^d$ space for fast ANN search on MapReduce. The Experimental results on a real large-scale text dataset show that the proposed ABS algorithm remained high-similarity and achieved good performances on ANN search.

For future work, we plan to compare with other popular ANN search methods and further explore more effective model for our scheme.

Acknowledgments. This work was supported by the National Natural Science Foundation of China (No.51077010) and by the Jilin provincial department of science and technology (No.20120338).

References

1. Indyk, P., Motwani, R.: Approximate nearest neighbors: towards removing the curse of dimensionality. In: Proceedings of the Thirtieth annual ACM Symposium on Theory of Computing, pp. 604–613. ACM Press, Dallas (1998)
2. Dean, J., Ghemawat, S.: MapReduce: Simplified Data Processing on Large Clusters. M. Communications, 107–113 (2008). ACM Press

3. Dean, J., Ghemawat, S.: MapReduce: A Flexible Data Processing Tool. M. Communications, 72–77 (2010). ACM Press
4. Thusoo, A., Sarma, J.S., Jain, N., Shao, Z., Chakka, P., Zhang, N., Anthony, S., Liu, H., Murthy, R.: Hive – a petabyte scale data warehouse using hadoop. In: Proceedings of the 26th International Conference on Data Engineering, pp. 996–1005. IEEE Press, Long Beach (2010)
5. Salakhutdinov, R., Hinton, G.: Semantic Hashing. J. Approximate Reasoning, 969–978 (2009). Elsevier Press
6. Liu, W., Wang, J., Ji, R., Jiang, Y.G., Chang, S.F.: Supervised hashing with kernels. In: Proceedings of the International Conference on Computer Vision and Pattern Recognition, pp. 2074–2081. IEEE Press, Providence (2012)
7. Wang, J., Kumar, S., Chang, S.F.: Semi-supervised hashing for scalable image retrieval. In: Proceedings of the International Conference on Computer Vision and Pattern Recognition, pp. 3424–3431. IEEE Press, San Francisco (2010)
8. Heo, J.P., Lee, Y., He, J., Chang, S.F., Yoon, S.E.: Spherical hashing. In: Proceedings of the International Conference on Computer Vision and Pattern Recognition, pp. 2957–2964. IEEE Press, Providence (2012)
9. Weber, R., Schek, H.J., Blott, S.: A quantitative analysis and performance study for similarity-search methods in high-dimensional space. In: Proceedings of the 24th International Conference on Very Large Data Bases, pp. 194–205. Morgan Kaufmann Press, New York (1998)
10. Heisterkamp, D.R., Peng, J.: Kernel vector approximation files for relevance feedback retrieval in large image databases. J. Multimedia Tools and Applications, pp. 175–189. Kluwer Academic Press (2005)
11. The Apache Software Foundation. Hadoop. http://hadoop.apache.org/
12. Shim, K.: MapReduce algorithms for big data analysis. J. PVLDB, pp. 2016–2017 (2012). VLDB Endowment Press
13. The Apache Software Foundation. Mahout. http://mahout.apache.org/
14. Deng, D., Li, G., Hao, S., Wang, J., Feng, J.: Massjoin: a mapreduce-based algorithm for string similarity joins. In: Proceedings of International Conference on Data Engineering, pp. 340–351. IEEE Press, Chicago (2013)
15. Wang, Y., Metwally, A., Parthasarathy, S.: Scalable all-pairs similarity search in metric spaces. In: Proceedings of the 19th ACM SIGKDD International Conference on Knowledge Discovery and Data Mining, pp. 829–837. ACM Press, Chicago (2013)
16. Lu, W., Shen, Y., Chen, S., Ooi, B.C.: Efficient Processing of k Nearest Neighbor Joins Using MapReduce. J. PVLDB, pp. 1016–1027 (2012). VLDB Endowment
17. Zhang, C., Li, F., Jestes, J.: Efficient parallel kNN joins for large data in MapReduce. In: Proceedings of the 15th International Conference on Extending Database Technology, pp. 38–49. ACM Press, Berlin (2012)
18. Kllapi, H., Harb, B., Yu, C.: Near neighbor join. In: Proceedings of International Conference on Data Engineering, pp. 1120–1131. IEEE Press, Chicago (2014)
19. Metwally, A., Faloutsos, C.: V-SMART-Join: A Scalable MapReduce Framework for All-Pair Similarity Joins of Multisets and Vectors. J. PVLDB, pp. 704–715 (2012). VLDB Endowment
20. Okcan, A., and Riedewald, M.: Processing Theta-Joins Using MapReduce. In: Proceedings of the 2011 ACM SIGMOD International Conference on Management of data, pp. 949–960. ACM Press, Athens (2011)

21. Kim, Y., Shim, K.: Parallel top-k similarity join algorithms using MapReduce. In: Proceedings of International Conference on Data Engineering, pp. 510–521. IEEE Press, Washington (2012)

22. Vernica, R., Carey, M.J., Li, C.: Efficient parallel set-similarity joins using MapReduce. In: Proceedings of the 2010 ACM SIGMOD International Conference on Management of data, pp. 495–506. ACM Press, Indianapolis (2010)

23. Lin, J.: Brute force and indexed approaches to pairwise document similarity comparisons with MapReduce. In: Proceedings of the 32nd International ACM SIGIR Conference on Research and Development in Information Retrieval, pp. 155–162. ACM Press, Boston (2009)

24. Lewis, D.D.: Reuters-21578 Text Categorization Test Collection. http://www.daviddlewis.com/resources/testcollections/reuters21578/

Linguistic Summaries of Graph Datasets Using Ontologies: An Application to Semantic Web

Lukasz Strobin[✉] and Adam Niewiadomski

Insitute of Information Technology, Lodz University of Technology,
ul. Wólczaúska 215, 90-924 Lodz, Poland
800337@edu.p.lodz.pl, adam.niewiadomski@p.lodz.pl

Abstract. This paper presents a new approach to performing linguistic summaries of graph datasets with the use of ontologies. Linguistic summarization is a well known data mining technique, aimed to discover patterns in data and present them in natural language. So far, this method has been applied only to relational databases. However amount of available graph datasets with associated ontologies is growing fast, hence we have investigated the problem of applying linguistic summaries in this scenario. As our first contribution, we propose to use an ontological class as subject of a summary, showing that its class taxonomy has to be used to properly select objects for summarization. Our second contribution is an extension to a summarizer, by analysis of set of ontological superclasses. We then propose extensions to quality measures T_1 and T_2, measuring informativeness of a summary in the context of ontological class taxonomy. We also show that our approach can create more general summarizations (higher in class taxonomy). We verify our proposals by performing linguistic summarization on Semantic Web, which is a vast distributed graph dataset with several associated ontologies. We conclude the paper with showing the possibilities of future work.

1 Introduction

This paper focuses on performing linguistic summaries on graph datasets with associated ontologies, which has not been attempted before. We propose to rebuild and extend the notion of a subject and a summarizer, by including class taxonomies into problem analysis. For summary subjects we show that all subclasses of a given class have to be analyzed, while for the summarizer - all super classes, which leads to obtaining new information (more general summaries). Secondly, also based on class taxonomies, we propose extensions to quality measures T_1 and T_2.

In this paper we firstly introduce main concepts of linguistic summaries, formerly defined by Yager [1],[2], [3],[4], which are intended for relational databases only. These algorithms provide means of discovering general knowledge and complex patterns in data and presenting it in human-readable quasi-natural sentences. This form of data mining is especially suitable for very large datasets, like Semantic Web. The central notion of our approach is the usage of ontologies, with particular emphasis on class taxonomy. We show how analysis of sub- and superclasses

© Springer International Publishing Switzerland 2015
M. Núñez et al. (Eds.): ICCCI 2015, Part I, LNAI 9329, pp. 380–389, 2015.
DOI: 10.1007/978-3-319-24069-5_36

of a given class can lead to creation of new lingusitic summaries. First of all, in our approach we use an ontological class as a subject of a summary, e.g. 'artist'. Note however that 'is-a' relationship that indicates class membership is a transitive predicate, hence, in a general case, proper selection of summary subjects require analysis of all subclasses of a given subject class (since a 'writer' is also an 'artist'). On the other hand, for creation of summarizers, taking all superclasses of summarizer class (given that this class is a member of an ontology) can lead to creating new (more general) summaries.

This paper is organized as follows. In Section 2 we only remind the reader the main concepts of linguistic summaries. In section 3 we discuss how algorithms for linguistic summaries may be adopted for Semantic Web with the use of ontologies. The exact algorithm of generating summaries for Semantic Web is presented in section 3.5. Proposed algorithm adapts to the dataset, in which the set of summarizers is created dynamically. Section 3.4 shows how T_1 and T_2 quality measures may be extended with class taxonomy. We introduce the notion of summary on different level of generality (Degree of Summarizer Imprecision), depending on the class used in summary. Afterwards, in Section 4 we show the results of an experiment. In the end, in Section 5 we draw the conclusions and show the possibilities of future work.

2   Linguistic Summaries of Relational Databases

This chapter is only meant to remind the reader the main concepts of linguistic summaries, and for in-depth understanding the reader is asked to refer to [1], [2], [3], [4].

Consider the database D. The first form of a linguistic summary is presented by (1):

$$Q \quad P \quad are/have \quad S_j[T] \tag{1}$$

where Q is the *linguistic quantifier*; P is the *subject of the summary* (set of objects represented by the database tuples d_i); S_j is a *property of interest*, the so-called *summarizer* represented by a fuzzy or a crisp set (discrete set in particular).

The crucial part of the algorithm in the sense of Yager is the computation of the degree of truth T. The algorithm is strictly based on Zadeh calculus of linguistically quantified statements, and is computed as:

$$T_1 \; (Q \; P \; are/have \; S_j) \; = \mu_Q(\frac{r}{m}) \tag{2}$$

where

$$r = \sum_{i=1}^{m} \mu_{S_j}(d_i) \tag{3}$$

In typical applications the symbol μ_{S_j} is a membership function of d_i to fuzzy set S_j. However, S_j may also be a discrete set, e.g. *BORN IN Poland*, hence $S_j = \{Poland\}$. In this case membership value is given by (4) (this trivial formula is quoted in this paper, because it is a starting point to author's original contribution, see Def. 5).

$$\mu_{S_j}(d_i) = \begin{cases} 1 & \text{if } d_i \in S_j \\ 0 & \text{otherwise} \end{cases} \tag{4}$$

The *degree of truth* as calculated by (2) describes only one of the many aspects of a summary. Quality measure T_2 called *degree of imprecision* describes how imprecise the summarizer is, see (4). The meaning of this quality measure is as follows: the more general the statement is, the higher the value of the imprecision.

$$T2 = 1 - (\prod_{j=1}^{n} in(S_j))^{1/n} \tag{5}$$

where in is the inprecision of a fuzzy set [1].

In this paper we propose an interpretation of the notion of degree of imprecision for ontological classes, see (7) on page 386.

3   Linguistic Summaries of Graph Databases Using Ontologies

The first part of this section introduces a set of concepts and definitions from the field of ontologies. In the remainder of this section author's original contributions are presented - subject selection (using ontological subclasses), extensions of summarizer (using ontological superclasses) and extensions of quality measures T_1 and T_2 (using a complete class taxonomy).

3.1   Definitions Related to Ontological Classes Taxonomy

An ontology is an explicit specification of a conceptualization [7], which defines classes (types), properties of these classes, and taxonomies. In this paper we will denote an ontological class as c.

Consider the fragment of DBPedia ontology shown in figure 1. Say we want to summarize class 'writer'. Belonging to a class is defined by predicate 'rdf:type', and classes in an ontology are linked using predicate 'rdfs:subClassOf' which is transitive (see www.w3.org/TR/rdf-schema for rdf and rdfs reference). As a result, 'rdf:type' is a transitive property with respect to ontological class taxonomy.

Definition 1. *Subclasses of class c are classes, which are directly below class c in a given taxonomy. We denote this set of subclasses by Sub_c.*

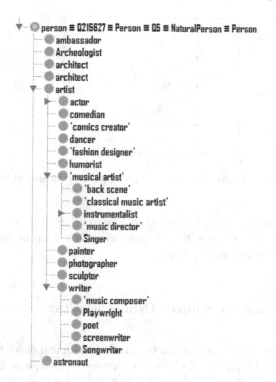

Fig. 1. Fragment of DBPedia ontology, class 'Person'

Example 1. For fig. 1, class *writer* has the following subclasses:
$Sub_{artist} = \{actor, comedian, comics \ creator, dancer, fashion \ designer,$
$humorist, musical \ artist, painter, photographer, sculptor, writer\}$.
Note that $Sub_{writer} \not\subseteq Sub_{artist}$, because we consider only direct subclasses.

Definition 2. *Subclasses of of n-th level of class c are classes, which are separated from class c by not more then n specialization relations. We denote this set by Sub_c^n.*
Note that $Sub_c = Sub_c^1$, $Sub_c^{n-1} \subseteq Sub_c^n$

Definition 3. *The complete set of subclasses of class c (all levels below class c) is denoted by Sub_c^∞.*

Example 2. Let's consider a class *artist* shown in fig. 1. For this class the following sets may be defined:
$Sub_{artist} = Sub_{artist}^1 = \{actor, comedian,' comicscreator', dancer,' fashion$
$designer', humorist,' musicalartist', painter, photographer, sculptor, writer\}$
$Sub_{artist}^2 = Sub_{artist}^1 \cup Sub_{actor}^1 \cup Sub_{musical \ artist}^1 \cup Sub_{writer}^1$

Analogously to the notion of subclass we define a superclass, and set of superclasses of n-th level - see definitions 1, 2, 3.

– artist
  • writer
  • poet
  • painter
  • photographer
– engineer
  • programmer
  • mechanical engineer
  • electronic engineer

Fig. 2. 'Occupation' value class taxonomy for example

Definition 4. *Superclasses of of n-th level of class c are classes, which are separated from class c by not more then n generalization relation from class c. We denote this set of classes by Sup_c^n.*

3.2   Building Summary Subject Using Class Taxonomy - Including Subclasses

As a subject of a summary, we propose to use an ontological class, with additional consideration of the hierarchy of classes. In a general case, a graph vertex does not specify all classes that is belongs to, but only the most specific one, that is - lowest in class taxonomy. Hence, in order to properly select objects for summarization, also all subclasses of given class have to be selected (see def. 1). Hence, when a linguistic summary of class $'c'$ is created, vertices not only of class $'c'$, but also all vertices of classes Sub_c^∞ have to be selected (see Def. 3), because each member of any of the classes Sub_c^∞ also belongs to class $'c'$.

Example 3. Say we want to create a summary of class 'writer'. We select all vertices of class *writer* but also - classes 'music composer', 'Playwrit', 'poet', 'screenwriter' and 'Songwriter'.

3.3   Building Summarizer Using Class Taxonomy - Including Superclasses

In the proposed method, set of summarizers S is created during selecting objects for summarization, so each triple that has predicate $rdf : type \in Sub_c^\infty$ (see Def. 2). Now for each attribute A_i (that is predicate label) its discrete value set is created based on the values of all retrieved triples. We denote this set of values as X_{A_i}. The attribute A_i may be a graph vertex that has its class ($'a_j rdf : type' = c_{A_i}$) that belongs to an ontology - may be the same or different then the subject ontology. In this case, due to transitivity of 'rdf:type' predicate, attribute value a_j also belongs to super classes of class c_{A_i}, hence a set of classes $Sup_{c_{A_i}}^n$ (see def. 4). In this case the set of summarizers is augmented by this set of super classes.

Example 4. Consider a simple ontology 'Occupations' that has the taxonomy shown in figure 2. Say we are creating linguistic summaries of class 'Person', and one of the atributes/predicates is $A_1 ='$ *occupation'* (A belongs to 'Occupations' ontology). Assume that in the considered dataset the set of values is $X_{Occupation} = \{writer, poet, painter\}$. In this case, in a regular approach, the summarizer $S_{Occupation}$ based on this attribute would have only three possible values $S_{Occupation} = \{writer, poet, painter\}$. However, by taking the set of super-classes of each class in $X_{Occupation}$ and adding to the summarizer set we obtain the following summarizer set: $S_{Occupation} = \{writer, poet, painter, artist\}$. Summarizing on more general attribute value *artist* may lead to extracting new knowledge from the data.

3.4 Ontological Extensions to Quality Measures

T_1 Extended by Class Taxonomy
Recall the summary truth value T_1, given in (1), and the notion of membership function for a discrete summarizer (4). Now consider an ontological class as a summarizer or a qualifier. In this case, the notion of 'being member of a class' may be extended. Since a class $poet \in Sub^{\infty}_{writer}$ (see fig. 1), hence (4) may be extended to have the form as in definition 5.

Definition 5.

$$\mu_c^{ont.}(d_i) = \begin{cases} 1 & if\ d_i \in \{c, S_c^{\infty}\} \\ 0 & otherwise \end{cases} \tag{6}$$

Example 5. Let's consider a summarizer (or a qualifier) 'is writer' and an object, which belongs to a class *poet*. Using a regular approach, so using equation 4 the obtained membership value is 0, while by using an extended approach, equation 6 evaluates to 1.

T_2 Extended by Class Taxonomy
Using (6) for evaluating the membership value of an attribute to a summarizer leads to generating more general summaries, when using classes higher in an ontology. For example, let's imagine a summarizer related to geographical locations, like cities, countries and continents. Assume also that each instance of a class (e.g. *Person*) has a property 'city of birth'. When formula 6 is used, we may create new summaries, extending by the notion of a city to a broader term, like a country. Then we may form a new summary, otherwise not possible (since each attribute specifies only a city), like 'average number of people that are tall are born in Europe'. However, such summaries are less precise - extreme case would be 'most people are born on Earth'. This summary is definitely true, however it is very imprecise.

Hence, we propose an analogous measure to a degree of imprecision for fuzzy sets - we call this notion degree of ontological class imprecision, see definition 6. Proposed formula describes the intuition that the imprecision of the class depends linearly on the number of classes that are below a given class in a taxonomy.

Definition 6. *By degree of ontological class imprecision we call the level of generality of a given class, and evaluated by using (7) in (5).*

$$in^{ont.}(S_j) = \frac{|Sub_{S_j}^{\infty}|}{|Sub_{S_j}^{\infty}| + |Sup_{S_j}^{\infty}|} \tag{7}$$

Example 6. Evaluated degrees of selected ontological classes imprecisions for ontology presented in figure 1 are shown below:

1. $in^{ont.}(Writer) = \frac{5}{5+5} = 0.5$
2. $in^{ont.}(Artist) = \frac{24}{24+4} = 0.85$
3. $in^{ont.}(Person) = \frac{173}{173+3} = 0.98$
4. In the top of the ontology there is a class Thing. For this class the degree of imprecision is equal to 1.

3.5 Generating Linguistic Summaries for Graph Databases - Complete Process with Example

The set of quantifiers Q is known beforehand, as well as the summary subject - an ontological class. We also know the set of ontologies that will be used for summaries - we denote this set of ontologies by T.

For universality of the method, we do not know the attributes, denoted by A, nor their set of values, denoted by X_A. Attributes and their values will be used as summarizers. Exact steps to be followed are listed below.

1. define the ontological class c that will be the subject of the summary
2. generate full set of subclasses for class c Sub_c^{∞} (see Def. 2)
3. query the database for objects of classes Sub_c^{∞} and their attributes (so vertices that are directly connected to them).
4. based on the queried data we create a set of attributes and their value sets: $A = \langle A_1, X_{A1} \rangle, \langle A_2, X_{A2} \rangle, ... , \langle A_i, X_{A_i} \rangle$
5. for each attribute A_i we check if it belongs to any considered ontology T, which means to check if it is an ontological class. If so, we take the full set of superclasses of this attribute value $Sup_{A_i}^{\infty}$ (see definition 4). Each of the superclasses may be used to form a more general summary.
6. we create a set of linguistic summaries using found attributes as summarizers - $X_{A_i} \cup Sup_{A_i}^{\infty}$
7. for each qualifier we calculate truth values $T_1 - T_2$

Example 7. Assume that the subject of a linguistic summary is on ontological class writer, and for the summary we will use an ontology GeoNames, which contains information about administrative classification of the world. Hence the set $T = \{GeoNames\}$.

1. $c = writer$
2. $Sub_{writer}^{\infty} = \{'musiccomposer', Playwright, poet, screenwriter, Songwriter\}$ (see fig. 1)

3. query the database for all objects of class *writer* and all subclasses - Sub^{∞}_{writer}

4. $A = \langle 'born', \{Paris, NewYork, Katowice, Amsterdam\}\rangle$,
   $\langle 'height', \{176cmv, 186cm, 190cm, 166cm\}\rangle$.

5. $'born' \in T.$ $Sup^{\infty}_{Paris} \cup Sup^{\infty}_{NewYork} \cup Sup^{\infty}_{Katowice}$
   $= \{France, USA, Poland, Europe\}$

6. set of summarizers $S = \{Paris, NewYork, Katowice, Amsterdam,$
   $France, USA, Netherlands, Poland, Europe\}$

7. calculation T_1, T_2 and T_{final} - an average of T_1, T_2

4 Application Example - Generating Linguistic Summaries of DBPedia as Part of Semantic Web

DBPedia (see [8], [9]) is an extraction of info boxes from Wikipedia articles into Semantic Web format - Resource Description Framework, RDF (see [10]). In short, RDF is a data format/model composed of triples (subject, predicate and object) that allow easy data integrations. Currently DBPedia contains over 4 milion objects in the main dataset, while it can be easily connected to other data sources using $owl : sameAs$ links available. DBPedia created its own multi-domain ontology which will be used for this experiment, but is also using several others - like subject categories (dcterms), Open Cyc, Wordnet, Freebase, UMBEL and YAGO2. We have implemented our system in Java using jFuzzy-Logic [11] and Apache Jena [12] for querying and processing DBPedia (using its SPARQL endpoint). We have used typical triangular definitions of qualifiers.

We have created summaries for a subclass of class Person - class Artists (96300 instances) - see table 1. Due to the nature of the data, there is an unusually large number of summarizers (in comparison to a typical relational database case) - 56. Due to that complexity, including compound summarizers is not directly feasible and we have not included them in the experiment. As can be seen from the table, some interesting patters may be found - for example that about half artists are musicians. A especially interesting summary is summary

Table 1. A small subset of obtained linguistic summaries of DBPedia for class Artists

No.	Summary	T_1	T_2	T_{final}
1	almost none artists have genre jazz[1]	0.86	0.63	0.75
2	almost none artists are born in France[1]	0.92	1.0	0.93
3	small number of artists are born in Europe	0.83	0.11	0.47
4	about half artists are musicians	0.53	0.55	0.54
5	almost none artists are comics creators	0.78	1.0	0.89
6	almost all artists born in Russia are actors	1.0	0.72	0.96
7	small number of artists that play piano are singers	1.0	1.0	1.0
8	about half of artists with genre Soul music are singers	0.98	1.0	0.99
9	about half of artists with genre Soul music are guitarists	0.93	1.0	0.97

number 3 from table 1 - summarizer 'born in Europe' has been obtained by using summarizer generalization described in section 3.3 [1]. Note also the much lower value of T_2 for this summarizer, which indicates that this summary is very general.

5   Conclusions and Future Work

In this paper we have presented a novel approach to linguistic summarization of graph datasets with the use of ontologies. We have rebuild the notion of subject summary by including ontological class taxonomy. Also, we have shown that when an ontological class is used as a summarizer, it is possible and useful also to include the class taxonomy into analysis, since we may obtain summaries that cannot be directly computed with typical approaches, that is more general summaries. Creating such summaries (e.g. summarizing on continental level, while only the information about a country in directly available) leads to finding new dependencies in data. We have also extended the T_1 and T_2 quality measures, also by including class taxonomy into analysis. By quality measure T_2 we are able to determine the informativeness of a summary. We have proven our approach by generating linguistic summaries for a small subset of DBPedia.

Further work may be focused on taking attributes of attributes into account (vertices with distance of 2 edges from summary subjects). As an example consider a subject type 'movie'. A common attribute of this type may be 'director', who also has its properties (like country of birth, age). Another direction of research continuation could be leveraging Linked Data nature - incorporating other ontologies and databases. Since Semantic Web is based on Linked Data, we may also use other ontologies and information sources to create new summaries. For instance, DBPedia is interconnected to DBTune (music database), Eurostat (statistical information), LinkedMDB (movies database), LinkedGeoData (geographical database), GeoSpecies (various information about species) and many others.

Authors of this paper have already conducted research from the area of acquiring knowledge from graph databases, so far focused on path analysis using artificial intelligence [5],[6]. A comprehensive work on extracting knowledge from graph databases using artificial intelligence is being prepared.

References

1. Yager, R.R.: A new approach to the summarization of data. Inf. Sci. **28**(1), 69–86 (1982)
2. Yager, R.R.: Linguistic summaries as a tool for database discovery. In: FQAS, pp. 17–22 (1994)

---

[1] Since France is not an ontological class, we used a political taxonomy of the world to derive imprecison of summarizer 'France'. For musical genres, like jazz, we used the genre taxonomy given in [13].

3. Yager, R.R., Ford, K.M., Cañas, A.J.: An approach to the linguistic summarization of data. In: Bouchon-Meunier, B., Yager, R.R., Zadeh, L.A. (eds.) Bouchon-Meunier, IPMU. LNCS, vol. 521. Springer, Heidelberg (1990)
4. Yager, R.R.: On linguistic summaries of data. In: Knowledge Discovery in Databases, pp. 347–366 (1991)
5. Strobin, U., Niewiadomski, A.: Evaluating semantic similarity with a new method of path analysis in RDF using genetic algorithms. Journal of Applied Computer Science
6. Strobin, L., Niewiadomski, A.: Recommendations and object discovery in graph databases using path semantic analysis. In: Rutkowski, L., Korytkowski, M., Scherer, R., Tadeusiewicz, R., Zadeh, L.A., Zurada, J.M. (eds.) ICAISC 2014, Part I. LNCS, vol. 8467, pp. 793–804. Springer, Heidelberg (2014)
7. Gruber, T.R.: A translation approach to portable ontology specifications. Knowl. Acquis. 5(2), 199–220 (1993)
8. Lehmann, J., Isele, R., Jakob, M., Jentzsch, A., Kontokostas, D., Mendes, P.N., Hellmann, S., Morsey, M., van Kleef, P., Auer, S., Bizer, C.: DBpedia - a large-scale, multilingual knowledge base extracted from wikipedia. Semantic Web Journal (2014)
9. Auer, S., Bizer, C., Kobilarov, G., Lehmann, J., Cyganiak, R., Ives, Z.G.: DBpedia: a nucleus for a web of open data. In: Aberer, K., Choi, K.-S., Noy, N., Allemang, D., Lee, K.-I., Nixon, L.J.B., Golbeck, J., Mika, P., Maynard, D., Mizoguchi, R., Schreiber, G., Cudré-Mauroux, P. (eds.) ASWC 2007 and ISWC 2007. LNCS, vol. 4825, pp. 722–735. Springer, Heidelberg (2007)
10. Candan, K.S., Liu, H., Suvarna, R.: Resource description framework: metadata and its applications. SIGKDD Explor. Newsl. 3(1), 6–19 (2001)
11. Cingolani, P., Alcalá-Fdez, J.: jfuzzylogic: a robust and flexible fuzzy-logic inference system language implementation. In: FUZZ-IEEE, pp. 1–8. IEEE (2012)
12. Seaborne, A.: Jena, a Semantic Web Framework, November 2010
13. Garcia, J., Barbedo, A., Lopes, A.: doi:10.1155/2007/64960 research article automatic genre classification of musical signals

Movie Summarization Using Characters Network Analysis

Quang Dieu Tran[1], Dosam Hwang[1], and Jason J. Jung[2(✉)]

[1] Department of Computer Engineering, Yeungnam University,
Gyeongsan 712-749, Korea
{dieutq,dosamhwang}@gmail.com
[2] Department of Computer Engineering, Chung-Ang University,
Seoul 156-756, Korea
j2jung@gmail.com

Abstract. Movie summarization focuses to obtain as much as possible of information as a shorter movie clip does, that but it keeps the content of the original and presents to the audience the faster way for understanding the movie. In this paper, we propose a co-occurrence characters network analysis for movie summarization based on discovery and analysis movie storytelling. Experiments on 17 movies in the Star War series, the Lord of the Ring series and Harry Porter series with more than 2000 minutes of movies play time and the evaluated results are compared to IMDb and IMSDb database. Our results show that proposed method has outperformed the conventional approaches in terms of the movie summarization rate.

Keywords: Movie analysis · Movie storytelling · Move segmentation · Social network analysis · Movie summarization

1 Introduction

Nowadays, many movies are published every year and the numbers has continue to rise rapidly. Understanding of movies has becomes a wide research field as discovering contents, stories, and relationships among characters from the movies are challenging and complicated. Numerous approaches are used to aid users to gain a better understanding of movies such as recommendation systems [2,3], recognition systems [6] that using a number of techniques to explore knowledge on them.

Movie analysis focus on analyzing and discovering contents of the movie. The purpose of this research is provided the best way for move to be understood. When the audience watches a movie, he/she usually analyzes the movie storytellings and the relationships among characters. Archetypal of the characters exists as a form of storytelling shorthand. Characters in the movie known as "Main Character(s)" or "Protagonist(s)" - the heroes or who have main purpose or a chief proponent and principal driver of the effort to achieve the story's

© Springer International Publishing Switzerland 2015
M. Núñez et al. (Eds.): ICCCI 2015, Part I, LNAI 9329, pp. 390–399, 2015.
DOI: 10.1007/978-3-319-24069-5_37

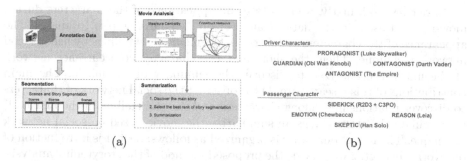

Fig. 1. Proposed Frameworks and Characters Archetypes. (a)Proposed Frameworks, (b)The Characters Archetypes in the Star War Episode IV. A New Hope

goal, "Antagonists" - characters who are opposite of the heroes and try to stop the mission of the heroes, "Reason" - characters who are calm, cold collected and "Emotions" characters who are frenetic, disorganized, and driven by feeling, "Sidekicks" - characters who support protagonist and "Skeptic" - characters who are opposite to Sidekicks, "Guardian" - characters who strongly support protagonist and "Contagonists" - characters who are opposite to Guardian [5]. One of the movie normally contain all types of characters and the biggest task is to identify who belongs to which group. It is trivial to start thinking about the main character (known as "protagonist" or "hero"). A main character is the player who makes the audience experiments the story of the movie first hand and the minor characters, who support the main character to tell the story. Extracting characters to analyze relationships among them is one of the methods to retrieve useful information from the movie. Fig. 1(b) illustrates the relationships belong to characters in the movie and their roles in the storytelling. Identifying this fact is the most important work in movie analysis.

Movie summarization is one of movie analysis research which aims in considering to summarize the full video to a significantly shortened version to provide a condensed and succinct representations of the story of the movie through a combination, analyzing and processing of movie frame, shot, scenes, segmentation and textual description. The easiest summarization method is to analyze sample directly from movie such as key frames, shots and scenes during movie playing time. The cognitive approaches extract the feature of movie such as audio, visual and text features to identify the importance of key frame or scenes in the movie. Some proposed methods use low-level features that are visual-based, audio-based, text-based, audio-visual, visual-textual, audio-textual, motion features and hybrid segmentation approaches [1] that are considered to video/movie segmentation based on scenes detection and classification. Some others cognitive methods are based on events extraction and analyzing in the movie. The non-cognitive methods extract a video/movie summarization via user feedback [2]. Almost of these concept are considered to emotion detect and recognize (eg. [3,4,6]).

In our method, firstly, we annotate the appearance of the characters during movie play. Based on the data of annotation, our next step is to analyze the appearance of the characters and discover the movie storytellings by using social network technique. Then, the shots and the character co-occurrence network of the movie are analyzed to discover the summary of the movie. The main contributions of this research are: 1) to construct the social network of characters to discover relationships among characters and the storytelling in the movie. 2) to formulate the storytellings an summarize the movie based on social network technique. The rest of our paper is organized as follows: Sect. 1 is introduction of our works, in Sect. 2 we provide the proposed method of the storytelling analysis, in Sect. 3 we describe the experiment results, some discussions and conclusions are given in Sect. 4.

2   Movie Storytellings Analysis

2.1   Shot Segmentation

Movie normally is a set of frames in which one frame should contains the sub-story of the movie. In the frame, the characters appear and/or occurrence to each other. In this research, we only consider to the shot in which, the characters are appeared or not. To represent the shots in the movie based on co-occurrence of the characters, we define the signal of the character in the movie.

Definition 1 (Signal of a Character). *Signal of a character in the movie is the time in which the character appear in the shot describes as follows*

$$\wp_{ci}(t) = \begin{cases} 1, & \text{if the character } c_i \text{ appear in the time } t \\ 0, & \text{otherwise.} \end{cases} \tag{1}$$

The appearance of the characters in the movie should be measured by cross-added measurement of the signal of the characters as follows

Definition 2 (Signal of the Characters). *Signal of the characters in the movie is cross-added measurement of the signal of a characters as follows*

$$C(t) = \sum_{i=1}^{n} (\wp_{ci}(t)) \tag{2}$$

where $\wp_{ci}(t)$ is the signal of a character in the movie.

A shot in the case of considering to characters appearance is the time that the characters appear and have some activities in the movie. In the case of the appearance of the characters during movie playing time, a shot in the movie should be measured by the integral of characters appearance signal as follows

$$\Gamma_{ij} = t_j - t_i, \text{ if } \int_{t_i}^{t_j} C(t)dt \geq 1 \tag{3}$$

where Γ_{ij} is i^{th} a shot in the movie and t_i, t_j are times that characters appear in the movie.

By identifying the shot, the movie is converted into the shot sequence during movie playing time as follows

$$\Gamma = \{\Gamma_1, \Gamma_2, ..., \Gamma_n\} \tag{4}$$

In order to determine the appearance of the characters in the movie. The distance between the shots in the movie should be measured as follows

$$\Re_{ij} = t_j - t_i, \; if \int_{t_i}^{t_j} C(t)dt = 0 \tag{5}$$

where $\Re_{ij}$ is distance between two shots in the movie. t_i, t_j are the time that the characters do not appear in the movie.

We formulate the sequence of the distance in the shots in the movie as a set of distance as follows

$$\Re = \{\Re_1, \Re_2, ...\Re_m\}. \tag{6}$$

2.2 Story Segmentation

By using shot detection, we segment the storytellings by using average of shot distance in the movie as follows

$$\Gamma_\Re = \frac{\sum_{i=1}^m (\Re_i)}{|\Re|} \tag{7}$$

where $|\Re|$ is total elements the shot distance $\Re$ and $m = |\Re|$ are total shots of the movie.

A shot should be discover if the distance between two shots is bigger than average of a shot distance that described in Eq. 5. The storytellings detection then use to discover sub-storytellings of the movie and create a movie summarization version. The storytellings detection in this case are segmented by shot and the appearance of the characters.

Let M is movie with a set of $\Im$ shot and a set of $\Re$ distances, sub-storytellings of the movie should be measure by follows

$$\Upsilon_{ij} = \sum_i \sum_j (\Delta_i \Im_i \Re_j | if \; \Re_j \geq \Gamma_\Re) \tag{8}$$

where $\Im_i$ is the ith shot, Δ_i is the social score of the characters in the shot $\Im_i$ in the movie, $\Gamma_\Re$ is the average distance of the shot and $\Re_k$ is the distance of the shot $\Im_i$. In this case, Δ_i should be measure in the social network analysis stage.

We use a characters social network to analysis and discover the main character and the main storytelling of the movie. To construct character social network, we annotate the appearance and co-occurrence of the characters in the movie during movie playing time. In our method, the social network of characters is a weighted graph describes as follows

Definition 3 (Characters Network). *The characters network in the movie is a weighted graph as follows*

$$G = \langle U, V \rangle \tag{9}$$

where $U = \{c_1, c_2, ..., c_n\}$ *is represented as a set of characters in a movie,* $V = \{v_{ij}|$ *where* v_{ij} *is relationship strength between* c_i *and* c_j *if character* c_i *and character* c_j *have relationships}, is represented a set of relationships between the characters.*

Suppose that a movie has n characters that are annotated, V is represented the weight of relationships among characters. V also is related matrix as follows

$$V = [v_{ij}]_{n \times n} \tag{10}$$

where element v_{ij} is

$$v_{ij} = \begin{cases} w_{ij}, & \text{if the } i^{th} \text{ character have a relationship} \\ & \text{with a } j^{th} \text{ character as weight } w_{ij} \\ 0, & \text{otherwise.} \end{cases} \tag{11}$$

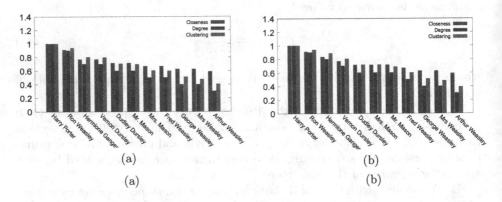

(a)

(b)

Fig. 2. The Social Network Measurement from the movie Harry Porter and the Chamber of Secrets. (a)Total time appear together, (b)Number of Co-Occurrence

In movie analysis, the challenge task is how to determine main character (or protagonist). We use a social network of characters co-occurrence to solve this by using characters network. Characters network applies the Clustering Coefficient, Closeness Centrality and Degree measurement to evaluate the importance of characters in the movie. In the analysis stage, social network measurement values will be used to divide characters in to two main communities: Main characters class - the characters who are leading of social network and Minor characters class - the characters who are supporting to the main characters in the movie.

Table 1. The main character in the movie

ID	Movie	IMDB	Proposed
MV1	Episode I. The Phantom Menace	Qui Gon Jinn	Qui Gon Jinn
MV2	Episode II. Attack of the Clones	Obi Wan Kenobi	Obi Wan Kenobi
MV3	Episode III. Revenge of the Sith	Obi Wan Kenobi	Obi Wan Kenobi
MV4	Episode IV. A New Hope	Luke Skywalker	Luke Skywalker
MV5	Episode V. The Empire Strikes Back	Luke Skywalker	Luke Skywalker
MV6	Episode VI. Return of the Jedi	Luke Skywalker	Luke Skywalker
MV7	The Lord of Rings I. Fellowship of the Ring	Frodo Baggins	Frodo Baggins
MV8	The Lord of Rings II. The Two Towers	Frodo Baggins	Frodo Baggins
MV9	The Lord of Rings III. The Return of the King	Frodo Baggins	Frodo Baggins
MV10	Harry Porter I. Harry Potter and the Philosopher's Stone	Harry Porter	Harry Porter
MV11	Harry Porter II. Harry Potter and the Chamber of Secrets	Harry Porter	Harry Porter
MV12	Harry Porter III. Harry Potter and the Prisoner of Azkaban	Harry Porter	Harry Porter
MV13	Harry Porter IV. Harry Potter and the Goblet of Fire	Harry Porter	Harry Porter
MV14	Harry Porter V. Harry Potter and the Order of the Phoenix	Harry Porter	Harry Porter
MV15	Harry Porter VI. Harry Potter and the Half-Blood Prince	Harry Porter	Harry Porter
MV16	Harry Porter VII. Harry Potter and the Deathly Hallows (Part 1)	Harry Porter	Harry Porter
MV17	Harry Porter VII. Harry Potter and the Deathly Hallows (Part 2)	Harry Porter	Harry Porter

2.3 Summarization

Movie summarization is focused to contain as much as possible of information as a shorter movie clip and present to the audience the fastest way to understand the movie. Our research, belongs to the summarization strategy, firstly, we detect the movie storytellings based on the scenes detection and segmentation. Then we use social network technique to analyze and discover the main story of the movie. Fig. 1(a) illustrates the frame work of proposed method to summary the movie. In our approach, the annotation data are used to the shot segmentation and analysis. In the shot segmentation, the annotation data are used to segment the shots and the shot's distance in the movie. In analysis, the annotation data are used to analyze and extract a characters network in the movie by using social network techniques. Results then use to discover main storytellings and exploit a summarization version of the movie.

Suppose that M is the movie with a set of scenes $\Im$, a set of scenes distance $\Re$, a set of characters C, and a relationship matrix E. The main storytelling of the movie could be measure by Eq. 8 with Δ, the characters story measurement as follows

$$\Delta(c_i) = E(c_i)f(c_i) + C(c_i)f(c_i) + \Psi(c_i)f(c_i) \tag{12}$$

where $E(c_i), C(c_i), \Psi(c_i)$ are Clustering Coefficient, Closeness Centrality and Degree measurements, $f(c_i)$ is signal function of character c_i in the movie and $E(c_i), C(c_i), \Psi(c_i)$ are greater than average values.

3 Results and Discussion

We selected 17 movies from the Star Wars series: *Star Wars series: MV1, MV2, MV3, MV4, MV5, MV6,* The Lord of Rings series: *The Lord of Rings series: MV7, MV8, MV9* and Harry Porter series: *Harry Porter series: MV10, MV11, MV12, MV13, MV14, MV15, MV16, MV17* to extracting characters network. To summarize the move, we extract the characters network by performing technique with centralities measurement and social network analysis and representation. Table 1 show results of characters analysis, by using our method, the main character (or protagonist) in the movie is detected. These results are matched to IMDB ranking list.

Fig. 2 illustrates the centralities of the characters in the movie Harry Porter and the Chamber of Secrets. The results shown that the main characters in the movie have high value of centrality will hold the importance role in the movie. Combining the centralities of these results by, we found that Harry Porter had

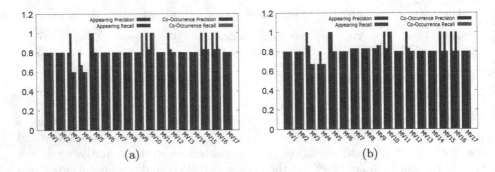

Fig. 3. Characters Classification's Precision and Recall. (a)Main Characters, (b)Minor Character

Table 2. Sub-Storytellings Segmentation

ID	Movie	Script	Proposed
MV1	Star Wars I. The Phantom Menace	182	175
MV2	Star Wars II. Attack of the Clones	160	153
MV3	Star Wars III. Revenge of the Sith	241	231
MV4	Star Wars IV. A New Hope	496	492
MV5	Star Wars V. The Empire Strikes Back	279	340
MV6	Star Wars VI. Return of the Jedi	136	109
MV7	The Lord of Rings I. Fellowship of the Ring	144	139
MV8	The Lord of Rings II. The Two Towers	145	142
MV9	The Lord of Rings III. The Return of the King	231	228
MV10	Harry Porter I. Harry Potter and the Philosopher's Stone	35	34
MV11	Harry Porter II. Harry Potter and the Chamber of Secrets	107	105
MV12	Harry Porter III. Harry Potter and the Prisoner of Azkaban	152	150
MV13	Harry Porter IV. Harry Potter and the Goblet of Fire	68	68
MV14	Harry Porter V. Harry Potter and the Order of the Phoenix	106	102
MV15	Harry Porter VI. Harry Potter and the Half-Blood Prince	142	139
MV16	Harry Porter VII. Harry Potter and the Deathly Hallows (Part 1)	161	161
MV17	Harry Porter VII. Harry Potter and the Deathly Hallows (Part 2)	207	203

Table 3. Summarization Results

ID	Movie	Length	Version 1	Version 2
MV1	Episode I. The Phantom Menace	136mins	43.11mins	33.43mins
MV2	Episode II. Attack of the Clones	142mins	26.04mins	22.04mins
MV3	Episode III. Revenge of the Sith	140mins	32.35mins	35.62mins
MV4	Episode IV. A New Hope	121mins	47.10mins	39.96mins
MV5	Episode V. The Empire Strikes Back	124mins	27.61mins	21.26mins
MV6	Episode VI. Return of the Jedi	134mins	34.82mins	28.47mins
MV7	The Lord of Rings I. Fellowship of the Ring	178mins	44.12mins	39.58mins
MV8	The Lord of Rings II. The Two Towers	179mins	45.23mins	41.21mins
MV9	The Lord of Rings III. The Return of the King	201mins	55.63mins	50.26mins
MV10	Harry Porter I. Harry Potter and the Philosopher's Stone	152mins	50.12mins	45.23 mins
MV11	Harry Porter II. Harry Potter and the Chamber of Secrets	161mins	55.78mins	49.68mins
MV12	Harry Porter III. Harry Potter and the Prisoner of Azkaban	142mins	45.32mins	39.58mins
MV13	Harry Porter IV. Harry Potter and the Goblet of Fire	157mins	53.54mins	48.26mins
MV14	Harry Porter V. Harry Potter and the Order of the Phoenix	138mins	44.63mins	39.86mins
MV15	Harry Porter VI. Harry Potter and the Half-Blood Prince	153mins	48.96mins	42.16mins
MV16	Harry Porter VII. Harry Potter and the Deathly Hallows (Part 1)	146mins	44.36mins	38.58mins
MV17	Harry Porter VII. Harry Potter and the Deathly Hallows (Part 2)	130mins	38.56mins	31.36mins

highest values. In the IMDb database ranking list of cast and crew of this episode, Harry Porter is main character who tell the main story of the movie.

Fig. 3 illustrates precision and recall of our proposed method, by using total time appearing together, Main characters class accomplished a precision of 86% and a recall of 82% on average. The Minor characters class achieves a precision of 87% and a recall of 82% on average. In other approach, by counting number of characters co-occurrence , Main characters class accomplished a precision of 81% and a recall of 82% on average. The Minor characters class achieves a precision of 83% and a recall of 80% on average. Compare to other approaches, RoleNet [6] is a method to extract the roles in the movies. This method uses face recognition strategy to detect the co-occurrence of the characters on the scenes. If two characters co-occurrence in the scenes, a relationship between them is created and the weight of this relationship is measured. Social network analysis technology is used to analyze and discover the story of characters in the movie. RoleNet automatically classifies major roles, supporting roles and identify characters belong to its. RoleNet is a weighted graph with the nodes illustrate the character in the movie, the edges instance the relationships among characters and the weights are the number of characters' co-occurrence in the frames. There are limitation in this method occurred by the image processing accuracy is not enough. Its not enough to discover the main character in the movie.

To evaluate the sub-storytelling segmentation, we parsed the scenes from script of the movie by using movie scripts database (IMSDb). By using movie

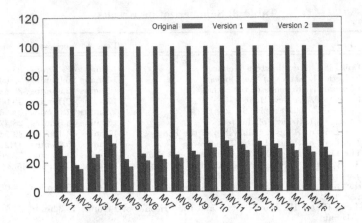

Fig. 4. Movie Summarization Results

script, all of sub-storytellings of the movie will be parsed and compared to the proposed method. Table 2 shows the sub-storytellings segmentation compared to IMSDb database, these results have some incorrect rate because that in the some case, the characters do not appear. We then use the storytellings measured by Eq. 8 as a summarization strategy, in our proposed method, the main storytellings of the movie depends on Δ, if $\Delta(c_i)$ belongs to protagonist, Υ_i is a candidate scene of selecting to summary. We also use the measurement of the characters in the classification step to discover the main storytellings of the movie. When a candidate scene is selected, we measure the score of scenes using the score of the characters, if the average of centralities belong to characters is greater than Γ_S, the scene is selected to a summary version. After that, we have truncated all of scenes which have length less than 10 seconds in the Version 1 and truncated all of scenes which have length less than 20 seconds in the Version 2 of summarization because that in the normal, most films averaged 2 to 4 minutes per scene and many scenes ran from 4 minutes or more. Results of the summarization is illustrated in Table 3, this result shows our proposed method can summarize the movie as a main storytellings based on the appearance of the main character (known as protagonist) in the movie more efficiently than other. With our approached, we can compress the movie as overall 75.33% rate to the summarization version as illustrates in Fig. 4 Movie Summarization Results. These results are more efficient compare to [7], this method accomplished with 65% of summarization level.

4   Conclusion

Movie summarization is focused to obtain as much as possible of information as a shorter movie clip and present to the audience the fastest way for understanding of the movie. In our research, we have proposed the method to analyze and discover main movie storytelling based on the characters appearance and co-occurrence using social network analysis technique. The movie summarization follows the main storytellings, which depends on the main character of the movie. In the experimental results, we have compared it to the IMDb and IMSDb database and discovered that our proposed method has provided more efficient results. Experimental results have shown that our proposed method provides good solution for the movie summary with compress rate accomplished 75.33%. Despite of having better results of movie summarization, our proposed method still has some limitation and weaknesses. And it takes time to annotate and it does not consider the emotion or behavior of the characters in the movie, respectively. These limitations need to be considered in future works, that will consider to other approaches such as combining the characters appearance and movie scripts analysis, through which, movie summarization process will be more efficient in performance.

Acknowledgments. This work was supported by the National Research Foundation of Korea (NRF) grant funded by the Korea government (MSIP) (NRF-2014R1A2A2A05007154). Also, this research was supported by the MSIP (Ministry of Science, ICT and Future Planning), Korea, under the ITRC (Information Technology Research Center) support program (IITP-2015-H8501-15-1018) supervised by the IITP (Institute for Information & Communications Technology Promotion).

References

1. Del Fabro, M., Böszörmenyi, L.: State-of-the-art and future challenges in video scene detection: a survey. Multimedia systems **19**(5), 427–454 (2013)
2. Jung, J.J.: Attribute selection-based recommendation framework for short-head user group: An empirical study by movielens and imdb. Expert Systems with Applications **39**(4), 4049–4054 (2012)
3. Jung, J.J., You, E., Park, S.-B.: Emotion-based character clustering for managing story-based contents: a cinemetric analysis. Multimedia tools and applications **65**(1), 29–45 (2013)
4. Park, S.-B., Oh, K.-J., Jo, G.-S.: Social network analysis in a movie using character-net. Multimedia Tools and Applications **59**(2), 601–627 (2012)
5. Phillips, M.A., Huntley, C.: Dramatica: a new theory of story. Screenplay Systems (1996)
6. Weng, C.-Y., Chu, W.-T., Wu, J.-L.: Rolenet: Movie analysis from the perspective of social networks. IEEE Transactions on Multimedia **11**(2), 256–271 (2009)
7. Zhu, X., Wu, X., Fan, J., Elmagarmid, A.K., Aref, W.G.: Exploring video content structure for hierarchical summarization. Multimedia Systems **10**(2), 98–115 (2004)

A Mobile Context-Aware Proactive Recommendation Approach

Imen Akermi[1,2](✉) and Rim Faiz[2]

[1] IRIT, Paul Sabatier University, Toulouse, France
imen.akermi@irit.fr
[2] IHEC LARODEC, University of Carthage, Tunis, Tunisia
rim.faiz@ihec.rnu.tn

Abstract. The Proactive Context Aware Recommender Systems aim at combining a set of technologies and knowledge about the user context not only in order to deliver the most appropriate information to the user need at the right time but also to recommend it without a user query. In this paper, we propose a contextualized proactive multi-domain recommendation approach for mobile devices. Its objective is to efficiently recommend relevant items that match users' personal interests at the right time without waiting for users to initiate any interaction. Our contribution is divided into two main areas: The modeling of a situational user profile and the definition of an aggregation frame for contextual dimensions combination.

Keywords: Context modeling · Context-aware recommendation · User modeling · Proactive recommendation

1 Introduction

The development of mobile devices equipped with persistent data connections, geolocation, cameras and wireless capabilities allows current context-aware recommender systems (CARS) to be highly contextualized and proactive. They provide users with relevant information when it is most needed at the right time without waiting for the user to initiate any query. There are several context aware systems that attempted to meet the challenge of providing the right information at the right time without the interference of the user in a mobile environment. However,this requires good modeling of the dimensions of the context and especially the modeling of the user profile. Indeed, as mentioned by [1], several dimensions of context, such as location, time, users activities, needs, resources in the nearbies, light, noise, movement, etc., have to be managed and represented which requires a big amount of information and are time consuming. Besides, the incorporation of too many context dimensions generate complex context models. On the other hand, context models integrating few dimensions are unable to figure out the whole user context.

We propose, in this paper, a proactive context-aware recommendation approach that integrates the modeling of a situational user profile and the definition of an aggregation frame for contextual dimensions combination.

© Springer International Publishing Switzerland 2015
M. Núñez et al. (Eds.): ICCCI 2015, Part I, LNAI 9329, pp. 400–409, 2015.
DOI: 10.1007/978-3-319-24069-5_38

The paper is organised as follows. We provide in section 2 an overview about the related work. Section 3 presents the proposed approach. We describe in section 4 the experiments done within the TREC 2014 Contextual Suggestion Track. We finish in section 5 with our conclusions and thoughts for future work.

2 Related Work

The user profile modeling covers broad aspects such as the cognitive, social and professional environment to determine the user intentions during a search session [1]. The user profile is an important dimension considered within context modeling. Indeed, context is defined as a set of dimensions that describe and/or infer user intentions and perception of relevance. Those dimensions cover situations related to factors such as location, time and the current application. Work in context-aware recommendation makes use of one or all of these dimensions to describe the user and integrate him forward in the various phases of the recommendation process. Proactive recommendation systems (PRSs) as described in [2], intend to sort among a large quantity of documents, the information that is most likely to be relevant to the user needs, and recommend that information without user requests. Several systems have been developed to support proactive recommendation. Various approaches relied on the user's past or actual behavior history to determine the user interests. Behavior history includes Web browsing history/clicks ([2]); previous visiting behaviors for location based systems ([4,5]) and previous reading patters for news recommendation systems ([6,7,8,9]). Other approaches dealt with user profiling from an activity centric angle. The common activities used to build the user profile in these systems might take the form of: Opened web pages or documents ([10,11,12]); Ongoing conversation or activity such as phone calls [13]; The social media activity of the user such as the user's tweet stream on Twitter ([14,15,16]). However, some approaches require that users express their interests and input keywords or tags which is, most of the time, inconvenient in a mobile environment since it entails extra efforts from the user such as tagging, searching, or clicking. Mobile systems can help keep track of user's activities, preference and location. Besides, many context aware systems dealt with user profiling from an activity centric angle. Nevertheless, we cannot reduce the user profile to some activities. One can simply open a document to work on without being related to it in any way or have a conversation about an issue that he/she is not concerned to know any recommendation about it. There are also the domain dependency issue. In fact, many of the actual contextualized systems are domain dependent (tourism, movie, news ...) and have specific context dimensions to apply according to the domain. However, most of them rely almost on the same context combination which includes location, time and user preferences with a slight difference on how to approach this information. Our approach tries to deals with these issues by integrating the modeling of a situational user profile and the definition of an aggregation frame for contextual dimensions combination.

3   Proposed Approach

We propose a multi-domain proactive context-aware recommendation approach that recommends the right item that match users personal interests at the right time without waiting for users to initiate any interaction or activity.

3.1   Context Modeling

We consider the context as a two-dimension representation
context=(profile,location).
These dimensions' instantiations form a situation S where recommendation is needed. We define in the following sections the different component of the context modeling process.

The User Profile. The user profile (UP) model is defined by two features, *Demographic data*; the user related information such as name,age,etc, C;the user's interests related to specific weighted categories :
$profile = \{C_i, w_i\}; i = 1..n$
A category is a set of weighted terms that are associated to the user preferences and to the interest that he expresses towards this particular category :
$C_i = \{t_j^{(i)}, w_j^{(i)}\}; j = 1..m$
The categories are predefined using the Open Directory Project Dmoz[1].

Location. The location is inferred according to GPS coordinates (*latitude and longitude*). Those GPS coordinates are not the only features that we can consider when defining a location. Indeed, as discussed by [16], there are different ways to characterize the location of the mobile user, *Absolute position*; *Relative* (next to, ...); *A Place name*; *A named class* that represents the type of the place, eg. museum, school, ... We consider the absolute position and the named class representation to characterize the spatial dimension of the user. We define two levels of the location dimension: *The actual location*, that refers to the user's actual location at a given time; *The user's related locations*, the places related to the daily life of the user (work, home, ...). The actual location can be recovered using several tools such as Geonames[2] or the Social Network Foursquare[3] which assign a location category to a given GPS coordinates.

3.2   The Recommendation Process

Type of Information to Recommend. We consider that the recommendation process entails a pack of situations which reflect a specific area of interest

---

[1] http://www.dmoz.org/docs/en/about.html
[2] http://www.geonames.org/
[3] https://fr.foursquare.com/

characterized by the instantiations of the context dimensions. The possible situations are organized within a knowledge database. A situation is then represented by two specific dimensions: the actual location (D_l) and the user's category of interests related to that location.

These values are used to assess the current situation need in information and the category type of the information to recommend. Our key idea is that the user's need in information changes according to the user's actual location. For instance, a user might want to check the news once he is at work or he might want to visit a specific shop if he's found to be near a mall.

The category of interest of the information to recommend is inferred from the current situation.

Information Extraction. In order to retrieve the information to recommend, a query q is formulated as : *q=(latitude, longitude, category_of_ interest)*
The query is sent to a geo-based service. The query result is represented by a set of items $I : I = \{i_1, ?, i_n\}$.
An item is defined as a weighted terms vector $i_j = \{t_k^j, w_k^j\}; j = 1..n, k = 1..p$
We filter out from the set I, the items suiting best the user's preferences by calculating a relevance score.
The relevance of an item with respect to the category of interest entails two components: the topic and the location relevance. The topic relevance assess to which degree the user preferences related to the given category are related to an item and is calculated as :

$$Topic_{rel}(VC_i, It) = \frac{\sum_{j=1}^{n} VC_i^j * It_j}{\sqrt{\sum_{j=1}^{n}(VC_i^j)^2} * \sqrt{\sum_{j=1}^{n}(It_j)^2}} \qquad (1)$$

Where:
VC_i: the preferences keywords vector related to category C_i
It: the item keywords vector
The location relevance is only used in case where the user has to move to get to the suggested item. It is expressed by a score measuring the accessibility to the actual place's location and is calculated as the distance between 2 GPS coordinates corresponding to the user's current location and the suggested item location: ($P1(lat1, long1)$ et $P2(lat2, long2)$):

$$accessibility = R * c \qquad (2)$$

Where:
R: The earth radius=6,371Km
$c = 2 * atan2(\sqrt{a}, \sqrt{(1 - a)})$
$a = sin^2((lat2 - lat1)/2) + cos(lat1) * cos(lat2) * sin^2((long2 - long1)/2)$
The overall relevance for a result is calculated as :

$$Rel = \alpha * Topic_{rel}(C_i, It) + (1 - \alpha) * accessibility \qquad (3)$$

The results are ranked according to their overall relevance scores. The recommendation process is summarized as follows :

```
Input: Profile{{Ci, wi}^{i=1..n}}
       Ci = {(t^i_j, w^i_j); j = 1..m}
       Situation {UP, Dl}
       Situation Knowledge Database (KB)
Case of Dl
       item type(s)← get(KB, Dl)
       For each item type
          I← get(service, type, Dl)
          For each i ∈ {I}
             Compute topic-relevance of i
             If item type is accessibility sensitive
                compute geo-relevance of i
             End If
             Compute the overall relevance of i :
                R(i) ← f(topic − relevance(i), geo − relevance(i))
          End For
       End For
```

4   Experiments

We evaluated our approach using the TREC 2014 *Contexual Suggestion Track* task. We present in this section, a general description of the task, then we expose the obtained results.

4.1   The TREC 2014 Contexual Suggestion Track

This task offers an evaluation platform for search techniques that depend highly on the context and the user interests. The input to this task consist of a set of profiles, a set of sample suggestions (a set of venues evaluated by the profiles) and a set of contexts. Each profile corresponds to a single user, and indicates the preference of the user with respect to the set of suggestions. For example, one suggestion could be a recommendation to have a beer at the Dogfish Head Alehouse. The profile describes the negative or the positive preference of the user regarding the set of suggested venues.

Profiles Processing. The profiles are constructed using the list of the suggested venues evaluated by the user. Each suggestion is evaluated according to two ratings: a rating for the venue's title and description and a rating for the venue's website. The profile should indicate which venues a user likes or does not like. The ratings are fixed on a five-point scale based on how interesting a venue would be for the user if he was visiting the city the venue was in:
4, Strongly interested; 3, Interested; 2, Neutral; 1, Disinterested; 0, Strongly disinterested; -1, Website didn't load or no rating given

The suggestions (venues) representation :

```
id,title,description,url
1,Fresh on Bloor,"Our vegan menu ...",www.freshrestaurants.ca
```

The user's ratings :

```
id,attraction_id,description,website
1,1,1,0
```

In order to define the user's thematic profile, we identify for each suggested venue its category using Google Places API[4]. A profile is then expressed as a set of weighted categories under which there are terms set related to the liked suggestions :$profile = \{C_i, w_i\}; i = 1..n$

The categories are represented by a set of weighted terms extracted from the suggestions' descriptions. For each profile, the weight assigned to a particular category takes into account the two ratings of the suggested venues that were rated by users. One rating for the venue's title and description and the other one is for the venue's website.

$$weight(category) = \frac{\sum_{\forall s \in C} R_{td} + R_w}{N_p} \tag{4}$$

Where:

$\forall s \in C$: for each suggestion s belonging to this category C

R_{td}: The venue's title and description rating

R_w: The venue's website rating

N_p: the number of suggestions belonging to this category

Contexts Processing. The TREC task defines context according to GPS coordinates $context=(latitude, longitude)$. For each context, we gather possible venues by querying two geo-based services: Google Places and Foursquare. The query is modeled as:

$Query=\{(latitude, longitude), category\}$

A venue (query result) is modeled as an object having specific attributes and belongs to a specific category:

$venue=\{name, url, description, accessibility, category\}$

$accessibility$: Represents the distance separating the venue from the specified location. The venue accessibility is measured as the distance between 2 GPS coordinates(see formula number 2).

---

[4] https://developers.google.com/places/documentation/?hl=fr

(Profiles, Suggestions) Matching. The selection process of interesting places for each context-profile pair is summarized as follows:

```
For each pᵢ ∈ P
    For each cⱼ ∈ Cx
        - Calculate, for each suggestion s ∈ Cxⱼₛᵤₘₘₑₛₜᵢₒₙₛ
          the relevance(Formula 3)
        - Normalize the relevance scores of the suggestions
          between 0 and 1
        - Extract the suggestions which relevance score
          is ≥ 0,5
          and that belongs to the categories of interest
          that are appreciated by the profile
          category_weight_normalised ≥ 0.5
```

Where:

P: The Profiles set

Cx: The Contexts set

$Cx_{j_{suggestions}}$: The set of suggestions related to context C_j

s: suggestion $\in Cx_{j_{suggestions}}$

4.2   Results

To measure the geographic relevance of the venues that we suggested for each context, we extracted a venues' sample V_g which involves the intersection of our venues collection with the venues that have been evaluated geographically in the TREC task for each context. We obtained $|V_g| = 4802$ venues across all contexts. Among these places, 4644 were evaluated geographically relevant which implies a total geographical precision equal to 0.97 and is calculated as:

$$geo\_relevance = \frac{Nb\_Geo\_Relevant\_Venues}{|V_g|} \tag{5}$$

To measure the profile relevance, there were two alternatives to note. A first alternative is to consider for each context, the intersection of our venues' collection with the venues provided by each run[5], however, this intersection almost gave the empty set. We opted for an intermediate solution of considering the intersection of our venues' collection with the union of the venues that each run has proposed across all profiles in order to get the suggested venues ratings. The cardinality of this intersection is $|V_p| = 889$ venues.

Then we measured for each venue the level of interest that it has requested from the profiles based on the number of profiles that have evaluated this venue

---

[5] A run represents the set of venues proposed by a team participating in the TREC 2014 Contextual Suggestion Track.

Table 1. Profile relevance (NBInV is the number of venues that were rated as interesting by the profiles; NbTotV is the total rated venues that were suggested by the run; Prec. stands for the precision)

id	run	NbInV	NbTotV	Prec.	id	run	NbInV	NbTotV	Prec.
1	BJUTa	352	1495	0,24	20	SScore	244	1496	0,16
2	BJUTb	327	1495	0,22	21	SScoreImp	254	1496	0,17
3	BUPT_01	81	671	0,12	22	tueNet	129	1497	0,09
4	BUPT_02	78	704	0,11	23	tueRforest	138	1497	0,09
5	cat	318	1496	0,21	24	UDInfo_1	201	1495	0,13
6	choqrun	196	1465	0,13	25	UDInfo_2	360	1495	0,24
9	dixlticmu	367	1496	0,25	26	uogTrBun	272	1495	0,18
10	gw1	58	1466	0,04	27	uogTrCsL	199	1496	0,13
11	lda	163	1496	0,11	28	waterlooA	260	1497	0,17
14	RAMAR2	260	1497	0,17	29	waterlooB	286	1497	0,19
15	RUN1	229	1494	0,15	30	webis_1	269	1477	0,18
16	run_DwD	181	1496	0,12	31	webis_2	230	1474	0,16
17	run_FDwD	262	1496	0,18					

as relevant compared to the total number of evaluations of the same venue. We obtained an average precision of 0,56 that is calculated as :

$$profile\_relevance = \frac{\sum_{v=1}^{|V_p|}(nb\_profiles_v/tot\_nb\_profiles_v)}{|V_p|} \tag{6}$$

where:

$nb\_profiles_v$: is the number of profiles that evaluated the venue v as interesting.
$tot\_nb\_profiles_v$: is the total number of profiles that evaluated the venue v.

In order to compare approximately the result that we obtained with the other participants'results, we also applied this method to the other runs in order to measure the profile relevance. Indeed, for each run we calculated the number of the venues that were rated interesting by the profiles compared to the total rated venues that were suggested by the run (see table 1).

As we can notice from Table 1, the profile relevance for each run does not exceed 0.25. This is explained by the fact that for each run, there is a large gap between the number of profiles that have judged a venue (belonging to a given context) as relevant and the number of the total judgments for the given venue.

The results that we obtained using our approach are promising and show that the use of the categories classification of a profile's preferences implies better thematic relevance compared to the profile interests. These results also indicate that the choice of the parameters that we have set such as the radius used for the definition of the premises of the venues for a given context, are effective. However, this evaluation has only indicted a part of our approach. Indeed, the time notion and the current activity of the user are not incorporated in the TREC task whereas these dimensions are considered in our approach.

5   Conclusion

The fundamental purpose of Context-Aware Recommender Systems consists in combining the user's context and environment in a same infrastructure to better characterize the user information needs in order to improve the recommendation process.We proposed a proactive context-aware recommendation approach that can help users deal with information overload problem efficiently by recommending relevant items that match users? personal interests at the right time without waiting for users to initiate any interaction or activity. More specifically, our contribution is divided into two main areas: The modeling of a situational user profile and the definition of an aggregation frame for social contextual dimensions combination. Actually, we are planning to participate in the RecSys 2015 Challenge *YOUCHOOSE* in order to validate our approach by incorporating, this time, the user's current activity and the time notions within the experiments.

References

1. Mizzaro, S., Vassena, L.: A social approach to context-aware retrieval. World Wide Web **14**(4), 377–405 (2011)
2. Melguizo, M.C.P., Bogers, T., Deshpande, A., Boves, L., van den Bosch, A.: What a proactive recommendation system needs - relevance, non-intrusiveness, and a new long-term memory. In: ICEIS (5), pp. 86–91 (2007)
3. Li, W., Eickhoff, C., de Vries, A.P.: Want a coffee?: Predicting users' trails. In: Proceedings of the 35th International ACM SIGIR Conference on Research and Development in Information Retrieval. pp. 1171–1172. ACM, New York, NY, USA (2012)
4. Pu, Q., Lbath, A., He, D.: Location based recommendation for mobile users using language model and skyline query. International Journal of Information Technology & Computer Science (IJITCS) **4**(10), 19–28 (2012)
5. IJntema, W., Goossen, F., Frasincar, F., Hogenboom, F.: Ontology-based news recommendation. In: Proceedings of the 2010 EDBT/ICDT Workshops. pp. 16:1–16:6. ACM, New York (2010)
6. Arora, A., Shah, P.: Personalized News Prediction and Recommendation. Ph.D. thesis, Stanford University (2011)
7. Athalye, S.: Recommendation System for News Reader. Ph.D. thesis, San Jose State University (2013)
8. Dumitrescu, D.A., Santini, S.: Improving novelty in streaming recommendation using a context model. In: CARS 2012: ACM RecSys Workshop on Context-Aware Recommender Systems (2012)
9. Prekop, P., Burnett, M.: Activities, context and ubiquitous computing. Comput. Commun. **26**(11), 1168–1176 (2003)
10. Dumais, S., Cutrell, E., Sarin, R., Horvitz, E.: Implicit queries (iq) for contextualized search. In: Proceedings of the 27th Annual International ACM SIGIR Conference on Research and Development in Information Retrieval, pp. 594–594. ACM, New York (2004)
11. Karkali, M., Pontikis, D., Vazirgiannis, M.: Match the news: A firefox extension for real-time news recommendation. In: Proceedings of the 36th International ACM SIGIR Conference on Research and Development in Information Retrieval, pp. 1117–1118. ACM, New York (2013)

12. Popescu-Belis, A., Yazdani, M., Nanchen, A., Garner, P.N.: A speech-based just-in-time retrieval system using semantic search. In: Proceedings of the 49th Annual Meeting of the Association for Computational Linguistics: Human Language Technologies: Systems Demonstrations, pp. 80–85. Association for Computational Linguistics, Stroudsburg (2011)
13. Phelan, O., McCarthy, K., Bennett, M., Smyth, B.: On using the real-time web for news recommendation & #38; discovery. In: Proceedings of the 20th International Conference Companion on World Wide Web. pp. 103–104. ACM, New York (2011)
14. De Francisci Morales, G., Gionis, A., Lucchese, C.: From chatter to headlines: Harnessing the real-time web for personalized news recommendation. In: Proceedings of the Fifth ACM International Conference on Web Search and Data Mining, pp. 153–162. ACM, New York (2012)
15. O'Banion, S., Birnbaum, L., Hammond, K.: Social media-driven news personalization. In: Proceedings of the 4th ACM RecSys Workshop on Recommender Systems and the Social Web, pp. 45–52. ACM, New York (2012)
16. Dobson, S.: Leveraging the subtleties of location. In: Proceedings of the 2005 Joint Conference on Smart Objects and Ambient Intelligence: Innovative Context-aware Services: Usages and Technologies, pp. 189–193. ACM, New York (2005)

A Method for Profile Clustering Using Ontology Alignment in Personalized Document Retrieval Systems

Bernadetta Maleszka[✉]

Department of Information System, Wroclaw University of Technology,
Wybrzeze Wyspianskiego 27, 50-370 Wroclaw, Poland
Bernadetta.Maleszka@pwr.edu.pl

Abstract. User modeling is crucial aspect of personalized document retrieval systems. In this paper we propose to use ontology-based user profile while ontological structures are appropriate to represent dependencies between concepts in user profile. A method for clustering set of users is proposed. As a similarity measure between ontological profiles we present a novel approach using ontology alignment methods. To avoid "cold-start problem" we developed method for profile recommendation for a new user.

Keywords: Ontology-based user profile · User preference · Ontology alignment · Personalization

1 Introduction

Modeling user interests is more and more important issue in modern personalized retrieval systems. Information about the user is usually stored in some form of user profile. An accurate representation of a profile is crucial to the performance of personalized search. The most important question during profile building is how to build an accurate profile and particularly how to identify major versus minor concepts of interest to the users, and how to represent the profile [7].

User profiles may be built explicitly, by asking users questions, or implicitly, by observing their activity. User profile designer should take into account that user interests change over time. This implies that most user profiles are constructed using implicit methods, as such approach gives potential chance to adapt over time and reflects changing user interests.

The next aspect of building a user profile is its structure. The simplest models assume a set of words or vectors of keywords but they don't contain information about relations between words or concepts [9]. More significant profiles have hierarchical structure (sometimes weighted hierarchical structure) or ontological structure. In hierarchical user profile a relation of generalization – specification is considered, while an ontology can contain more relations and levels of semantic. In both cases one needs the domain structure which is used as a basis of user profile.

© Springer International Publishing Switzerland 2015
M. Núñez et al. (Eds.): ICCCI 2015, Part I, LNAI 9329, pp. 410–420, 2015.
DOI: 10.1007/978-3-319-24069-5_39

An ontological approach to user profiling has been proven successful in addressing the cold-start problem in recommender systems, since it allows for propagation from a small number of initial concepts to other related domain concepts by exploiting the ontological structure of the domain [2].

Important part of user profiling is adaptation. User interests change over time. It implies that user profile should be modified according to user's current interests. There are various methods of user profile adaptation. In many system one can find spreading activation methods, while a more and more popular method is to build long-term and short-term profiles and try to merge them.

In this paper we propose to use an ontology-based user profile. Each user interest should be presented as a ontology concept (keyword) and instances (relevant document described by this keyword). To cluster user profiles and determine representative profile of each group we adapt ontology alignment methods proposed in paper [14]. Such approach allows a new user to start personalized retrieval when he registers into system without "cold-start problem".

The rest of the paper is organized as follows. In Section 2 we present a survey of methods to build and align ontology-based user profiles in personalized document retrieval system. The model of documents set and user profile are presented in Section 3. The proposition of using ontology alignment method for collaborative recommendation is described. There is also presented an idea of evaluation methodology. In the last Section 4 we gather the main conclusions and future works.

2   Related Works

Approaches to user modeling range from demographic methods, through relevance feedback and methods of modeling users in mobile environments [15]. In this paper we focus on ontology-based user profiling. In this section we describe popular formalism of user model, ways to gather information about user behaviuor, methods of ontology-based profile adaptation and methodology of evaluations.

Hoppe et al. [6] consider three key issues: analysis of resources, user profiling and the scale of gathered information. A crucial step in building user profile is the analysis of the web resources viewed by the user. Profile in this context is only a machine-interpretable representation of the user information needs. The background knowledge about relations between user characteristics has to be extended with knowledge that can be deduced from the user's specific trace. One should remember that the user can generate high volume of heterogeneous data that should be interpreted. Due to this fact methods for profile adaptation should be efficient and scalable.

2.1   Modelling User Preferences

A user model should consist of knowledge about the user preferences which determine the user's interactions.

Ontology is often used as a basis for the construction of a user model in personalized systems, e.g. information delivery systems [12], Intelligent Tutoring Systems [4], etc.

A specific kind of ontology is a hierarchical ontology. It is represented by a directed acyclic graph. A node models a concept and its label reflects the meaning of the concept. Relationships between the nodes are usually represented by *narrow-than* or *broader-than* relations in the graph. The hierarchical ontology classifies the concepts at each level and proceeds from generalized to specialized concepts [8].

2.2   Information Acquisition and User Profile Adaptation

In the process of building accurate user profile, machine learning or soft computing techniques are often used. Their aim is to extract knowledge from data. User model is used as a wrapper for the entire process [15].

Hierarchy of concepts is used to represent course content in an e-learning system [19]. The authors proposed two different approaches for acquisition of users knowledge requirements about courses. At the beginning the user has to take part in interactive question-answer session. Additionally, the system gathers historical session logs and analyzes them to determine users requirements. In the second approach the system collects information about users reading behavior while browsing e-documents. Combining these two approaches allows to create an accurate user profile.

Monitoring the users browsing behaviours was also proposed by Zhang et al. [20]. The system tracks the user (by analyzing usage logs) and adapts the user ontology. Similar approach can be found in [3]. Chen et al. consider the previous browsing and reading behavior of users (i.e., usage history log). They also developed a mechanism for domain ontology adaptation for personalized knowledge search and recommendation to adapt a suitable domain ontology. Adaptive domain ontology can satisfy the future requirements of users, thereby promoting the reuse value of domain ontology.

Domain ontology is also used in the system proposed by Sieg et al. [17]. The process of building ontological user profiles is based mainly on assigning interest scores to existing concepts in a domain ontology. User profiles consists of annotated specializations of a preexisting reference domain ontology. A spreading activation algorithm is used for maintaining the interest scores in the user profile based on the users ongoing behavior.

Cena et al. [2] propose an approach of propagating user interests in a domain ontology, starting from the user behavior in the system. User interests are recorded for the classes of the domain ontology. Each ontological user profile is an instance of the reference ontology where every domain object in the ontology has an interest value associated to it. This model considers user interests, which are received directly from user feedback, and calculates the value of propagated interests. The main idea of propagation is based on *IS-A* relations. If a user is interested in a keyword from class C, he may be also interested in a class which is more general. He might be also interested in some more specific area of class C.

The approach presented by Tang et al. [18] assumes two types of interest profiles: explicit profiles and implicit profiles, during construction of user interest profiles. The explicit profile is determined based on relating users interest-topic relevance factors to users interest measurements of these topics computed by a conventional ontology-based method, and the implicit profile is acquired on the basis of the correlative relationships among the topic nodes in topic network graphs.

Razmerita et al. [16] present an ontology-based user model which integrates three ontologies: user ontology that includes different user's characteristics and their relationships, domain ontology that captures the domain or application specific concepts and their relationships, and log ontology which represents the semantics of the user interaction with the system.

User profile can be adapted using information about his activity or information about activities of other users. Also contextual information can improve system's performance by combining data mining and ontologies. Eyharabide et al. [5] take advantage of data mining techniques enriched with the semantics gathered within an ontology to produce highly relevant user profiles. The aim of their system is to improve the user profile with contextual information. Profile is enriched semantically by an ontological representation of context to dynamically adapt the attribute selection according to the user's preferences. The main objective of this adaptation method is the identification of patterns and finding the context attributes that frequently occur together. Ontology is used to find common semantics to those context attributes.

2.3 Personalized System Evaluation Approaches

When the user profile is determined we need to find out if it is accurate for user needs. Ontology-based user profile in paper [7] is created without the user interaction. The authors propose algorithms which evaluate profile, followed by investigations into ranking concepts in the profile by their importance and, ultimately, the improvement of the profile by removing unimportant concepts.

Sieg et. al [17] performed experiments and has shown that re-ranking the search results based on the interest scores and the semantic evidence in an ontological user profile successfully provides the user with a personalized view of the search results by bringing results closer to the top when they are most relevant to the user.

In this paper we focus on building ontology-based profile and method for determining groups of user profiles. We propose to use ontology alignment method to find similar users.

3 Model of Personalized Document Retrieval System

In [10] the authors proposed the following schema of collaborative recommendation system (Fig. 1). It consists of module for storing user profiles, module of determining representative profile of group and new user classification and adaptation module.

The module for storing user profiles consists of database with user profiles and a method for profile clustering used to determine groups of users. In every group of users a representative profile is determined using various tools from knowledge integration area. The last module refers to a new user – when he comes to the system, a representative profile is recommended for him. The system observes his activities and adapts his profile according his behaviour.

In this paper we consider only module of user profiles on this schema (Fig. 1). We propose ontological structure of user profile and a method of determining groups of users based on the ontology alignment approach. The authors of [14] presented a framework for multi-attribute ontology alignment. The main idea of our clustering model is to determine alignments of each pair of user profiles and to find similar profiles. A function of ontology alignment is used as similarity measure between user profiles.

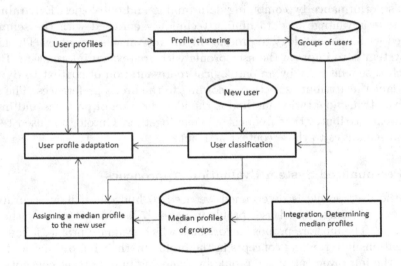

Fig. 1. Schema of Personalized Document Retrieval System

3.1   Model of Document

We consider the following definition of a set of documents.

$$D = \{d_i : i = 1, 2, \ldots, n_d\} \tag{1}$$

where n_d is a number of documents and each document d_i is described in the following way:

$$d_i = \{(t_j^i, w_j^i) : t_j^i \in T \wedge w_j^i \in [0.5, 1), j = 1, 2, \ldots, n_d^i\} \tag{2}$$

where t_j^i is the index term coming from assumed set of terms T, w_j^i is the appropriate weight and n_d^i is a number of index terms that describe document d_i.

3.2 Model of User Profile

We assume that the user profile is an ontology. The definition of ontology is taken from paper [14].

Ontology-based user profile is defined as a triple:

$$O = (C, R, I) \tag{3}$$

in which by C we denote a finite set of concepts, by R a finite set of relations between concepts $R = \{r_1, r_2, ..., r_n\}$, $n \in N$ and $r_i \subset C \times C$ for $i \in [1, n]$ and by I a finite set of instances.

Assuming the existence of a finite set A of all possible attributes and a finite set V of their valid valuations such that $V = \bigcup_{a \in A} V_a$, where V_a is a domain of an attribute a in real world we will call a pair (A, V).

The structure of a concept c is defined as a triple:

$$c = (Id^c, A^c, V^c) \tag{4}$$

where Id^c is its unique identificator, A^c is a set of attributes assigned to c and V^c is a set of domains of attributes from A^c defined as $V^c = \bigcup_{a \in A^c} V_a$.

Every ontology that meets the following criteria:

1. $\forall_{c \in C} A^c \subseteq A$
2. $\forall_{c \in C} V^c \subseteq V$

will be called (A, V)-based.

An instance i taken from the set I is defined as:

$$i = (id, A_i, v_i) \tag{5}$$

where id is its identificator, A_i is a set of attributes describing the instance i and v_i is a function with a signature $v_i : A_i \rightarrow \bigcup_{a \in A_i} V_a$, which assigns to attributes from the set A_i specific values taken from their domains.

Assuming the existence of a concept $c = (Id^c, A^c, V^c)$ we say that a member of the set I $i = (id, v_i, A_i)$ is its instance only if:

1. $A^c \subseteq A_i$
2. $\forall_{a \in A_i \cap A^c} v_i(a) \in V^c$

In the user profile model we would understand a concept as a keyword (or keywords which are synonyms) that the user is interested in, relations between concepts are relations between the keywords (e.g. relation "is-a-kind" or generalization – specification relation) and the instance is a document that is described using a particular keyword.

3.3   Method of User Profile Clustering

We assume the collaborative recommendation system which is presented in Fig. 1. Here we consider a method for profile clustering using ontology alignment approach on concept and instance level.

Aligning ontologies on the concept level is defined as follows:

For given two (A, V)-based ontologies O and O', one should determine a function $\lambda_c(c, c')$ that represents the degree to which concept c from the ontology O can be aligned to concept c' from the ontology O'.

The aim of our system is to determine a representative profile using alignment between two users' profiles. On the concept level we compare sets of keywords which users are interested in. We can define a function $\lambda_c^{(i,j)}(c_i, c_j')$ in the following way:

$$\lambda_c^{(i,j)}(c_i, c_j') = \begin{cases} 1 & \text{if } c_i \text{ and } c_j' \text{ are synonyms} \\ \beta & \text{if } c_i \text{ and } c_j' \text{ are connected by a relation in reference ontology} \\ 0 & \text{in other case.} \end{cases}$$

$$(6)$$

where β is a constant value which refers the strength of relation between two concepts (keywords) in reference ontology, eg. $\beta = 0.75$ for "is-a" relation; c_i is a i-th concept in ontology O, $i \in \{1, \ldots, m\}$ and c_j' is a j-th concept in ontology O', $j \in \{1, \ldots, l\}$.

Then we can calculate the value of function $\lambda_c(c, c')$ as:

$$\lambda_c(c, c') = \frac{\sum_{i,j} \lambda_c^{(i,j)}(c_i, c_j')}{k \cdot l}$$

$$(7)$$

Aligning ontologies on the instance level is a more sophisticated procedure. We propose not to compare every pair of two profiles but only such pairs that the value of their alignment on the concept level is greater than assumed threshold η. In this way we can decrease the number of comparisons on instance level.

The comparison of two profiles on the instance level is conducted as follows. First we should determine which instances should be taken into account. We consider only those concepts which have value of $\lambda_c^{(i,j)}(c_i, c_j')$ greater than the assumed threshold γ.

Instance of a concept in a single user profile contains a document that the user has assigned as relevant for his information needs. Each document has a list of keywords (eq. 2) and can be assign to many concepts. To compare two instances of the same concept in different profiles we check if document in both instances are the same.

Let us consider instances of two user profiles: i and i'. Let D_c be a set of instances (documents that are relevant for information needs of the user) connected with concept c.

We propose the formula 8 to calculate similarity of concepts of two profiles on instances level.

$$\lambda_i(i_{c1}, i'_{c2}) = \frac{\text{card} \{x : x \in D_{c1} \cap D'_{c2}\}}{\min(\text{card } D_{c1}, \text{card } D'_{c2})} \tag{8}$$

where card D_c is a cardinality of set D_c.

Similarity of two ontology-based user profiles on instance level is calculated as average value of instance similarities.

$$\lambda_i(i, i') = \frac{\sum_{j,k} \lambda_i(i_{cj}, i'_{ck})}{j \cdot k} \tag{9}$$

Merging two ontology-based user profiles is defined in the following way:

For given two (A, V)-based ontologies O_1 and O_2, one should determine a new (A, V)-based ontology O that represents the common part of concept c_1 from the ontology O_1 and concept c_2 from the ontology O_2 and common part of instance i_1 from the ontology O_1 and instance i_2 from the ontology O_2.

We can merge two concepts c and c' from ontologies O and O' when the following conditions are satisfied:

1. $\lambda_c(c, c') \geq \alpha_c$
2. $\lambda_i(i, i') \geq \alpha_i$

where α_c and α_i are parameters from interval $(0, 1)$ that should be tuned in experimental evaluations.

A method for user profiles clustering. uses prepared methods of aligning ontology-based profiles on concept and instance levels.

In the system we have N users that have ontology-based profiles described by formula 3. We use hierarchical clustering algorithm to obtain assumed number of groups of similar user profiles [1]. To calculate similarity of user profiles we use methods of ontology alignment on concept and instances level.

Algorithm for user profiles clustering is presented below.

3.4   Idea of Evaluation Procedure

We propose to adapt ontology alignment model to an application of document recommendation to users of collaborative environment based on their preferences.

To evaluate the personalized document retrieval system presented in Fig. 1 we need to prepare methods for each module: storing user profiles, module of determining representative profile of group and new user classification and adaptation module.

The idea of experiments is as follows. In the system we have a set of users (their profiles and demographic data). We cluster the users based on their profiles using methodology presented in Section 3 and determine a representative profile of each group of users. When a new user is coming to the system he is asked

---

Algorithm 1. Algorithm for user profiles clustering.

Input: A set of N ontology-based user profile

Output: A set of K users profiles clusters.

1. Assign $k \leftarrow 0$.
2. Assign each profile to its own cluster (each cluster contains one element).
3. Calculate $N \times N$ matrix of similarities on instance level between each pair of clusters using eq. 9.
4. **while** $k < N - K$ **do**
   - (a) Find the most similar clusters (the greatest value of similarities).
   - (b) Merge the clusters using ontology alignment method.
   - (c) Calculate the similarities between new cluster and each of the old clusters.
   - (d) $k \leftarrow k + 1$.

---

to provide his demographic data. Using the method for rough classification [13] we can determine an appropriate group for a new user and recommend him a non-empty profile. The new user can obtain personalized results just in the first session in the system. The system should take into account user current activities and adapt user profile according to his current information needs. When each user profile is evolving, the clustering procedure should be performed to determine appropriate groups of similar users.

In this paper we have presented methods for clustering user profiles and determining representative profile of a group using ontology alignment approach. Experimental evaluation will be performed to tune system parameters but to check the quality of the system, we need to develop methods for each part of the system.

4 Summary and Future Works

In this paper the author considers ontology-based user profile in a personalized document retrieval system. Methods for comparing user profiles using ontology alignment on concept and instance level are presented. To divide the whole set of users in the system, a standard hierarchical clustering approach is proposed. When a new user is coming to a group a non-empty profile is recommended for him.

In the future we plan to develop methods for every module and to evaluate the presented model of personalized document retrieval system in simulated and real environment.

Acknowledgments. This research was partially supported by Polish Ministry of Science and Higher Education.

References

1. Borgatti, S.P.: How to Explain Hierarchical Clustering. Connections **17**(2), 78–80 (1994)
2. Cena, Federica, Likavec, Silvia, Osborne, Francesco: Propagating User Interests in Ontology-Based User Model. In: Pirrone, Roberto, Sorbello, Filippo (eds.) AI*IA 2011. LNCS, vol. 6934, pp. 299–311. Springer, Heidelberg (2011)
3. Chen, Y.J., Chu, H.C., Chen, Y.M., Chao, C.Y.: Adapting domain ontology for personalized knowledge search and recommendation. Information & Management **50**, 285–303 (2013)
4. Dicheva D., Aroyo L.: An approach to intelligent information handling in web-based learning environments. In: Proceedings of ICAI 2000 (2000)
5. Eyharabide, V., Amandi, A.: Ontology-based user profile learning. Applied Intelligence **36**, 857–869 (2012). doi:10.1007/s10489-011-0301-4
6. Hoppe, Anett, Roxin, Ana, Nicolle, Christophe: Dynamic, behavior-based user profiling using semantic web technologies in a big data context. In: Demey, Yan Tang, Panetto, Hervé (eds.) OTM 2013 Workshops 2013. LNCS, vol. 8186, pp. 363–372. Springer, Heidelberg (2013)
7. Jayanthi, J., Jayakumar, K.S., Surendran, S.: Generation of ontology based user profiles for personalized web search. In: Proceedings of 3rd International Conference on Electronics Computer Technology (ICECT) (2011). doi:10.1109/ICECTECH.2011.5942090
8. Khan, S., Safyan, M.: Semantic matching in hierarchical ontologies. Journal of King Saud University - Computer and Information Sciences **26**, 247–257 (2014)
9. Mianowska, B., Nguyen, N.T.: Tuning User Profiles Based on Analyzing Dynamic Preference in Document Retrieval Systems. Multimedia Tools and Applications (2012). doi:10.1007/s11042-012-1145-6
10. Maleszka, M., Mianowska, B., Nguyen, N.T.: A method for collaborative recommendation using knowledge integration tools and hierarchical structure of user profiles. Knowledge-Based Systems **47**, 1–13 (2013)
11. Maleszka B.: Methods for User Personalization in Document Retrieval Systems Using Collective Knowledge. Ph.D. thesis. Wroclaw University of Technology (2014)
12. Middleton, S.E., Alani, H., Roure, D.: Exploiting synergy between ontologies and recommender systems. arXiv preprint cs/0204012, (2002)
13. Nguyen, N.T.: Rough Classification - New Approach and Applications. Journal of Universal Computer Science **15**(13), 2622–2628 (2009)
14. Pietranik, M., Nguyen, N.T.: A multi-attribute based framework for ontology aligning. Neurocomputing **146**, 276–290 (2014)
15. Potey, M., Sinha, P.K.: Review and analysis of machine learning and soft computing approaches for user modeling. International Journal of Web & Semantic Technology (IJWesT) **6**(1), 39–55 (2015)
16. Razmerita, L., Angehrn, A., Maedche, A.: Ontology-based user modelling for knowledge management systems. In: Brusilovsky, Peter, Corbett, Albert T., de Rosis, Fiorella (eds.) UM 2003. LNCS, vol. 2702, pp. 213–217. Springer, Heidelberg (2003)
17. Sieg, A., Mobasher, B., Burke, R.: Learning Ontology-Based User Profiles: A Semantic Approach to Personalized Web Search. IEEE Intelligent Informatics Bulletin **8**(1), 7–18 (2007)

18. Tang, X., Zeng, Q.: Keyword clustering for user interest profiling refinement within paper recommender systems. The Journal of Systems and Software **85**, 87–101 (2012)
19. Zeng, Q., Zhao, Z., Liang, Y.: Course ontology-based users knowledge requirement acquisition from behaviors within e-learning systems. Computers & Education **53**(3), 809–818 (2009)
20. Zhang, Hui, Song, Yu., Song, Han-tao: Construction of ontology-based user model for web personalization. In: Conati, Cristina, McCoy, Kathleen, Paliouras, Georgios (eds.) UM 2007. LNCS (LNAI), vol. 4511, pp. 67–76. Springer, Heidelberg (2007)

User Personalisation for the Web Information Retrieval Using Lexico-Semantic Relations

Agnieszka Indyka-Piasecka[1(✉)], Piotr Jacewicz[2], and Elżbieta Kukla[1]

[1] Department of Information Systems, Wrocław University of Technology,
Wybrzeże Wyspiańskiego 27 50-370, Wrocław, Poland
{agnieszka.indyka-piasecka,elzbieta.kukla}@pwr.edu.pl
[2] Faculty of Computer Science and Management, Wrocław University of Technology,
Wybrzeże Wyspiańskiego 27 50-370, Wrocław, Poland

Abstract. This contribution presents a new approach to the representation of user interests and preferences at information retrieval process on the Web. The adaptive user profile includes both interests given explicitly by the user, as a query, and also preferences expressed during relevance valuation process, so to express field independent translation between terminology used by the user and terminology accepted in some field of knowledge. Building, modifying, expanding (by semantically related terms) and using procedures for the profile are presented. Experiments concerning the profile, as a personalization mechanism of Web retrieval system, are presented and discussed.

Keywords: Personalized web retrieval · User profile adaptation · plwordnet · Lexico-semantic relations

1 Introduction

In today's World Wide Web reality, the common facts are: increasingly growing number of documents, high frequency of their modifications and, as consequence, the difficulty for users to find important and valuable information. These problems caused that much attention is paid to helping user in finding important information on Internet Information Retrieval (IIR) systems. Individual characteristic and user needs are taken under consideration, what lead to system personalization. System personalization is usually achieved by introducing user model into information system. User model might include information about user preferences and interests, attitudes and goals [3], knowledge and beliefs [6], personal characteristics [7], or history of user interaction with a system [12]. User model is called user profile at the domain of Information Retrieval (IR). The profile represents user information needs, such as interests and preferences and can be used for ranking documents received from the IR system. Such ranking is usually created due to degree of similarity between user query and retrieved documents [8], [14]. In Information Filtering (IF) systems, user profile became the query during process of information filtering. Such profile represents user information needs, relatively stable over a period of time [1]. User profile

© Springer International Publishing Switzerland 2015
M. Núñez et al. (Eds.): ICCCI 2015, Part I, LNAI 9329, pp. 421–429, 2015.
DOI: 10.1007/978-3-319-24069-5_40

also has been used for query expansion, based on explicit and implicit information obtained from the user [5].

The main issue, at the domain of user profile for IR, is a representation of user information needs and interests. Usually user interests are represented as a set of keywords or a n–dimensional vector of keywords, where every keyword's weight (at the vector) represents the importance of a keyword in the description of user interests [8], [9], [16]. The approaches with more sophisticated structures representing knowledge about user preferences are also described: stereotypes – the set of characteristics of a prototype user of users groups, sharing the same interests [4], or semantic net, which discriminates subject of user interests by underlining the main topic of interests [2].

The approaches to determine user profile can be also divided into a few groups. The first group includes methods where interests are stated explicitly by the user in specially prepared questionnaires or during answering a set of standard questions [4], [8], or by an example snippet of text, written by the user [11]. The second group are these approaches, where user profile is determine during the analysis of terms frequency in the user queries directed to IR system [8]. Behind these methods is the assumption that the user interest, represented by a term, is higher as the term is more frequently used in the queries. Analysis of the queries supported by genetic algorithms [13], reinforcement learning [17] or semantic nets [2] are extensions to this approach. The third group of approaches includes methods, in which the user evaluates retrieved documents. Additional index terms, obtained from documents and assessed as interesting by the user, are introduced to the user profile and expanded the user interests description [4], [8].

Most of research, concerning user modeling for IR, acquire user information needs expressed directly by the user of the IR system. Serious difficulties of a user in expressing his real information need are frequently neglected. As well the fact is ignored that user usually does not know precisely, which words he should use to describe his interests and to receive valuable documents from IR system. User formulates his query with his *subjectively chosen words*, which may occur to be not very correct and popular at the retrieval domain. So the main idea of the *adaptive user profile*, presented in our approach, is to join user subjective words with the objective description of retrieval domain. The connection between terminology used by the user and terminology accepted at some field of knowledge might be considered as a kind of *translation*, which describes the meaning of words used by the user in a context fixed by relevant documents. At the *adaptive user profile* the *translation* is described by assigning to the user *query pattern* a *subprofile*, created during the *relevance valuation process* for retrieved documents. From relevant documents the objectively "proper" vocabulary of the retrieval domain could be identified.

We claim at these contribution that user can express own preferences by relevance valuation of retrieved documents. The valuation process is not a burdensome task for the user, while he only points out several documents, as an example of the most relevant documents. The user neither assigns any relevance values to the retrieved documents nor evaluate all documents from the answer of IR system.

2   User Profile

The IR system we define as four elements: set of documents D, user profiles P, set of queries Q and set of terms in dictionary T. The retrieval function is $\omega: Q \rightarrow 2^D$. Retrieval function returns a set of documents, which is the answer to query q. Set $T = \{t_1, t_2, ..., t_n\}$ contains terms from all documents, which have been indexed by Web IR system and is called dictionary. By D_q we describe a set of relevant documents among retrieved documents D_q' for query q: $D_q' = \omega(q,D)$ and $D_q \subseteq D_q'$.

Therefore, the *user profile* $p \in P$ is a set of pairs:

$$p = \{\langle s_1, sp_1 \rangle, \langle s_2, sp_2 \rangle, ..., \langle s_l, sp_l \rangle\} \tag{1}$$

where: s_j – user query pattern, sp_j – user subprofile (one user query pattern indicates only one user subprofile univocally).

For profile p we define function π, which maps: user query q, the set of retrieved relevant documents D_q and the previous user profile p_{m-1} into new user profile p_m. The function π determines the profile modifications. Thus, the profile is the following multi-attribute structure: $p_0 = \varnothing$, $p_m = \pi(q_m, D_q, p_{m-1})$. For the user profile we define a set of *user subprofiles SP* (presented below).

The user profile is created on the basis of *user verification of the documents* retrieved by Web IR system. During verification the user points out a sample of few documents which he considers the most relevant to his interests.

The user query pattern s_j is a Boolean statement, the same as user query q: $s_j = r_1 \wedge r_2 \wedge r_3 \wedge ... \wedge r_n$, where r_i is a term: $r_i = t_i$, a negated term: $r_i = \neg t_i$ or logical one: $r_i = 1$ (for terms which do not appear at the question). The user query pattern s_j is assigned to subprofile and is connected to only one subprofile.

The user subprofile $sp \in SP$ is defined as n-dimensional vector of terms weights, (the terms are from relevant documents): $sp_j^{(k)} = (w_{j,1}^{(k)}, w_{j,2}^{(k)}, w_{j,3}^{(k)}, ..., w_{j,n}^{(k)})$, where SP – set of subprofiles, n – number of terms in dictionary T: $n = |T|$, $w_{j,i}^{(k)}$ – weight of significant term tz_i in subprofile after k-th subprofile modification.

The terms from dictionary T are an indexing terms at Web IR system. These terms are indexing documents retrieved for query q and belong to relevant documents.

The weight of significant term tz_i in subprofile is calculated as following:

$$w_{j,i}^{(k)} = \frac{1}{k}((k-1) w_{j,i}^{(k-1)} + wz_i^{(k)}) \tag{2}$$

where: k – number of retrievals made with using the subprofile, i – index of term in the dictionary T, j – index of subprofile, $w_{j,i}^{(k)}$ – weight of significant term tz_i in subprofile after k-th modification of subprofile[1], subprofile is indicated by pattern s_j (i.e. after k-th document retrieval with the use of this subprofile), $wz_i^{(k)}$ – weight of

---

[1] The weight, called *cue validity*[10], is calculated according to a frequency of term tz_i in relevant documents retrieved by Web IR system at k-th retrieval and a frequency of this term in whole documents of the collection.

significant term tz_i after k–th selection of term t_i. The significant terms selection and weighting algorithms have been presented in [18].

3   Modification of User Profile

The main idea of the presented approach is to join the user subjective words with the objective description of retrieval domain. The *adaptive user profile* expresses the connection between terminology used by the user and terminology accepted at some field of knowledge. This connection might be considered as a kind of translation describing the meaning of words used by the user in a context fixed by relevant documents. Translation is described by assigning to user query pattern s_j a subprofile ('translation') created during the process of significant terms tz_i selection from relevant documents of an answer. We assume the following designations: q – the user query, D_q' – the set of documents retrieved for the user query q, $D_q \subseteq D$, D_q – the set of documents pointed by the user as relevant documents among the documents retrieved for user query q, $D_q \subseteq D_q'$.

As it was described above, user profile p_m is the representation of the user query q, the set of relevant documents D_q and the previous (former) user profile p_{m-1}. After every retrieval and verification of documents made by the user, the profile is modified. The modification is performed according to the following procedure: $p_0 = \varnothing$, $p_m = \pi(q_m, D_q, p_{m-1})$ where p_0 – the initial profile, this profile is empty, p_m – the profile after m–times the user has asked different queries and after each retrieval the analysis of relevant documents was made.

Traditionally, a user profile is represented by one n–dimensional vector of terms describing user interests. User interests change and so should the profile. Usually changes of a profile are achieved by modifications of weights in the vector. After appearance of queries from various domains, modifications made for this profile can lead to an unpredictable state of the profile. By the unpredictable state we mean a disproportional increase of weights of some terms in the vector representing the profile that might not reflect a real increase of user interests in the domain represented by these terms. The weights of terms could be growing, because of high frequency of these terms at whole collection, regardless of the domain of actual retrieval.

The representation of a profile as single vector could also cause ambiguity during the use of such profile for query modification. At certain moment of user retrieval history, a query refers to only one domain of user interests. To use the profile mentioned above, a single vector form, we need a mechanism of choosing from single vector of terms, representing various interests of the user, only these terms that are related to a domain of current query. To obtain this information, knowledge about relationship between terms from a query and a profile, and between terms in profile is needed. In literature, this information is obtained from a co-occurrence matrix created for a collection of documents [15] or from a semantic net [9]. One of disadvantages of these approaches is that two mentioned above structures, namely a user profile and a structure representing term dependencies, should be maintained and managed for each user and also that creating the structure representing term relationships is difficult for so diverging and frequently changing environment as the Web.

The above-mentioned problems are not encountered at the adaptive user profile p created at the presented approach. After singular retrieval, only weighs of terms from the subprofile identified by pattern s_j (identical to users' query) are modified, not all weighs of all terms in the profile. Similarly, when the profile is used to modify user query, the direct translation between current query q and significant terms from the domain associated with the query is used. In profile p, between single user query pattern s_j and single subprofile sp_j a kind of mapping exists representing this translation.

The user profile is created during a period of time – during sequence of retrievals at Web IR system. There could appear a problem how many subprofiles should be kept in the user profile. We have decided that only subprofiles that are frequently used for query modifications should not be deleted. If a subprofile is frequently used, it is important for representing user interests.

The modification of the user subprofile sp is made when, from the set of relevant documents pointed out by the user among retrieved documents, some significant term tz_i is determined as described in [18]. The weight $w_{j,i}^{(k)}$ (2) is modified for these terms and only in one appropriate subprofile identified by the user query pattern s_j. During each retrieval cycle a modification is applicable to only one subprofile and for all significant terms tz_i obtain during the k–th selection of significant terms from the relevant documents retrieved for query q, which was asked k–th time. If the modification took place for significant terms tz_i for every subprofile in user profile, it would cause disfigurement of significant terms importance for single question.

The selected significant terms are further processed with the use of wordnet build for Polish language – plWordNet [19]. A wordnet is dictionary-like lexico-semantic resource that describes lexical meaning of words. Every wordnet, following the Princeton WordNet [20], includes network of synsets – sets of near-synonyms that are linked by lexico-semantic relations such as hypernymy, hyponymy, meronymy, holonymy etc. Synsets represent distinct lexical meanings.

We assume that the use of words lexically related to significant terms enriches the description of user interest domain. So at a subprofile of adaptive user profile, for each significant term we are selecting direct hypernyms, direct hyponyms and all synonyms to expand subprofile. If a word from newly selected set of hyponyms have already been among significant terms at subprofile, hyponyms of this word are not added to subprofile. Introducing further hyponyms for such word could affect unpredictable narrowing of modified query (presented below).

4   Application of User Profile

The user profile contains terms selected from relevant documents. These terms are good discriminators distinguishing relevant documents among the other documents of the collection and also represent the whole set of relevant documents.

The application of user profile p is performed during each retrieval for a user query q. One of the main problems is the selection of significant terms tz_i for query modification. Not all significant terms from a subprofile will be appropriate to modify the next user query, because the query becomes too long.

The user asks new query q_j to Web IR system, new pattern s_j and a subprofile identified by this pattern are added to the profile. The subprofile is created as a result of relevant documents analysis. If user asks next query q_k and this query is identical to previous query q_j, the given query is changed by user profile. The modified query is asked to Web IR system, retrieved documents are verified by the user and the subprofile in user profile is brought up to date. After each use of the same query as query q_j, the subprofile identified by the pattern s_j represents user interests, described at the beginning by the query q_j, even better. Each retrieval, with the use of the subprofile identified by the pattern s_j, leads to query narrowing, a decrease in the number of answer documents, an increase in the number of relevant documents.

Adaptive user profile can be used for query modification if pattern s_j, existing in the profile, is *identical* or *similar* to the current query q_i. For example, for queries: $q_a = t_1 \wedge t_2 \wedge t_3 \wedge t_4$, $q_b = t_1 \wedge \neg t_2$, following patterns: $s_1 = t_1 \wedge t_2 \wedge t_3 \wedge t_4$, $s_2 = t_1 \wedge \neg t_2$ are identical to queries q_a, q_b, respectively. For the same query q_a patterns: $s_2 = t_2 \wedge t_4$, $s_4 = t_1 \wedge t_2$, $s_5 = t_1 \wedge t_3$, $s_6 = t_2$ are similar to query q_a.

If pattern s_j is identical to current user query q_i, the query q_i is replaced by the best significant terms tz_i from subprofile identified by pattern s_j. The weights of these terms are over $\tau_{profile}$ – a dynamically calculated threshold. If user profile consists of more than one pattern similar to current user query q_i, all significant terms tz_i from all subprofiles identified by these patterns are taken under consideration. In such case significant terms weights from all subprofiles identified by similar patterns are summed. The n–dimensional vector $R=(r_1, r_2, ..., r_n)$ is created. The ranking list of all significant terms is created. If the weight r_i is over $\tau_{profile}$ dynamic threshold, significant term tz_i will be introduced to replace current user query q_i.

5   Experiments

The *adaptive user profile* was implemented as a part of Web IR system - the *Profiler* module. User profile has been applied as a retrieval personalisation mechanism. User query modification is performed during user interaction with Web IR system (i.e. the verification process). After verification, the Profiler automatically asks the modified query to Web IR system and presents new answer to the user.

The experiments were performed in two directions. Firstly, we aimed to establish all parameters (i.e. thresholds) for the Profiler module. Then we were verifying the usefulness of multi-attribute adaptive profile, i.e. our goal was to confirm the increase of number of relevant documents retrieved at every retrieval cycle and the decrease of number of document at following answers.

With the Apache Solr, the open source solution for search engine, adopted for Polish language and the Profiler (user profiling module) testing environment has been established. The real users of Web IR system (group of 13 persons) have been asked to evaluate testing sets of documents consisting of relevant documents that represent their real interests. Three types of testing sets were used: *dense*, *loose* and *mixed sets of documents.* The dense sets of relevant documents consist of documents describing only one precise domain of user interests - the strongly similar to each other documents were assessed by users. The loose sets of documents consist of no similar to each other documents originating from various domains of user interests. The mixed sets of documents consist of subsets of closely related documents, identifying one precise domain of interests and a number of documents from disjoint domains of user interests.

The experiments were arranged as a simulation of user behaviour during the information retrieval at Web IR system. The user interests could came from any domain of knowledge but the user query did not always reflects the vocabulary commonly used at this knowledge. The fact of posting by the user not a very precise query was simulated by the process of producing the number of 50 random queries selected randomly among the words of dictionary T. Each random query was asked to Web IR system. If in the answer there were relevant documents from testing sets of documents, the randomly generated query was modified – the significant terms replaced the preliminary query. Further, the modified query was automatically asked to Web IR system and another relevant documents were found from testing sets of documents. Thus the usage of testing sets of documents simulates the process of answer verification made by the user.

Each stage of above described cyclic process we describe as *iteration*. Iterations were repeated until all relevant documents from the dense sets of relevant documents were found or no changes in number of relevant documents were observed (for the loose and mixed sets of relevant documents).

For every random query at experiment: the number of all retrieved documents D'_q, effectiveness $\%DR$ (the effectiveness is defined as a percent of relevant documents retrieved by the modified queries from the set of all relevant documents in the testing sets) and precision $Prec_m$ (as a standard precision in IR for the first $m=10, 20, 30$ documents in the answer) were calculated at every iteration. Above mentioned measures has been chosen to calculate retrieval improvement for the proposed method.

Table 1. Measures of retrievals improvement made during experiments.

	Percent of modified queries			Effectiveness %DR			
	improvement in precision $Prec_m$	no improvement in precision $Prec_m$	partial improvement in precision $Prec_m$	75 -100%	50 -75%	25 -50%	0 -25 %
dense sets	85 %	9 %	6 %	57 %	10 %	17 %	16 %
loose sets	64 %	15 %	21 %	0 %	6 %	66 %	28 %
mixed sets	61 %	15 %	24 %	9 %	36 %	28 %	27 %

The retrievals made for the dense sets of relevant documents confirmed the assumption that for most of the modified queries the retrieval results were better in comparison to preliminary query. For over 85% of preliminary queries, $Prec_m$ measure was increasing at every iteration of query modification (Table 1). The effectiveness measure %*DR* of the proposed approach shows that over 75% of all relevant documents from testing dense sets were found for over 57% of asked queries. The number of all documents retrieved and returned as the answer diminishes with every iteration.

The experimental retrievals arranged for the loose sets of relevant documents showed that all subsequently modified queries (originating from the same preliminary query) have always been focusing on the same domain of user interests. This property was assured by the method of adaptive profile creation, modification and application and verified by experiments.

In case of the loose sets of relevant documents, for more then 64% of preliminary queries, $Prec_m$ measure rose with each iteration of query modification. The number of all retrieved documents diminishes as well. For the rest of the queries in the loose sets of relevant documents, the measured parameters were worse, because only a single document has been selected by Web IR system as the answer for each subsequent modification of the query. No similar documents were found because the loose testing sets consist of no similar to each other documents taken from disjoint domains of user interests.

6   Conclusions

In this contribution we propose a new and universal approach to the representation of user interests and preferences. The *adaptive user profile* includes both interests given explicitly by the user, as a query, and also preferences expressed by relevance valuation of retrieved documents. The achievement of the profile is the ability to express field independent translation between user terminology and terminology accepted in some field of knowledge. This universal translation is supposed to describe the meaning of words used by user (i.e. user query) in context fixed by the retrieved documents (i.e. user subprofile). During retrievals at Web IR system the user is supported by the adaptive user profile, performing more precise document retrievals during every query reformulation. The query modification procedure determines that subsequent retrievals return to the user the set of retrieved documents that definitely meets user information needs.

Acknowledgments. This work was supported by the European Commission under the 7th Framework Programme. Coordination and Support Action, Grant Agreement Number 316097, ENGINE - European research center of Network intelliGence for INnovation Enhancement (http://engine.pwr.wroc.pl/).

References

1. Gentili, G., Micarelli, A., Sciarrone, F.: Infoweb: An adaptive information filtering system for the cultural heritage domain. Applied Artificial Intelligence 17(8–9), 715–744 (2003)
2. Casoto, P., Dattolo, A., Omero, P., Pudota, N., Tasso, C.: Accessing, analyzing, and extracting information from user generated contents. Handbook of Research on Web 2(3.0), 312–328 (2009)

3. Pazzani, M.J., Billsus, D.: Content-based recommendation systems. In: Brusilovsky, P., Kobsa, A., Nejdl, W. (eds.) Adaptive web 2007. LNCS, vol. 4321, pp. 325–341. Springer, Heidelberg (2007)
4. Paliouras, G., Papatheodorou, C., Karkaletsis, V., Spyropoulos, C.D.: Discovering user communities on the Internet using unsupervised machine learning techniques. Interacting with Computers **14**(6), 761–791 (2002)
5. Nanas, N., Uren, V., De Roeck, A.: Building and applying a concept hierarchy representation of a user profile. In Proc. of the 26th Annual International ACM SIGIR Conference on Research and Development in Information Retrieval, pp. 198–204. ACM (2003)
6. Bull, S., Mabbott, A., Abu-Issa, A.S.: UMPTEEN: Named and anonymous learner model access for instructors and peers. Int. Journal of Artificial Intelligence in Education **17**(3), 227–253 (2007)
7. Kumar, V., Greer, J., McCalla, G.: Assisting online helpers. International Journal of Learning Technology **1**(3), 293–321 (2005)
8. Daniłowicz, C.Z.: Modelling of user preferences and needs in Boolean retrieval systems. Information Processing and Management **30**(3), 363–378 (1994)
9. Davies, N.J., Revett, M.C.: Networked information management. BT Technology Journal **25**(3–4), 285–298 (2007)
10. Goldberg, J.L.: CDM: An Approach to Learning in Text Categorization. International Journal on Artificial Intelligence Tools **5**(1 and 2), 229–253 (1996)
11. Indyka-Piasecka, A., Piasecki, M.: Adaptive translation between user's vocabulary and internet queries. In: Proc. of the IIS IPWM 2003, pp.149–157. Springer (2003)
12. Danilowicz, C., Indyka-Piasecka, A.: Dynamic user profiles based on boolean formulas. In: Orchard, B., Yang, C., Ali, M. (eds.) IEA/AIE 2004. LNCS (LNAI), vol. 3029, pp. 779–787. Springer, Heidelberg (2004)
13. Jeapes, B.: Neural Intelligent Agents. Online and CDROM Rev. **20**(5), 260–262 (1996)
14. Maglio, P.P., Barrett, R.: How to build modeling agents to support web searchers. In: Proc. of the 6th Int. Conf. on User Modeling, pp. 5–16. Springer (1997)
15. Moukas, A., Zachatia, G.: Evolving a multi-agent information filtering solution in amalthaea. In: Proc. of the Conference on Agents, Agents 1997. ACM Press (1997)
16. Salton, G., Bukley, C.H.: Term-Weighting Approaches in Automatic Text Retrieval. Information Processing and Management **24**(5), 513–523 (1988)
17. Seo, Y.W., Zhang, B.T.: A reinforcement learning agent for personalised information filtering. In: Int. Conf. on the Intelligent User Interfaces, pp. 248–251. ACM (2000)
18. Indyka-Piasecka, A.: Using multi-attribute structures and significance term evaluation for user profile adaptation. In: Jędrzejowicz, P., Nguyen, N.T., Hoang, K. (eds.) ICCCI 2011, Part I. LNCS, vol. 6922, pp. 336–345. Springer, Heidelberg (2011)
19. Piasecki, M., Szpakowicz, S., Broda, B.: A Wordnet from the Ground Up. Oficyna Wydawnicza Politechniki Wrocławskiej (2009)
20. Fellbaum, C. (ed.): WordNet – An Electronic Lexical Database. The MIT Press (1998)

Formal Models and
Simulation

Formal and Computational Model
for A. Smith's Invisible Hand Paradigm

Tadeusz (Tad) Szuba[1(✉)], Stanislaw Szydlo[2], and Pawel Skrzynski[1]

[1] Dept. of Applied Computer Science, AGH University, Cracow, Poland
{szuba,skrzynia}@agh.edu.pl
[2] Faculty of Management, AGH University, Cracow, Poland
sszydlo@zarz.agh.edu.pl

Abstract. We present probably the first formal theory of A. Smith's Invisible Hand paradigm (ASIH). It proves that it is not only an idea, that is often conflicting with established governing methods, but something real, for which a formal theory can be built. This should allow for the creation of new tools for market analysis and prediction. It claims, that a market has another dimension of computational nature, which itself is a complete programmable computer; however quite different from a digital computer. There, unconscious meta-inference process of ASIH is spread on the platform of brains of agents and the structure of a market. This process is: unconscious, distributed, parallel, non-deterministic with properties of chaotic systems. A given thread of ASIH emerges spontaneously in certain market circumstances and can vanish when the market situation changes. Since this computer is made up of brains of agents, conclusions of this inference process affect the behavior of agents and therefore the behavior of the entire market. For a description of ASIH, a molecular model of computations must be used. Our theory shows that ASIH is much more "universal" than expected, and is not only restricted to market optimization and stabilization, but can also act as a "discoverer" of new technologies, (technical market optimization). Also rules of social behavior can be discovered (social market optimization). ASIH can be considered as Collective Intelligence (CI) of a market, similarly to Collective Intelligence of an ant hill.

Keywords: Macroeconomic · A. smith's invisible hand (ASIH) · Collective intelligence (CI) · Computational model · Formal theory · Simulation model · Unconscious meta-inference process

1   Introduction

Secondary school textbooks define Economics as *"a study of human behavior"*. However, behavior is driven by individual "mental calculations and inferences" moderated by interaction with other market agents (individuals and organizations). A question can be raised, if it is possible to build a formal and computational model of global mental and computational processes in brains of agents, over the given market? Let's tentatively name this as Collective Intelligence [7] of a Market (CIM).

© Springer International Publishing Switzerland 2015
M. Núñez et al. (Eds.): ICCCI 2015, Part I, LNAI 9329, pp. 433–442, 2015.
DOI: 10.1007/978-3-319-24069-5_41

An ant colony is composed of primitive beings (compared to humans), considered by scientists as automata, displaying astonishing computational abilities as a colony. Ant colonies use specific computational methods (ant colony optimization algorithms), on the platform of moving agents, who are using pheromones for mutual information transfer. Some computational methods used by ants are still not decoded, e.g. how architecture of an ant hill is calculated and optimized. It is certain, that ants are unconscious of this optimization, because their brains are too primitive. We should respect the computational power of an ant colony; individually intelligent and stronger dinosaurs have become extinct, despite originating from a similar time in Earth's history.

Adam Smith (1723 -1790); a Scottish moral philosopher, pioneer of political economy, with just two documented statements, had managed to seed theoretical economics with the term „Invisible Hand" - which is a force controlling and optimizing markets beyond our consciousness. It means that many economists were subconsciously aware of symptoms of the Invisible Hand however, conscious macroeconomic thinking was not able to prove the existence of the Invisible Hand. Thus, since the advent of this term, it remained labeled as a phenomenon or metaphor [3], [4], [5]. Such collective, mass belief must be treated with respect, and an attempt must be taken to formalize this phenomenon/metaphor.

A. Smith had also provided a hint on how to begin formalization of the Invisible Hand almost 260 years ago. This is also a hint on how to formalize a market-agent into market-automata.

Let us assume, that the Invisible Hand for a given market is a symptom of existence of another dimension of a market, which is of computational nature, where computational processes (calculations + inferences) take place. They are spontaneous, not continuous, parallel, nondeterministic/chaotic[1]. They run on the platform of agent's brains, and have control nature. This is a set of entry assumptions (or axioms[2]) for our theory. This will be scientifically justified (this is this paper's main task) in such way, that a complete formal simulation model of very special, however programmable computer, functioning in this dimension, will be given. This computer has a different architecture compared to digital computers. Let's name this computer ASIH-computer.

A similar idea has been also proposed in much wider sense by Stephen Wolfram (author of Mathematica), who following an idea of Edward Fredkin[3], has concluded in his book "A new kind of science" [9] that the universe itself would then be an automaton (cellular automation), like a giant computer.

In the next, second part of this paper it will be demonstrated, that following A. Turing's methodology and A. Smith's hint, a market agent can be formalized into the concept of market automata, very similar in his nature to the famous Turing Machine. The difference is, that the Turing Machine has been designed to formalize "a mathematical agent". Market automata representing market agents as computational and inferring elements in ASIH-computer are displacing around the market (most of

---

[1] E.g. "butterfly effect" can take place.

[2] Axioms to build certain theory.

[3] Edward Fredkin: professor at Carnegie Mellon University, pioneer of digital physics.

them). Thus the standard model of computations, which is the basis of a digital computer cannot be used. For this, in the third part of the paper, a molecular model of computations will be used to assemble market automata into a global computational system i.e. The ASIH-computer. In the fourth part, we will discuss how computations are performed in an ASIH-computer, since an operating system is not applicable in a distributed and chaotic system. Moreover, it is difficult to express a business way of thinking with the help of 0/1 Boolean algebra. Thus, the concept of computations based on inference logic (similar to predicate calculus) and driven by abstract *value*, will be given. Part five will demonstrate, on the basis of a very simple example, how the ASIH-computer works. As an example, a 15th century Mediterranean quasi-free market of spice trade has been selected. It was an extremely fascinating moment in European history, because Arab taxes and Dardanelle sea route cut-off by Ottoman Turks had "pushed" Europeans toward great geographical discoveries and rapid technological and economic development. This resulted in the emergence of today's modern Europe, oriented toward industry and trade. It resembles Renaissance in art. Computations in an ASIH-computer emerge spontaneously, when any market impulse happens. The above case is a perfect example. The paper will end with conclusions.

2   Model of Market Agent as Computational Element in an ASIH-Computer

Let's quote Adam Smith's probably best known remark [6]:

> *„It is not from the benevolence of the butcher, the brewer or the baker, that we expect our dinner, but from their regard to their own self-interest. We address ourselves, not to their humanity but to their self-love, and never talk to them of our own necessities but of their advantages."*

This remark allows us to convert a human-agent intro inferring-automata, oriented for his profit, in the same way as Alan Turing converted a human-clerk into Turing Machine [8] - computational automata to investigate what a human can calculate. This allows us to reduce the market agent into an automata with a very limited thinking horizon (narrowed down to only self-interest). Speaking more precisely, it allows us "to make him" a computational element (a processor). Such an agent is deprived of humanity[4] (what A. Smith states directly), but not from intelligence in terms of planning, problem-solving, discovering analogies or generalizations. Business thinking makes agent logically sensitive only to his profit or loss. In our concept all agent's behavior is based on handling of market goods (irrelevant of what nature) in terms of logic (logic description of goods) and abstract *value*[5] assign to market objects. We can derive more from the aforementioned remark, even though it is not a direct statement: a butcher, brewer and baker are settled agents, but the agent-narrator is moving, driven by necessity or will to do business. However, if a butcher's, brewer's and

---

[4] Humanity will require a lot of extra computations which we cannot afford at this moment.
[5] Abstract *value* can be instantiated with help of concept of money (what Phoenicians did).

baker's self-interests are high, they will visit the narrator offering goods they produce. This significantly simplifies the formalization of a market agent. Fig. 1 demonstrates the assumed structure of a single agent. Please note, the similarity to a Turing Machine, which has emerged on its own, quite unintentionally.

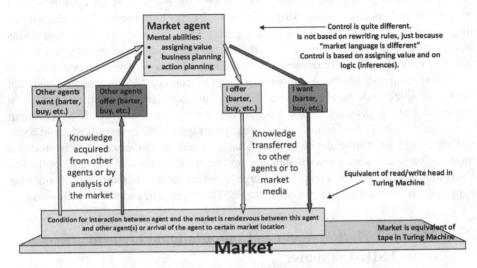

Fig. 1. Market agent as basic computational element in ASIH-computer. On right side there are references to Turing Machine.

It is possible to make and simulate this model, however several advanced programming tools were necessary to be used and interfaced (see part 5) into one simulation system.

3   Molecular Model of Computations as Frame for ASIH-Computer

Standard Model of Computations (SMC) based on von Neumann architecture (Princeton architecture) assumes, that elements for processing, data storage and control are "settled" e.g. on a silicon chip, and interconnections are fixed. This way, the architecture of a computer is defined. Only data and instructions are moving (are transferred). However, if we will consider today's global network as a computer, the SMC model of computations should be questioned. The reason for this is, that traveling human-agents carrying mobile computers (smartphones) can vanish from one network location and emerge in another network location. Airports with Wi-Fi facilities are perfect examples. This can be named as rendezvous of traveler's computer with the world www network at any new location. In such a case, both processing elements, as well as data, are moving. Such displacements are highly random/chaotic in character. Rendezvous can take place both between traveler's mobile computer and network at certain network location, as well as mutual rendezvous between two computers (processors). Since the involved live agents provide mobility and hardware processor(s), for the purity of definitions, it is better to encapsulate (pair) them into one "concept" of *information_molecule* (IM). Figure 2 depicts, with the help of Feynman diagrams, the idea of Molecular Model of Computations (MMC) in two-dimensional spacetime.

The MMC model has until now only one, but famous implementation, i.e. Adleman's DNA computer [1]. In 1994 Adleman had demonstrated, that a computer based on this concept can be built, is programmable and is able to solve a Hamiltonian problem. The interesting thing about this implementation of the MMC model, is that the DNA molecule is a processor, data and an agent at the same time, encapsulating all three. In MMC model of computations, following properties are fundamental:

- Processors, data storage are encapsulated in various ways in *information_molecules* (IM). From a formal point of view, an IM hierarchy can be defined, e.g. to define company as a structure of agents;
- IM are displacing in Computational Space (CS). Metrics for this space can be Cartesian or abstract (e.g. for case of network). Displacements can be of different nature, e.g. in DNA-computer they are based on Brownian motion. In an ASIH-computer, IM representing business agents can intentionally travel to business destinations (market locations). The nature of CS depends on implementation. It can be chemical reactor, network, market, etc.;
- Inference process take place when given IM rendezvous with other IM at rendezvous distance or given IM rendezvous with specific location in CS (allowing or firing inference because of his properties);
- In a MMC model there are no entry assumptions on nature of inferences in case of rendezvous. It is open for definition or specific implementation.

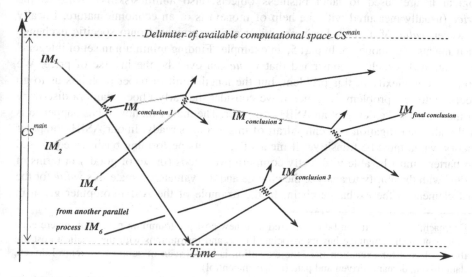

Fig. 2. Basic idea of Molecular Model of Computations (MMC). CS spacetime has 2 dimensions.

4   Computations Driven by Logic and Abstract Value

When analyzing an MMC model of computations, it is immediately visible how inferences there are uncontrollable and vulnerable to so-called combinatorial explosions. This is also visible in Adleman's DNA computer [1]. Adleman tackled combinatorial explosion of emerging local sub-solutions of a traveling-salesman problem with the help of a sequence of manual filtrations of DNA molecules which owned required logic properties. It is amazing that nature has found an alternative and powerful method for the need of an ASIH computer: calculations are driven by abstract *value* + ability of information molecules (agents) to displace in computational space. What is natural in economy, is not yet natural in theory of computations. Some researchers have theoretically noticed that controlling computations (in digital computer) with concept of *value* is feasible[6] [2]. Yet, no such digital computer has been built. Combinatorial explosion is profitable for a market because it works as a generator of all possible solutions for business activity, however possible solutions which are not labeled at the end with *profit*, will be filtered-out. Thus in the ASIH model of computations, calculations/inferences are controlled and thus driven by an abstract root parameter: *value*, and derived parameters *cost, loss, profit*. *Value* is used to label business objects or actions, whereas parameters *cost, loss, profit* are used to label actions only. It is assumed that the agent is able to label any given business object or action with these subjective parameters on the basis of fixed business rules, on the basis of his private considerations or past experience. Thus, such an assignment can be quite different from the point of view of different agents. Moreover, it can be temporary. Most probably it is also an element of an individual as well as group learning (acquiring knowledge). *Value*[7] in ASIH-computer theory is not a price: *value* is an abstraction of logical nature[8] used to label business objects (also abilities/skills); whereas the *price* (usually measured with the help of money) is of an economic nature. Locally however, *value* of a given business object can be projected onto specific exchange instruments, e.g. money p. In part 5, an example "Finding minimum in set of integers" is mentioned, which demonstrated that *value* can even be the inverse of price. We note the complexity of this problem, but the length limit of paper prohibits us to go deeper into this problem. In general, we consider the early Phoenician era discovery of money to be the doing of an ASIH computer. The theory on how it has happened is still under investigation - an entry draft of this theory is ready. In our ASIH-computer theory, *value* must be transitive. It means that agents performing business e.g. based on barter - must be able to directly compare two goods (or set of goods) in terms of *value*, with the ability to analyze the change and to evaluate/estimate this *value* for the last element in the exchange chain. In the example of the ASIH-computer given in

---

[6] <u>Eberbach</u>: ...In fact, it can be considered as a new cost programming paradigm, where each statement of the language has its associated cost, and instructions with the smallest cost are those selected for execution (i.e., the execution is cost-driven in contrast to control-driven, data-driven, demand-driven and pattern-driven control)...

[7] The difference between value and price is also explained and dissected in the book: K. Marx: Das <u>Kapital</u>.

[8] Predicate calculus style inference rules are applicable.

this paper, *value* is expressed with the help of money on the basis of a simple decision table. Individual *costs* (of actions) are based on the length of a trip of a 15th century agent and local taxes (if any). When defining the resale price, small individual over-head for mediation is added (profit intermediary). Observation of our simulation model demonstrates, that *value* is an astonishingly powerful and universal computational mechanism, which we cannot fully comprehend at this very moment... Most computational algorithms from SMC model of computations (if not all, no proof as yet[9]) can be probably redefined to MMC by applying *value* abstraction, and will probably still function properly. Labeling with *value, cost, loss, profit* must be done in such a way that advancing computations provides *profit* for the involved elements (agents, processors, etc.) with acceptable *profit* for those who finish computations. It must be done dynamically by agents. It can be assumed, that there is *profit* in every step or over several steps (with planned *loss* for some steps). The above discussed *situation*, will get more complicated if we *include* relations between objects; beyond the fundamental, business question if $\exists$ (exists) somebody, somewhere, wanting to sell and if $\exists$ another agent somewhere wanting to buy. Now suppose that agents are "more intelligent" i.e. they can not only use integer calculus to calculate money, *profit*, trip *costs*, etc. but they can also analyze logical relations (e.g. technological) between business objects they trade. Let this ability will be on the propositional calculus level[10]. An example is given below.

Example (Discovering-Fire)
Suppose that the case of fire discovery is formally defined by set of facts and rules, using propositional calculus as given below. Let us also assume (for the moment) that fire is temporarily not necessary (because no agent at given market wants to get *heat*). Therefore all items in the set below are labeled by agents with *value=low*.

$$
\left\{
\begin{aligned}
&1.1 \quad \langle tinder. \rangle^{value=low}, 1.2 \quad \langle fire\_striker. \rangle^{value=low}, 1.3 \quad \langle flint. \rangle^{value=low}, 1.4 \quad \langle fuel. \rangle^{value=low} \\
&1.5 \quad \langle tinder \wedge fire\_striker \wedge flint \xrightarrow{making\ small\ fire} temporary\_fire. \rangle^{value=low}, \\
&1.6 \quad \langle temporary\_fire \wedge fuel \xrightarrow{making\_permanent\_fire} permanent\_fire. \rangle^{value=low} \\
&1.7 \quad \langle permanent\_fire \xrightarrow{heating} heat. \rangle^{value=low}
\end{aligned}
\right\}
\tag{1}
$$

Now suppose, that all of the above facts and rules are randomly spread (as business items: goods, skills or knowledge) among agents and treated in the same way as other business items are. Some items will exist in parallel in different locations. Agents will not trade above items (e.g. they will continue other, more profitable trading, e.g. spices), because nobody is expressing the will to get even one item from the list. We even can say that this thread of inference is possible, but dormant. Moreover, its components are spread around market in brains of different agents, perhaps far away from

---

[9] Under research.

[10] We should not assume, that businessman are philosophers, thus propositional calculus as simplified version of 1st order predicate calculus can be used.

each other. Now, let a strong demand for heat emerge (e.g. outbreak of cold weather), expressed as change of *value* assignment (even local):

$$\langle\,?\,heat.\rangle^{value=veryhigh} \tag{2}$$

Assuming that agents are be able to process *value* according to given below inference rules (*a*, *b*, *c* are any business items):

$$\frac{a,a \rightarrow b^{\,value=high}}{a^{\,value=high}} \quad , \quad \frac{a \rightarrow b, b \rightarrow c^{\,value=high}}{a^{\,value=high} \rightarrow c^{\,value=high}} \tag{3}$$

In our example the system of agents after a certain time will discover fire in the background of doing usual business. This is probably (currently under investigation) the gateway to proper analysis on how in certain situations an ASIH-computer can act as a universal inventor in terms of new technologies, which can also overturn the present state of market as a consequence. It can be considered as mentioned in abstract (technical optimization). Assignment of *value* label to objects and its importance for computational process is well visible in mentioned in part 5 example on how ASIH-computer can solve monkey-and-banana problem. Firing computations is based on assigning *value=high* to banana hanging from the ceiling (palm in our example). Humans at the age of 3 are able to perceive the concept of money. Most probably, monkeys do not perform any form of trade; however concept of *value* works … We can therefore risk the following statement; that perhaps sometimes, the real inventor of scientific and technological progress is not always one, "a devoted researcher or scientist", but "the anonymous world of business".

5   Example of ASIH-Computer

A very simplified model of an ASIH-computer for spice trade quasi free market in the 15th century has been built, to analyze its behavior in terms of what it can "calculate". That particular "moment" in the history of Europe is fascinating from macroeconomic point of view and thus has been chosen as the object of experiments.

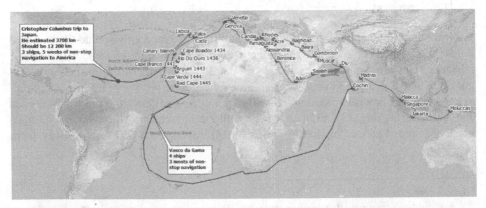

Fig. 3. Structure of the market in terms of computational nodes and data transfer routes.

An ASIH-computer for such a market should be turned on like a digital computer, which must also be provided with power, an operating system and application software must be loaded. In this case, it means that the spice market must be activated and stabilized i.e. agents should start to behave rationally i.e. to do business on Lisbon ↔ Moluccas spice route. This resembles a quasi-free market, because Arabs were monopolizing spice trade in the Levant, intermediating, creating barriers for Europeans and Indians.

Until now, 3 problems and their calculations in the ASIH computer – built for the above market – have been tested. Since the language of economy is different than 0/1 calculus, it was necessary to translate all 3 problems into an economic form. For example "Finding minimum in set of integers" has been translated into "cheapest product will eliminate more expensive ones". In case of "monkey and banana" it was necessary to generate a distributed version, where knowledge, resources and skills were distributed around Portuguese, Italian, Arab and Indian "monkeys". The third problem - mentioned before – was the problem of fire discovery. The above problems were "injected" in random into the brains of agents, in such way, that no one owned a complete description of the problem. The ASIH-computer managed to solve them. The current page limit does not allow us to go into detail.

5.1   Structure of Simulation Model

At the lowest level of our model, the QGIS system is used to allow editing of all necessary geographical data, basic for the 15th century spice trade. Important cities (trade centers, with their geometric properties and relation), important sea routes and a map of the then-known (15th century) world have been defined, with the help of QGIS. QGIS is powerful, user friendly Open Source Geographic Information System (GIS). A basic structure and behavior of agents has been programmed in the Mason system. Mason is a fast discrete-event multiagent simulation library core in Java, designed for large custom-purpose Java simulations. It is a joint effort by George Mason University's Evolutionary Computation Laboratory and the GMU Center for Social Complexity. Since agents interact and move in a geographical environment, it was necessary to use GeoMason also. This is a geospatial support for MASON, also designed at George Mason University. As mentioned in the introduction, business items are labeled with logic expressions (predicate calculus) and abstract value. This SWI-Prolog (University of Amsterdam) was used to equip agents with an inference engine. For this, JPL (Java-Prolog interface) was necessary. The size of the model is more than 5500 lines of code as counted in the Eclipse environment. Constructing a set of procedures for a single agent in our model, while keeping consistency in-between databases: *WantsToSell*, *WantsToBuy*, *OtherWantsToSell*, *OtherWantsToBuy* appeared astonishingly difficult and complex in our model. Every single rendezvous and contact between two agents, resulting with information exchange and a real transaction, forced reshuffling of all these databases. Moreover, the flow of time forced keeping only up-to-date records in *OtherWantsToSell*, *OtherWantsToBuy* databases.

6   Conclusions

An ant colonies, beside their specific economic activity, breeding and the provision of safety, display astonishing computational abilities, which optimize their behavior. We name this "ant colony optimization algorithms". However, a single ant is a small, primitive being with about $250 \cdot 10^3$ neurons in its brain, which classifies it as automata. A human brain has $86 \cdot 10^9$ neurons. Adam Smith was the first to discover (as an economist and philosopher) that a human market is optimized unconsciously (to us), much like an ant colony. He referred to this as a phenomenon, which he has termed the „Invisible Hand". This can be formally explained with the help of a theory, which assumes that every market has an additional, invisible dimension of a computational nature. This dimension creates a complete, programmable computer, which is self-programming. Results of computations in this computer (ASIH-computer) affects the behavior of agents and therefore the whole market. We perceive this as a kind of control or optimization. From this theory emerges a conclusion that Adam Smith's Invisible Hand is far more "universal and powerful" than expected, because it is not only restricted to market optimization and stabilization, but can also act as a "discoverer" of new technologies, (technical market optimization). Also rules of social behavior can likely be discovered (social market optimization).

Theory on the existence of an ASIH-computer unveils chances for quite new macroeconomic models and tools for future market analysis and prediction. Comparing to CGE, DSGE or agent-based ACE macroeconomic models, they will focus on the hidden computational nature of a given market as its basic property, paying only restricted attention to observable elements and aspects of the market.

References

1. Adleman, L.M.: Molecular computation of solutions to combinatorial problems. Science **266**(5187), 1021–1024 (1994)
2. Eberbach, E.: The S-calculus process algebra for problem solving: A paradigmatic shift in handling hard computational problems. Theor. Comput. Sci. **383** (2007)
3. Joyce, H.: Adam Smith and the invisible hand. Millennium Mathematics Project, Plus Magazine (2001)
4. Kennedy, G.: Adam Smith and the Invisible Hand: From Metaphor to Myth. Econ Journal Watch **6**(2), 25 (2009)
5. Samuels, W.J., Johnson, M.F., Perry, W.H.: Erasing the Invisible Hand: Essays on an Elusive and Misused Concept in Economics, p. 358, hardcover. Cambridge University Press (2011)
6. Smith, A.: An Inquiry into the Nature and Causes of the Wealth of Nations (1776). http://www2.hn.psu.edu/faculty/jmanis/adam-smith/wealth-nations.pdf
7. Szuba, T.: Computational Collective Intelligence. Wiley Series on Parallel and Distributed Computing, Wiley and Sons, NY (2001)
8. Turing, A.M.: On computable numbers, with an application to the entscheidungsproblem. In: Proceedings of the London Mathematical Society, 2, vol. 42, pp. 230–265 (1937)
9. Wolfram, S.: A New Kind of Science, vol. 1, p. 197. Wolfram Media (2002)

Comparison of Edge Operators for Detection of Vanishing Points

Dongwook Seo, Danilo Cáceres Hernández, and Kang-Hyun Jo[⊠]

School of Electrical Engineering, University of Ulsan, Ulsan, Korea
{seodonguk,danilo}@islab.ulsan.ac.kr, acejo@ulsan.ac.kr
http://islab.ulsan.ac.kr

Abstract. This paper describes a comparative study of edge operators for the task of detecting dominant vanishing points in the image. Segmentation of line is required in order to detect the vanishing points. Three edge operators such as Sobel, Canny, and LoG (Laplacian of Gaussian) are used in order to compare that edge has influence on detecting the vanishing points. Most of line segments are obtained based on edge detection. The vanishing points are estimated by MSAC (m-estimator sample consensus) based algorithm. First, lines are extracted from edge images produced by edge operators. Second, vanishing points are obtained. The results of line segments and vanishing points detection are compared and discussed. The comparison is carried out based on the result of implementation on images with different buildings.

Keywords: Sobel · Canny · LoG · Man-made and vanishing points

1 Introduction

The 3D reconstruction of urban scenes from multiple views including street-level photographs and aerial images has received significant interest over time. The reconstructed urban models are used in various fields such as entertainment industry (movie and game), digital mapping (Google Earth and Microsoft Bing Maps), urban planning, training and simulation, and so on [6]. Urban environment consists of many objects, such vehicles, streets, parks, traffic signs, vegetation and buildings. We focus on the reconstruction of 3D geometric models of a building with multiple facades. The building has a rich geometric structure as a representative artifacts. A man-made environment has two properties: (1) many lines in the scene are parallel and (2) a variety of edges in the scene are orthogonal to each other. There are many lines on the building regions. The line segments are gathered with two or three dominant directions orthogonal to each other. In the 2D image plane, the parallel lines meet in a common point which is called a vanishing point [2,7–9]. It provides strong cues for extracting an information about the 3D structure of a building. Therefore, the reconstruction of a building is significantly simplified by the detection of vanishing points.

In this paper, we focus on the comparison of the effects of edge operators to detect the vanishing points. The algorithm of detection of vanishing points

© Springer International Publishing Switzerland 2015
M. Núñez et al. (Eds.): ICCCI 2015, Part I, LNAI 9329, pp. 443–452, 2015.
DOI: 10.1007/978-3-319-24069-5_42

based on extracting the line starts by detecting the edges in the 2D image plane. Therefore, the results of detection of vanishing points are affected by the detected edges in the image. Sobel [3], Canny [1], and LoG [5] are used to compare the results of detection of vanishing points under the influences of edges. MSAC based method is applied to detect the vanishing points in the image [10,11].

This paper is organized as follows: Section 2 presents the vanishing point. Section 3 is the explanation of edge operators. We discuss the detection of vanishing points in section 4. Finally, section 5 conclude this paper.

2   Vanishing Point

The parallel lines in the 3D space are projected onto lines in the 2D image plane that intersect in a common point, which is called the vanishing points. These points are finite (real) or infinite on the image plane. Vanishing points which lie on the same plane in the 3D space are defined by a line in the image, the so-called vanishing line.

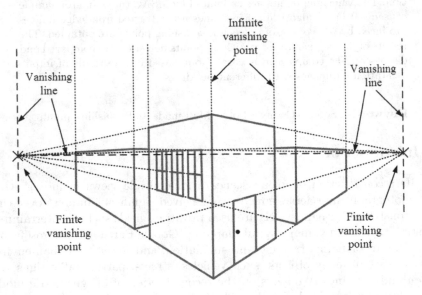

Fig. 1. The three vanishing points and lines of buildings, where a vanishing point at infinity is depicted by a direction on the image plane

The building has a rich geometric structure of a man-made object. There are many lines in the man-made object. In addition, the lines are distributed into more than one principle direction in 3D space. The principle directions are orthogonal to each other. Therefore, the vanishing point provides a powerful cue to extract 3D information from the image. It means that the understanding and reconstruction of a man-made object are able to be simplified by detecting vanishing points.

3   Edge Operators

Sobel. The Sobel operator represents differentiation for performing a 2-D spatial gradient measurement on an image. In general, the operator uses a pair of 3×3 kernels which are convolved with the original image to compute an approximation of the derivatives as follows:

$$G_x = \begin{bmatrix} -1 & 0 & 1 \\ -2 & 0 & 2 \\ -1 & 0 & 1 \end{bmatrix} * I \ \text{ and } \ G_y = \begin{bmatrix} -1 & -2 & -1 \\ 0 & 0 & 0 \\ 1 & 2 & 1 \end{bmatrix} * I \tag{1}$$

where I is an input image and $*$ is the 2D convolution operation. G_x and G_y are two images which at each point contain the horizontal and vertical derivative approximations. The resulting gradient approximations are given by

$$G = \sqrt{G_x^2 + G_y^2} \tag{2}$$

If G is more than some thresholding value, the point is an edge. The angle of orientation of the edge is defined as follows:

$$\theta = \arctan\left(\frac{G_y}{G_x}\right) \tag{3}$$

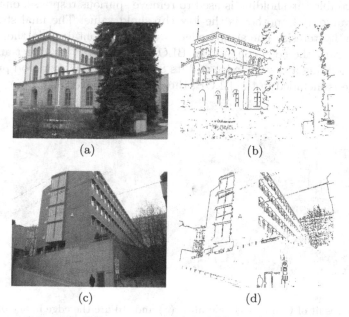

(a)   (b)

(c)   (d)

Fig. 2. The results of Sobel edge operator. (a) and (c) are the input image. (b) and (c) are the edge image.

Canny. The Canny operator is an optimal edge detector that uses a multi-stage algorithm to detect a wide range of edges in images. The process of Canny edge detection algorithm consist of 5 steps:

1. Smoothing: apply a Gaussian filter to smooth an image in order to get rid of a noise.
2. Finding gradients: find the intensity gradients of an image.
3. Non-maximum suppression: apply non-maximum suppression to remove spurious response to edge detection.
4. Double thresholding: apply a double threshold to determine the potential edges.
5. Edge tracking by hysteresis: finalize the detection of edges by suppressing all the other edges that are weak and not connected to strong edges by keeping track of hysteresis.

Each step is described briefly as follows. First a Gaussian blur is used to remove the noise which has a great influence on the results of edge detection. The next step is to use Sobel to find edge gradient in the image. The third step is non-maximum suppression to find weak edges that parallel to strong edges and eliminate them. If the edge strength of the current pixel is greater than other pixels in the mask, the value is preserved. Otherwise, the value is suppressed. After the non-maximum suppression method, edge pixels in the image are quite accurate to present real edge. However, there are still some edge pixels included by noise. Double thresholding is used to remove spurious response, one is a high threshold value and another is the low threshold value. The final step is edge tracking by hysteresis. The strong edge is certainly contained in the final edge image. To track the edge connection, BLOB (Binary Large Object-analysis) is applied by looking at a weak edge and its 8-connected neighborhood pixels. The strong edge is included in the BLOB, the other is removed.

(a)   (b)

Fig. 3. The result of Canny edge operator. (a) and (b) are the edge image of Fig. 2 (a) and 2 (c), respectively.

LoG. This edge detector combines Gaussian filtering with the Laplacian, which is called LoG (Laplacian of Gaussian). The edge pixels in the image are detected by finding the zero-crossing of the 2nd derivative of the image intensity [5]. However, the 2nd derivative is very sensitive to noise. In order to detect accurately edge pixels, the noise is filtered out before edge detection. A Gaussian filter is used to get rid of noise in the image. The 2D LoG function is defined as follows

$$LoG = \nabla^2 G(x, y) = \left[\frac{x^2 + y^2 - 2\sigma^2}{\sigma^4} \right] e^{-\frac{x^2+y^2}{\sigma^2}}, \tag{4}$$

where the Gaussian is given by

$$G(x, y) = e^{-\frac{x^2+y^2}{\sigma^2}}. \tag{5}$$

Edge detection is done by finding the zero crossing. The detection criterion is the presence of a zero crossing in the second derivative with the corresponding large peak in the first derivative. To find a zero crossing it is possible to use a 3×3 mask that checks sign changes around a pixel. Fig. 5(a) shows a cross section of 1D LoG. LoG mask is an approximation to the shape of $\nabla^2 G$ in Fig. 4.

0	0	-1	0	0
0	-1	-2	-1	0
-1	-2	16	-2	-1
0	-1	-2	-1	0
0	0	-1	0	0

Fig. 4. The example of LoG mask

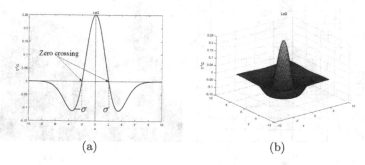

(a)   (b)

Fig. 5. (a) Cross section showing zeros crossing in 1D. (b) $\nabla^2 G$ shown in 2D space. Vertical axis corresponds to intensity.

(a)   (b)

Fig. 6. The result of LoG edge operator. (a) and (b) is the edge image of Fig. 2 (a) and 2 (c), respectively.

4   Vanishing Points Detection

This section describes the detection of vanishing points. We apply the method proposed by Thrinh et al. [11] to detect the vanishing points. This method is used to estimate the vanishing points based on MSAC [10].

The first step of the detection of vanishing points is segmentation of line in the image. In order to detect the lines, the computation of image edge is required. We used the three edge operators (Sobel, Canny, and LoG) to extract edges in the image. The line segment as a part of the edge is satisfied two constraints as:

1. The number of connected edge pixels is larger than thresholding value T_1.
2. All distance from the outer points to the shored line which lies through the end points must be less than a certain length T_2.

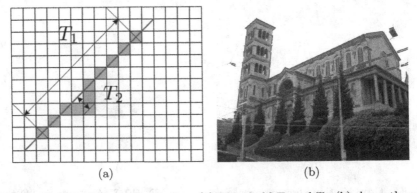

(a)   (b)

Fig. 7. Illustration of line segmentation. (a) Threshold T_1 and T_2. (b) shows the example of line segments in the image. the red color is the detected lines. (best viewed in color)

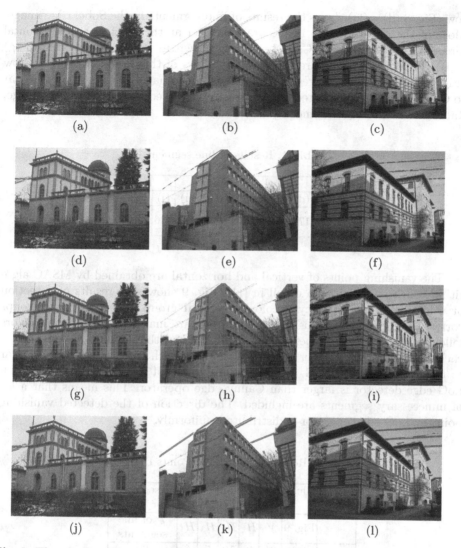

Fig. 8. The results of line segments with three edge operators. (a) – (c) are the input image in ZuBuD data set. (d) – (l) are the line segments with three different edge operators. Second Row: Soble. Third Row: Canny. Final Row: LoG. the red color is segmented lines.

The length and a number of line segments are controlled by T_1 and T_2 thresholding values. Fig. 7 (a) illustrates two thresholds. The example of line segments in the image shown in Fig. 7 (b). We use ZuBuD data set [4].

Illustration of line segments with three different edge operators is Fig. 8. (d) – (l) shows the results of the detected line segments. The red lines are part of the detected segments. We applied each edge operator to three images. The second row (Fig. 8 (d) – (f)) shows the results of line segments in the Sobel edge image. Most of the results in the image are appeared at the vertical and horizontal directions. The results in Canny edge are shown in the third row (Fig. 8 (g) – (i)). The line segments are detected regardless of the direction. The last row (Fig. 8 (j) – (l)) shows the results in the LoG edge image. The results are similar to Canny. However, the detection result is not affected by the direction of edge. Table 1 shows number of line segment.

Table 1. Results of line segments

		Sobel			Canny			LoG	
In Fig. 8	(d)	(e)	(f)	(g)	(h)	(i)	(j)	(k)	(l)
# of line segments	171	257	283	397	379	424	436	468	492

The vanishing points of vertical and horizontal are obtained by MSAC algorithm which is explained a detail in [11]. Fig. 9 shows the results of detection of vanishing points. The cyan color is vertical group. The horizontal groups corresponding to their vanishing point orders are marked with different colors following by red, green, blue, yellow. Table 2 shows the number of segments for each edge operator. Number of line segments for each vanishing point is similar to both of Canny and LoG. However, the number of segmented lines based on LoG edge detector is larger than Canny edge operator. This means that a lot of unnecessary segments are included. The direction of the detected vanishing point by LoG edge operator is distributed uniformly.

Table 2. Results of Vanishing Points Detection

	Clusters and number of line segments for each edge operator						
	Fig. 9	V	H_1	H_2	H_3	H_4	# of line segments
Sobel	(a)	21	48	19	5	6	99
	(b)	72	23	25	5	0	125
	(c)	53	38	16	8	0	115
Canny	(d)	111	89	15	8	8	231
	(e)	102	20	7	0	0	129
	(f)	85	89	32	9	0	215
LoG	(g)	94	74	22	13	13	216
	(h)	84	41	33	7	0	165
	(i)	76	73	61	18	9	228

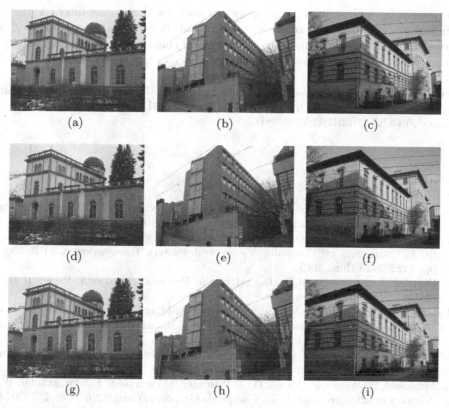

Fig. 9. Results from Sobel, Canny, and LoG edge images. The cyan color is vertical group. The red, green, blue, and yellow colors are horizontal groups from high to low priority. (best viewed in color)

5   Conclusion

This paper presented a comparative study of edge operators for the detection of vanishing points in the image. In order to detect vanishing points, segmentation of line based on edge is required. Three edge operators as Sobel, Canny, and LoG are used to extract edge in the image. Experimental results show that the results of detection of vanishing points are affected by edge operators. The number of line segments for detecting vanishing points by Sobel edge operator is smaller than other two operators. And then the number of the detected vanishing points is also small. The number of detected vanishing points by Canny and LoG is similar. However, the number of line segments based on LoG is larger than Canny edge operator. It means that the rate of detection of vanishing points by Canny is higher than others.

In this paper, we use only three kinds of operators for comparing the results of detection of vanishing points by the effects of edge detectors. It is not enough to compare the effects of the edge detector. Our future work will be compared the results by adding more edge operators such as the Gabor filter. Also, we compare only the number of the detected vanishing points. We would add a variety of comparative data, such as processing time, accuracy, and sensitivity, because it is not enough to evaluate the influence of the edge operators.

Acknowledgments. This research was supported by Basic Science Research Program through the National Research Foundation of Korea (NRF) funded by the Ministry of Education (NRF-2013R1A1A2009984).

References

1. Canny, J.: A computational approach to edge detection. IEEE Transactions on Pattern Analysis and Machine Intelligence **PAMI**–8(6), 679–698 (1986)
2. Furukawa, Y., Curless, B., Seitz, S., Szeliski, R.: Manhattan-world stereo. In: IEEE Conference on Computer Vision and Pattern Recognition. CVPR 2009, pp. 1422–1429, June 2009
3. Gonzlez, R.C., Woods, R.E.: Digital Image Processing, 3rd edn. Prenticel Hall (2007)
4. Hao Shao, T.S., Gool, L.V.: Zubud - zurich buildings database for image based recognition. Computer Vison Lab, Swiss Federal Institute of Technology, Technical Report No. 260 (2003)
5. Haralick, R.M., Shapiro, L.G.: Computer and Robot Vision, 1st edn. Addison-Wesley Longman Publishing Co., Inc, Boston (1992)
6. Musialski, P., Wonka, P., Aliaga, D.G., Wimmer, M., van Gool, L., Purgathofer, W.: A survey of urban reconstruction. Computer Graphics Forum **32**(6), 146–177 (2013)
7. Rother, C.: A new approach to vanishing point detection in architectural environments. Image and Vision Computing **20**(910), 647–655 (2002)
8. Schindler, G., Dellaert, F.: Atlanta world: an expectation maximization framework for simultaneous low-level edge grouping and camera calibration in complex man-made environments. In: Proceedings of the 2004 IEEE Computer Society Conference on Computer Vision and Pattern Recognition. CVPR 2004, vol. 1, pp. I-203–I-209, June 2004
9. Sinha, S., Steedly, D., Szeliski, R.: Piecewise planar stereo for image-based rendering. In: 2009 IEEE 12th International Conference on Computer Vision, pp. 1881–1888, September 2009
10. Torr, P., Zisserman, A.: Mlesac: A new robust estimator with application to estimating image geometry. Computer Vision and Image Understanding **78**(1), 138–156 (2000)
11. Trinh, H.H., Kim, D.N., Jo, K.H.: Facet-based multiple building analysis for robot intelligence. Applied Mathematics and Computation **205**(2), 537–549 (2008)

A Tolerance-Based Semantics of Temporal Relations: First Steps

Aymen Gammoudi[1,2], Allel Hadjali[2(✉)], and Boutheina Ben Yaghlane[3]

[1] ISGT, University of Tunis,
92, Boulevard 9 Avril 1938, 1007 Tunis, Tunisia
aymen.gammoudi@ensma.fr
[2] LIAS/ENSMA, 1 Avenue Clement Ader,
86960 Futuroscope Cedex, France
allel.hadjali@ensma.fr
[3] IHECT, University of Carthage,
IHEC-Carthage Presidence, 2016 Tunis, Tunisia
boutheina.yaghlane@ihec.rnu.tn

Abstract. Allen temporal relations is a well-known formalism used for modeling and handling temporal data. This paper discusses an idea to introduce some flexibility in defining such relations between two fuzzy time intervals. The key concept of this approach is a fuzzy tolerance relation conveniently modeled. Tolerant Allen temporal relations are then defined using the dilated and the eroded intervals of the initial fuzzy time intervals. This extension of Allen relation is investigated for the purpose of temporal databases querying thanks to the language *TSQLf* introduced in our previous works.

1 Introduction

Temporal information is often perceived or expressed in an imprecise/fuzzy manner. For example, periods of global revolutions are characterized by beginnings and endings naturally gradual and ill-defined (such as "well after early 20" or "to late 30"). Unfortunately, and to the best of our knowledge, there is not much work devoted to querying/handling fuzzy/imprecise information in a temporal databases context. On the contrary, in Artificial Intelligence field, several works exist to represent and handle imprecise or uncertain information in temporal reasoning (see for instance Dubois et al. 2003 [4], Schockaert S. and De Cock 2009 [10]).

As mentioned above, only very few studies have considered the issue of modeling and handling flexible queries over regular/fuzzy temporal databases. Billiet et al. [2] have proposed an approach that integrates bipolar classifications to determine the degree of satisfaction of records by using both positive and negative imprecise and possibly temporal preferences. But this approach is still unable to model complex temporal relationships and cannot be applied in historical temporal databases (for instance, the user may request one time period but reject a part of this period, when specifying the valid time constraint in the

© Springer International Publishing Switzerland 2015
M. Núñez et al. (Eds.): ICCCI 2015, Part I, LNAI 9329, pp. 453–462, 2015.
DOI: 10.1007/978-3-319-24069-5_43

query). Deng et al. [3] have proposed a temporal extension to an extended ERT model to handle fuzzy numbers. They have specified a fuzzy temporal query which is an extension from the TQuel language and they have introduced the concepts of fuzzy temporal in specification expressions, selection, join and projection. Tudorie et al. [11] have proposed a fuzzy model for vague temporal terms and their implication in queries' evaluation. Unfortunately, this approach does not allow to model a large number of temporal terms (such as: just after and much before). In [7], Galindo and Medina have proposed an extension of temporal fuzzy comparators and have introduced the notion of dates in Relational Databases (RDB) by adding two extra precise attributes on dates (VST, VET). Very recently, in [8] we have proposed an extension, named $TSQLf$, of SQLf language [9] by adding the time dimension. TSQLf language allows for expressing user queries involving fuzzy criteria on time. It is founded on the fuzzy extension of Allen temporal relations already proposed in [4].

Unfortunately, all the above approaches consider (fuzzy) temporal relations only between regular time intervals (i.e., their lower and upper bounds are crisp instants). While in real world applications, time intervals are often described by ill-defined bounds to better capture the vagueness inherent to the available pieces information. This paper inspired from the study of [4], is a step towards dealing with that issue. It introduces an extension of Allen temporal relations between fuzzy time intervals. This extension relies on a particular tolerance relation that allows associating a fuzzy time interval with two nested fuzzy time intervals (i.e., the dilated and the eroded intervals). Based on these two nested intervals, tolerant Allen temporal relations are defined that show some new idea about softness in expressing relations between temporal entities.

The paper is structured as follows. In Section 2, we provide some background on Allen temporal relation, fuzzy comparators and some fuzzy extension of Allen relations. In Section 3, tolerant Allen temporal relations between fuzzy time intervals are presented. Section 4 concludes the paper.

2   Background

The purpose of this section is manifold. We begin by recalling Allen temporal relation, then we recall some fuzzy comparators of interest. Finally, we present a fuzzy extension of Allen relations based on two particular fuzzy comparators expressing an approximate equality and a graded strict inequality. This extension operates only on regular time intervals. This section is mainly borrowed from [4].

2.1   Allen Temporal Relations

Allen [1] has proposed a set of mutually exclusive primitive relations that can be applied between two temporal intervals. These relationships between events are usually denoted by *before* ($\prec$), *after* ($\succ$), *meets* (m), *met by* (mi), *overlaps* (o), *overlapped by* (oi), *during* (d), *contains* (di), *starts* (s), *started by* (si), *finishes* (f), *finished by* (fi), and *equal* ($\equiv$). Their meaning is illustrated in Table 1

Table 1. Allen Relations

Relation	Inverse	Relations between bounds
A ≺ B	B ≻ A	b > a'
A m B	B mi A	a' = b
A o B	B oi A	b > a ∧ a' > b ∧ b' > a'
A d B	B di A	a > b ∧ b' > a'
A s B	B si A	a = b ∧ b' > a'
A f B	B fi A	a > b ∧ b' = a'
A ≡ B	B ≡ A	a ≐ b ∧ a' = b'

($A = [a, a']$ and $B = [b, b']$ are two time intervals where a and a' (respectively b and b') represent the two bounds of A (respectively B), with $a < a'$.

These relationships can be defined from three ordinary relations $<, =, >$ between bounds of the two intervals to describe relative position to each other. For example, the statement *A overlaps B* corresponds to $(b > a) \wedge (a' > b) \wedge (b' > a')$ as indicated in the Table 1. Allen [1] has provided a set of axioms describing the composition of the 13 relationships and an algorithm to infer new information. For example, from *A before B* and *B before C*, one can deduce *A before C*.

2.2 Fuzzy Absolute Comparators

We recall here two fuzzy comparators defined in terms of difference of values, that are of great of interest for our proposal.

An approximate equality between two values, here representing dates, modeled by a fuzzy relation E with membership function μ_E (E stands for "equal"), can be defined in terms of a distance such as the absolute value of the difference. Namely,

$$\mu_E(x, y) = \mu_L(|x - y|)$$

For simplicity, fuzzy sets and fuzzy relations are assumed to be defined on the real line. Approximate equality can be represented by
$\forall x, y \in \mathbb{R}$
$$\mu_E(x, y) = \mu_L(|x - y|) = max(0, min(1, \tfrac{\delta + \varepsilon - |x-y|}{\varepsilon})) =$$

$$\begin{cases} 1 & if\ |x - y| \le \delta \\ 0 & if\ |x - y| > \delta + \varepsilon \\ \frac{\delta + \varepsilon - |x-y|}{\varepsilon} & otherwise \end{cases}$$

where δ and ε are respectively positive and strictly positive parameters which affect the approximate equality. With the following intended meaning: the possible values of the difference $a - b$ are restricted by the fuzzy set $L = (-\delta, \delta, \varepsilon, \varepsilon)^1$. In particular $a\ E(0)\ b$ means $a = b$. Note that L is symmetrical and $L = -L$. Similarly, *a more or less strong inequality* can be modeled by a fuzzy relation G (G stands for "greater"), of the form

$$\mu_G(x, y) = \mu_K(x - y)$$

We assume $\rho > 0$, i.e. G more demanding than the idea of "strictly greater" or "clearly greater". $K = (\lambda + \rho, +\infty, \rho, +\infty)$ is a fuzzy interval which gathers all the values equal to or greater than a value fuzzily located between λ and $\lambda + \rho$. K is thus a fuzzy set of positive values with an increasing membership function. See Figure 1.

According to the values of parameters $\lambda + \rho$, the modality, which indicates how much larger than b is a, may be linguistically labeled by "slightly", "moderately", "much". In a given context, G(0) stands for '>'.

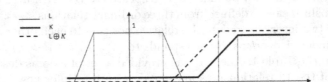

Fig. 1. Modeling "approximate equality" and "graded strict inequality"

2.3   Fuzzy Allen Relations

Using the fuzzy parameterized comparators E(L) and G(K), fuzzy counterparts of Allen relations have been established in [4]. The idea is that the relations which can hold between the endpoints of the intervals we consider may not be described in precise terms. Then, in approximate terms, only two distinct relations may hold between two dates a and b. Indeed, a date a can be "*approximately equal*" to a date b in the sense of E(L), or a can be "*clearly different from*" b in the sense of not E(L). This last relation corresponds to "*much larger*" in the sense of G(K) or "*much smaller*" in the sense of S(K^{ant})[2]

Let $A = [a, a']$ and $B = [b, b']$ be two time intervals. Table 2 summarizes the fuzzy Allen relations established (where $L=(-\delta, \delta, \varepsilon, \varepsilon)$ and $K = \bar{L} \cap [0, +\infty] = (\delta + \varepsilon, +\infty, \delta, +\infty)$ which is denoted L_+^c).

---

[1] where $[-\delta, \delta]$ (resp. $[-\delta - \varepsilon, \delta + \varepsilon]$) is the core (resp. support) of L.
[2] K^{ant} is the antonym of K defined by $\mu_{K^{ant}}(x) = \mu_K(-x)$.

Table 2. Fuzzy Allen Relations

Fuzzy Allen Relation	Definition	Label
A *fuzz-before(L)* B	b $G(L_+^c)$ a'	$fb(L)$
A *fuzz-meets(L)* B	a' E(L) b	$fm(L)$
A *fuzz-overlaps(L)* B	b $G(L_+^c)$ a $\land$ a' $G(L_+^c)$ b $\land$ b' $G(L_+^c)$ a'	$fo(L)$
A *fuzz-during(L)* B	a $G(L_+^c)$ b $\land$ b' $G(L_+^c)$a'	$fd(L)$
A *fuzz-starts(L)* B	a E(L) b $\land$ b' $G(L_+^c)$ a'	$fs(L)$
A *fuzz-finishes(L)* B	a' E(L) b' $\land$ a $G(L_+^c)$ b	$ff(L)$
A *fuzz-equals(L)* B	a E(L) b $\land$ b' E(L) a'	$fe(L)$

3 Tolerance-Based Semantics of Allen Relations

We discuss here the basis of a tolerance-based semantics of Allen relations between two (crisp/fuzzy) time intervals. Each *tolerant* Allen relation is expressed in terms of a class of traditional Allen relations applied on dilated and erode versions of A and B by means of a tolerance relation $E(L)$.

3.1 Dilation and Erosion Operations

Let us consider a fuzzy set A representing a time interval, and *an approximate equality* relation $E(L$. A can be associated with a nested pair of fuzzy sets when using the parameterized relation $E(L)$ as a tolerance relation [4]. Indeed,

- one can build a fuzzy set of temporal instants close to A such that $A \subseteq A^L$. This is the dilation operation,
- one can build a fuzzy set of temporal instants close to A such that $A_L \subseteq A$. This is the erosion operation.

Dilation Operation. Dilating the fuzzy set of temporal instants A by L will provide a fuzzy set A^L defined by

$$\mu_{A^L}(r) = sup_s min(\mu_{E[L]}(s,r), \mu_A(s)) \tag{1}$$

$$= sup_s min(\mu_L(r-s), \mu_A(s)) \tag{2}$$

$$= \mu_{A \oplus L}(r). \tag{3}$$

Hence,

$$A^L = A \oplus L,$$

where $\oplus$ is the addition operation extended to fuzzy sets [6]. A^L gathers the elements of A and the elements outside A which are somewhat close to an element in A.

Lemma 1. Given a fuzzy set of temporal instants A and a tolerance relation $E[L]$, we have $A \subseteq A^L$.

Proof. Obvious.

One can easily check that the fuzzy set of temporal instants A^L is less restrictive than A, but still semantically close to A. Thus, A^L can be viewed as a relaxed variant of A. In terms of t.m.f., if $A = (a, a', \alpha, \alpha')$ [3] and $L = (-\delta, \delta, \epsilon, \epsilon)$ then $A^L = (a - \delta, a' + \delta, \alpha + \epsilon, \alpha' + \epsilon)$, see Figure 2.

Erosion Operation. Let $L \oplus X = A$ be an equation where X is the unknown variable. Solving this equation has extensively been discussed in [5]. It has been demonstrated that the greatest solution of this equation is given by $\bar{X} = A \ominus (-L) = A \ominus L$ since $L = -L$ and where $\ominus$ is the extended Minkowski subtraction defined by [5]:

$$\mu_{A \ominus L}(r) = inf_s(\mu_L(r - s) \Rightarrow_{\mathbb{T}} \mu_A(s)) \tag{4}$$

where $\mathbb{T}$ is a t-norm, and $\Rightarrow_{\mathbb{T}}$ is the R-implication induced by $\mathbb{T}$ and defined by $\Rightarrow_{\mathbb{T}} (u, v) = sup\{\lambda \in [0, 1], \mathbb{T}(u, \lambda) \leq v\}$, for $u, v \in [0, 1]$. We make use of the same t-norm $\mathbb{T} (= min)$ as in the dilation operation which implies that $\Rightarrow_{\mathbb{T}}$ is the so-called Gödel implication.

Let $(E[L])_r = \{s, \mu_{E[L]}(s, r) > 0\}$ be the set of elements that are close to r in the sense of $E[L]$. Then, the above expression can be interpreted as the degree of inclusion of $(E[Z])_r$ in A. This means that r belongs to $A \ominus L$ if all the elements s that are close to r are A. Hence, the inclusion $A \ominus L \subseteq A$ holds. This operation is very useful in natural language to intensify the meaning of vague terms. Now, eroding the fuzzy set A by L results in the fuzzy set A_L defined by $A_L = A \ominus L$.

Lemma 2. Given a fuzzy set of temporal instants A and a tolerance relation $E[L]$, we have $A_L \subseteq A$.

Proof. Obvious.

The fuzzy set A_L is more precise than the original fuzzy set A but it still remains not too far from A semantically speaking. If $A = (a, a', \alpha, \alpha')$ and $L = (-\delta, \delta, \epsilon, \epsilon)$ then $A_L = A \ominus L = (a + \delta, a' - \delta, \alpha - \epsilon, \alpha' - \epsilon)$ provide that $\alpha \geq \epsilon$ and $\alpha' \geq \epsilon$). Figure 2 illustrates this operation. In the crisp case, $A \ominus L = [a, a'] \ominus [-\delta, \delta'] = [a + \delta, a' - \delta']$ (while $A \oplus Z = [a - \delta, a' + \delta']$).

One can easily check that the following proposition holds:

---

[3] with $[a, a']$ (resp. $[a - \alpha, a' + \alpha']$) represents the core (resp. support) of A.

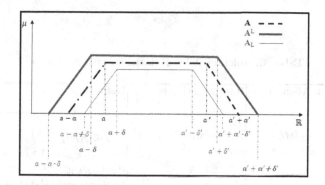

Fig. 2. Dilated and eroded time intervals of a fuzzy set of temporal instants A

Proposition 1. Using the t.m.f. of A^L and A_L given above, we have:

- $(A^L)_L = (A_L)^L = A$
- $(A^L)^L = A \oplus 2L$
- $(A_L)_L = A \ominus 2L$

3.2   Tolerant Allen Relations

Using the dilatation and erosion operations defined above, we can provide the basis for defining a tolerance-based extension of Allen relations. For instance, the tolerance-based temporal relation corresponding to *meet* Allen relation writes:

A *toler-meets(L)* B would correspond to A_L before B_L and A^L overlaps B^L,

The statement A *toler-meets(L)* B means that the temporal relation between the two (crisp/fuzzy) temporal intervals A and B is perceived as a variant of *meet* relation thanks to some tolerance expressed by the indicator L. This is a human perception which is often encountered in real world applications (such as in managing historical temporal data, planning and scheduling, natural language understanding, etc.).

The *toler-meet* relation gathers a class of Allen relations (i.e., *before* and *overlaps*) applied on the eroded and dilated time intervals corresponding to the original time intervals A and B. This boils down to compute the traditional Allen relations, *before* and *overlaps*, on fuzzy temporal intervals (since A^L and A_L are fuzzy sets as shown in the previous subsection).

Let $A = (a, a', \alpha, \alpha)$ be a fuzzy time interval with $\tilde{a} = (a, a, \alpha, 0)$ and $\tilde{a}' = (a', a', 0, \alpha')$ the two fuzzy bounds of validity of A. One can write $A = (\tilde{a}, a, a', \tilde{a}')$. Under this forms:

- A^L writes $(\tilde{a}^{(L)}, a - \delta, a' + \delta, \tilde{a}'^{(L)})$ where $\tilde{a}^{(L)} = (a - \delta, a - \delta, \alpha + \epsilon, 0)$ and $\tilde{a}'^{(L)} = (a' + \delta, a' + \delta, 0, \alpha' + \epsilon)$.
- A_L writes $(\tilde{a}_{(L)}, a + \delta, a' - \delta, \tilde{a}'_{(L)}))$ where $\tilde{a}_{(L)} = (a + \delta, a + \delta, \alpha - \epsilon, 0))$ and $\tilde{a}'_{(L)} = (a' - \delta, a' - \delta, 0, \alpha' - \epsilon))$.

Table 3. Tolerant Allen Relations (with $K = L_+^c$)

Tolerant Allen Relation	Interpretation	Definition
A $toler\text{-}meets(L)$ B	A_L before B_L $\wedge$ A^L overlaps B^L	$\mu_{\bar{b}_{(L)}G(K)\bar{a}'_{(L)}}(t,v)$ $\wedge$ $(\mu_{\bar{b}(L)G(K)\bar{a}(L)}(t,v)$ $\wedge$ $\mu_{\bar{a}'(L)G(K)\bar{b}(L)}(t,v)$ $\wedge$ $\mu_{\bar{b}'(L)G(K)\bar{a}'(L)}(t,v))$
A $toler\text{-}before(L)$ B	A^L toler-meets B^L	$\mu_{\bar{b}G(K)\bar{a}'}(t,v)$ $\wedge$ $\mu_{\bar{b}(2L)G(K)\bar{a}(2L)}(t,v)$ $\wedge$ $\mu_{\bar{a}'(2L)G(K)\bar{b}(2L)}(t,v)$ $\wedge$ $\mu_{\bar{b}'(2L)G(K)\bar{a}(2L)}(t,v)$
A $toler\text{-}overlaps(L)$ B	A_L toler-meets B_L	$\mu_{\bar{b}'(2L)G(K)\bar{a}'(2L)}(t,v)$ $\wedge$ $\mu_{\bar{b}G(K)\bar{a}}(t,v) \wedge \mu_{\bar{a}'G(K)\bar{b}}(t,v)$ $\wedge\ \mu_{\bar{b}'G(K)\bar{a}'}(t,v)$
A $toler\text{-}during(L)$ B	A^L toler-equals B_L	$(\mu_{\bar{b}G(K)\bar{a}(2L)}(t,v)$ $\wedge$ $\mu_{\bar{a}'(2L)G(K)\bar{b}'}(t,v))$ $\wedge$ $(\mu_{\bar{a}G(K)\bar{b}_{(2L)}}(t,v)$ $\wedge$ $\mu_{\bar{b}'_{(2L)}G(K)\bar{a}'}(t,v))$
A $toler\text{-}starts(L)$ B	A_L during B^L $\wedge$ A^L overlaps B_L	$(\mu_{\bar{a}_{(L)}G(K)\bar{b}(L)}(t,v)$ $\wedge$ $\mu_{\bar{b}'(L)G(K)\bar{a}'_{(L)}}(t,v))$ $\wedge$ $(\mu_{\bar{b}_{(L)}G(K)\bar{a}(L)}(t,v)$ $\wedge$ $\mu_{\bar{a}'(L)G(K)\bar{b}_{(L)}}(t,v)$ $\wedge$ $\mu_{\bar{b}'_{(L)}G(K)\bar{a}'(L)}(t,v))$
A $toler\text{-}finishes(L)$ B	A^L overlapped_by B_L $\wedge$ A_L during B^L	$(\mu_{\bar{a}(L)G(K)\bar{b}_{(L)}}(t,v)$ $\wedge$ $\mu_{\bar{b}'_{(L)}G(K)\bar{a}(L)}(t,v)$ $\wedge$ $\mu_{\bar{a}'(L)G(K)\bar{b}'_{(L)}}(t,v))$ $\wedge$ $(\mu_{\bar{a}_{(L)}G(K)\bar{b}(L)}(t,v)$ $\wedge$ $\mu_{\bar{b}'(L)G(K)\bar{a}'_{(L)}}(t,v))$
A $toler\text{-}equals(L)$ B	A^L contains B_L $\wedge$ A_L during B^L	$(\mu_{\bar{b}_{(L)}G(K)\bar{a}(L)}(t,v)$ $\wedge$ $\mu_{\bar{a}'(L)G(K)\bar{b}'_{(L)}}(t,v))$ $\wedge$ $(\mu_{\bar{a}_{(L)}G(K)\bar{b}(L)}(t,v)$ $\wedge$ $\mu_{\bar{b}'(L)G(K)\bar{a}'_{(L)}}(t,v))$

One can see that the relation *toler-meet(L)* is of a gradual nature and its satisfaction is a matter of degree (i.e., A and B satisfy *toler-meet(L)* with a degree $d \in [0,1]$ instead of $\{0,1\}$). According to the interpretation given above of *toler-meet(L)*, $d = min(d_1, d_2)$ where

- $d_1 = \mu_{\tilde{b}_{(L)}G(L_+^c)\tilde{a}'_{(L)}}(t, v)$ is the degree expressing the extent to which A_L is *before* B_L.
- $d_2 = min(\mu_{\tilde{b}(L)G(L_+^c)\tilde{a}(L)}(t, v), \mu_{\tilde{a}'(L)G(L_+^c)\tilde{b}(L)}(t, v), \mu_{\tilde{b}'(L)G(L_+^c)\tilde{a}'(L)}(t, v))$ is the degree expressing the extent to which A^L overlaps B^L.

Now, to compute the degrees d_1 and d_2, the following formulas can be used:
$$\mu_{\tilde{a}E(L)\tilde{b}}(t, v) = sup(min(\mu_{\tilde{a}}(t), \mu_{\tilde{b}}(v)), \mu_L(t\text{-}v))$$
$$\mu_{\tilde{a}G(L_+^c)\tilde{b}}(t, v) = sup(min(\mu_{\tilde{a}}(t), \mu_{\tilde{b}}(v)), \mu_{L_+^c}(t\text{-}v))$$

In a similar way, the tolerant counterparts of the *start* and *during* Allen relations write:
A *toler-starts(L)* B as A_L *during* B^L and A^L *overlaps* B_L,
and
A *toler-before(L)* B as A^L *toler-meets(L)* B^L.
In Table 3, we provide the tolerant counterparts of all Allen relations.

To reason on the tolerant Allen relations introduced, one can extend the composition rules established in the case of Allen relations. For instance, if A *toler-before(L_1)* B and B *toler-before(L_2)* C, then A *toler-before(L_1 $\oplus$ L_2)* C.

4 Conclusion

In this paper, we have defined the first steps to introduce the concept of tolerance-based of Allen temporal relations to manage time intervals with fuzzy bounds. The key notion of this extension is the dilation and erosion operations defined on fuzzy time intervals.

As for immediate future work, we first plan to establish the complete set of composition rules of the tolerant Allen relations for the purpose of reasoning and inference. Second, we plan also to incorporate such new variant of Allen relations and their reasoning in our language *TSQLf* to design and develop an advanced intelligent temporal database querying tool.

References

1. Allen, J.F.: Maintaining knowledge about temporal intervals. Comm. of the ACM 832–843 (1983)
2. Billiet, C., Pons, J.E., Matthé, T., De Tré, G., Pons Capote, O.: Bipolar fuzzy querying of temporal databases. In: Christiansen, H., De Tré, G., Yazici, A., Zadrozny, S., Andreasen, T., Larsen, H.L. (eds.) FQAS 2011. LNCS, vol. 7022, pp. 60–71. Springer, Heidelberg (2011)

3. Deng, L., Liang, Z., Zhang, Y.: A fuzzy temporal model and query language for fter databases. In: Eighth International Conference on Intelligent Systems Design and Applications, pp. 77–82 (2008)
4. Dubois, D., Hadjali, A., Prade, H.: Fuzziness and uncertainty in temporal reasoning. Journal of Universal Computer Science, Special Issue on Spatial and Temporal Reasoning 1168–1194 (2003)
5. Dubois, D., Prade, H.: Inverse operations for fuzzy numbers. In: Proc. IFAC Symp. on Fuzzy Info., Knowledge Representation and Decision Analysis, pp. 391–395 (1983)
6. Dubois, D., Prade, H.: Possibility theory. Plenum Press (1988)
7. Galindo, J., Medina, J.M.: Ftsql2: fuzzy time in relational databases. In: Proceedings of the 2nd International Conf. in Fuzzy Loggic and Technology, pp. 5–7 (2001)
8. Gammoudi, A., Hadjali, A., Yaghlane, B.B.: An intelligent flexible querying approach for temporal databases. In: Proc. of the 7th International Conference on Intelligent Systems IS 2014. Warsar, Poland (IS14), pp. 523–534 (2014)
9. Pivert, O., Smits, G.: On fuzzy preference queries explicitly handling satisfaction levels. Information Science 341–350 (2012)
10. Schockaert, S., De Cock, M.: Temporal reasoning about fuzzy intervals. Artificial Intelligence 1158–1193 (2008)
11. Tudorie, C., Vlase, M., Nica, C., Muntranu, D.: Fuzzy temporal criteria in database querying. Artificial Intelligence Applications 112 (2012)

Integer Programming Based Stable and Efficiency Algorithm for Two-sided Matching with Indifferences

Naoki Ohta[✉]

College of Information Science and Engineering, Ritsumeikan University,
1-1-1 Noji Higashi, Kusatsu, Shiga 525-8577, Japan
n-ohta@fc.ritsumei.ac.jp

Abstract. To make use of collective intelligence of many autonomous self-interested agents, it is important to form a team that all the agents agree. Two-sided matching is one of the basic approaches to form a team that consists of agents from two disjoint agent groups. Traditional two-sided matching assumes that an agent has totally ordered preference list of agents to be paired with. However, it is unrealistic to have a totally ordered list for a large-scale two-sided matching problem. Therefore, two-sided matching with indifferences is proposed. It allows indifferences in the preference list of agents. Two-sided matching with indifferences has two important characters weakly stable and Pareto efficiency. In this paper, we propose a new integer programming based algorithm "nucleolus" for two-sided matching with indifferences. This algorithm propose the matching which satisfies weakly stable and Pareto efficiency.

Keywords: Multi-agent system · Two-sided matching · Integer programming

1 Introduction

Forming a team of autonomous agents is an important capability in a multi-agent system in order to make use of collective capabilities of intelligent agents. An agent will not form a team if forming a team does not increase the agent's utility. Generally, an autonomous agent acts to maximize its utility.

Two-sided matching is one of the basic methods to form a team. In the two-sided matching problem, two disjoint agent groups exist. A member of one group is paired with one or multiple members of the other group. Each agent has a preference in agents of the other group to be paired with. A two-sided matching algorithm finds a matching that all agents are consented with. Two-sided matching is used in a variety of fields in a real world, such as matching interns with hospitals [1–4]. We can expect that two-sided matching is applied to resource allocation [5].

The Gale Shapley algorithm is one of the major algorithms to solve a two-sided matching problem [1,4]. This algorithm assumes that the agent has a

© Springer International Publishing Switzerland 2015
M. Núñez et al. (Eds.): ICCCI 2015, Part I, LNAI 9329, pp. 463–472, 2015.
DOI: 10.1007/978-3-319-24069-5_44

totally ordered preference list in the agents to be paired with. A matching found by this algorithm is known to be *stable*. The stable matching means that there is no pair of agents that collectively prefer another match.

The two-sided matching problem can also be solved by integer programming [6,7]. By properly defining constraints in the integer programming, a stable matching can be calculated. Although the matching found by the Gale Shapley algorithm is known to be the best matching to one group and the worst matching to the other group [8], various stable matchings can be obtained by using a different objective function when integer programming is used.

In the traditional two-sided matching that uses the Gale Shapley algorithm, it is assumed that all agents have totally ordered preference list on the agents in the other group. This assumption is somewhat unrealistic for a large scale two-sided matching. For example, let us suppose that there are 1000 job seekers and many companies that potentially offer jobs. When we consider this job matching problem as a two-side matching, it is very difficult for a company to determine the total order of preference in job seekers. To cope with this kind of issue, two-sided matching was extended to allow indifferences in the preference list of agents [9]. Three types of stability are identified for two-sided matching with indifferences [9]. In addition, "Pareto efficiency" is defined for a two-sided matching problem, based on the agents' preferences on the matching [10].

We can extend the algorithms for traditional two-sided matching such as Gale Shapley algorithm to apply to two-sided matching with indifferences. In addition to, integer programming based algorithm for two-sided matching with indifferences is proposed [11]. This algorithm proposes the matching which satisfies one type of stability. This paper contains some parts borrowed from [11]. [11] is one of the last author's papers presented at ICCCI 2014.

In this paper, we propose a new integer programming based algorithm "nucleolus" for two-sided matching with indifferences. This algorithm propose the matching which satisfies one type of stability and "Pareto efficiency"

The key idea of the proposed algorithm is based on the concept of "nucleous" [12] in the coalition game research.

The rest of this paper is organized as follows. First, we review the model of two-sided matching (Sect. 2). Next, we propose nucleolus and explain nucleolus's character (Sect. 3). Finally, we present a conclusion of this paper (Sect. 4).

2   Two-Sided Matching

2.1   Traditional Two-Sided Matching

In this section, we illustrate traditional two-sided matching. There are two finite and disjoint sets of agents L and R. In two-sided matching, a member of L is paired with one or multiple members of R, a member of R is paired with one or multiple members of L. Each agent p ($p \in L \cup R$) has capacity n_p. n_p is a natural number and represents the number of agents that can be matched from the other group. Therefore, $\forall l \in L$, $n_l \leq \|R\|$ and $\forall r \in R$, $n_r \leq \|L\|$ holds.

Every member of L has preference over R, and every member of R has preference over L.

Definition 1 (preference). *The preference over X is a total order on $X \cup \{\emptyset\}$.*

Assume that there is an agent p whose preference is denoted by $\succ_p$. If $a \succ_p b$ holds, p prefers agent a to agent b. In this report, we assume that all agents prefer pairing some agent to pairing no agent. It is illustrated that $\forall l \in L, \forall r \in R$, $r \succ_l \emptyset$ and $l \succ_r \emptyset$.

The target of Two sided Matching (L, R) is finding a matching.

Definition 2 (matching). *Given a two-sided matching (L, R), (L, R)'s matching m is a function, which satisfies the following conditions:*

- *Its domain is $R \cup L$.*
- *$\forall r \in R$, $m(r) \subseteq L$ holds.*
- *$\forall l \in L$, $m(l) \subseteq R$ holds.*
- *$\forall p \in L \cup R$, $\|m(p)\| \leq n_p$.*
- *If $\forall q \in m(p)$ holds, $p \in m(q)$ holds.*

Two-sided matching where $\forall p \in L \cap R$, $n_p = 1$ holds is called one-to-one matching, two-sided matching where $\forall l \in L, n_l = 1$ or $r \in R$, $n_r = 1$ holds, is called many-to-one matching, and two-sided matching where $\exists l, n_l > 1$ and $\exists r, n_r > 1$ hold, is called many-to-many matching. Given a matching m and an agent p. $m_i(p)$ is p's i-th best agent in $m(p)$ ($i \in \mathcal{N}$ and $i \leq n_p$ hold). If $\|m(p)\| \leq k \leq n_p$ holds, $m_k(p) = \emptyset$ holds. For example, if $a \succ_p b \succ_p c \succ_p d$, $m(p) = \{a, c\}$, and $n_p = 3$ hold, $m_1(p) = a$, $m_2(p) = c$, and $m_3(p) = \emptyset$ hold.

There are a number of algorithms to solve two-sided matching problem. These algorithms find a matching with a good property such as stable matching.

Definition 3 (blocking pair). *For matching m of given two-sided matching (L, R), the blocking pair (l, r) ($l \in L$, $r \in R$) is a pair of agents that satisfies the following conditions:*

- *$r \succ_l m_{n_l}(l)$ holds.*
- *$l \succ_r m_{n_r}(r)$ holds.*

If a blocking pair (l, r) exists for a matching m, l and r have an incentive to collectively deviate from the matching m. Therefore, a matching with a blocking pair is not stable.

Definition 4 (stable matching). *Given a two-sided matching (L, R), a matching m is stable when there is no blocking pair of m.*

The Gale Shapley algorithm [1] is a well-known algorithm for calculating stable matchings.

Definition 5 (Gale Shapley Algorithm). *The Gale Shapley algorithm calculates stale matching as follows: (We call this algorithm "L proposed". If transposed L to R and R to L, we call it "R proposed".)*

- Given matching m where $\forall p \in R \cup L$, $m(p) = \emptyset$.
- While there is an agent $l \in L$ who satisfies $\|m(l)\| < n_l$ and has not proposed all members of R, execute the following procedure:
  - l proposed agent r that satisfies $\forall \in R' \setminus \{r\}$, $r \succ_l r'$ where R' is a set of agents who have not been proposed by l.
  - r replies as follows:
    * If $\|m(r)\| < n_r$ holds, r accepts l's proposal. l joins $m(r)$ and r joins $m(l)$
    * If $\|m(r)\| = n_r$ and $\not\exists\, l' \in m(r), l \succ_r l'$ hold, r does not accept l's proposal.
    * If $\|m(r)\| = n_r$ and $\exists l' \in m(r), l \succ_r l'$ hold, r accepts l's proposal and drops $m_w(r)$. $m_w(r)$ is a member of $m(r)$. $m_w(r)$ satisfies that $\forall l' \in m(r) \setminus \{m_w(r)\}$, $l' \succ_r m_w(r)$ holds. $m_w(r)$ defects from $m(r)$, r defects from $m(m_w(r))$, l joins $m(r)$
- If there is no agent who is not paired, and who has not proposed all members of R, this algorithm returns the matching m

The Gale Shapley algorithm finds a stable matching that is best to members of proposing group [1].

This characteristics of the matching is desirable in such a case as assigning students to professor's seminars in a university where preferences of the students are usually placed before the professors'. However, the cases where the two groups are on equal basis, other characteristics are wanted.

Integer programming can find a stable matching that has other characteristics by defining a proper objective function. Before discussing the objective function, we show how a two-sided matching problem can be represented in the integer programming framework. There are $\|L\| \times \|R\|$ variables $x(l, r)$ ($\forall l \in L$, $\forall r \in R$) $x(l, r) \in \{0, 1\}$ holds. If $x(l, r)$ is 0, $l \notin m(r)$ and $r \notin m(l)$ hold. If $x(l, r)$ is 1, $l \in m(r)$ and $r \in m(l)$ hold.

The Constrains that are needed to find a stable matching in one-to-one matching can be defined as follows [6].

Definition 6 (constraints for stable matchings). *If variables $x(l, r)$ ($\forall l \in L, \forall r \in R$) satisfy the following conditions, the matching is stable.*

- $\forall r \in R$, $\sum_{r \in R} x(l, r) = n_r$ holds.
- $\forall l \in L$, $\sum_{l \in L} x(l, r) = n_l$ holds.
- $\forall l \in L$, $\{\sum_{r \in R} x(l, r)\} + \|R\| \times e(l) \geq n_l$
- $\forall r \in R$, $\{\sum_{l \in L} x(l, r)\} + \|L\| \times e(r) \geq n_r$
- $\forall l \in L$, $\forall r \in R$, $-x(l, r) + e(l) + e(r) < 2$
- $\forall l \in L$, $\forall r \in R$, $\forall l' \in L$, $\forall r' \in R$, if $r \succ_l r'$ and $l \succ_r l'$ hold, $-x(l, r) + x(l', r) + x(l, r') < 2$

Let us consider an example of an objective function. Assume that we define a function $r(p, q)$ to denote that q is the $r(p, q)$-th best agent for agent p. In addition, we define $r(p, \emptyset) = \|O\|$. If $p \in L$ holds, O is R. If $p \in R$ holds, O is L.

When we define an objective function as minimizing $(\sum_{l \in L, r \in R}\{r(l, r) + r(r, l)\} \times x(l, r)) + (\sum_{l \in L}(n_l - \|m(l)\|) \times r(l, \emptyset)) + (\sum_{r \in R}(n_r - \|m(r)\|) \times r(r, \emptyset))$, a stable matching that maximizes the sum of all agents' satisfaction can be found ($\|m(l)\|$ is equal to $\sum_{r \in R} x(l, r)$ and $\|m(r)\|$ is equal to $\sum_{l \in L} x(l, r)$).

2.2   Two-Sided Matching with Indifferences

In some applications, especially those with a large number of agents, the assumption that all agents have totally ordered preference is unrealistic. Two-sided matching was extended to allow an agent to have indifferent preference in agents in the other group [9].

Definition 7 (indifferent preference). *If there is a set of agents X, the indifferent preference over X is total order on SX, which is a group of sets of agents that satisfy the following conditions.*

- $\bigcup_{X' \in SX} X' = X$ *holds.*
- $\forall X', X'' \in SX,\ X' \cap X'' = \emptyset$ *holds.*
- $\exists \{\emptyset\} \in SX$

Assume that there is an agent p whose preference is represented as $\succ'_p$. $A \succ'_p B$ means that p prefers a member of A to a member of B. We will write it also as $a \succ_p b$ if $a \in A$ and $b \in B$ hold. In addition, if two agents p' and p'' are members of one group (for example A or B), p likes p' as well as p''. In this case, we will write $p' =_p p''$. If $a \succeq_p b$ holds, $a \succ_p b$ or $a =_p b$ holds. In the following, we write an agent p's preference as $(\{a, b\}, \{c, d, e\}, \{f, g\})$, which means that followings:

- $A = \{a, b\}$, $B = \{c, d, e\}$, $C = \{f, g\}$.
- $A \succ'_p B$, $A \succ'_p C$, and $B \succ_p C$.

In this report, we assume that all agents prefer pairing some agent to pairing no agent. It is illustrated that $\forall p \in L \cup R, \forall X \in SX \setminus \{\emptyset\}, X \succ'_p \{\emptyset\}$ and $\forall p, p' \in L \cup R,\ p' \succ_p \emptyset$ hold.

In two sided-matching with indifferences, there are two finite and disjoint sets of agents L and R. A member of L is paired with one or multiple members of R, and a member of R is paired with one or multiple members of L. Each agent p ($p \in L \cup R$) has capacity n_p. Every member of L has indifferent preferences over R, and every member of R has indifferent preference over L. The target of two-sided matching with indifferences (L, R) is finding a matching m.

Given a matching m and an agent p. $m_i(p)$ is p's i-th best agent in $m(p)$ ($i \in \mathcal{N}$ and $i \le n_p$ hold). If $\|m(p)\| \le k \le n_p$ holds, $m_k(p) = \emptyset$ holds. For example, if p's preference is $(\{a\}, \{b, c\}, \{d, e\}, \{f\})$, and $m(p) = \{a, d, e\}$ and $n_p = 4$ holds, $m_1(p) = a$, $m_2(p) = dore$, $m_3(p) = dore$, and $m_4(p) = \emptyset$ hold.

In traditional two-sided matching, stable condition is one of the most important condition. We can't define stable condition in indifferent two-sided matching.

Therefore, three types substitute conditions for stable condition is proposed [9].

Definition 8 (weakly stable). *If a pair of agent* (l, r) *does not exist that satisfies the following conditions for a given matching* m, m *is said to be weakly stable.*

- $r \notin m(l)$ *and* $l \notin m(r)$ *hold.*
- $r \succ_l m_{n_l}(l)$ *holds.*
- $l \succ_r m_{n_r}(r)$ *holds.*

Definition 9 (strongly stable). *If a pair of agent* (l, r) *does not exist that satisfies the following conditions for a given matching* m, m *is said to be strongly stable.*

- $r \notin m(l)$ *and* $l \notin m(r)$ *hold.*
- *One of the following conditions is satisfied.*
  - *The following conditions are satisfied.*
    - $r \succ_l m_{n_l}(l)$ *holds.*
    - $l \succeq_r m_{n_r}(r)$ *holds.*
  - *The following conditions are satisfied.*
    - $r \succeq_l m_{n_l}(l)$ *holds.*
    - $l \succ_r m_{n_r}(r)$ *holds.*

Definition 10 (super stable). *If a pair of agent* (l, r) *does not exist that satisfies the following conditions for a given matching* m, m *is said to be super stable.*

- $r \notin m(l)$ *and* $l \notin m(r)$ *hold.*
- $r \succeq_l m_{n_l}(l)$ *holds.*
- $l \succeq_r m_{n_r}(r)$ *holds.*

We can apply the algorithms of traditional two-sided matching to two-sided matching with indifferences. For example, Gale Shapley algorithm can search weakly stable matching and we can define the constraints for weakly stable matching, strongly stable matching and super stable matching.

However, These condition have matters. In two-sided matching with indifferences, there doesn't always exist strongly stable and super stable. On the other hand, there always exist weakly stable. However, weakly stable doesn't always satisfy the Pareto efficiency [10] (stable condition in traditional two-sided matching always exist and satisfy the Pareto efficiency).

We illustrate Pareto efficiency. Given two matchings m, m' and an agent p. If $\forall i \in \mathcal{N}(i \le n_p)$, $m_i(p) \succeq_p m_i'(p)$ and $\exists i \in \mathcal{N}(i \le n_p)$, $m_i(p) \succ_p m_i'(p)$ hold, we can say that p prefer the matching m to the matching m'. If p prefer a matching m to a matching m', we write $m >_p m'$. On the other hand, if $\forall i \in \mathcal{N}$, $m_i(p) =_p m_i'(p)$ holds, we can say that p likes the matching m as well as the matching m'. If p likes the matching m as well as the matching m', we write $m =_p m'$. If $m >_p m'$ or $m =_p m'$ holds, we can write $m \ge_p m'$.

Definition 11 (dominant matching). *A matching* m' *is dominant matching of* m *while they satisfy the following conditions:*

– $\forall p \in L \cup R$, $m' \geq_p m$ holds.
– $\exists p \in L \cup R$, $m' >_p m$ holds.

Definition 12 (Pareto efficiency). *A matching m satisfies Pareto efficiency while there doesn't exist dominant matching of m.*

If a matching m doesn't satisfy Pareto efficiency, there exist the matching m' that no agent oppose change the matching m into and there exist one or multiple agents who want to change the matching m into. Therefore, Pareto efficiency is one of the most important character in two-sided matching in indifferences.

There always exist one or multiple matchings which satisfy weakly stable and Pareto efficiency. We call "Pareto stability" these matching. To find a Pareto stability matching is important issue in the research field of two-sided matching with indifferences [10]. There are some algorithm to find a Pareto stability matching.

3 Nucleolus

In this section, we illustrate the proposing algorithm to find a Pareto stability matching named "nucleolus". This algorithm's idea is based on nucleolus in the research field of coalitional game which is popular solution concept in this field [12].

We assume that every agents have complaint about a matching and we can represent these complaint in integer.

Definition 13 (complaint). *In two-sided matching with indifferences, an agent p has complaint $c(m,p)$ about a matching m which satisfies $c(m,p) \in \mathbb{Z}$.*

Complaint is decided freely. However, if complaint is decided seriously, complaint has a character which is always satisfied.

Definition 14 (serious complaint). *If complaint c satisfies the following condition, c is serious.*

– *there exist two matchings m, m' and agent p. If $m >_p m'$ holds, $c(m',p) > c(m,p)$ holds.*
– *there exist two matchings m, m' and agent p. If $m =_p m'$ holds, $c(m',p) = c(m,p)$ holds.*

Nucleons's idea is similar to minimization of maximum complaint. Nucleolus is weakly stable matching which is minimized maximum complaint. If there exists multiple such weakly stable matching, nucleolus is minimized 2nd-maximum complaint in such matchings. If there exists multiple such weakly stable matching, nucleolus is minimized 3rd-maximum complaint in such matchings ....

Definition 15 (complaint vector). *In two-sided matching with indifferences, a matching m has complaint vector cv. $cv = \{c(m,p_1), \ldots c(m,p_k)) \ldots c(m,p_{\|L \cup R\|})\}$ holds. p_k is the agent who has k-th biggest complaint in $L \cup R$.*

Definition 16 (nucleolus). *In two-sided matching with indifferences, the nucleolus is weakly stable matching which has has the lexicographically minimum of complaint vector in all possible weakly stable matchings.*

Complaint vector's lexicographical order satisfies transitive relation. Therefore, in two-sided matching with indifferences, (one or multiple) nucleolus always exist.

If complaint is serious, nucleolus satisfies Pareto efficiency (Of course, nucleolus satisfies weakly stable. Therefore, if complaint is serious, nucleolus satisfies Pareto stable).

Lemma 1. *In two-sided matching with indifferences, there exists a matching m which satisfies weakly stable. A matching m' which is dominant matching of m satisfies weakly stable.*

Proof. We assume that m' doesn't satisfy weakly stable. m' has blocking pair (l, r). From definition 8, $l \succ_r m_{n_r}(r)$ and $r \succ_l m_{n_l}(l)$ hold. Because m' is dominant matching of m, $m' \geq_l m$ and $m' \geq_r m$ hold. It means that $m'_{n_r}(r) \succeq_r m_{n_r}(r)$ and $m'_{n_l}(l) \succeq_l m_{n_l}(l)$ hold. Therefore $l \succ_r m'_{n_r}(r)$ and $r \succ_l m'_{n_l}(l)$ hold. We can say that if m' doesn't satisfy weakly stable, m doesn't satisfy weakly stable. Therfore, m' is weakly stable.

Lemma 2. *In two-sided matching with indifferences, there exists a matching m and a matching m' which is dominant matching of m. If complaint is serious, the complaint vector of m' is lexicographically less than the complaint vector of m*

Proof. $\forall p \in L \cup R$, $m' \geq_p m$ and $\exists p \in L \cup R$, $m' >_p m$ hold because m' is dominant matching of m. From definition 14, $\forall p \in L \cup R$, $c(m, p) \geq c(m', p)$ and $\exists p \in L \cup R$, $c(m, p) > c(m', p)$ hold. Therefore, the complaint vector of m' is lexicographically less than the complaint vector of m.

Theorem 1. *If complaint is serious, nucleolus satisfies Pareto efficiency.*

Proof. Let's assume that nucleolus matching m doesn't satisfy the Pareto efficiency. m has matching m' which is dominant matching of m. From lemma 1, m' satisfies is weakly stable. From lemma 2, the complaint vector of m' is lexicographically less than the complaint vector of m. Therefore, if m doesn't satisfy the Pareto efficiency, m is not nucleolus. It means that the nucleolus matching m is Pareto efficiency.

While we can calculate complaint from liner function for $x(l, r)$, we can calculate nucleolus to use integer programming.

The following are examples of complaint which can calculate from liner function for $x(l, r)$ ($\|m(l)\|$ is equal to $\sum_{r \in R} x(l, r)$ and $\|m(r)\|$ is equal to $\sum_{l \in L} x(l, r)$).

– In the case where agent evaluates from all members of his pairs:
  - $\forall l \in L$, $c(l) = (\sum_{r \in R} x(l, r) \times r(l, r)) + (n_l - m(l)) \times r(l, \emptyset)$
  - $\forall r \in R$, $c(r) = (\sum_{l \in L} x(l, r) \times r(r, l)) + (n_r - m(r)) \times r(r, \emptyset)$

- In the case where agent regard his worst pair as important:
  - $\forall l \in L,\ c(l) = (\sum_{r \in R} x(l,r) \times 2^{r(l,r)-1}) + (n_l - m(l)) \times 2^{r(l,\emptyset)-1}$
  - $\forall r \in R,\ c(r) = (\sum_{l \in L} x(l,r) \times 2^{r(r,l)-1}) + (n_r - m(r)) \times 2^{r(r,\emptyset)-1}$
- In the case where agent regard his best pair as important:
  - $\forall l \in L,\ c(l) = -(\sum_{r \in R} x(l,r) \times 2^{\|R\|-r(l,r)})$
  - $\forall r \in R,\ c(r) = -(\sum_{l \in L} x(l,r) \times 2^{\|L\|-r(r,l)})$

Definition 17 (Integer programming to calculate nucleolus). *Nucleolus is calculated from following integer programming*

- *Minimize* $\sum_{1 \le k \le \|L \cup R\|} M^{\|L \cup R\|-k} \times \epsilon(k)$
- $\forall l \in L,\ \sum_{r \in R} x(l,r) \le n_l$
- $\forall r \in R,\ \sum_{l \in L} x(l,r) \le n_r$
- $\forall l \in L,\ \{\sum_{r \in R} x(l,r)\} + \|R\| \times e(l) \ge n_l$
- $\forall r \in R,\ \{\sum_{l \in L} x(l,r)\} + \|L\| \times e(r) \ge n_r$
- $\forall l \in L,\ \forall r \in R,\ -x(l,r) + e(l) + e(r) < 2$
- $\forall l \in L,\ \forall r \in R,\ \forall r' \in R,\ if\ r \succ_l r'\ holds,\ -x(l,r) + x(l,r') + e(r) < 2$
- $\forall l \in L,\ \forall r \in R,\ \forall l' \in L,\ if\ l \succ_r l'\ holds,\ -x(l,r) + x(l',r) + e(l) < 2$
- $\forall l \in L,\ \forall r \in R,\ \forall l' \in L,\ \forall r' \in R,\ if\ r \succ_l r'\ and\ l \succ_r l'\ hold,\ -x(l,r) + x(l',r) + x(l,r') < 2$
- $\forall k \in \mathcal{N}(i \le \|L \cup R\|)$
  - $\forall p \in L \cup R,\ c(p) - \epsilon(k) - C \times s(k,p) \ge 0$
  - $\sum_{p \in L \cup R} s(k,p) = k - 1$

M is a number which is more than possible maximum complaint

4 Conclusion

Two-sided matching is one of the main approaches to form a team in a multi-agent system. The matching is found based on the agent's preferences in the agents to be paired with. For a large scale multi-agent system, handling indifferences in the agent's preferences is important.

In this paper, we propose a new integer programming based algorithm "nucleolus" for two-sided matching with indifferences. The matching which is proposed by nucleolus satisfies important character Paretostable. Nucleolus propose the matching that minimizes the maximum of agent's complaint. By defining an apposite complaint, a Pareto stable matching with proper characteristics can be found.

Future works include extending the proposed algorithm to satisfy incentive compatibility.

References

1. Gale, F., Shapley, L.S.: College Admissions and the Stability of Marriage. American Mathematical Monthly **69**, 9–15 (1962)

2. Roth, A.E.: The Evolution of the Labor Market for Medical Interns and Residents: A Case Study in Game Theory. Journal of Political Economy **92**, 991–1016 (1984)
3. Roth, A.E.: The National Residency Matching Program as a Labor Market. Journal of American Medical Association **275**(13), 1054–1056 (1996)
4. Roth, A.E., Sotomayor, M.A.O.: Two-sided Matching: A study in Game-Theoretic Modeling and Analysis. University Press, Cambridge (1990)
5. Haas, C., Kimbrough, S.O., Caton, S., Weinhardt, C.: Preference-based resource allocation: using heuristics to solve two-sided matching problems with indifferences. In: Altmann, J., Vanmechelen, K., Rana, O.F. (eds.) GECON 2013. LNCS, vol. 8193, pp. 149–160. Springer, Heidelberg (2013)
6. Gusfield, D., Irving, R.W.: The Stable Marriage Problem: Structure and Algorithms. MIT Press, Cambridge (1989)
7. Vate, J.H.V.: Linear Programming Brings Marital Bliss. Oper. Res. Lett. **8**, 147–153 (1988)
8. McVitie, D., Wilson, L.B.: The Stable Marriage Problem. Communications of the ACM **14**, 486–490 (1971)
9. Irving, R.W.: Stable Marriage and Indifference. Discrete Applied Mathematics Archive **48**(3), 261–272 (1994)
10. Kamiyama, N.: A New Approach to the Pareto Stable Matching Problem, Mathematics of Operations Research (2013). (published online in Articles in Advance October 24, 2013 http://dx.doi.org/10.1287/moor.2013.0627)
11. Ohta, N., Kuwabara, K.: An integer programming approach for two-sided matching with indifferences. In: Hwang, D., Jung, J.J., Nguyen, N.-T. (eds.) ICCCI 2014. LNCS, vol. 8733, pp. 563–572. Springer, Heidelberg (2014)
12. Schmeidler, D.: The Nucleolus of a Characteristic Function Game. Journal of Applied Mathematics **17**, 1163–1170 (1969)

Neural Networks, SMT and MIS

Design Analysis of Intelligent Dynamically Phased Array Smart Antenna Using Dipole Leg and Radial Basis Function Neural Network

Abhishek Rawat[1(✉)], Vidhi Rawat[2], and R. N. Yadav[3]

[1] Department of Electrical Engineering, Institute of Infrastructure Technology Research
and Management (IITRAM), Ahmedabad, India
arawat@iitram.ac.in
[2] Department of Electronics and Instrumentation Engineering,
Samrat Ashok Technological Institute, Vidisha, India
[3] Department of Electronics and Communication Engineering,
Maulana Azad National Institute of Technology, Bhopal, India

Abstract. The smart antennas are the antenna arrays with smart signal processing algorithms used to track and locate the antenna beam on the target. The idea of smart antennas is to use base station antenna patterns that are not fixed, but adapt to the current radio conditions. The DPA smart antenna may simplify the problem of design and implementation in comparison of adaptive or switched lobe method because of their capabilities of interference suppression, easy design and implementation. A simple DPA based smart antenna using dipole leg is proposed in this paper which makes it suitable for practical implementation without any compromise in its performance, thus avoiding the need for a high cost adaptive array. Side lobe magnitude and number of sidelobe are reduced by adding null point to the design of array signal in conventional Fourier series window method techniques. A control algorithm based on RBFNN technique is proposed to control the variations in shape and interference which provides the performance of DPA smart antenna equivalent to adaptive array technique. It effectively controls the directivity pattern variations without sacrificing the spectral performance of antenna array.

Keywords: Dynamically phased array smart antenna · Radial basis neural network (RBFNN) · Half wave dipole · Fourier series design method

1 Introduction

Smart antenna[1-4] are emerging as a viable alternative over the conventional antenna array systems in different wireless applications due to automatic pattern synthesis across the whole range, low side lobe stress, electronics steering and beam forming of the spectrum. During the last two decades, smart antenna have been researched, developed and implemented in civil and defense applications such as mobile and satellite communication, fault finding and tolerance with the real time wireless systems, direction of arrival (DOA) estimation, beam forming etc. With the remarkable

© Springer International Publishing Switzerland 2015
M. Núñez et al. (Eds.): ICCCI 2015, Part I, LNAI 9329, pp. 475–484, 2015.
DOI: 10.1007/978-3-319-24069-5_45

development and advances in array design methods and digital signal processor (DSP) devices, high speed semiconductor devices, various topologies of smart antenna are developed. Smart antenna system technologies include intelligent antennas, dynamic phased array, digital beam forming, adaptive antenna systems, and others but broadly smart antenna systems are customarily categorized, however, as switched beam, dynamically phased array (DPA) and adaptive array systems. Among these, adaptive array smart antennas are proved to be effective in mobile and satellite applications. But this system is complex and costly. Design and control scheme for smart antenna decides its performance. Switched beam and Adaptive array techniques have been widely used in smart antenna. Adaptive array technique gives better radiation performance even at higher interference environment, improved directivity utilization, reduced side lobe losses, higher intended signal reception for the same array index, and reduced interference and power losses. However, the complexity involved in determination of desired, and interference signal, DOA estimation, beam forming is evident from the work reported in literature. On the other hand, switched beam smart antennas are easy to implement and does not require any computation extensive algorithm, but its performance is degraded in presence of interference. This encouraged to explore the strength of simple dynamically phased array (DPA) smart antenna techniques to improve the performance of smart antenna array without any complexity involved as in the case of adaptive array technique.

DPA smart antenna is proven technology now-a-days in medium & high range wireless applications. Despite of its several advantages such as side lobe reduction and achieving higher directive capabilities without mathematical computation of array devices, required beam shape generation, electronics steering and consideration of mutual induction which are the main problems with design of array antenna. Keeping this in mind, a DPA smart antenna using half wave dipole is designed and investigated for improved radiation pattern quality. It successfully eliminates the side lobe level from the direction of interference. Further, the "DPA smart antenna" technique is developed to reduce the side lobe losses of radiation pattern with maintaining the particular beam width and directivity within permissible limits. Extensive simulation and experimental results are obtained to investigate the performance of developed antenna system under different conditions. The viability of proposed DPA smart antenna design are ascertained by extensive simulation results and verified experimentally.

This paper is organized as follows. In Section 2, the DPA smart antenna presented. In Section 3, the theory of the RBFNN and its application to the smart antenna problem are reported. Section 4 discusses Fourier series design method. Section 5, describe simulation details with detail result analysis and sections 6 finally conclude the paper.

2   DPA Smart Antenna

The DPA smart antenna is an antenna which controls its own pattern by means of feed-back or feed-forward control, and it performs gain enhancement for desired signals and suppression for interfering signals. The array forms a beam by activating certain omni directional elements in the array which have a multiplying effect to form a beam. This beam can then be 'steered' or pointed in the direction of a device by phasing the transmission of the signal in the elements and adjusting the gain on each antenna element. A DPA steers the created beam at the desired device. As the beams are formed digitally the same array of elements can target beams at multiple devices on multiple frequencies. It could be an active area of academic research and development.

The DPA smart antenna [5] achieved optimum gain but the interference not suppressed completely but up to certain extent where its effect is almost negligible. In this approach smart antennas communicate directionally by forming specific antenna-beam patters. They direct their main lobe, with increased gain, in the direction of the user, and they direct nulls or very small side lobe in directions of interference. Since the behavior of antenna array is nonlinear in nature and changes rapidly as any parameters (spacing, feeding current, phase and frequency) changes so the optimization techniques like artificial neural network (ANN) based synthesis method could be able to calculate the voltages that must be applied to the elements of an array, or taking into account coupling effects between them and their real radiating properties. It is performed without any increase of the complexity from the designer's point of view, improving the efficiency and accuracy of non-coupling traditional methods (based on array factor information or equivalent current reconstruction). The fast operation of trained optimization techniques like neural network makes this method suitable for real-time implementation. In this work DPA smart antenna using Fourier series and trained with RBFNN based design proposed here.

3   Radial Basis Neural Network

Techniques employed in synthesis of smart antenna arrays vary from complex analytical methods to iterative numerical methods. The drawback of these techniques is that they usually work with the array factor information and do not consider interaction between array elements. This causes certain error in resultant radiation pattern. The complexity of this approach is extremely high and it is usually disregarded. A neural network (NN) [13-14] based optimization techniques can avoid complexity establishing a relation between the desired radiation patterns and feeding detail like voltage spacing etc in the real antenna.

RBFNN's [6-8] are a member of a class of general purpose method for approximating nonlinear mappings since the DOA problem and Beamforming is of nonlinear nature. Unlike the back propagation networks which can be viewed as an application of an optimization problem, RBFNN can be considered as designing neural networks as a curve fitting (or interpolation) problem in a high-dimensional space. RBF neural networks consist of neurons which are locally tuned. An RBF network can be regarded as a feed forward neural network with a single layer of hidden units, whose responses are the outputs of radial basis functions. The input of each radial basis function of an RBF neural network is the distance between the input vector (activation) and its center (location). Since the radial basis neural networks are excellent candidates for selecting relevant features in pattern recognition problems, so it could be successfully applied for DPA smart antenna designing.

A typical RBF network is a two layer network having input layer of the dimension of training patterns, hidden Layer of up to p locally tuned neurons centered over receptive fields for non-linear, local mapping and output layer that provides the response of the network. Each hidden unit output zj is obtained by calculating the "closeness" of the input x to an n-dimensional parameter vector μ_j associated with the ith hidden unit. Receptive fields center on areas of the input space where input vectors lie, and serve to cluster similar input vectors. If an input vector (x) lies near the center of a receptive field (μ), then that hidden node will be activated. The output layer is a layer of standard linear neurons and performs a linear transformation of the hidden node outputs. This layer is equivalent to a linear output layer in a MLP, but the weights are usually solved for using a least square algorithm rather trained for using back propagation. The output layer may, or may not, contain biases.

4   Fourier series Design Method

Arrays of antennas are used to direct radiated power towards a desired angular sector. The number, geometrical arrangement, and relative amplitudes and phases of the array elements depend on the angular pattern that must be achieved. Once an array has been designed to focus towards a particular direction, it becomes a simple matter to steer it towards some other direction by changing the relative phases of the array elements, a process called steering or scanning. For uniformly spaced arrays, fourier series design methods are identical to the methods for designing FIR digital filters in digital signal processing (DSP), such as window-based and frequency-sampling designs. In fact, these methods were first developed in antenna theory and only later were adopted and further developed in DSP. This method is based on the inverse discrete-space Fourier transforms of the array factor. The one-dimensional equally-spaced arrays are usually considered symmetrically with respect to the origin of the array axis. This requires a slight redefinition of the array factor in the case of even number of array elements. Consider an array of N elements at locations xm along the x-axis with element spacing d. The array factor will be

$$A(\varphi) = \sum_m a_m e^{jk_x x_m} = \sum_m a_m e^{jkx_m \cos\varphi} \tag{1}$$

Where $k_x = k\cos\varphi$, $x_m = md$.

$$A(\phi) = a_0 + \sum_{m=1} a_m e^{jm\phi} + a_{-m} e^{-jm\phi} \tag{2}$$

Then the corresponding inverse function will be

$$a_m = \frac{1}{2\pi} \int_{-\pi}^{\pi} A(\varphi) e^{-jm\varphi} \, d\varphi \tag{3}$$

In general, a desired array factor [5][6] requires an infinite number of coefficients a_m to be represented exactly. Keeping only a finite number of coefficients in the Fourier series introduces unwanted ripples in the desired response, known as the Gibbs phenomenon. Such ripples can be minimized using an appropriate window, but at the expense of wider transition regions. So the Fourier series method may be summarized as a desired response, say Ad(φ), pick an odd or even window length, for example N = 2M+1, and calculate the N ideal weights by evaluating the inverse transform:

$$a_d(m) = \frac{1}{2\pi} \int_{-\pi}^{\pi} A(\varphi) e^{-jm\varphi} \, d\varphi \qquad (4)$$

$$m = 0, \pm 1, \ldots, \pm M$$

Then, the final weights are obtained by windowing with a length-N window w(m):

$$a(m) = w(m)ad(m), \qquad (5)$$

This method is convenient only when the required integral can be done exactly. The detail of desired and interference signal for five dipole leg at 860 MHZ are given in Table 1 and their correspondent feeding detail are given in Table 2. The corresponding radiation pattern is shown in Fig. 1.

Table 1. Details of desired and interference signal

Angle	Power(dB)	Angle	Power(dB)
-90	-50	10	-50
-80	-50	15	-50
-70	-50	20	-50
-60	-50	25	0
-50	-50	30	0
-40	-50	35	0
-30	-50	40	-50
-20	-50	45	-50
-10	-50	50	-50
-5	-50	55	-50
0	-50	60	-50
5	-50	65	-50

5   Simulation Setup and Result Analysis

For the experimental setup, design prototype of DPA smart antenna has been prepared in MATLAB simulink environment using linear 5 element half wave dipole antenna at 860 MHz. In the first phase, array of different beam shape from 0 to 60 deg. and restrict side or minor lobe up to -50dB level is designed and collect the data for the difference of five degree. In second phase RBFNN trained and tested for data set generated by the first phase. Once proper training and testing completed, RBFNN can predict the feeding values for given number of element and scan angles. The RBFNN model is applied for the training of DPA antenna. The data generated in first phase is used for design, trained and testing of RBF neural model. The performance of neural model is strictly dependent on the training parameters, so training parameters adjusted to achieve required accuracy. Once the neural model properly trained it can provide accurate feeding detail for required shape of beam lobe without any complicated manipulations. The performance improves as epoch and neurons advances as shown in Table 5. The simulation and experimental results shows that proposed model can predict, feeding detail for required radiation pattern of required shape and size. It can be achieved by using this neural model without any mathematical computation. The detail of RBFNN is given in Table 4.

Table 2. Feeding details for desired and interference signal.

Dipole leg	Applied voltage	Phase in radian
1.	0.913229394554452	-3.12397316904215
2.	0.952513181471378	-1.56195203151085
3.	1	0
4.	0.952513181471378	1.56195203151085
5.	0.913229394554452	3.12397316904215

Table 3. Design parameters and specifications.

PARAMETERS	SPECIFICATIONS
ARRAY ELEMENT	DIPOLE LEG
APPLIED FREQUENCY	860 MHz
NUMBER OF ELEMENTS	5
SPACING BETWEEN ELEMENTS	0.5
RADIUS FOR SUMMATION OF FIELD CONTRIBUTIONS	999 METER

Table 4. Detail of RBFNN model.

Total number of neuron	150
Number of hidden layer	01
SSE	2.3547e-007
Performance goal	0.00
Spread of radial function	6

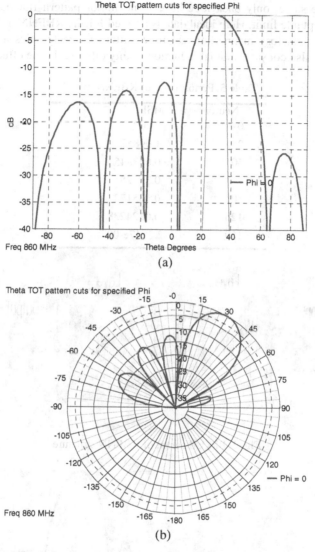

Fig. 1. Radiation pattern for five element at 860 MHz (a) The Linear plot of radiation pattern, (b) The Polar Plot of radiation pattern.

RBFNN play key role to convert ordinary DPA to smart one. The detail data set for 5 element data set applied on RBFNN, set its parameters and trained it. The block diagrams of RBFNN are shown in Fig 2 and detail are given in Table. 4. After successful training, when performance goal is achieved, it is tested for some unknown data. The training details are feeding values, number of elements and scan at particular scan angle. The design parameters and training performances for RBFNN are shown in Table. 3 and 5. The Fig. 3 compares real time pattern with predicted pattern from RBFNN model. The comparison shows that main lobe and nearby side lobe pattern are the same, only back lobe have some minor pattern fluctuation which is under the acceptable limit. Here ideal data is applied for the RBFNN training and it is performs in satisfactory manner. If real time pattern characteristics are applied for training it will also consider mutual induction or any other real time effect.

Table 5. Performance of RBFNN model.

Neurons	MSE
0	2.602
25	1.09852
50	0.653845
75	0.331255
100	0.135172
125	0.0124238
150	2.3547e-007

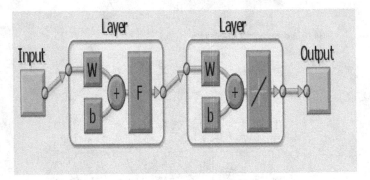

Fig. 2. Block diagram of RBFNN structure.

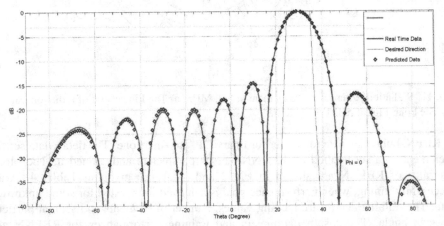

Fig. 3. Radiation pattern comparision for real time and RBFNN predicted for 5 elements at 30 degree.

Table 6. Feeding detail for radiation pattern of 5 element at 30 degree scan angle.

Element	Applied Voltage in volt. (Real time)	Applied Voltage In volt. (Predicted from RBFNN)	Phase in Radian (Real time)	Phase in Radian (Predicted from RBFNN)
1	0.7664	0.7764	0.03467	0.0367
2	0.8501	0.8501	1.5970	1.5705
3	0.9132	0.9332	-3.1239	-3.123
4	0.9525	0.95251	-1.5619	-1.5618
5	1	1	0	0
6	0.9525	0.95251	1.5619	1.5618
7	0.9132	0.93322	3.1239	3.1239
8	0.8501	0.8501	-1.5970	-1.5705
9	0.7664	0.7764	-0.0346	-0.0367

6   Conclusion

In this approach DPA smart antenna can estimate and design required major beam lobe in a particular scan angle. At the same time it maintains the minor or side lobe level within the limits which have no interference effect on proposed communication system. The soft prototype for 5 elements at 860MHz have been design and tested in MATLAB simulink environment. Fourier series based array lobe control has been very effectively controlled using neural based techniques. The obtained results are in the close confirmation with simulation results and are equivalent to DPA smart antenna to solve these issues with advantage of not using any mathematically computation extensive techniques. Since the most of the beam shape need for mobile towers are not so complicated so fourier series design method could used successfully for DPA smart antenna design applicable for mobile towers. The proposed design method DPA smart antenna is case sensitive and becoming less complex, and can be involve combinations with processing in time domain. To implementation DPA smart antenna, there should always be a reasonable compromise between the amount of information about radio channels in different domains to be exploited at the receiver and the expected level of improvements. The possibility to obtain more detailed information related to the radio channel is restricted by the signal processing algorithms and antenna element. Feeding hardware system, data transmission speed, and user mobility is highly dependent on the radio interface type and parameters. The proposed methodology is only for linear array of dipole antenna and provides satisfactory result up to 60 degree scan angle. But it could be applied for different array element and scan angle can be increased by using planer array set up.

The ANN are capable to convert ordinary DPA antenna in to smart one. Basically it learn from its past data set and predict effective data for unknown condition. It can also work on real time system and can update its weights continually. This network provides command DSP processor and hardware system for appropriate feeding detail according to the system requirements. The design of DPA smart antenna is very specific to the applications, and may vary case to case. For the accuracy of this system initially broad data collected for specific element, operational frequency and pattern are required. RBFNN trained and tested with data set and connected to the feeding hardware systems to accumulate specific feeding detail for particular direction which is instructed form DOA estimator. A specific feedback system is also planted for in far field distance to monitor its performance and provide necessary updation. These data are used to reset RBFNN weight time to time to improve its performance.

References

1. Alexiou, A., Haardt, M.: Smart antenna technologies for future wireless systems: trends and challenges. IEEE Commun. Mag. **42**(9), 90–97 (2004)
2. Chryssomallis, M.: smarty antennas. Antennas and Propagation Magazine, IEEE **42**(3), 129–136 (2000)
3. Array Comm: Smart Antenna Systems, May 2000. http://www.webproforum.comlsmar_anl and http://www.arraycomin.com, IntclliCell@technology
4. Compton, R.T.: Adaptive Antcnnas: Conccpts and Perfonnance. Prentice-Hall, Englewood Cliffs (1988)
5. Stevanovi´c, I., Skrivervik, A., Mosig, J.R.: Smart Antenna Systems for Mobile Communications. Technical report, Laboratoired'Electromagn´etisme et d'Acoustique, January 2003
6. Haykins, S.: Neural Networks: A Comprehensive Foundation. IEEE Press/IEEE Computer Society Press, New York (1994)
7. Rawat, A., Yadav, R.N., Shrivastava, S.C.: Neural Modeling of 15 Element Dynamic Phased Array Smart Antenna. In: IEEE International Conference on Advances in Computing, Control, and Telecommunication Technologies, ACT 2009, Trivendrum, India, pp. 45–49, December 28, 2009
8. El Zooghby, A.H., Christodoulou, C.G., Georgiopoulos, M.: Performance of Radial-Basis Function Networks for Direction of Arrival Estimation with Antenna Arrays. IEEE Trans. Antennas Propag. **45**(11), 1611–1617 (1997)
9. Rawat, A., Yadav, R.N., Shrivastava, S.C.: Neural Network Application in Smart Antenna Arrays: A Review. AEU - International Journal of Electronics and Communications **66**(11), 903–912 (2012)

On the Accuracy of Copula-Based Bayesian Classifiers: An Experimental Comparison with Neural Networks

Lukáš Slechan and Jan Górecki[✉]

Department of Informatics, SBA in Karvina, Silesian University in Opava,
Karvina, Czech Republic
{o130079,gorecki}@opf.slu.cz

Abstract. In this work, we compare three classifiers in terms of accuracy. The first is a copula-based Bayesian classifier based on elliptical and Archimedean copulas. The remaining two are Naive Bayes and Neural Networks. Such a comparison, particularly for the recently proposed Archimedean copula-based Bayesian classifiers, hasn't been reported in the literature. The results show that copula-based Bayesian classifiers are a viable alternative to Neural Networks in terms of accuracy while keeping the models relatively simple.

Keywords: Copula · Elliptical copula · Archimedean copula · Bayesian classification · Neural networks

1   Introduction

In machine learning, classification is the problem of identifying to which of the set of categories a new observation belongs, on the basis of a training set of data containing observations whose category membership is known. Learning relationships between random variables is a decisive task in the field of knowledge discovery and data mining. The dependence between the observed variables can be studied by means of copulas. Copulas are distribution functions with standard uniform univariate margins and are widely used particularly for studying dependence between continuously distributed random variables; for more details on copulas, e.g., see [12]. The word copula comes from the latin copula, which means "a bond or a link" and was first used by Sklar [15]. Despite the fact that a large part of the success of copulas is attributed to finance [14], copulas are more and more adopted in data mining [9], [7], hydro-climatic and water-resources [4], [10], gene analysis [18] or cluster analysis [3].

In this paper, we consider copula-based Bayesian classifiers (CBCs) based on elliptical copulas (ECs) and Archimedean copulas (ACs). These classifiers are experimentally compared on 8 real-world datasets with other two commonly known classifiers - Neural Networks (NN), which are appreciated for their high accuracy but which however produce complex and thus low-understandable models, and Naive Bayes (NB), which, as a special case of CBCs assuming the independence copula for the variables, produces well-understandable however less-accurate models. The research reported in this paper follows the research presented in [7], where CBCs are

© Springer International Publishing Switzerland 2015
M. Núñez et al. (Eds.): ICCCI 2015, Part I, LNAI 9329, pp. 485–493, 2015.
DOI: 10.1007/978-3-319-24069-5_46

compared with Classification and regression trees, Random forests and Support vector machines. Our comparison complements that work with a comparison of CBCs to NN.

The paper is structured as follows. Section 2 summarizes some needed theoretical concepts including ECs and ACs. Section 3 recalls CBCs. Section 4 presents the results of an experimental comparison of the above-mentioned classifiers based on real-world datasets and Section 5 concludes.

2   Preliminaries

2.1   Copulas

Definition 1. A d-dimensional copula C: $[0,1]^d \rightarrow [0,1]$ is a function which is a multivariate distribution function with uniform univariate margins on the interval $[0,1]$.

Copulas establish a connection between multivariate distributions and their univariate margins, see the following theorem.

Theorem 1 (Sklar's Theorem). Let F be a d-dimensional multivariate distribution function with univariate margins $F_1, ..., F_d$. Then there exists a copula C: $[0,1]^d \rightarrow [0,1]$ such that

$$F(x_1, ..., x_d) = C\big(F_1(x_1), ..., F_d(x_d)\big) \tag{1}$$

holds for all $(x_1, ..., x_d) \in \overline{\mathbb{R}}^d$, where $\overline{\mathbb{R}} = \mathbb{R} \cup \{-\infty, +\infty\}$. Such a function C is uniquely determined, if $F_1, ..., F_d$ are all continuous. Conversely, if C is a copula and $F_1, ..., F_d$ are univariate distribution functions, then the function F given by (1) is a multivariate distribution function with margins $F_1, ..., F_d$.

Hence, Sklar's Theorem allows to model the univariate margins and the copula of a multivariate distribution function separately, which allows to build very flexible class of multivariate distribution models. There are many parametric copula families available, which usually have parameters that control the strength of dependence. Some popular parametric copula models are recalled below.

2.2   Some Parametric Copula Families

Basic information about copula families are presented, e.g., in [12]. In this paper, we use two families of ECs and three families of ACs.

Elliptical Copulas

ECs are based on existing multivariate elliptical distributions and are derived directly using Sklar's Theorem. The Gaussian copula is based on the multivariate normal distribution and the Student's t-copula is based on the multivariate Student's t-distribution. Formally, a Gaussian copula is given by

$$C_\Sigma^{Ga}(u_1, ..., u_d) = \Phi_\Sigma\big(\phi^{-1}(u_1), ..., \phi^{-1}(u_d)\big), \tag{2}$$

where ϕ is the cumulative distribution function (CDF) of the univariate normal distribution and ϕ_Σ is the CDF of the multivariate normal distribution with a correlation matrix Σ. A Student's t-copula is given by

$$C_{\nu,\Sigma}^t(u_1,...,u_d) = t_{\nu,\Sigma}(t_\nu^{-1}(u_1), ..., t_\nu^{-1}(u_d)), \tag{3}$$

where t_ν is the CDF of the univariate Student's t-distribution with ν degrees of freedom and $t_{\nu,\Sigma}$ is the CDF of the multivariate Student's t-distribution with a correlation matrix Σ and ν degrees of freedom.

2.3 Archimedean Copulas

Definition 2. An *Archimedean generator* (simply, *generator*) is a continuous, non-increasing function ψ : $[0, \infty] \to [0,1]$, which satisfies $\psi(0) = 1, \psi(\infty) = \lim_{t\to\infty} \psi(t) = 0$ and which is strictly decreasing on $[0, \inf\{t \mid \psi(t) = 0\}]$.

Definition 3. Any d-copula C is called AC, if it admits the form

$$C(\mathbf{u}) := C(\mathbf{u}; \psi) := \psi(\psi^{-1}(u_1) + \cdots + \psi^{-1}(u_d)), \mathbf{u} \in [0, 1]^d, \tag{4}$$

where ψ is a generator and $\psi^{-1}: [0,1] \to [0,\infty]$ is defined by $\psi^{-1}(s) = \inf\{t \, \psi(t) = s\}, s \in [0, 1]$.

To derive an explicit form of an AC, we need explicit generators. For the construction of CBCs described in Section 3, we use the three popular families of ACs presented in Table 1. For other families of ACs, e.g., see [9].

Table 1. The three considered one-parametric Archimedean copula families with the corresponding parameter ranges and forms.

Family	θ	$\psi(t)$
Clayton (C)	$(0, \infty)$	$(1 + t)^{-1/\theta}$
Frank (F)	$(0, \infty)$	$-\log(1 - (1 - e^{-\theta})\exp(-t))/\theta$
Gumbel (G)	$[1, \infty)$	$\exp(-t^{1/\theta})$

3 Construction of Copula-Based Bayesian Classifiers

We briefly recall some basics for Bayesian classifiers and a way how copulas could be integrated into them; see also [7], [14].

Let $\Omega = \{\omega_1, ..., \omega_m\}$ be a finite set of m classes. The problem of classification is to assign each x from the variable space $\mathbb{R}^d$ a class from Ω. A Bayesian classifier is said to assign x to the class ω_i, if

$$g_i(x) > g_j(x) \qquad \text{for all } j \neq i, \tag{5}$$

where $g_i\colon [0,\infty)^d \to \mathbb{R}, i = 1, \ldots, m$ are called *discriminant functions* that are defined by

$$g_i(x) = \mathbb{P}(\omega_i|x) = \frac{f(x|\omega_i)\mathbb{P}(\omega_i)}{\sum_{j=1}^m f(x|\omega_j)\mathbb{P}(\omega_j)}, \tag{6}$$

where $f\colon \mathbb{R}^d \to [0,\infty)$ is a probability density function and $\mathbb{P}(\omega_i)$, $i = 1, \ldots, m$ are the prior probabilities of the classes from Ω. Since any monotonically increasing function $Q\colon \mathbb{R} \to \mathbb{R}$ keeps the classification unaltered, the discriminant functions can be simplified by $g_i := Q \circ g_i$ with $Q(t) = \ln\left(t\sum_{j=1}^m f(x \mid \omega_j)\mathbb{P}(\omega_j)\right)$ from (6) to

$$g_i(x) = \ln f(x|\omega_i) + \ln \mathbb{P}(\omega_i). \tag{7}$$

Provided F given by (1) is an absolutely continuous multivariate distribution function with margins $F_1, \ldots, F_d$, the pdf f of F can be expressed by

$$f(x_1, \ldots, x_d) = c\big(F_1(x_1), \ldots, F_d(x_d)\big) \prod_{k=1}^d f_k(x_k), \tag{8}$$

where $c(u_1, \ldots, u_d) = \frac{\partial^d C(u_1, \ldots, u_d)}{\partial u_1 \ldots \partial u_d}$ denotes the density of the copula $C(u_1, \ldots, u_d)$ and f_k denotes the density of $F_k, k = 1, \ldots, d$. Using (8), $f(x|\omega_i)$ can be rewritten to

$$f(x|\omega_i) = c(F_1(x_1|\omega_i), \ldots, F_d(x_d|\omega_i)|\omega_i) \prod_{k=1}^d f_k(x_k|\omega_i), \tag{9}$$

and thus (7) turns to

$$g_i(x) = \ln(c(F_1(x_1|\omega_i), \ldots, (F_d(x_d|\omega_i)|\omega_i))) + \sum_{k=1}^d \ln(f_k(x_k|\omega_i)) + \ln(\mathbb{P}(\omega_i)). \tag{10}$$

Hence, the discriminant function g_i is composed of three ingredients: the conditional copula density, the conditional marginal densities and the prior probability of the class ω_i. Note that these ingredients do not impose any restrictions on each other.

4   Experiments

4.1   Design on the Experiments

In this subsection, we evaluate the accuracy of 8 classifiers. Five of them, considering different underlying families of copulas, are CBCs, where two of them are based on ECs and three of them are based on ACs. More precisely, we consider:

- **Elliptical Copula-Based Bayesian Classifiers**. For these classifiers, it is assumed that $\hat{C}(\cdot|\omega_i)$ is a Gaussian (denoted as ECBC(G)) or Student's t-copula (ECBC(t)), respectively. The computation $\hat{C}(\cdot|\omega_i)$ is implemented by Matlab's Statistics and machine learning toolbox function `copulafit` with the parameter `family` set to the value `Gaussian` or `t copula`, respectively.
- **Archimedean Copula-Based Bayesian Classifiers**. For these classifiers, it is assumed that $\hat{C}(\cdot|\omega_i)$ is a Clayton (denoted as ACBC(C)), Gumbel (ACBC(G)) or

Frank (ACBC(F)) copula, respectively. The copula parameter is estimated by inversion of pairwise Kendall's tau, e.g., see [4], [5], [7].

The estimates $\hat{F}_1(\cdot \mid \omega_i), \dots, \hat{F}_d(\cdot \mid \omega_i)$ of $F_1(\cdot \mid \omega_i), \dots, F_d(\cdot \mid \omega_i)$ in (10) are computed in the same way for all above-mentioned classifiers using the Kernel smoothing function `ksdensity` with the parameter `function` set to `cdf`.

These CBCs are compared with the following classifiers available in Matlab:

- **Naive Bayes** (denoted by NB). For the classifier $\hat{C}(\cdot \mid \omega_i), i = 1, \dots, m$ is independence copula. For the following tt is implemented by function `fitNaiveBayes` and it is training phase, we set the parameter `Distribution` to the value `normal` (this classifier is denoted by NAIVE(N)) or `kernel` (denoted by NAIVE(K)).
- **Neural Networks** (denoted by NN). It is a two-layer feed-forward network with hidden and softmax output neurons. The classifier is implemented by the function `patternnet` with the training function set to the default, i.e., the scaled conjugate gradient back-propagation is used. The parameter `hiddenLayerSize` is set to one of the values 5, 10 and 15 based on a 10-fold cross-validation.

Note that if the reader is interested in a comparison of CBCs to other types of classifiers, e.g., to Classification and regression trees, Random forests or Support vector machines, such a comparison is reported in [7].

In summary, we evaluate these 8 classifiers on 8 commonly known datasets from UCI-dataset repository [2], namely on Iris (4 variables, 3 classes), BankNote (4 variables, 2 classes), Seeds (7 variables, 3 classes) and BreastTissue (9 variables, 4 classes), Wine (13 variables, 3 classes), and two datasets from the KEEL-dataset repository[1], namely Hayes-Roth (5 variables, 3 classes) and Appendicitis (7 variables, 2 classes). The eighth dataset, which is a results of a recent real-world application in catalysis [11], we the Catalysis dataset. We selected these datasets in order to all considered classifiers could be applicable for each of them.

The accuracy computation for a given classifier and dataset is based on a 10-fold cross-validation and repeated 10 times. All computations were performed in Matlab on a PC with Intel Core i3-3220 CPU @ 3.30 GHz, 8GB RAM.

4.2   Results of the Experiments

The accuracy of the classifiers computed for the selected datasets is shown in Fig. 1.

It can be observed in Fig. 1, that there is not a top winning classifier on all chosen datasets. It thus confirms the "No Free Lunch Theorem" [17]. However, one can observe that there are classifiers which scored higher more often than others. This observation is supported by rankings of the classifiers shown in Table 2. Each of the classifiers is ranked according to its averaged accuracy (1 - the highest, 8 – the lowest).

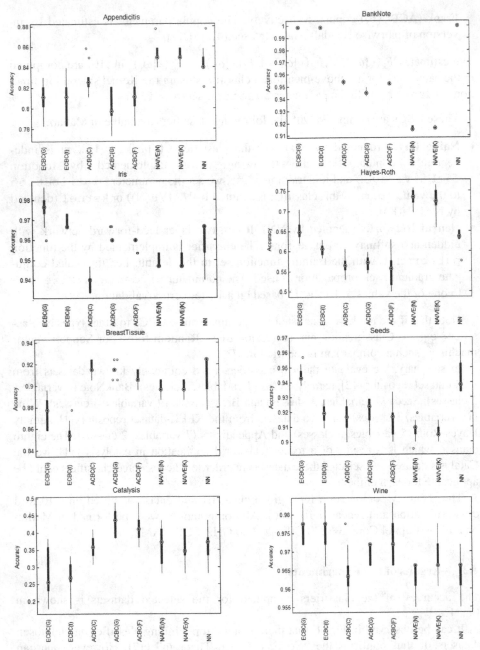

Fig. 1. The accuracy (boxplots) of the classifiers computed on the Appendicitis, BankNote, Iris, Hayes-Roth, BreastTissue, Seeds, Catalysis and Wine datasets.

Table 2. The rankings of the classifiers according to the averaged accuracy on a given dataset.

Classifier	ECBC(G)	ECBC(t)	ACBC(C)	ACBC(G)	ACBC(F)	NAIVE(N)	NAIVE(K)	NN
Appendicitis	5	7	4	8	5	1	2	3
BankNote	3	2	4	6	5	8	7	1
Iris	1	2	8	4	5	7	6	3
Hayes-Roth	3	5	7	6	8	1	2	4
BreastTissue	7	8	2	3	4	5	5	1
Seeds	1	4	5	3	6	7	8	2
Catalysis	8	7	5	1	2	3	6	4
Wine	1	2	8	4	3	7	5	6
average rank	3.625	4.625	5.375	4.375	4.75	4.875	5.125	3

Observing the averages of the ranks – the average rank row in Table 2 - three groups of classifiers can be distinguished:

— The highest-ranked classifiers – NN (average rank = 3) and ECBC(G) (3.625);
— The middle-ranked classifiers – ACBC(G) (4.375), ECBC(t) (4.625), ACBC(F) (4.75) and NAIVE(N) (4,875);
— The lowest-ranked classifiers – NAIVE(K) (5.125) and ACBC(C) (5.375).

These results show that CBCs, particularly ECBC(G), could be considered competitive with the best in average performing NN. It should be also mentioned that, even if the CBCs based on ACs perform in average worse than the highest-ranked classifiers (this result corresponds to [7]), these classifiers, namely ACBC(G) and ACBC(F) are the best performing classifiers on the Catalysis dataset and the second best performing classifiers on the BreastTissue dataset, and hence could also be considered as a viable and simple alternative to NN.

5   Conclusion

In this work, elliptical and Archimedean copula-based Bayesian classifiers are experimentally compared to Neural Networks and Naive Bayes in terms of accuracy for the first time. The results based on 8 real-world datasets have shown that copula-based Bayesian classifiers are, in terms of accuracy, a viable alternative to highly accurate Neural Networks while keeping the models relatively simple.

In further research, we would like to extend the research presented here by involving other copula-based Bayesian classifiers based on, e.g., hierarchical Archimedean copulas, pair copulas, etc. These families of copulas, for which serious researching effort can be recently observed, e.g., see [6], [8], [16], overcome some restrictions of elliptical and Archimedean copulas, are flexible but more computationally demanding, see, e.g., the discussion concerning the computation of hierarchical Archimedean copula density functions in high dimensions in [7]. Nevertheless, in our opinion, bringing these families into a play could substantially increase the accuracy of copula-based Bayesian classifiers while still keeping the models relatively simple.

Also, the considered copula-based Bayesian classifiers, despite ranked lower than the neural network based classifier in the averaged accuracy, outperform it on several datasets. In the light of this fact, it would be desirable to consider, e.g., the relation between the datasets features and the ranks, or which are the subclasses of problems for which the copula-based classifiers perform better than other classifiers.

Acknowledgement. The research in this work has been funded by the project SGS/21/2014.

References

1. Alcalá, J., Fernández, A., Luengo, J., Derrac, J., García, S., Sánchez, L., Herrera, F.: Keel data-mining software tool: Data set repository, integration of algorithms and experimental analysis framework. Journal of Multiple-Valued Logic and Soft Computing **17**, 255–287 (2010)
2. Bache, K., Lichman, M.: UCI machine learning repository (2013). http://archive.ics.uci.edu/ml
3. Cuvelier, E., Noirhomme-Fraitur, M.: Clayton copula and mixture decomposition. In: Applied Stochastic Models and Data Analysis, ASMDA 2005. Brest (2005)
4. Genest, C., Favre, A.: Everything you always wanted to know about copula modeling but were afraid to ask. Hydrol. Eng. **12**, 347–368 (2007)
5. Górecki, J., Holeňa, M.: Structure determination and estimation of hierarchical archimedean copulas based on Kendall correlation matrix. In: Appice, A., Ceci, M., Loglisci, C., Manco, G., Masciari, E., Ras, Z.W. (eds.) NFMCP 2013. LNCS, vol. 8399, pp. 132–147. Springer, Heidelberg (2014)
6. Górecki, J., Holeňa. M.: An alternative approach to the structure determination of hierarchical Archimedean copulas. In: Proceedings of the 31st International Conference on Mathematical Methods in Economics (MME 2013), pp. 201–206. Jihlava, Czech Republic (2013)
7. Górecki, J., Hofert, M., Holeňa, M.: An Approach to Structure Determination and Estimation of Hierarchical Archimedean Copulas and Its Application in Bayesian Classification. Journal of Intelligent Information Systems 1–39 (2015). Springer Science+Business Media New York
8. Górecki, J., Hofert, M., Holeňa, M.: On the consistency of an estimator for hierarchical Archimedean copulas. In: Proceedings of the 32nd International Conference on Mathematical Methods in Economics, pp. 239–234. Olomouc, Czech Republic (2014)
9. Holeňa, M., Ščavnický, M.: Application of copulas to data mining based on observational logic. In: ITAT: Information Technologies Applications and Theory Workshops, Posters, and Tutorials, North Charleston: CreateSpace Independent Publishing Platform, Donovaly, Slovakia (2013)
10. Kao, S.C., Govindaraju, R.S.: Trivariate statistical analysis of extreme rainfall events via plackett family of copulas. Water Resour. Res. **44** (2008)

11. Moehmel, S., Steinfeldt, N., Engelschalt, S., Holena, M., Kolf, S., Baerns, M., Dingerdissen, U., Wolf, D., Weber, R., Bewersdorf, M.: New catalytic materials for the high-temperature synthesis of hydrocyanic acid from methane and ammonia by high-throughput approach. Applied Catalysis A: General **334**(1), 73–83 (2008)
12. Nelsen, R.B.: An Introduction to Copulas, 2nd edn. Springer, New York (2006)
13. Rank, J.: Copulas: From Theory to Application in Finance. Risk Books, London (2006)
14. Sathe, S.: A novel Bayesian classifier using copula functions (2006). arXiv preprint cs/0611150
15. Sklar, A.: Fonctions de répartition a n dimensions et leurs marges. Publ. Inst. Stat. Univ. Paris **8**, 229–231 (1959)
16. Stöber, J., Joe, H., Czado, C.: Simplified pair copula constructions - limitations and extensions. Journal of Multivariate Analysis **119**, 101–118 (2013)
17. Wolpert, D.H.: The supervised learning no-free-lunch theorems. In: Soft Computing and Industry, pp. 25–42. Springer (2002)
18. Yuan, A., Chen, G., Zhou, Z.C., Bonney, G., Rotimi, C.: Gene copy number analysis for family data using semiparametric copula model. Bioinform. Biol. Insights **2**, 343–355 (2008)

Automating Event Recognition for SMT Systems

Emna Hkiri[✉], Souheyl Mallat, Mohsen Maraoui, and Mounir Zrigui

Faculty of Sciences of Monastir, Monastir, Tunisia
emna.hkiri@gmail.com

Abstract. Event Named entity Recognition (NER) is different from most past research on NER in Arabic texts. Most of the effort in named entity recognition focused on a specific domains and general classes especially the categories; Organization, Location and Person. In this work, we build a system for Event named entities annotation and recognition. To reach our goal we combined between linguistic resources and tools. Our method is fully automatic and aims to ameliorate the performance of our machine translation system.

1 Introduction

Information extraction (IE) as defined in the message understanding conferences (MUC), aims to analyze natural language texts and extract useful information from a particular domain or application. IE is used in processing text for Arabic systems such as search engines, clustering, classification, text mining systems Question and answering, information retrieval, text summary, indexation, and classification (Cimiano et al, 2004) ,(Cohen et al, 2009).

The main task that we tackle in this paper is to develop a system to extract events and test the applicability of our method to Arabic texts. Arabic language is considered difficult to control in the NLP. It is a semitic language and presents specific morphological, syntactic, phonetic and phonological features(Hkiri et al, 2013). To overcome these challenges, we propose an event system detection in order to improve the performance of SMT system. Our system used a set of unclassified Arabic news websites articles, and generates a set of events with their attributes. In this work we adopted the Automatic Content Extraction model definition to extract and annotate Event named entities. The ACE defines the event as an action, a process in which participants are connected(George, 2004).

The rest of the paper is organized as follows: Section (2) deals with the definition of the event and reviews related work on event detection. In section (3), we present our method for the automatic extraction of events. In section (4), we discuss the main elements of our proposed system. In section (5), we presents our experiments and discusses the results. Finally, we conclude our work highlight the future work.

2 State of Art

The problem of extracting information from text acquired in recent years considerable attention. But while the problem of detecting and extracting named entities is relatively

© Springer International Publishing Switzerland 2015
M. Núñez et al. (Eds.): ICCCI 2015, Part I, LNAI 9329, pp. 494–502, 2015.
DOI: 10.1007/978-3-319-24069-5_47

well understood, the event extraction remains a problem far from being solved tasks because an event mention can be expressed by several linguistic expressions and several sentences (Béatrice, 2012)(Aymen, 2007).

The same definition of event varies depending on the domain of application: software engineering, history, philosophy and linguistics ...etc.

Extracting event attracted a number of works, mostly restricted to English (Gabriel, 2006) or French (Gabriel,2008), (Ludovic, 2011). This domain of research is widespread, but it is still very little studied for the Arabic language because it has several unique features that do not exist in other languages (Soraya, 2013) (Mallat,2013). For other languages different approaches have proposed different models in the definition, which can be summarized in two main flows: The approach or TimeML standard and the ACE Model.

The model ACE (Automatic Content Extraction) defines the event as an action, a trial at which the participants are connected. Events are represented according to their attributes and their participants. ACE event can have a number of participants, and each participant is characterized by a role (agent, object, source, target).The event attributes are its types (destruction, creation, transfer, movement, interaction) and the modality of an event is (real and not real). Several systems and projects were founded on this model such as the Aone and Ramos-Santacruz approach, which is interested in events and relations between them; it is based on patterns for the labeling. The idea is to search find all informations about the event and to fill the grid corresponding to it for example the event "bying" their REES tool do extract from the text informations about the time, place, the seller and the buyer.. (Aone and Ramos-Santacruz , 2000).

The approach of Ahn is founded on machine learning. It is composed of 4 modules, each one is generated by TiMBL tool these modules are trigger identification; events anchors; attributes identification and the last one corresponds to the coreference of the events (Ahn, 2006).

The same principle of modules is used in Beatrice approach for French events extraction, The module of anchors identification is based on the prepositions or determinants and the events could be nouns, verbs, adjectives or adverbs. The identification of these anchors is based on classification of words in the texts. The basic indices of the classification are morpho-syntactic categories of words, lexical features (lemma, form, depth in the dependency tree), features issued from WordNet , words surrounding the event ,etc...

The approach or TimeML standard (Roser, 2006) is based on three fundamental concepts: time, event and relations. The TimeML project was created with the aim of improving Q-A systems to deal with questions of temporal order on the entities and events. TimeML adopts a large conception of events. The <event> tag classifies an annotated event by the TimeML ontology, the latter includes different classes of events (aspectual, intentional action, intentional state, perception, and state case).

Parent used TimeML guide to manually annotate the English texts. Their model is based on patterns and syntactic analysis. The author is interested in adverbs, verbs, adjectives and nouns events. In their approach, they are based on words of action verbs and syntactic rules to annotate nouns events which depend on temporal prepositions. These prepositions are initially issued from TimeML guide. (Parent et al., 2008).

The approach of Sauri in (Saurí et al., 2005) is focused on the disambiguation of nouns that may have event interpretation using SemCor corpus and TimeBank1.2. He extracted 25 sub-tree of the semantic network, which contains essentially nouns referring to events. This approach is also based on a statistical module and Wordnet for event annotation. If the process of events search in Wordnet is worthless they pass to use their baysien classifier. This last is founded on a set of rules learned from the SemCor corpus. The evaluation of their results is done by comparison with the tags of the TimeBank1.2

In our work we are interested in the EVITA system (Sauri, 2005). This system is part of the project TARSQI (Verhagen, 2005) respecting the TimeML specification, EVITA is a tool for event detection for the English language. In this system, we consider verbs, nouns and adjectives as events. These textual elements are considered the most meaningful. In this system for each type of "trigger", an extraction method is associated particularly considering the grammatical and textual features.

EVITA system tests first if the category of a word is a verb, noun or adjective and then classifies it among the types of predefined events. These processes are complemented by the identification of the various arguments whose objective is the identification of event participants. The identification is made by detecting pairs between the trigger and the other entities of the sentence.

Detection of pairs is realized based on indices such as the morpho-syntactic category of the two elements, the type of event, the type of entity or the dependency relations in the pair.

3   Proposed Method

In our work, as already mentioned, we were inspired by the EVITA system. EVITA considers verbs, nouns and adjectives as the most meaningful events. For us, we judged that only verbs and nouns as events while adjectives as less significant

Our event extraction system was built using GATE General Architecture for Text Engineering). Our system identifies predefined named entities (names of "people", "places" "organization", and "dates"), and the relations between the entities and the defined events. The extraction of an event consists of the discovery of links between the "trigger" of the event and its arguments/attributes. The extraction of the link is established based on a syntactic analysis "dependency analysis» and of extraction rules exploiting this analysis. To implement this task we used JAPE transducer (JAVA Annotation Pattern Engine) provided by Gate Toolkit

Our event detection system is composed of four main phases summarized below:

The first phase is lemmatization; we first start by cutting the Arabic text into words, and then we assign to each word its lemma.

The second phase is the identification of triggers; it is done by the use of gazetteers. These lists are composed of verbal lemmas for verbal triggers and nominal lemmas for nominal triggers. This phase is realized as the following steps:

We compare each lemma to the list of triggers (list of gazetteers).

If matching, we annotate the corresponding lemmatized word as an event trigger

In the third phase, as in the EVITA system we associated the trigger to the class that it represents. At this stage of our work we do not deal with polysemous triggers, we just developed gazetteers for monosemtic triggers.

The fourth phase includes the identification of participants and their semantic roles. For identification, the parser is used to extract the main dependencies "subject", "object", "preposition", "agent ». The purpose of the semantic analysis is the assignment of semantic roles to participants extracted in the previous phase.

4   System Implementation

4.1   Work Environment

4.1.1   GATE

Most of existing tools for text engineering, if not all, were not originally developed for the treatment of the Arabic language. Most tools were built for English, French or other languages as ACABIT, LEXTER, FASTER, ANA, EXIT(Roch, 2004). Some tools can support or be modified to treat Arabic as Nooj and GATE. We chose the latter for the implementation of our system. GATE[1] (General Architecture for Text Engineering) is a Java open source platform dedicated to textual engineering; it appeared to us well suited for the development of our system. It is a toolkit for natural language processing and very useful for information extraction.

By a system of plugins GATE provides to users a variety of modules dedicated to textual analysis. The most commonly used are tokenizers (segmenters), Part Of Speech Taggers (morpho-syntactic taggers), Gazetteers (lexicons) and transducers (JAPE).

Named entities extracted by GATE correspond to person names, organization, location, dates, etc.. To realize new annotation Gate permits to load new resources and plugins. Also, it permits to combine and parameter them within the same treatment chain.

4.1.2   External Resources

For the Arabic language, Gate does not have a satisfactory number of instances in his predefined Gazetteers of named entities. For example, the Gazetteers_Personne is composed only of two types (female_names.lst and male_names.lst) whereas for English almost 20000 entries, classified in nine different types (person_ambig.lst, person_female_cap.lst, person_female.lst, person_full.lst, person_male_cap.lst, person_male.lst, person_relig.lst, person_sci.lst, person_spur.lst).

To overcome, this problem of deficiency in the Gazetteers, we have used external resources to enrich Gate predefined Gazetteers, which consist of five different types; all built manually using our corpus and web resources.

Person gazetteer: a list of 3257 complete names of people found in Wikipedia and other websites. These names are normalized and formalized to end with a list of 3000 names (first and final names).

---

[1] http://gate.ac.uk/

Organization gazetteer: we collected 2400 names of organizations.

Location gazetteer: we enrich it by Arabic Wikipedia, taking the page labeled "countries of the world in Arabic" to retrieve location names. It consists of 2500 of continents, cities, countries... etc.

Date Gazetteers: we did enrich this predefined gazetteer by 910 new entities.

Event Gazetteers: Gate as mentioned above do not do the annotation of event and have no predefined Gazetteer for it, that's why we did create a new one, composed of verbal and nominal list of triggers. These lists are collected from our domain corpus.To enrich it we added to their extracted lemmas their synonyms using Arabic Wordnet and as well as the argument prepositions structures (for verbs) by the Arabic dictionary. This final Gazetteer contains 700 entries.

Table 1. Enrichment of Gate Gazetteers

Gate	predefined entries	enriched entries	Overall
Person	1700	3000	4700
Organization	96	2400	2496
Location	485	2500	2985
Date	84	910	994
Event	0	700	700

4.2 Implementation of the Method

The implementation of our method in GATE platform has necessitated the use of additional Gazetteers and tools; therefore the installation of new plugins.

For the first phase of lemmatization, we used morphological analyzer (GATE Morphological Analyser). This analyzer has allowed us to obtain, for each word in the Arabic text, the associated lemma. These lemmas are then used in the next phase.

The second phase of identifying triggers is performed using resources Flexible Gazetteer, which allowed us to compare the tokens of the Arabic text to lemmas in both created Gazetteers (one for verbal lemmas and the other for nominal lemmas).

In the third phase, we classified the verb and noun lemmas into subclasses of the "Event" class (Attack, Military_Operation, Crash, Shooting, Damage, Bombing, Death, Kidnapping, War, Injure).

In the fourth phase, for the identification of participants, sentences were cut with the Noun Phrase Chunker and VP Chunker resources. For syntactic analysis, Stanford Parser was used and configured to process the Arabic language.

In the last phase, we pass to develop our linguistic rules with JAPE transducer (Java Annotation Patterns Engine). The role of the transducer is to identify named entities (Person, Organization, Location, and Date). So in our case, JAPE [2] executes the developed rules to extract the various arguments of the event. Writing these rules is followed by a test phase which aims to detect annotation errors and therefore the correction of these rules.

---

[2] http://gate.ac.uk/sale/tao/splitch8.html#chap:jape

Here we present an example of some predefined event classes, that are annotated by our tool: death: مقتل attack: الهجوم InjureEvent: إصابة Military_ Status: ضربةعسكرية War الحربالأهلية / الحربالعالميةالثانية

5   System Evaluation

We used MILTcorp to test our system. MILTcorp is annotated especially for the Event extraction task. This corpus is related to our domain research; the military domain. Its articles are collected from news websites and news wire like (aljaziraa, al_arabia, al_manar, France24 (Arabic) also from electronic journals(al-Quds).

Table 2. Data collection

	Test corpus
Number of sentences	1650
Number of words	45350
Number of words/sentences	24

The evaluation of our system is as follows:

The texts of test corpus are all manually annotated.

After we have annotated automatically these texts with our system.

- We have established a comparison between these two annotations with Corpus Quality Assurance. This Gate tool makes the comparison between two annotations on a corpus.

- The result of the comparison leads to an overall F - measure equal to 0.48, which uses the recall and precision (accuracy).

Precision measurement is defined by the percentage of entities found by the system and which are correct, and the recall is defined as the ratio between the numbers of found correct entities by the number of entities extracted from the reference articles.

$$\text{Precision} = \frac{\text{number of found correct entities}}{\text{number of entities found}} \tag{1}$$

The overall evaluation of entities extracted by our system compared to reference articles is based on the use of F-measure. This measure combines the precision and the recall, it is defined as following (we have used the $F_\beta = 1$; the precision and the recall are weighted equally):

$$F - \text{mesure} = \frac{(\beta^2 + 1) * \text{Precision} * \text{Recalll}}{\beta^2 * \text{Precision} + \text{Recall}}$$

Table 1 shows results for different entities annotated by Gate (using only the basic predefined gazetteers without the enriched gazetteers)

Table 3. Gate baseline results

NE	Precision	Recall	F-mesure
Date	51%	29%	36 ,97%
Person	33%	15%	20 ,62%
Organization	52%	33%	40,73%
Location	59%	44%	50,40%
Event	0%	0%	0%
Overall	39%	24,2%	29,67%

Table3 show the results obtained by using   the enriched gazetteers (person, organ-ization, location, date) and our method of event extraction.

Table 4. Results with event, date, person, Organization, location gazetteers

Named entities	Precision	Recall	F-measure
Date	84.03%	77.19%	80,64%
Person	83.04%	79.34%	81,14%
Organization	78.6%	62.11%	69,38%
Location	86.14%	77.42%	81,54%
Event	70%	35%	46,66%
Overall	80,36%	66,21%	71,87%

From the table above, we see the effect of the annotation of events on the annota-tion of named entities extracted by Gate. We obtain an improvement of accuracy of 39% to 80,36 % same with the recall from 24,2 % to 66,21 %.

These results are satisfying according to limitations due to the complexity of Ara-bic sentences and the lack of adequate tools for the treatment of the Arabic language.

In the next future, we plan to complete the event Gazetteers (verbal and nominal) and the disambiguation of polysemous triggers also to increase the size of our corpus to obtain a higher performance of the system.

6   Conclusion and Perspectives

In this paper we have presented an integrated system to detect events in Arabic texts. The event identification was performed through several stages; data collection, pre-processing, classification and identification of participants and their semantic roles. Extensive experiments were conducted to evaluate the effectiveness of the proposed system using our Arabic Corpus. This system can be generalized for the purposes of decision making enrichment which can be implemented in many areas such as infor-mation intelligence or crises management. In future work we look to improve results and compare with other works in the domain of event detection. We aim to annotate different types of relations between the named entities in order to improve the per-formance of our proposed system.

References

1. Ahn, D.: The stages of event extraction. In: ARTE 2006: Proceedings of the Workshop on Annotating and Reasoning about Time and Events, Morristown, NJ, USA, pp. 1–8, July 2006. Association for Computational Linguistics (2006)
2. Aone, C., Ramos-Santacruz, M.: REES: a large-scale relation and event extraction system. In: Proceedings of the Sixth Conference on Applied Natural Language Processing, ANLC 2000, Stroudsburg, PA, USA, pp. 76–83. Association for Computational Linguistics (2000)
3. Elkhlifi, A., Faiz, R.: Machine learning approach for the automatic annotation of the events. In: FLAIRS Conference 2007, pp. 362–367 (2007)
4. Bethard, S., Martin, J.H.: Identification of event mentions and their semantic class. In: Proceedings of the 2006 Conference on Empirical Methods in Natural Language Processing (EMNLP 2006), Sydney, pp. 146–154 (2006)
5. Borsje, J., Hogenboom, F., Frasincar, F.: Semi-Automatic Financial Events Discovery Based on Lexico-Semantic Patterns. International Journal of Web Engineering and Technology 6(2), 115–140 (2010)
6. Capet, P., Delavallade, T., Nakamura, T., Sandor, A., Tarsitano, C., Voyatzi, S.: A risk assessment system with automatic extraction of event types. In: Intelligent Information Processing IV, IFIP International Federation for Information Processing, vol. 288, pp. 220–229. Springer, Boston (2008)
7. Cimiano, P., Staab, S.: (2004): Learning by Googling. SIGKDD Explorations Newsletter 6(2), 24–33 (2004)
8. Cohen, K.B., Verspoor, K., Johnson, H.L., Roeder, C., Ogren, P.V., Baumgartner Jr., W.A., White, E., Tipney, H., Hunter, L.: High-precision biological event extraction with a concept recognizer. In: Workshop on BioNLP: Shared Task Collocated with the NAACL-HLT 2009 Meeting, pp. 50–58. Association for Computational Linguistics (2009)
9. Frasincar, F., Borsje, J., Levering, L.: A Semantic Web-Based Approach for Building Personalized News Services. International Journal of E-Business Research 5(3), 35–53 (2009)
10. Parent, G., Gagnon, M., Muller, P.: Annotation d'expressions temporelles et d'événements en français TALN 2008, Avignon, Juin 9–13, 2008
11. Doddington, G., Mitchell, A., Przybocki, M.: The automatic content extraction (ACE) program tasks, data, and evaluation. In: Proceedings of LREC 2004, pp. 837–840 (2004)
12. Emna, H., Mallat, S., Zrigui, M.: Automatic translation of Arabic text based on ontology. In: Proceedings of Fifth International Conference on Web and Information Technologies, ICWIT2013 (2013)
13. Kamijo, S., Matsushita, Y., Ikeuchi, K., Sakauchi, M.: (2000): Traffic monitoring and accident detection at intersections. IEEE Transactions on Intelligent Transportation Systems 1(2), 108–118 (2000)
14. Mallat, S., Zouaghi, A., Hkiri, E.: Zrigui, M, Method of lexical enrichment in information retrieval system in arabic. International Journal of Information Retrieval Research IJIRR 3(4), 35–51 (2013)
15. Jean-Louis, L.: Approches supervisées et faiblement supervisées pour l'extraction d'événements et le peuplement de bases de connaissances. Thèse, Université Paris Sud - Paris XI, Decembre 2011

16. Gabriel, P., Gagnon, M., Muller, P.: Annotation d'expressions temporelles et d'événements en français. In: Béchet, F. (ed.) Traitement Automatique des Langues Naturelles (TALN), Avignon, 09/06/08–13/06/08, (support électronique) (2008). http://www.atala.org/. Association pour le Traitement Automatique des Langues (ATALA) (2008)
17. Roche, M., Heitz, T., Matte-Tailliez, O., Kodratoff, Y., EXIT: Un système itératif pour l'extraction de la terminologie du domaine à partir de corpus spécialisés (2004)
18. Sauri, R., Littman, J., Knippen, B., Gaizauskas, R., Setzer, A., Pustejovsky, J.: TimeML Annotation Guidelines Version 1.2 (2006)
19. Saurí, R., Knippen, R., Verhagen, M., Pustejovsky, J.: Evita: a robust event recognizer for QA systems. In: Proceedings of HLT/EMNLP 2005, pp. 700–707 (2005)
20. Zaidi–Ayad, S.: Une plateforme pour la construction d'ontologie en arabe: Extraction des termes et des relations à partir de textes (Application sur le Saint Coran). These, Université Badij Mokhtar, Annaba (2013)
21. Verhagen, M., Mani, I., Saurí, R., Knippen, R., Littman, J., Pustejovsky, J.: Automating temporal annotation with TARSQI. In: Proceedings of the ACL (2005)
22. Wei, C.P., Lee, Y.H.: Event detection from Online News Documents for Supporting Environmental Scanning. Decision Support Systems **36**(4), 385–401 (2004)
23. Yakushiji, A., Tateisi, Y., Miyao, Y.: Event extraction from biomedical papers using a full parser. In: 6th Pacific Symposium on Biocomputing, pp. 408–419 (2001)

Deriving Consensus for Term Frequency Matrix in a Cognitive Integrated Management Information System

Marcin Hernes[✉]

Wrocław University of Economics, ul. Komandorska 118/120, 53-345, Wrocław, Poland
marcin.hernes@ue.wroc.pl

Abstract. An unstructured knowledge processing in integrated management information systems is increasingly becoming a major challenge, mainly due to the possibility to obtain better flexibility and competitiveness of the organization. However, the most prevailing phenomenon is a conflict in unstructured knowledge. As an example may serve opinions of users about a given product offered by online shops. Some users may have positive opinions; others negative ones, while some of them may not have any opinion about a given product. Therefore this paper focus on developing a consensus deriving method for resolving conflict of unstructured knowledge of text documents represented by Term Frequency Matrix in integrated management information system.

Keywords: Multiagent systems · Unstructured knowledge · Knowledge conflicts · Term frequency matrix · Consensus method

1 Introduction

Contemporary economy forces company managers to make complex operational, tactical, yet most of all, strategic decisions that influence the future of the organization. Those who actually make decisions in a company, are usually exposed to risk and doubt, because they cannot foresee the consequences of their decisions or their predictions have very low probability [11]. Therefore, the entire decision-making process is very complicated.

The process of decisions making employs decision support systems as well as multiagent integrated management information systems [1]. The example of such system is Cognitive Integrated Management Information System (CIMIS) presented at work [9], composed of a few to more than a dozen of cognitive agents. The cognitive agent is an autonomous object which has a specified aim, is able to communicate with other agents, takes action and reacts to changes of the environment in which it operates and it is capable of understanding the real meaning of the observed phenomena and economic processes taking place in the organization environment [16]. Multiagent systems allow to gather up-to-date knowledge quickly, process it and provide the decision maker with all possible acceptable decisions (decisions that meet the decisions maker's conditions) or optimum decisions (those acceptable decisions

© Springer International Publishing Switzerland 2015
M. Núñez et al. (Eds.): ICCCI 2015, Part I, LNAI 9329, pp. 503–512, 2015.
DOI: 10.1007/978-3-319-24069-5_48

which are most accurate in terms of assessment criteria assumed by the decision maker). However, the final decision is made by the decision maker who also takes responsibility for the results of the decision. Multi-agent integrated management information systems significantly shorten the time necessary to make a decision, because they relieve the decision maker of the task of knowledge processing and they are able to draw decision-based conclusions and react properly, following the conclusions. Therefore they can suggest various new solutions to the decision maker.

The biggest problem currently, however, turns out to be the processing of unstructured knowledge in systems of this kind. Note that knowledge contained in integrated management information systems is generally structuralized and the systems employ various methods for processing structuralized knowledge. However, in contemporary companies, unstructured knowledge is essential, mainly due to the possibility to obtain better flexibility and competitiveness of the organization. Therefore, unstructured knowledge supports structuralized knowledge to a high degree. It is mainly stored in natural language, so it is processed with symbols (not numbers), e.g. users' opinions on a forum. Generally speaking, there are text databases that contain various types of text files, such as newspaper articles, e-books, e-mail, websites and all sorts of text files. The documents describe certain phenomena that occur in the real world, in the environment where a given organization operates. Text files are not internally structured in any way, the knowledge that they contain is non-structuralized or structuralized to a small degree. It is important to say that text files are often a source of significant and useful knowledge.

Simultaneously, one must note that the most prevailing phenomenon is a conflict in unstructured knowledge. As an example may serve opinions of users about a given product offered by online shops. Some users may have positive opinions; others negative ones, while some of them may not have any opinion about a given product. Moreover, opinions or reviews of the same product found in other online shops may be totally different. Another example may be documents created by employees, describing actions or phenomena taking place in an organization. The same actions or phenomena may be described differently by different employees, and even more differently if described by third parties (customers, suppliers). It is extremely difficult to resolve conflicts of this kind properly. However, it is also very important, since it can improve the operation of integrated management information systems and, consequently, help the organization that employs the system become more flexible and competitive.

The aim of this paper is to develop a consensus deriving method for resolving conflict of unstructured knowledge of text documents represented by Term Frequency Matrix, in integrated management information system.

This paper is organized as follows: the first part shortly presents the state-of-the-art in considered field; next, an unstructured knowledge processing agents running in Cognitive Integrated Management Information System (CIMIS) are described; a developed consensus method is presented in the last part of paper.

2   Related Works

Representation of the test documents is relates with results of document retrieval process [2, 15]. Text documents are often represented in databases according to key words contained in them or ontology (symbol representation of knowledge). With such representation, however, it is very difficult to compare documents, especially measuring distance between documents, however distance is understood here as the degree of documents' similarity [17].

Increasingly to represent text documents are used for semantic nets, and especially the topic maps. In work [10] it was found that the semantic nets allows to record information ontology and data taxonomy, structured semantically and at the same time it allows for knowledge mapping (both structured and unstructured) on a wide variety of hierarchical dependencies exist between economic concepts and semantic (the concept of this type include, inter alia, to text documents in the field of management and economics).

An alternative approach to text document representation is an approach based on vector representation of a document (representation of numeric knowledge). The basic idea of vector representation boils down to the fact that any given document is represented in the form of a vector of frequency of key words appearance, also called index terms [3].

Knowledge conflicts occur when different attributes are assigned to the same world objects, or different values are assigned to the same attributes (features) [12].

After analysing some sources covering the topic [4], [17] one can draw conclusions that there are no generally accepted methods of solving conflicts of unstructured knowledge represented by symbols, which is related to the problem of processing this type of knowledge. At present, hybrid methods of processing unstructured knowledge are used, in the course of which knowledge is first structured and then symbolically processed (for example with the use of expert systems or genetic algorithms), or transformed into numerical representation and then numerically processed (for example with the use of neural network or fuzzy logics systems). In both cases, in order to extract knowledge, documents may be subject to data exploration [18]. Methods such as machine learning [7], or rules, on the basis of which identification (annotation) of pieces of texts on a given topic is made are often used in text documents analysis.

In the face of the presented problems, in the economic practice, transformations "in the other direction" are also performed – unstructured data is transformed into documents saved in a natural language, and then a "manual" analysis of the documents is performed. However, this approach is not too effective as it is labour-intensive and time-consuming, and the turbulent economic reality forces decision-makers to make decision in near real time.

The works on the subject present various methods of resolving conflicts of knowledge, especially structuralized knowledge. However, they contain some minor flaws. Negotiation methods [6], for example, guarantee reaching the desired compromise between sides of a conflict, yet it is done at the expense of increased communication between system components, which obviously affects the speed of processing. Whereas, deduction-computing methods (e.g. methods based on game theory, Newtonian mechanics, methods originating from operational research, from sociology and behavioral studies, selection methods or consensus methods [12], [5]) do not influence the speed of processing too much. However, apart from the consensus methods, they do not guarantee a good compromise. User, however, expects the system to be efficient (often to operate at a rate approximate to real time) and to resolve conflicts of knowledge effectively. For this reason, as means to resolve conflicts of knowledge, it is better to employ consensus methods. Unlike other methods, they guarantee good compromise and are a much better choice than negotiation methods, because they do not require substantial processor capacity or increased communication between system components, therefore allowing the system to operate at a rate approximate to real time.

Using the consensus method to solve unstructured knowledge conflicts will enable the system to present its user with one, reliable version of representation of a set of text documents describing the same object or phenomenon, which will result in eliminating doubts which decision-makers might otherwise have when making quick decisions.

3   Unstructured Knowledge Processing Agents in CIMIS

The CIMIS system is dedicated mainly for the middle and large manufacturing enterprises operating on the Polish market (because the user language, at the moment, is a Polish language). The CIMIS consist of following sub-systems: fixed assets, logistics, manufacturing management, human resources management, financial and accounting, controlling, CRM, business intelligence. This system have been detailed described in [9]. In this paper the unstructured knowledge processing agents, running in the frame of CRM subsystem, will be shortly described.

The four main types of LIDA architecture [16] agents run in CIMIS in order to perform the text analysis process:

- Document retrieval agents,
- Information extraction agents,
- Text analysis agents,
- Conflict resolving agent.

Document retrieval agents search and retrieve, from the internet sources, the documents according to users' needs. Each agent run on the basis of different document retrieving method. Next, the information extraction agents extract only valuable information from documents (for example the advertisements are removed from text document). Each agent implements different information extraction method. The text analysis agent performs a natural language processing. Each agent run on the basis of different deep analysis method[1].

---

[1] Document retrieval agents, information extraction agents and text analysis agents are detailed described in [7]

The conflict resolving agent performs using a consensus method [15]. This agent determines the one, reliable version of representation of a set of text documents describing the same object or phenomenon.

The text document in CIMIS are represented by Term Frequency Matrix (TFM) and by "slipnet" – semantic net with nodes and links activation level [11]. This paper focus on consensus method for TFM.

The set of M stored text documents can be represented in the form of a N word frequency matrix referred to as Term Frequency Matrix, whose TFM$[d_i, t_i]$ element represents the number of appearance of the keyword t_i (where: $1 < i < N$) in document d_i (where: $1 < i < M$). Any given document d_i is represented in the form of a keywords appearance frequency vector. The element TFM$[d_i, t_i]$ is called the t_i word weight in d_i document. In the simplest Boolean representation, words weights in a document vector may have only two values: 0 or 1 (table 1). If the weight of t_i word in a d_i document equals 1 means that t_i word appears in a given d_i document, however, if t_i word does not appear in d_i document, the weight of t_i word in d_i document equals 0. It needs to be noted that the Boolean representation of documents specifies only whether a key word appears or does not appear. Consequently, for example a document in which a keyword appears once equals a document in which the same key word appears many times. It shows that Boolean representation of documents serves as their representation based on key words.

Table 1. The Boolean TFM

	t_1	t_2	t_3	t_4	t_5
d_1	1	0	1	1	0
d_2	1	0	1	1	1
d_3	1	1	0	1	0
d_4	0	1	1	1	0

A set of key words used in document representation approach in the form of the TFM matrix may be very large. The matrix may also be used in unstructured knowledge representation in integrated management information systems.

4   Deriving Consensus

The general meaning of the term consensus refers to an agreement. A consensus of a certain set (profile) of text documents may constitute a new document (hypothetical one) created on the basis of documents contained in the profile.

Driving consensus consists of three basic stages. In the first stage, one needs to determine a method of text documents representation. In this paper it has been assumed that the documents are represented in the form of binary vectors of frequency using the TFM matrix. The next step requires defining the function of calculating distance between individual variants. The third stage involves developing consensus deriving algorithms, i.e. determining a representation of a set of documents (profile) where the distance between the representation (consensus) and individual documents of the profile (stored in CIMIS) is minimal (according to various criteria). It needs to be noted that the profile is not made up of all text documents stored in the system, but documents connected with each other thematically, for example one profile may consist of documents containing opinions of users about a given $p1$ product, whereas another profile may consist of documents containing opinions of users about a $p2$ product.

The formal definition of a profile of the text documents is as follows:

Definition 1. *Set of N index terms* $T = \{t_1, t_2, \ldots, t_N\}$ *is given, where* $t_i = \{0,1\}$.

A profile $D = \{d_1, d_2, \ldots, d_M\}$ *is called set of M text documents described with the use of the index terms frequency vectors of finite set* T, *such, that:*

$$d_1 = \left\langle t_{1(d_1)}, t_{2(d_1)}, \ldots, t_{N(d_1)} \right\rangle$$

$$d_2 = \left\langle t_{1(d_2)}, t_{2(d_2)}, \ldots, t_{N(d_2)} \right\rangle$$

$$\cdots\cdots\cdots\cdots$$

$$d_M = \left\langle t_{1(d_M)}, t_{2(d_M)}, \ldots, t_{N(d_M)} \right\rangle, \tag{1}$$

where $t_{i(d_x)}$ *denotes value of the index term* t_i *at the document* d_x.

The main advantage of vector representation of documents over keywords-based representation is the possibility of defining measures of distance between documents or user's queries, which is the essence of the second stage of the process of deriving consensus. If a vector document representation has been defined, documents of similar topic should be characterized by a similar frequency of occurrence of the same keywords. With a vector document representation at hand, any document may be interpreted as a point in T-dimension space whose dimensions match individual keywords.

Consequently, in order to evaluate distances between documents, and between documents and a query, one may use any measures used to measure distances in the Euclidean space. The measures always satisfy metric conditions (distance function is metrics). Adopting the Euclidean distance is connected with problems resulting from the great influence changes in the scale of coordinates have on results of grouping elements of space features, for example keywords (to avoid the problem the features value space has to be normalized, which results in a greater complexity of the distance calculation algorithm). Apart from the known measures of distance used in multidimensional Euclidean spaces, for the need of information searching systems a lot of specific measures have been devised, such as the cosine distance or Hamming distance [8]. In the article, in order to calculate distance, the metric measure of Hamming distance will be used. It has been claimed that the advantage of such a type of distance is the lack of influence of independent increase of coordinates on its value [8].

Hamming distance between two text strings of equal length is the number of positions in which their corresponding symbols are different. In other words, it measures the minimum number of substitutions required to change one string into the other, or the minimum number of errors that could have transformed one string into the other.

Hamming distance meets all metric and formal conditions, and in case of binary vectors, it is defined in the following way:

Definition 2. *Let d_1, d_2 denote index terms frequency vectors characterized the text documents, then:*

$$\omega(d_1,d_2) = \sum_{i=1}^{N}\left[d_1[i] \otimes d_2[i]\right] \tag{2}$$

where $d_x[i] = \{0,1\}$ ($i = \{1,\dots,M\}$) denotes a value of the ith index term of vector d_x, while the symbol $\otimes$ denotes an Exclusive disjunction, operation means by which we obtain the following values:

$$\left[d_1[i] \otimes d_2[i]\right] = 0 \quad \Leftrightarrow \quad d_1[i] = d_2[i], \tag{3}$$

$$\left[d_1[i] \otimes d_2[i]\right] = 1 \quad \Leftrightarrow \quad d_1[i] \neq d_2[i]. \tag{4}$$

Using presented definition of distance, the consensus function can be defined (third stage of consensus deriving). Generally speaking a consensus is a function that minimizes the sum of distances to all the elements of profile. The work [12] states, that a better function of the distance, due to the fact that a greater uniformity of consensus is function to minimize the sum of squared distances to all elements of the consensus profile (a consensus is more even to all the profile elements). Deriving of consensus using this type of function is an NP-complete problem. With regard to the representation of text documents using binary index terms frequency vectors this function is defined as follows:

Definition 3. *Let D profile is given, and W denotes a set of all valuations of terms indexing T. The consensus of D is called following function:*

$$Con(D) = \left\{c \in W : \omega^2(c,D) = \min(\sum_{i=1}^{M}\omega(c,d_i)^2)\right\}. \tag{5}$$

An algorithm which derives consensus according to a function specified in definition 3 is a heuristic algorithm (as we are dealing with NP-complete problem) and it looks as follows:

Algorithm 1
Data: A profile $D = \{d_1, d_2, \dots, d_M\}$ consist of M frequency vectors characterized the text documents.
Result: Consensus $c = Con(D) = \langle t_1, t_2, \dots, t_n \rangle$ according D.
START
1: Let $j := 1$.
2: $s = \sum_{i=1}^{M} d_i[j]$.
3: If $s = M/2$ then $c[j] = random$.

If $s < M/2$ then $c[j] = 0$.

If $s > M/2$ then $c[j] = 1$.

4: If $j<N$ then $j=j+1$, go to: 2.

   If $j=N$ then go to: 5.

5: $o = \omega^2(c, D)$.

6: Let $j := 1$.

7: $c[j] = \neg c[j]$.

8: If $\omega^2(c, D) < o$ then $o = \omega^2(c, D)$.

   If $\omega^2(c, D) > o$ then $c[j] = \neg c[j]$.

9: If $j<N$ then $j=j+1$, go to: 7.

If $j=N$ then END.

END.

The complexity of the algorithm amounts to $O(NM)$.

5   Experiments

In order to verify the algorithm, a prototype of an agent determining consensus has been implemented and tested. The aim of the test was to verify the efficiency of the consensus algorithm.

The following assumptions have been adopted:

- a profile consist of 50 randomly selected documents,
- the consensus has been calculated, using an optimal algorithm (which involves determining the consensus by checking all the possible solutions), and the consensus based on the heuristic algorithm presented in this paper,
- after the above-mentioned calculations, the distance between a profile and consensus designated by the two algorithms have been calculated,
- 100 calculations related to different profiles have been performed.

On the basis of results of the research experiment performed by using 100 profiles of text documents represented by TFM it has been state, that in 96 cases consensus derived according to heuristic algorithm was in line with consensus derived by the optimal algorithm (heuristic algorithm compatibility level is 96%). Consensus according to the optimal algorithm was calculated about 15 second, while the consensus heuristic algorithm in about 2 seconds. Therefore, the heuristic algorithm, developed in this paper, characterize a higher performance then performance of an optimal algorithm.

6   Conclusions

The consensus method may appear useful in solving the knowledge conflicts due to the fact that each party to such a conflict is taken into account in the consensus and each one "loses" the least. If possible, each party contributes to the consensus due to the fact that a consensus constitutes the representation of all parties to a conflict. If, for example, there are various descriptions of a given phenomenon in an integrated management information system, using the consensus method, on the basis of the descriptions, one can determine one variant which is then presented to a user. The variant does not have to be one of the descriptions from the system. It can be a totally new variant created on the basis of descriptions existing within the system. Thanks to this, all descriptions of a given phenomenon can be taken into account. The sort of action allows also to shorten the time it takes to determine target descriptions (users do not have to analyse individual descriptions and wonder which one to choose – the system will perform the task automatically for a user), and to reduce the risk of selecting the worst description (as all descriptions are taken into consideration in the consensus). Consequently, the process of managing an organization may be realized more quickly and more efficiently.

It also needs to be noted that there are conflicting situations in which it is impossible to use the consensus method. For example if one is assessing documents which describe tasks performed by employees, connected with executing business-related process, and if some of the employees perform their tasks inadequately, their descriptions should not be taken into account in the consensus. It is a problem connected with profile's susceptibility to the consensus, characterized for example in [5, 12].

An important problem constitutes also a representation of text documents with the use of the TFM matrix, which is an attempt at partial formalization of unstructured knowledge, which is why the author have started research on solving conflicts of knowledge when text documents are represented with the use of ontology, with particular attention to the "slipnet". Additionally, in the paper the main emphasis has been placed on economic and IT related aspects of processing knowledge, however, further research will also include social and psychological aspects of the problem.

Acknowledgement. This research was financially supported by the National Science Center (decision No. DEC-2013/11/D/HS4/04096).

References

1. Badica, C., Ganzha, M., Gawinecki, M., Kobzdej, P., Paprzycki, M.: Towards trust managament in an agent-based e-commerce system – initial consideration. In: Zgrzywa, A. (ed.) Conference Multimedia and Network Information Systems MISSI 2006. Oficyna Wydawnicza PWr, Wrocław (2006)
2. Baeza-Yates, R., Ribeiro-Neto, B.: Modern information retrieval, vol. 463. ACM press, New York (1999)
3. Bush, P.: Tacit Knowledge in organizational knowledge. IGI Global, Hershey, New York (2008)
4. De Long, D., Seemann, P.: Confronting conceptual confusion and conflict in knowledge management. Organizational Dynamics **29**(1) (2000)
5. Duong, T.H., Nguyen, N.T., Jo, G.-S.: A method for integration of wordnet-based ontologies using distance measures. In: Lovrek, I., Howlett, R.J., Jain, L.C. (eds.) KES 2008, Part I. LNCS (LNAI), vol. 5177, pp. 210–219. Springer, Heidelberg (2008)

6. Dyk, P., Lenar, M.: Applying negotiation methods to resolve conflicts in multi-agent environments. In: Zgrzywa, A. (ed.) Conference Multimedia and Network Information Systems MISSI 2006. Oficyna Wydawnicza PWr, Wrocław (2006)

7. Frank, E., Bouckaert, R.R.: Naive Bayes for text classification with unbalanced classes. In: Fürnkranz, J., Scheffer, T., Spiliopoulou, M. (eds.) PKDD 2006. LNCS (LNAI), vol. 4213, pp. 503–510. Springer, Heidelberg (2006)

8. Hamming, R.W.: Error detecting and error correcting codes. Bell System Technical Journal 29(2) (1950)

9. Hernes, M.: A cognitive integrated management support system for enterprises. In: Hwang, D., Jung, J.J., Nguyen, N.-T. (eds.) ICCCI 2014. LNCS, vol. 8733, pp. 252–261. Springer, Heidelberg (2014)

10. Hofstadter, D.R., Mitchell, M.: The copycat project: a model of mental fluidity and analogy-making. In: Hofstadter, D. (ed.) The Fluid Analogies Research group, Fluid Concepts and Creative Analogies, chapter 5. Basic Books (1995)

11. Kubiak, B.F.: Knowledge and intellectual capital – management strategy in polish organizations. In: Kubiak, B.F., Korowicki, A. (eds.) Information Management. Gdansk University Press, Gdańsk (2009)

12. Nguyen, N.T.: Metody wyboru consensusu i ich zastosowanie w rozwiązywaniu konfliktów w systemach rozproszonych. Oficyna Wydawnicza PWr, Wrocław (2002)

13. Rosenfeld, A., Kraus, S.: Modeling Agents Based on Aspiration Adaptation Theory. Journal of Autonomous Agents and Multi-Agent Systems (JAAMAS) 24, 221–254 (2012)

14. Sadilek, A., Kautz, H.: Location-Based Reasoning about Complex Multi-Agent Behavior. Journal of Artificial Intelligence Research 43, 87–133 (2012)

15. Sliwko, L., Nguyen, N.T.: Using Multi-agent Systems and Consensus Methods for Information Retrieval in Internet. International Journal of Intelligent Information and Database Systems 1(2), 181–198 (2007)

16. Snaider, J., McCall, R., Franklin, S.: The LIDA framework as a general tool for AGI. In: Schmidhuber, J., Thórisson, K.R., Looks, M. (eds.) AGI 2011. LNCS, vol. 6830, pp. 133–142. Springer, Heidelberg (2011)

17. Vlas, R.E., Robinson, W.N.: Two rule-based natural language strategies for requirements discovery and classification in open source software development projects. Journal of Management Information Systems 28(4) (2012)

18. Zhang, C., Zhang, X., Jiang, W., Shen, Q., Zhang, S.: Rule-based extraction of spatial relations in natural language text. International Conference on Computational Intelligence and Software Engineering (2009)

Author Index

Printed in the United States
By Bookmasters